Chemistry in Space

NATO ASI Series

Advanced Science Institutes Series

A Series presenting the results of activities sponsored by the NATO Science Committee, which aims at the dissemination of advanced scientific and technological knowledge, with a view to strengthening links between scientific communities.

The Series is published by an international board of publishers in conjunction with the NATO Scientific Affairs Division

A Life Sciences **B Physics**	Plenum Publishing Corporation London and New York
C Mathematical and Physical Sciences **D Behavioural and Social Sciences** **E Applied Sciences**	Kluwer Academic Publishers Dordrecht, Boston and London
F Computer and Systems Sciences **G Ecological Sciences** **H Cell Biology**	Springer-Verlag Berlin, Heidelberg, New York, London, Paris and Tokyo

Series C: Mathematical and Physical Sciences - Vol. 323

Chemistry in Space

edited by

J. M. Greenberg

Laboratory Astrophysics,
University of Leiden,
Leiden, The Netherlands

and

V. Pirronello

Department of Physics,
University of Calabria,
Rende, Italy

Kluwer Academic Publishers

Dordrecht / Boston / London

Published in cooperation with NATO Scientific Affairs Division

Proceedings of the NATO Advanced Study Institute on
Chemistry in Space
Erice, Italy
May 10–20, 1989

ISBN 0–7923–0987–1

Published by Kluwer Academic Publishers,
P.O. Box 17, 3300 AA Dordrecht, The Netherlands.

Kluwer Academic Publishers incorporates the publishing programmes of
D. Reidel, Martinus Nijhoff, Dr W. Junk and MTP Press.

Sold and distributed in the U.S.A. and Canada
by Kluwer Academic Publishers,
101 Philip Drive, Norwell, MA 02061, U.S.A.

In all other countries, sold and distributed
by Kluwer Academic Publishers Group,
P.O. Box 322, 3300 AH Dordrecht, The Netherlands.

Printed on acid-free paper

TABLE OF CONTENTS

PREFACE

This volume contains the lectures presented at the first course of the International School of Space Chemistry held in Erice (Sicily) from May 10 to May 20 at the 'E. Majorana Centre for Scientific Culture'. The course was attended by 57 participants from 11 countries.

The recognition by Professor A. Zichichi that space chemistry is one of the important and rapidly growing scientific disciplines with many and varied applications provided the stimulation to initiate this new school.

Historically, the study of chemistry in space had its major origins in comets, the solar nebula and circumstellar envelopes before the interstellar medium achieved its current prominence. A remarkably rapid development in interstellar chemistry was precipitated by the discovery of formaldehyde in the late 1960's made possible by the new radio observational techniques. A four atom molecule in interstellar space was indeed a surprise considering that only a short time earlier there were still arguments about the existence of the simplest of all molecules - the hydrogen molecule. The application of ion-molecule reactions to interstellar cloud chemistry provided a rich variety of new possibilities which were, however, continuously under pressure to keep pace with radio-astronomical discoveries of more and more complex molecules. But, at the same time that the deluge of discoveries of interstellar molecules began there were initiated a number of experimental and theoretical attempts to bring into play the influence of the small solid particles (interstellar dust grains) towards creating new pathways in the chemical processes in interstellar space. The accretion of atoms and molecules on the dust and the solid state chemistry induced by ultraviolet photons or other energetic particles appeared to provide a means of creating even more complex molecules than could ever be identified in the gas. The full cycle of bringing the interstellar chemistry back to the solar system, particularly via the protosolar nebula and comets has been an inevitable consequence of these developments.

In this volume, we start with the chemical evolution of the galaxy in which nuclear processes are the means of producing the atomic species which provide the basis for all the interesting and complex chemistry discussed in the later chapters. The relative, as well as absolute, abundances of the elements are key parameters in determining the chemical processes in all regions of space. The role cosmic ray particles play in their interactions with the interstellar medium is discussed. The basic aspects of the atomic and ionic reactions in interstellar clouds, circumstellar regions and the most energetic regions around young hot stars are considered in detail by world experts in these fields and the observational results are a convincing proof of the complexity of the processes and the interactions. Although much of the same basic theory applies throughout, the widely varying physical environments - densities, temperatures, amounts and kinds of energetic radiation, as well as the relative abundances of the key atomic species (e.g., C:O ratios) - are sufficiently different to produce very different end products. The theoretical interpretation of the observations are needed to probe the nature and the physical evolution of the galactic regions starting with the most diffuse clouds, passing through the phases of dense clouds which condense to form protostars and subsequent evolution induced by star formation. Evolved stars are the sources not only of another phase in the chemical cycling, they are largely responsible for producing a population of solid particles (predominantly silicates) which are the nuclei or cores on which the atoms and molecules accrete to provide the basis for the bulk of the solid state chemistry in space.

Supernovae are well recognized as an important phase of galactic evolution but it is also shown here how molecular processes can provide a new probe of the evolution of the supernova itself.

The discovery of a new kind of large molecule or small particle - depending on the point of view - has stimulated a great deal of study on such examples as polycyclic aromatic hydrocarbons (PAH's). The answer to where and how such molecules are formed must ultimately lead to a new understanding of either circumstellar or interstellar chemistry or both.

The modelling of interstellar dust is constrained by basic observations of its optical properties and infrared spectroscopy. Laboratory analog studies of chemical and spectroscopic properties of accreted ices mantles on the dust as well as the physical and chemical interactions with the gas driven by ultraviolet radiation show how the photon induced solid state reactions including explosion/desorption of solid volatiles influence the abundances of the gas phase molecules. Laboratory studies of the effects of energetic particles on ices of astrophysical significance in the solar system and interstellar space are qualitatively similar to the ultraviolet studies with one significant difference, namely, the directly induced volatile desorption. One of the most important consequences of energetic processing of solids is the formation of complex organic residues. Infrared observations have confirmed that such material is as abundant as the silicates on which they are the mantles.

A major contribution to observational information about gas and dust in all the regions in our galaxy as well as external galaxies will come from the planned space observatories. The Infrared Space Observatory (ISO) has been singled out here but already the Cosmic Background Explorer (COBE) is revealing new and exciting results on the distribution and temperature of matter in space.

The understanding of how to account for the variety of objects in the solar system and particularly the most primitive examples in the form of asteroids, comets and their debris involves a study of the protosolar nebula of which a substantial introduction is presented.

Finally, the last four chapters take up the chemistry and physics of comet nuclei, comet volatiles and comet dust. Classically, the major information about comet nuclei was deduced from the atomic and molecular fluorescence radiation in the coma. Ultraviolet observations are presented and it is shown how their theoretical interpretation provides a major tool in describing the molecule, ion and radical distributions and abundances in comet comae. The in-situ measurement of comet Halley molecules by space probes showed that the spatial distribution of some of the observed species including CO and CN arose not from molecules which came off directly from the nucleus but rather from molecules which evaporated from the dust. This could be readily understood in terms of the fact that the Giotto and Vega analyses revealed that the comet dust consisted in large part of complex organic molecules some of which are volatile at the temperatures reached during the observations. The connection between interstellar dust and comet dust has been theoretically tested using all the available observational constraints. The use of the PUMA-1,2 atomic analyses and the molecular inferences deduced -both mineralogical and organic - of the dust in the coma of comet Halley, tell us about the solid components of comets. When combined with the studies of the volatiles the composition of comets is more clearly telling the story of how comets were formed in the protosolar nebula than was possible before from volatile studies alone. It is interesting to note that, to the extent that the interstellar dust model of comets is correct, *total* knowledge of the chemical and morphological structure of comets would give information not only about conditions in the protosolar nebula but in the sources of the cloud from which it evolved. Could comets be a window to the universe?

The school was fortunate in being able to benefit from having lecturers who so successfully guided us through the wide variety of interconnected subjects. And it must be noted that the participants made significant contributions, as well. We are grateful to the Scientific Affairs Division of the North Atlantic Treaty Organization for the generous support to the Institute and we owe a special debt of gratitude for the hospitality and the support given by the staff at the 'Ettore Majorana Centre for Scientific Culture'.

J. Mayo Greenberg
University of Leiden
Laboratory Astrophysics
Niels Bohrweg 2
2300 RA Leiden
The Netherlands

Valerio Pirronello
Università della Calabria
Dipartimento di Fisica
87036 Roges di Rende
Italy

1. MINH
2. JENNISKENS
3. MEHMET
4. LEPP
5. LAUREIJS
6. DESERT
7. BALM
8. HENDERSON
9. SCHWARTZ
10. CHLEWICKI
11. SCHMELCHER
12. SACK
13. GREENBERG
14. BLUM
15. TIMMERMANN
16. IRVINE
17. RIST
18. SHAPIRO
19. GREENHOUSE
20. MUTLU
21. RUZICKA
22. BREUKERS
23. HUTSON
24. LEWIS
25. AVERNA
26. FEUCHTGRUBER
27. FOMENKOVA
28. MADDEN
29. MUKHIN
30. LEE
31. GRECHINISKII
32. HAGE
33. GRAF
34. HARWIT
35. PETROVIC
36. MARQUETTE
37. HOWE
38. KOEMPE
39. FELDMAN
40. CECCHI-PESTELLINI
41. BRUNI
42. PIRRONELLO
43. FOING
44. EHRENFREUND-FOING
45. FERRARA
46. EVLANOV
47. CASSAM-CHENAI
48. LANZAFAME
49. PRADHAN
50. OMONT

List of Participants

Santi AIELLO
University of Florence
Dept. of Physics
Via L. Pancaldo, 3/45
50127 FIRENZE, Italy

Daniela AVERNA
Università di Catania
Istituto di Astronomia
Viale Andrea Doria, 6
95125 CATANIA, Italy

Simon BALM
University of Sussex
School of Chemistry
and Molecular Sciences
Brighton, BNI 9QJ, UK

Jean BENIT
Vanderbilt University
Dept. of Physics and Astronomy
P.O BOX 1807 B
NASHVILLE, TN 37235, USA

Jürgen BLUM
MPI für Kernphysik
Dept. of Cosmophysics
P.O. Box 103980
6900 HEIDELBERG, FRG

Raymond BREUKERS
University of Leiden
Laboratory Astrophysics
Niels Bohrweg 2
2300 RA LEIDEN, Netherlands

Gabriella BRUNI
ISM-CNR
Via Castagnoli 1
40126 BOLOGNA, Italy

Patrick CASSAM-CHENAI
Ecole Normale Superieure
Lab. Radioastr. millimetrique
24, Rue Lhomond
75005 PARIS, France

Cesare CECCHI-PESTELLINI
Università di Firenze
Dip. di Astronomia
Largo E. Fermi, 5
FIRENZE, Italy

Gregor CHLEWICKI
B50 Aerospace
Postbus 8280
3503 RG UTRECHT, Netherlands

Alexander DALGARNO
Harvard Univ.
Center for Astrophysics
60, Garden Street
Cambridge MA 02138 USA

Francois-Xavier DESERT
Laboratory of Astrophysics
Huygens Laboratorium
Niels Bohrweg 2
2300 RA LEIDEN, Netherlands

Pascale EHRENFREUND-FOING
Groupe de Physique des Solides
University Paris VII
2, Place Jussieu,
75251 PARIS, France

Evgeney EVLANOV
Space Research Institute
Profsoyuznaya 84/32
117810 MOSCOW, USSR

Paul-D. FELDMAN
Johns Hopkins University
Dept. of Physics and Astronomy
21218 BALTIMORE MD, USA

Andrea FERRARA
Università di Firenze
Dip. di Astronomia
Largo E. Fermi, 5
50125 FIRENZE, Italy

Helmut FEUCHTGRUBER
MPI für Physik und Astrophysik
Inst. für Extraterr. Physik
Giessenbachstrasse
8046 GARCHING, FRG

Bernard FOING
Institut d'Astrophys. Spatiale
Dept. of Astrophysics BP 10
91371 VERRIERES, France

Marina FOMENKOVA
Space Research Inst.
Academy of Sciences
Profsoyuznaja 84/32
117810 MOSCOW, USSR

R. GENZEL
MPI für Physik und Astrophysik
Inst. für Extraterr. Physik
Giessenbachstrasse
8046 GARCHING, FRG

Urs GRAF
MPI für Physik und Astrophysik
Inst. für Extraterr. Physik
Giessenbachstrasse
8046 GARCHING, FRG

Alexey GRECHINSKII
Inst. of Space Research
Academy of Sciences of USSR
Profsoyuznaya, 84/32
117810 MOSCOW, USSR

J. Mayo GREENBERG
University of Leiden
Laboratory Astrophysics
Niels Bohrweg 2
2300 RA LEIDEN, Netherlands

Matthew GREENHOUSE
National Air and Space Museum
Smithsonian Institution
20560 WASHINGTON, USA

Joniek HAGE
University of Leiden
Laboratory Astrophysics
Niels Bohrweg 2
2300 RA LEIDEN, Netherlands

Martin HARWIT
National Air and Space Museum
Washington, DC 20560, USA

James HENDERSON
University College London
Dept. of Physics and Astronomy
Gower St.
LONDON WC1E 6BT, UK

Yasushiro HIRAHARA
University of Tokio
Dept. of Chemistry
Hongo, Bunkyo-Ken
TOKIO, Japan

David Aitken HOWE
University of Manchester
Inst. of Science and Technology
M60 IQD MANCHESTER, UK

Melinda HUTSON
University of Arizona
Dept. of Planetary Science
TUCSON, AZ 85271, USA

William IRVINE
University of Massachusetts
FCRAO
619 Lederle GRC
AMHERST MA, 01003 USA.

Peter JENNISKENS
University of Leiden
Laboratory Astrophysics
Niels Bohrweg 2
2300 RA LEIDEN, Netherlands

Carsten KOEMPE
NRC of Canada
Herzberg Inst. of Astrophysics
100 Sussex Drive
OTTAWA K1L OR6, Canada

Giuseppe LANZAFAME
Università di Catania
Istituto di Astronomia
Viale A. Doria 6
95125 CATANIA, Italy

Rene J. LAUREIJS
Lab. for Space Research
Dept. of Astronomy
Landleven 12
9700 AV GRONINGEN, Netherlands

Alain LEGER
University of Paris VII
GPS Tour 23, 4 pl. Jussieu
75251 PARIS cédex 05, France

Young Bok LEE
Laboratory Astrophysics
Huygens Laboratorium
Niels Bohrweg 2
2300 RA LEIDEN, Netherlands

Stephen LEPP
Center for Astrophysics
60 Garden Street
Cambridge MA 07138, USA

John LEWIS
University of Arizona
Dept. of Planetary Sciences
TUCSON AZ 85721, USA

Suzanne MADDEN
University of Massachusetts
FCRAO Dept. of Astronomy
AMHERST MA 01003, USA

Jean Baptiste MARQUETTE
CNRS-Lab. Aerothermique
4ter route des Gardes
92190 MEUDON, France

Francesca MATTEUCCI
MPI fur Astrophysik
K. Schwarzschild Str. 1
8045 MUNICH, West Germany

Calis MEHMET
Anadolu University
Dept. of Chemistry
26470 ESKISEHIR, Turkey

Celia MENDOZA-GOMEZ
Huygens Laboratorium
Laboratory Astrohysics
Niels Bohrweg 2
2300 RA LEIDEN, Netherlands

Young MINH
FCRAO, Univ. of Massachusetts
Dept. of Astronomy
LGRT 619B
AMHERST MA 01003, USA

Lev MUKHIN
Space Research Institute
Profsoyuznaya 84/32
117810 MOSCOW, USSR

Yildiz MUTLU
Middle East Technical University
Dept. of Physics
06531 ANKARA, Turkey

Alain OMONT
Joseph Fourier University
Observatoire de Grenoble
Cerno BP53X
38041 GRENOBLE-CEDEX, France

Nada PETROVIC
Institut Ruder Boskovic
Dept. of Physical Chemistry
Bijenicka 54
41000 ZAGREB, Yugoslavia

Valerio PIRRONELLO
Università della Calabria
Dipartimento di Fisica
87036 RENDE, Italy

Atul D. PRADHAN
CFA-Dept. of Physics
60 Garden Str.
02138 CAMBRIDGE, USA

Claire RIST
Groupe Astrophysique
Dept. of Astrophysics
Rue de la Piscine
38000 GRENOBLE, France

Alex RUZICKA
University of Arizona
Dept. of Planetary Sciences
TUCSON AZ 89721, USA

Norbert SACK
University of Virginia
Dept. of Nuclear Engineering
and Engineering Physics
Charlottesville, VA USA

Peter SCHMELCHER
University of Heidelberg
Dept. of Theoretical Chemistry
Inf 253
6900 HEIDELBERG, FRG

Maurice M. SHAPIRO
University of Maryland
Dept. of Physics and Astronomy
205 Yoakum Pkwy, Apt. 1720
ALEXANDRIA VA, USA

Joel SCHWARTZ
Dept. of Planetary Science
170-25 Caltech
91125 PASADENA, USA

Ralf TIMMERMANN
MPI für Kernphysik
Dept. of Cosmophysics
P.O. 103980
6900 HEIDELBERG, FRG

Chemical Enrichment of the Interstellar Medium

Francesca Matteucci
Max Planck Institut für Astrophysik
K.Schwarzschild Str.1, D-8046 Garching bei München
and Istituto di Astrofisica Spaziale,C.N.R.
Frascati, Italy

The aim of these lessons is to show that there is evidence for chemical evolution of the interstellar gas, as proven by the spectroscopic analysis of stars of all ages in the solar neighbourhood, of interstellar molecules and of HII regions both in our Galaxy and in external ones.

Then we will review the most common ideas about stellar evolution and nucleosynthesis in order to understand where and how the most abundant chemical elements do form. In fact, apart from ^4He and some other light elements such as D, ^3He, and ^7Li, the synthesis of which is usually thought to have occurred during the Big Bang, all the other known chemical elements have been entirely produced in stellar interiors.

Stars sinthetize a mix of chemical elements and return some of the newly created material to the interstellar medium (ISM) so that the next generation of stars will contain a higher mass fraction of the synthetized elements than earlier generations. The way trhou which stars eject their processed material into the ISM can be quiescent (mass loss by stellar wind) or explosive (novae and supernovae), according to their initial mass.

The principal ingredients to build up a model of Galactic chemical evolution will be presented and discussed together with some model results for our Galaxy. The so-called cosmic abundances of matter, namely the solar system abundances, and their relative ratios will be presented and discussed in the light of observational and theoretical results.

Finally, some general conclusions on the most important issues in Galactic chemical evolution will be given together with a discussion about the still open problems and future prospects.

Observational Constraints on Abundances

a) Solar Vicinity

In this section we will briefly review the current status of abundances of field stars of different iron content together with the distribution of G–dwarfs relative to their iron content and the age–iron abundance relation as deduced from stars of all ages.

J. M. Greenberg and V. Pirronello (eds.), Chemistry in Space, 1–41.

Table I

ABUNDANCES IN STARS IN THE SOLAR VICINITY ($[el/H] = log(el/H)_* - log(el/H)_\odot$)

$[el/Fe]$	$[Fe/H]$	Type of stars	References
$[^{12}C/Fe] = -0.1$	$-2.0 \leq [Fe/H] \leq 0$	DW and SDW	Lambert (1988)
$[^{12}C/Fe] = +0.2$	$[Fe/H] = -2.5$		
$[^{14}N/Fe] = -0.25 \pm 0.15$	$-2.5 \leq [Fe/H] \leq -1.0$	DW and SDW	Lambert (1988)
$[^{14}N/Fe] = 0$	$[Fe/H] = 0$		
$[^{16}O/Fe] = +0.4 \pm 0.1$	$-2.6 \leq [Fe/H] \leq -1.0$	DW and G	Lambert (1988)
$[^{16}O/Fe] = -0.5\,[Fe/H]$	$-1.0 \leq [Fe/H] \leq 0$		
$[\alpha/Fe] = +0.45 \pm 0.1$	$-2.6 \leq [Fe/H] \leq -1.0$	DW and G	Lambert (1988)
$[\alpha/Fe] = 0$	$[Fe/H] = -0.4 \pm 0.2$		
$[^{32}S/Fe] = +0.6$	$[Fe/H] \leq -1.0$	DW	François (1987)
$[Fegroup/Fe] \simeq 0$	the whole range	G and DW	Lambert (1988)
$[Ni/Fe] > 0$ (?)	$[Fe/H] < -2.0$	G	Luck + Bond (1983, 1985)
odd-Z elements			
$[Al/Fe] = +0.4$	$[Fe/H] < -1.0$	DW	Abia et al. (1988)
$[Al/Fe] = 0$	over the whole range	DW	François (1986)
$[Al/Mg] = -0.4$	$[Fe/H] = -2$	DW	François (1988)
$[Al/Mg] = -1.2$	$[Fe/H] \leq -2.0$	DW	Magain (1988)
$[Na/Mg] = -0.6$	$[Fe/H] < -1.95$	DW and G	Lambert (1988)
$[Na/Fe] = 0$	over the whole range		
s-process elements:			
$[Sr, Y, Ba/Fe] < 0$	$[Fe/H] < -2.3$	DW	Magain (1988)
$[Zr/Fe] > 0$	$[Fe/H] < -2.3$	DW	Magain (1988)
$[Ba/Fe] < 0$	$[Fe/H] < -1.5$	G	Barbuy et al. (1985)
r-process elements:			
$[Eu/Fe] = +0.6$	$[Fe/H] < -2.3$	DW	Magain(1988)
$[Eu/Ba] = +0.9$	$[Fe/H] < -1.6$	DW and G	Gilroy et al. (1988)
$[Eu/Ba] = 0$	$[Fe/H] > -1.0$	DW and G	Gilroy et al. (1988)

It is worth noting that for many years the relative iron abundance $|Fe/H|$ (e.g. $|Fe/H| = log(Fe/H)_* - log(Fe/H)_\odot$), which is one of the most accessible elements to measure, was considered as representative of the global metal content Z (e.g. the sum of all the element heavier than ^4He), which is dominated by oxygen. This assumption implies a linear proportionality between the abundance of iron as a function of time and that of oxygen as a function of time. In recent years, the finding of an overabundance of oxygen and other heavy elements with respect to iron in metal poor stars has shown that the hypothesis of a linear proportionality between the evolution of iron and that of oxygen is not correct. As a consequence of this, iron should not be identified with the metal content.

Concerning the abundances in field stars there have been many reviews on the subject in the recent years (see for example, Gustafsson 1987; Gehren 1988; Lambert 1988; Wheeler, Sneden and Truran 1989), where quantitative intercomparisons of the abundace results from different investigations and discussion of the basic techniques of abundance determination can be found. Here, we will only mention results concerning the elements which are most relevant to the subsequent discussion.

In table I the ratios of several elements with respect to iron (column 1) are shown. In column 2 the range of relative iron abundance in which these ratios have been measured is reported. In column 3 we indicate the types of stars observed, namely dwarfs (DW), giants (G) and subgiants (SG). Finally, in column 4 there are the references.

Most of the results are taken from the review by Lambert (1988) where a synthesis of all the available observations is given.

As one can see from table I, elements such as C and N seem to be marginally solar in the whole iron abundance range, although the results for very metal poor stars ($|Fe/H| < -2.0$) are still uncertain.

For oxygen and the so-called α–elements (Mg, Si, S, Ca and perhaps Ti) it seems to exist a reasonably well established overabundance with respect to iron in unevolved stars ($|Fe/H| < -1.0$). It is worth noting that the $|Fe/H| = -1.0$ corresponds also to the transition between halo and disk phase, as recognized from kinematical arguments (Gilmore and Wyse, 1986).

The situation of odd–Z elements such as Al and Na is still very uncertain especially in unevolved stars, so that no firm conclusion on the existence of the odd–even effect predicted by Arnett (1971) can be drawn.

The iron–peak elements (V, Cr, Zn, Mn, Ni) generally seem to have a solar ratio with respect to iron in the whole range of iron abundances.

The s–process elements are those elements formed as a consequence of slow neutron capture, where for slow one intends that the time which elapses between two consecutive neutron captures is large compared to the time for a β decay. Conversely, if this time is small compared to the time for β decay we speak of formation of r–process elements. Observations suggest that s–process elements are more underabundant in unevolved stars than the r–process elements. In particular the s–process elements such as Ba, Y and Sr seem to be underabundant with

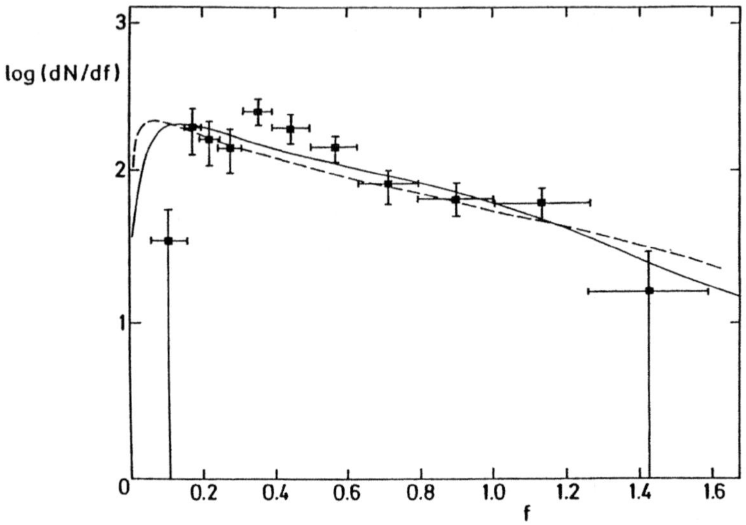

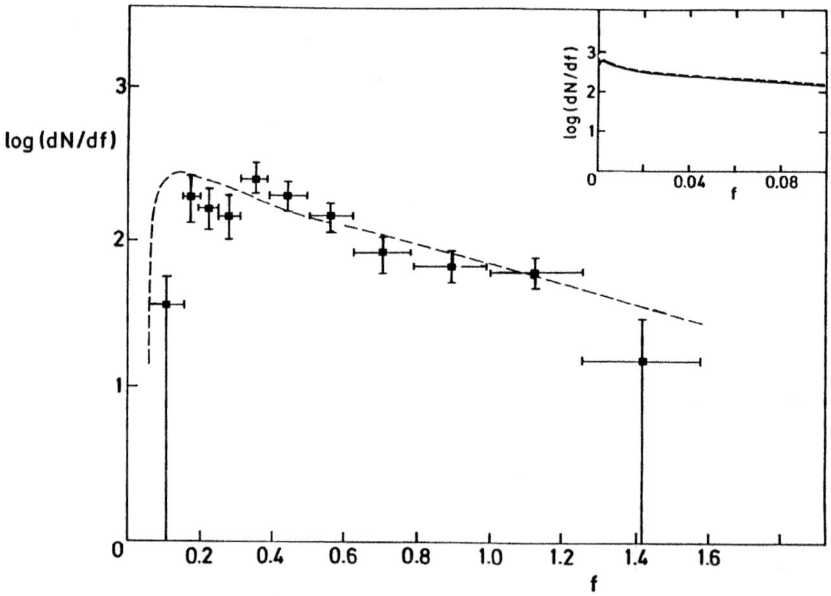

Fig.1a, b. G–dwarf distribution. In the abscissa is plotted $f = 10^{|Fe/H|}$, where $|Fe/H|$ is the iron abundance relative to the sun. The data are from Pagel (1989). Fig 1a shows the model predictions for models assuming primordial infall and different exponents of the surface gas density in the star formation law. In particular, the dashed line refers to k=1.1 and the continuous line to k=2.0.

Fig.1b shows the prediction of a model with enriched infall (as described in the text) and k=1.1. Predictions for the halo and disk are shown separately.

respect to iron in unevolved stars, their ratios becoming solar at a not yet clear iron abundance (between -2.3 and -1.5).

Concerning the r–process elements there is indication of an overabundance of Eu with respect to iron in metal poor stars.

The nucleosynthesis of s and r–process elements is not yet well established so that any definitive statement about it should be regarded with caution.

In the following we will refer mostly to the CNO elements and to the α–elements for which both observations and theory are reliable.

What is clear from the previous discussion is that the solar system abundance ratios are not constant in time and this is because they depend on the different lifetimes of the stellar progenitors of each element and on the modalities of their nucleosynthesis. This will appear more clear when we will discuss in detail the processes of stellar evolution and nucleosynthesis as functions of the initial stellar mass.

The distribution of the G dwarfs as a function of the iron abundance is one of the most important constraints for a model of chemical evolution.

The most recent compilation of data is from Pagel (1989) where a very extensive discussion of the G-dwarf distribution is given. I will only recall that the "G-dwarf problem" is the deficit of metal deficient stars in the solar neighbourhood relative to the simple model of Galactic evolution. In Fig. 1 the data from Pagel (1989) are presented and compared with some model predictions, which will be discussed later.

The distribution of Fig. 1 indicates that the stars with iron abundance lower than 0.3 times the solar value are less than 10 per cent of the total number of stars in the solar vicinity and that those with iron abundance lower than 0.1 times the solar value are about 3 per cent of the total.

The age-$|Fe/H|$ relation (Twarog, 1980; Nissen et al., 1985; Carlbergh et al., 1985) is shown in Fig. 2, where the most recent data are reported. As one can see from Fig.2 , Twarog's and Nissen et al's relations are generally in agreement with each other suggesting a strong increase of the iron abundance during the first two billion years, and a very shallow growth for the remaining time. On the other hand, Carlbergh et al.'s distribution looks much flatter than the others during the first 2 Gyr, although it seems to be biased toward metal poor stars.

It should be also noticed that great uncertainties are present in the determination of stellar ages from stellar evolutionary tracks. As an example of this, the age of the disk estimated by Carlbergh et al. is $\simeq 18Gyr$, whereas that estimated by Twarog is $\simeq 13Gyr$.

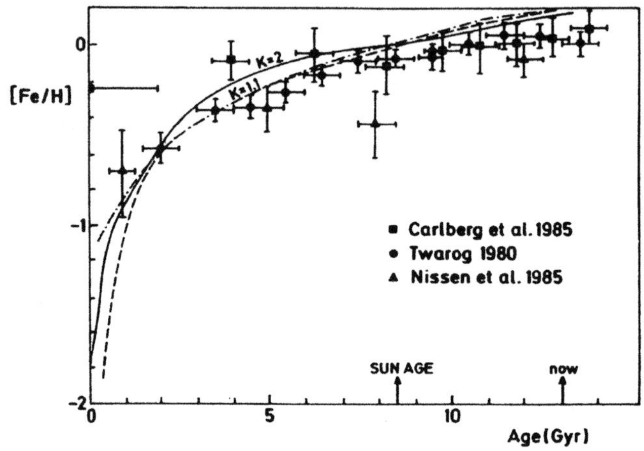

Fig.2. Age–iron relation. Model predictions for k=1.1 and k=2.0 are shown.

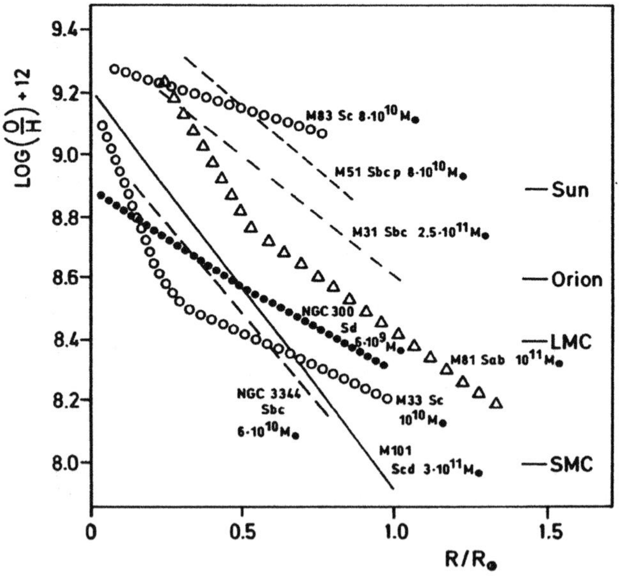

Fig.3. Examples of abundance gradients in the interstellar gas in spirals, deduced from HII regions spectra. From Edmunds (1989).

b) The Galactic disk

In Table II we present a compilation of data for the solar system, the interstellar medium and the Galactic centre.

The solar system abundances shown in Table II are representative of the so-called "cosmic abundances", and will be taken as a reference for comparison of nucleosynthesis and Galactic chemical evolution results.

Values for these abundances are usually taken from the Sun, and especially from chondritic meteorites which are regarded as samples of the primitive solar nebula.

Elemental abundances should in general be accurate by a factor of two, while isotopic ratios are known with much more precision. In fact, within the solar system at least, variations of isotopic ratios of more than 0.1% are regarded as unusual.

Abundances in the interstellar medium are mostly measured in HII regions from emission lines seen in the optical, in the radio and more recently also in the ultraviolet and in the infrared wavelengths. The reason for using the abundances derived from HII regions resides in the fact that the depletion of the gas phase abundances by condensation into grains is less severe than in most of the HI gas. However, Mg, Si and Fe are still depleted by large factors.

i) Abundance gradients of CNO elements

Radio and optical observations of HII regions (Shaver et al., 1983), in a galactocentric distance range 3.5-13.5 Kpc, suggest the existence of O and N abundance gradients.

Oxygen gradients from HII regions are found also in external galaxies, as shown in Fig.3.

Concerning ^{12}C, only measures of the isotopic ratio $^{12}C/^{13}C$ are available. Observations of CNO isotopes ($^{13}C,^{15}N,^{17,18}O$) come mainly from radio and millimetric spectroscopy of interstellar molecules made from CNO atoms.These species are especially interesting for the following reasons:

i) The CNO isotopic ratios are measured in various astrophysical sites : solar system, some red giants, ISM in the solar vicinity and Galactic center.

ii) They are synthetized by different nucleosynthetic processes (H and He-burning) occurring at various stages of stellar evolution both during quiet and explosive phases, so that their distribution represents a good test for the theory.

It is clear from table II that, although there are big uncertainties in the absolute value of the ratio of $^{13}C/^{12}C$ in the ISM, there is relative enrichment with time of ^{13}C with respect to ^{12}C. The same is true for the enrichment of ^{14}N with respect to ^{15}N and for ^{18}O with respect to ^{17}O and ^{16}O.

ii) Abundance gradients of the other elements

In Table II also the available data for Ne, Mg, S, Si and Fe are reported.

In particular, the gradient of S appears to be significantly flatter than the other

TABLE II COMPILATION OF SOLAR SYSTEM AND INTERSTELLAR ABUNDANCES

Isotopic Species	Solar System		Local ISM	Disk-Gradient [dex kpc^{-1}]	Galactic Center		Observed Lines
$^{2}H/H$	$3.6 \left(^{+1.0}_{-1.4}\right) \cdot 10^{-5}$	(5)	$\sim 2\ 10^{-5}$ $5\ 10^{-6}$	positive	?	(6, 15) (17)	molecular (HCN, DCN atomic (H, D)
$^{4}He/H$	0.05-0.10	(4)	~ 0.10	-0.015 -0.020 -0.01 ± 0.008	?	(7) (8) (11)	radio recomb. (HeII) optical (HeII)
$^{12}C/H$	$4.7\ (\pm 1.1) \cdot 10^{-4}$ $4.2 \left(^{+1.3}_{-1.0}\right) \cdot 10^{-4}$	(3) (2)	$(3 \cdot 10^{-4})$	< 0	?	(8)	optical (CII)
$^{12}C/^{13}C$	89±2	(1)	75 (±5) 43±4 47.3 (+5.5 -4.4)	~ 0.05	20±5	(9, 18, 19) (24) (25)	molecular (CO, H_2CO) optical
$^{16}O/H$	$8.3\ (\pm 2) \cdot 10^{-4}$ $6.9\ (\pm 1.3) \cdot 10^{-4}$	(3) (2)	$5\ (\pm 1) \cdot 10^{-4}$	-0.08 (±0.015) -0.07	?	(10) (11)	radio recomb. (HII) optical ([OII], [OIII])
$^{16}O/^{18}O$	490±25	(1)	~ 400	positive	~ 200	(12, 19, 20)	molecular (H_2CO, CO)
$^{18}O/^{17}O$	5.4	(1)	3.7	≤ 0.01	3.5 (± 0.2)	(13)	molecular (CO)
$^{14}N/H$	$1.0\ (\pm .2) \cdot 10^{-4}$ $.9\ (\pm .3) \cdot 10^{-4}$	(3) (2)	$4 \cdot 10^{-5}$	-0.09 (±0.015) -0.095 (±0.03)	?	(11) (16)	optical infrared ([NII])
$^{14}N/^{15}N$	272	(1)	400	(negative)	~ 1000	(14, 21)	molecular (CO)
$^{20}Ne/H$	$8.67 \cdot 10^{-5}$	(1′)	$7.9\ 10^{-5}$	?	?	(11)	optical ([NeII]) and infrared
$^{24}Mg/H$	$3.1 \cdot 10^{-5}$	(1′)	$3.8\ 10^{-5}$	?	?	(28)	
$^{28}Si/H$	$3.46 \cdot 10^{-5}$	(1′)	$3.7\ 10^{-5}$	?	?	(28)	
$^{32}S/H$	$1.78 \cdot 10^{-5}$	(1′)	$1.15\ 10^{-5}$	-0.01 ± 0.02	?	(11)	optical [SI], [SII], [SIII] near infrared [SIII]
$^{36}Ar/H$	$3.3 \cdot 10^{-6}$	(1′)	$2.6\ 10^{-6}$	-0.06 ± 0.015	$6.6 \cdot 10^{-6}$	(24, 11)	far infrared optical
$^{40}Ca/H$	$2.27 \cdot 10^{-6}$	(1′)	$2.29 \cdot 10^{-6}$	?	?	(28)	[ArII], [ArIII]
$^{56}Fe/H$	$3.09 \cdot 10^{-5}$	(1′)	$5.0\ 10^{-6}$	-0.05 ÷ -0.10		(22, 23, 11, 27)	UBV, DDO photometry optical [FeIII]

References from (1) to (21) in Güsten (1986)
(1′) Cameron (1982)
(22) Janes (1979)
(23) Panagia + Tosi (1981)
(24) Hawking + Jura (1987)
(25) Crane + Heygi (1988)
(26) Lester et al. (1983)
(27) Peimbert (1979)
(28) Meyer (1985)

elements, although the data relative to S are limited to the external regions of the disk.

Recent measurements of S gradients in external galaxies (Evans, 1987; Vilchez et al., 1988) seem also to indicate the same trend. The gradient of Ne may also be relatively flat, at least in the Milky Way.

Abundance gradients can also be measured from stellar photometry as well (for a critical discussion about gradients derived this way see Grenon, 1987). In particular, from photometry of K giants and open clusters, the gradient of iron, namely of $|Fe/H|$, can be derived.

Spectroscopic analyses of early B-type stars (Gehren et al., 1985) show very flat gradients for O and N in the galactocentric distance range 8-18 Kpc. However, the observed stellar sample is very limited and probably does not allow firm conclusions to be drawn.

Other constraints on the evolution of the Galactic disk

a) Surface gas density distribution.

Many uncertainties exist in the determination of the total surface gas density (HI+H_2) distribution along the Galactic disk, as is shown in Fig.4, where data collected from various sources by Lacey and Fall (1985), together with data by Bhat et al. (1984), Li et al. (1982) and Talbot (1980) are reported.

The HI component is usually measured from the 21cm emission. Uncertainties arise from the estimate of the optical depth in the solar neighbourhood and, outside it from the assumed rotation curve. Generally, the HI surface gas density is found to be approximately constant over most of the disk.

The H_2 component is obtained from z-integrated CO emissivity. A conversion factor has to be assumed and usually this factor is considered to be independent of the position in the disk, although it may not be so because of the existence of abundance gradients, except in the study of Bhat et al. (1984). These authors have also claimed evidence from the cosmic gamma rays that the surface mass density of molecular hydrogen in the Galaxy has been overestimated by radio observations (see Fig.4).

Generally, the H_2 distribution is found to vary exponentially along the Galactic disk, with a distribution very similar to that of the stellar component.

b) Variations in the Star Formation Rate.

The variations of the present time star formation rate (SFR) with galactocentric distance can be inferred by a variety of radio observations, under the assumption of an initial mass function (IMF). These observations are: the distribution of pulsars, the distribution of Lyman continuum photons and the distribution of supernova (SN) remnants. The absolute value of the inferred SFR is very uncertain because it depends on the assumptions made on the IMF as well as on the mass range for the progenitors of pulsars, and SNe.

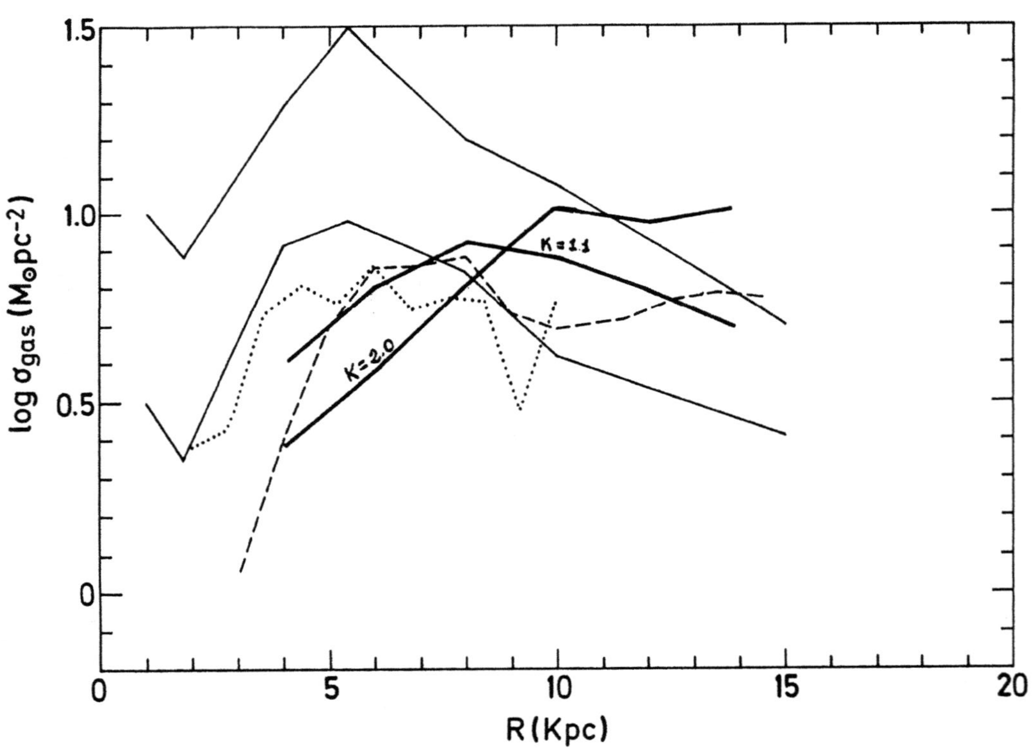

Fig.4. Total gas $(HI + H_2)$ distribution along the Galactic disk. The ticker lines represent model predictions for k=1.1 and k=2.0. The data are represented by the other lines. Most of the data lie between the two continuous lines (Lacey and Fall, 1985). The dashed line represents data from Bhat et al.(1984) and Li et al. (1982), whereas the dotted line represents data from Talbot (1980).

A big collection of data can be found in Lacey and Fall (1985): as a general trend, the SFR is exponentially increasing toward the inner Galactic regions with perhaps a peak at 8 Kpc. The current value of the SFR in the solar neighbourhood, although very uncertain, is given in the range 3-7 $M_{\odot} pc^{-2} Gyr^{-1}$, under the assumption of a constant IMF (Tinsley, 1980; Miller and Scalo, 1979). If variations in the IMF are taken into account this range can be 3-13$M_{\odot} pc^{-2} Gyr^{-1}$ (Güsten and Mezger, 1983).

c) Supernova rates.

It is very difficult to measure the SN rates in our Galaxy, because of the paucity of historical SNe and the difficulty of estimating a total blue luminosity for the Galaxy. van den Bergh et al. (1987) have derived SN rates (type Ia, Ib and II) for a sample of spiral galaxies in the Shapley-Ames catalog and suggested the following rates: 0.3h^2SNu for type Ia SNe, 0.4h^2SNu for type Ib SNe and 1.1h^2SNu for type II SNe, where SNu is SNe$10^{-10 L_{B\odot} 100yr^{-1}}$ and $h = H_o/100 km sec^{-1} Mpc^{-1}$.

van den Bergh (1988) computed the corresponding rates for our Galaxy under the assumption of a total Galactic blue luminosity of $\simeq 2.0\, 10^{10} L_{B\odot}$. The resulting rates are: 0.6$h^2 100yr^{-1}$ for type Ia SNe, 0.8$h^2 100yr^{-1}$ for type Ib SNe and 2.2$h^2 100yr^{-1}$ for type II SNe. Before the study of van den Bergh et al. the only existing estimates for SN rates in galaxies were those of Tamman (1982). These rates are larger by a factor of 3-3.5 than those estimated by van den Bergh et al. Very recent data from Cappellaro and Turatto (1988) seem to confirm that the Tamman SN rates may have been overestimated.

A very recent preprint from Evans et al. (1989) confirms the previous rates found by van den Bergh et al. (1987), although they are slightly different: $\simeq$ 0.28h^2SNu for type Ia's, $\simeq$ 0.27h^2SNu for type Ib's and $\simeq$ 1.04h^2SNu for type II's.

Nucleosynthesis and Stellar Evolution

In order to understand the history of chemical enrichment of the ISM one has to know how stars evolve thru the nucleosynthetic processes occurring in their interiors and how they restore the newly created elements into the ISM.

As is well known the life history of a star can be described as a sequence of nuclear burnings mostly occurring in its central core. This sequence stops with the formation of nuclei of ^{56}Fe, since the subsequent nuclear fusion reactions are endoenergetics. In Table III we give a schematic summary of the main hydrostatic burnings in stellar interiors in the order in which they are occurring: H, He, C, Ne, O and Si. In Table IV are shown the main explosive burnings, when a shock wave passes through the stellar mantle, in the order in which they are occurring: Si, O, Ne, C, He. As we will see more in detail later, only stars with initial mass greater than 10–12$M_{\odot}$ undergo all six nuclear burnings ending up with the so-called onion–skin structure. In order to better understand the effects of explosive nucleosynthesis we show in Figs. 5a,b the presupernova structure of a star of initially 20$M_{\odot}$ as well as the structure of the same star after the passage of a shock wave (occurring during the supernova event) where some explosive nucleosynthesis has taken place (from Thielemann et al. 1989).

As one can see, most of the changements in chemical composition occur because of Si–burning taking place inside the core and involve not more than a mass of $\simeq 1.65M_{\odot}$. Concerning chemical enrichment, it is worth noting that only the mass zones beyond $1.6M_{\odot}$ will be ejected and therefore the majority of the products of Si–burning will be retained into the compact remnant (neutron star).

Stellar Evolution: single stars

- $M \leq 0.08M_{\odot}$. These stars, if they exist, never ignite H ("brown dwarfs"). They do not contribute to galactic enrichment but only to lock up material. Their lifetimes are of the order of several Hubble times, where a Hubble time is $\simeq 15 - 20Gyr$.

- $0.08 < M/M_{\odot} < 0.5$. These stars ignite H in their center but their He–core becomes degenerate before getting hot enough to burn. They never ignite He and end up as He–white dwarfs. They contribute only to lock up material and their lifetimes are again much greater than a Hubble time.

- $0.5 < M/M_{\odot} < M_{He_F}$. These objects ignite He in a degenerate core (He–flash) and end their lives as C–O white dwarfs. The mass M_{He_F} depends on the treatment of convection. Classic stellar models, namely with no overshooting from the convective core, give a value of $\simeq 2.2M_{\odot}$. Classically, the extension of a stellar convective core is set at the layer where the acceleration of the fluid elements is zero. However, zero acceleration point does not coincide with zero velocity point, which should be the real limit of the convective core. Consideration of this effect leads to "overshooting" (for a more extensive and detailed description of this phenomenon see Chiosi, 1987). The value of M_{He_F} obtained by considering overshooting in

Table III

MAIN HYDROSTATIC BURNINGS IN STELLAR INTERIORS

HYDROGEN ⟶	^{4}He (p-p chain or CNO cycle) and ^{13}C,^{14}N,^{17}O
HELIUM ⟶	^{12}C,^{16}O and ^{18}O,^{22}Ne,$^{25,26}Mg$ light-s processing in core He-burning up to $A \simeq 90$; strong-s processing up to Pb and Bi in He-shell flashes
CARBON ⟶	^{20}Ne and ^{23}Na,^{24}Mg
OXYGEN ⟶	^{28}Si,^{31}P,^{32}S,....^{27}Al ^{30}Si,^{34}S,^{35}Cl
SILICON ⟶	^{56}Ni or ^{56}Fe

Table IV

EXPLOSIVE BURNINGS

SILICON —	a) complete b) incomplete
	a) $^{56}Ni \rightarrow {}^{56}Fe, {}^{58}Ni$ $^{57}Ni \rightarrow {}^{57}Fe, {}^{54}Fe$ $^{55}Co \rightarrow {}^{55}Mn, {}^{61,62}Zn \rightarrow {}^{61,62}Ni$ $^{60}Zn \rightarrow {}^{60}Ni, {}^{40}Ca$ $^{44}Ti \rightarrow {}^{44}Ca, {}^{48}Cr \rightarrow {}^{48}Ti$; $^{52}Fe \rightarrow {}^{52}Cr$
	b) $^{56}Ni \rightarrow {}^{56}Fe, {}^{54}Fe$ $^{52}Fe \rightarrow {}^{52}Cr, {}^{55}Co \rightarrow {}^{55}Mn$ $^{57}Ni \rightarrow {}^{57}Fe, {}^{48}Cr \rightarrow {}^{48}Ti$ and, if not consumed during explosion, $^{28}Si, {}^{32}S, {}^{36}Ar, {}^{40}Ca$
OXYGEN ⟶	$^{28}Si, {}^{32}S, {}^{36}Ar, {}^{40}Ca$ $^{34}S, {}^{38}Ar$ (Trace of ^{56}Ni; and ^{54}Fe)
NEON ⟶	$^{16}O, {}^{24}Mg, {}^{28}Si, {}^{32}S, {}^{30}Si$ $^{27}Al, {}^{29}Si$ ($^{26}Al, {}^{22}Na$ are mainly produced in this burning)
CARBON ⟶	$^{46}Ca, {}^{36}S$
HELIUM ⟶	^{56}Ni

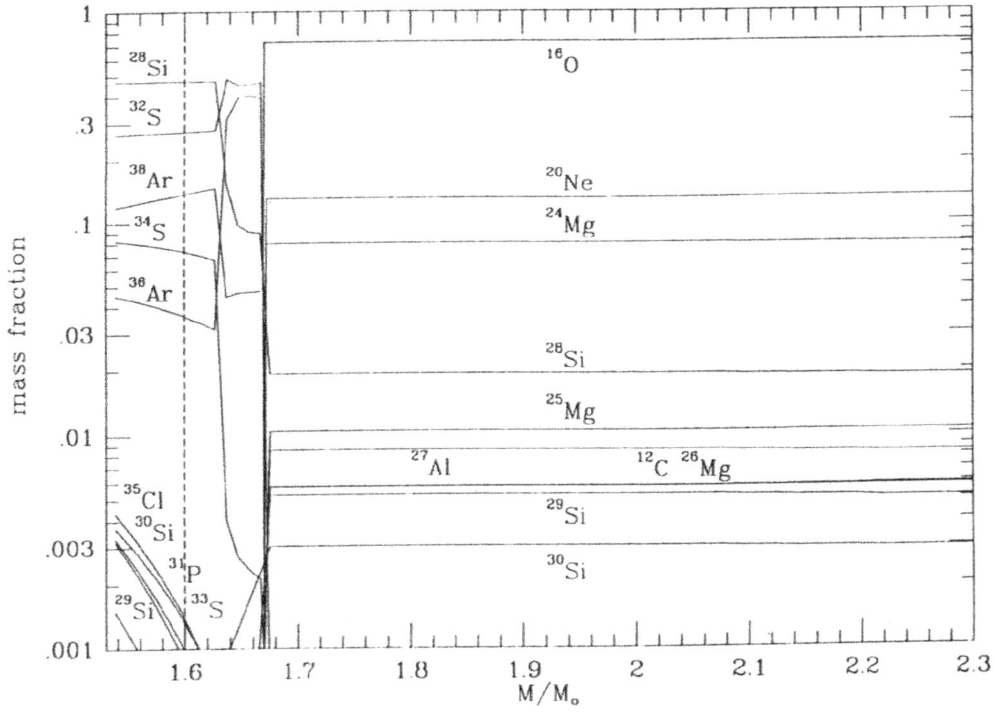

Fig.5a. Mass fractions of major nuclei resulting from hydrostatic burning stages prior to core collapse for a $20M_{\odot}$ star in Main Sequence.

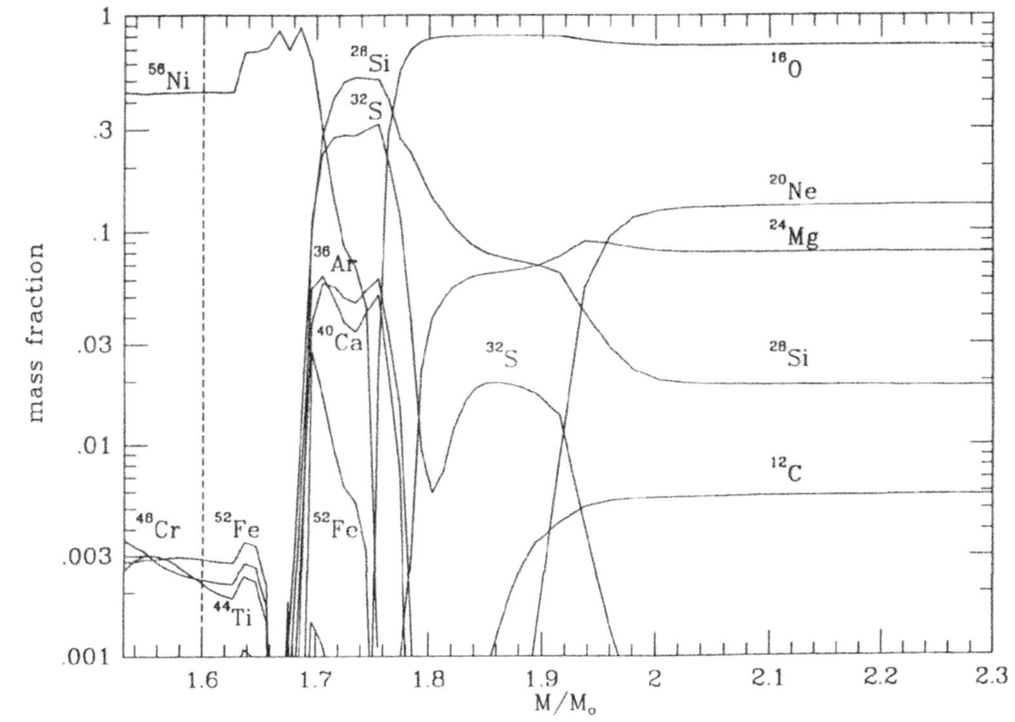

Fig.5b. Mass fractions of a few major nuclei resulting from explosive processes after the passage of the Supernova shock front for the same star.

stellar models is lower than the classical value ($1.85M_\odot$, Maeder and Meynet, 1989, hereafter referred to as MM89).

Stars in this mass range contribute to galactic enrichment through stellar wind and planetary nebula ejection only if their mass is greater than $1M_\odot$. They contribute to the enrichment in He, ^{14}N and perhaps s–process elements (see Gallino et al. 1988). Their lifetimes are ranging from $>> 15Gyr$ to 0.5–1Gyr.

- $M_{He_F} < M < M_w$.(M_w is the limiting mass for the formation of a C–O white dwarf).Stars in this mass range ignite He non–degenerately but develop an electron degenerate C–O core. They end their lives as C–O white dwarfs and contribute to the enrichment in He, C, ^{13}C, ^{17}O and s–process elements, through stellar wind and planetary nebula ejection.

-$M_w < M < M_{up}$.(M_{up} can be defined as the limiting mass for the formation of an electron degenerate C–O core or as the minimum initial mass for an off–center C ignition proceeding up to central C exhaustion). Unless $M_w \simeq M_{up}$ (depending on mass loss and overshooting), these stars should ignite C in a degenerate core, when its mass reaches the Chandrasekhar limit ($\simeq 1.4M_\odot$). As a consequence of this, an explosion should occur which will destroy completely the star (Type I 1/2 Supernovae, Iben and Renzini, 1983)

Current observational estimates give $M_w \simeq 8M_\odot$ (Koester and Reimers, 1985; Weidemann, 1987). On the other hand, recent estimates of M_{up} from stellar models containing mass loss and overshooting, suggest $M_{up} = 6.6M_\odot$ (MM89) whereas in classical models $M_{up} = 8 - 9M_\odot$ (Becker and Iben, 1979).

Therefore, the existence of type I 1/2 Supernovae seems very unlikely, at least at a solar metallicity, to which those estimates of M_{up} refer. The lifetimes of stars in this mass range are between $\simeq 0.5 - 1Gyr$ and several 10^7 years.

Stars in the mass range $M_{He_F} - M_{up}$ eject their nuclearly processed material into the ISM mostly during the asymptotic giant branch (AGB) phase, by means of mass loss by stellar wind and during the planetary nebula (PN) phase.

Episodes of dredge–up, namely when the external convection penetrates inward, in conjunction with thermal pulses, bring to the stellar surface freshly synthetized material (H and He-burning products). For example, during the third dredge up some fresh Carbon can be brought to the surface. Envelope burning will then tranform this Carbon into ^{14}N and ^{13}C, which will be "primary" elements (Renzini and Voli, 1981).

We define "primary" an element which is synthetized inside the core of a star starting from H and He, whereas we define "secondary" an element which needs the presence of some heavy elements (heavier than He) to be synthetized. In other words, a secondary element can never be formed in the very first generation of stars. Usually ^{14}N is a secondary element because is synthesized during the CNO cycle at the expenses of the preexisting C and O. Also secondary are ^{13}C and ^{18}O.

Stars in this mass range, if in binary systems, can give rise to Type I Supernovae and contribute substantially to iron enrichment, as we will see in the next sections.

-$M_{up} < M/M_\odot < 10-12$. These stars ignite C non degenerately and those which have He–cores between 2.2 and 2.5$M_\odot$ ignite oxygen in a degenerate Ne–O core. For those with He–cores from 2.5 to 3 $M_\odot$ all six burning stages are ignited non degenerately and an Fe–core in hydrostatic equilibrium is eventually formed (Woosley, 1987).

The Main Sequence masses of these stars in classical stellar models is 8–12 $M_\odot$, with overshooting is 6.6–10$M_\odot$ (MM89).

They end their lives as type II Supernovae and contribute mostly to galactic enrichment in He and only very little in heavier elements. For this reason those stars have often been indicated as possible progenitors for the Crab Nebula. Their lifetimes are of the order of several 10^7 years.

-$10-12 < M/M_\odot < M_{u_{SNII}}$. These stars are responsible for producing the bulk of heavy elements such as O, Ne, Mg, Si, S, Ar, Ca and r–process elements. They end their lives as Type II Supernovae, still possessing a H–rich envelope, and eject most of the nuclearly processed material throu the explosion. Some of the material, mostly He, and to a lesser extent C, ^{22}Ne and N, are restored to ISM also by means of stellar wind.

-$M_{u_{SNII}} < M/M_\odot < 100$. $M_{u_{SNII}}$ is the limiting mass for the formation of Wolf–Rayet stars and should be around 40$M_\odot$ (MM 89). Wolf–Rayet stars may still explode, but lacking an hydrogen envelope, their light curves and spectra would be different from type II Supernovae.

Type II Supernovae after explosion leave a compact remnant which can be a neutron star or a black hole. The limiting mass for the formation of black holes is very uncertain. Woosley (1987) suggests that stars with original mass greater than 20$M_\odot$ may leave black holes as remnants. The mass of the neutron star is usually assumed to be $\simeq 1.4M_\odot$.

The explosion mechanism in massive stars has not yet been understood so it is highly uncertain how the shock wave forms and propagates through the outer layers. Elements heavier than Si may experience, as we have seen before, considerable modification as the shock wave moves through, but also by accreting into the incipient neutron star before the explosion starts (see Woosley, 1987 for a detailed discussion on the possible explosion mechanisms). The lifetimes of these stars are of the order of million years.

-$M > 100M_\odot$. Following the exhaustion of He, the central regions contract and the star goes directly to oxygen burning. For a certain value of the He–core ^{16}O is not ignited in a stable fashion. This instability is known as "pair–instability" and occurs because a large portion of the energy from gravitational contraction goes , at a temperature of $\simeq 2\,10^9$ K, into creation of pairs (e^+, e^-) with the consequent reduction of the adiabatic index below 4/3. This kind of explosion destroys completely the star. From the point of view of galactic enrichment, these objects would mostly contribute to ^{16}O. Recent models with overshooting and mass loss indicate a mass range for these pair–creation Supernovae of 100–200$M_\odot$ (MM89), in agreement with previous calculations. The lifetimes of these stars are

$\leq 10^6$ years.

-**Supermassive objects** should collapse directly to black holes or suffer total disruption due to explosive H–burning. The minimum initial mass for such objects is $4.1\,10^2$ (Appenzeller and Tscharnuter, 1973).

Masses $M > 7.510^5$ end up as massive black holes (Appenzeller and Fricke, 1972). Masses in the range $4.110^2 - 7.510^5$ suffer total disruption during H–burning. The results are very sensitive to initial stellar metal content and the explosion is possible only for solar metal content. If they exist, these objects would produce mostly He and trace of ^{15}N and ^{7}Li (Woosley et al., 1984).

Stellar Evolution: binary systems

In this section we will not discuss in detail the evolution of all binary systems but only of some particular systems, which are relevant to Galactic chemical enrichment. In particular, we will discuss some possible evolutionary scenarios leading to the formation of Type I Supernovae, which are very important contributors to the enrichment in iron.

Carbon–deflagration in C–O white dwarfs whose mass exceeds the Chandrasekhar limit can reproduce the majority of the observed features in type Ia Supernovae (the classical type I SNe to be distinguished from other categories of peculiar ones called type Ib and Ip). In the explosion the right amount of ^{56}Ni is produced, which decayment into Co and Fe powers the light curve of SNI. Besides, the deflagration model (deflagration means that the flame velocity is subsonic as opposed to detonation where the flame velocity is supersonic) allows the formation of intermediate mass elements (from C to Si) which are observed in type Ia spectra.

Theoretical models computed by Nomoto et al. (1984) and Thielemann et al. (1986) for such a configuration predict $0.86M_\odot$ of Fe–group elements, $0.58M_\odot$ of which are originally in the form of ^{56}Ni, plus $0.27M_\odot$ of Si–Ca elements. These results are generally in agreement with other studies (Woosley et al. 1984), although the proportion between Fe–peak elements and intermediate mass elements varies due to the different assumptions made on the velocity of the deflagration flame.

Possible configurations for C–deflagration of a C–O white dwarfs are:

i) **A degenerate C–O white dwarf plus a red(super)giant companion.** When the red(super)giant fills its Roche lobe, C–deflagration occurs due to accretion of H–rich material on to the white dwarf, which leads the white dwarf mass to exceed the Chandrasekhar limit. This model was originally proposed by Whelan and Iben (1973). The expected realization frequency of this kind of configuration, considering the limits on the initial white dwarf mass, primordial separation of the binary system and mass ratio, is $\simeq 0.003$ Supernovae yr^{-1} (Iben and Tutukov, 1984), consistent with the observed rates. In fact, from Evans et al (1989) rates, one can infer a type Ia Supernova rate in our Galaxy in the range 0.0014 - 0.0056 Supernovae yr^{-1}, depending if one assumes an Hubble constant of 50 or 100 $Kmsec^{-1}Kpc^{-1}$, respectively.

The clock for the explosion of these systems is given by the lifetime of the secondary star (namely the initially less massive, in this case the red(super)giant). Since this star can have an initial mass in the range $0.8 - M_{up}$, this model can explain the existence of type Ia Supernovae in elliptical galaxies where star formation is likely to have stopped several billion years ago.

However, this kind of configuration should more probably give rise to cataclismic systems and Nova outbursts. In fact, the occurrence of H and He flashes prevents the dwarf from retaining matter accreted between flashes, unless the accretion rate is within a relatively narrow interval.

There are two classes of Nova outbursts, the so-called "dwarf" Novae, which can be modeled as the consequence of instabilities in the accretion disk surrounding a white dwarf in a cataclismic variable system, and "classical" Novae which are also thought to be accreting white dwarfs but powered during outburst by a thermonuclear instability rather than the sudden release of gravitational energy.

From the point of view of galactic enrichment Novae give a negligible contribution when compared to Supernovae, being the matter ejected in a nova outburts of the order of $\simeq 10^{-5}$ to $10^{-4} M_{\odot}$.

Truran (1985) estimated the enrichment rate from novae to be $\simeq 0.001 M_{\odot} yr^{-1}$ whereas that from Supernovae $\simeq 0.04 M_{\odot} yr^{-1}$.

The nuclear reactions involved in a nova outburst are those appropriate to the burning of hydrogen at high temperatures.

ii) **Double degenerate model.** Two degenerate C–O white dwarfs which merge after gravitational wave radiation, can give rise to a C–deflagration when the total mass of the system reaches the Chandrasekhar limit.

This scenario has been proposed by Iben and Tutukov (1984,1985) and Webbink (1984) and it has been the most favoured up to now.

The clock for the explosion in this case is given by the sum of the lifetime of the primordially less massive component plus the gravitational time delay. The mass range of the progenitor stars is $5 - M_{up}$ (Iben and Tutukov, 1984; Tornambè and Matteucci, 1987) but the gravitational time delays can be as long as several Hubble times. Again, type Ia Supernovae in elliptical galaxies can be explained.

Peculiar Type I Supernovae: type Ib. They have a luminosity at the maximum lower by a 1.5 to 2.0 magnitudes with respect to type Ia. On the other hand, both type Ia and type Ib show an exponential decay in their light curves which could be explained by a Ni–Co–Fe decay, but type Ib require a smaller amount of Ni. Another difference among type Ia and type Ib is that these latter are found only in spirals whereas type Ia are found in all type of galaxies.

Observational frequency estimates (Panagia 1987; van den Bergh et al. 1987) indicate that a fraction from 30 to 60 per cent of all type I Supernovae are of type b.

Possible evolutionary scenarios (still very uncertain) leading to the formation of type Ib Supernovae are:

i) **Explosion of a Wolf–Rayet star.** This model has been proposed by Gaskell et al. (1986) and Filippenko and Sargent (1986). Some problems may arise with the frequency of type Ib if Wolf–Rayet stars have as progenitors masses greater than $40M_\odot$. In fact, the number of these stars is too small to explain the observed frequency of type Ib. Another problem with this model arises from the fact that the final configuration of a Wolf–Rayet star before explosion cannot explain the observed structure of the ejecta of type Ib Supernovae (see Branch and Nomoto, 1986).

ii) **He–off–center detonation of a C–O white dwarf.**

This model has been proposed by Iben et al. (1987) and Tornambè and Matteucci (1987), and consists of a system of a C–O white dwarf exploding by He–detonation after merging, due to gravitational wave radiation with an He–nondegenerate star.

The clock for the explosion is given by the lifetime of the He–star ($\geq 2.3M_\odot$) plus the gravitational time delay, which is not longer than 810^8 years for this special configuration (see Tornambè and Matteucci, 1987, for more details).

The He–off center detonation in a $1.4M_\odot$ white dwarf should produce almost only iron–peak elements. In particular $\simeq 0.3M_\odot$ of ^{56}Ni should be created, which is the right amount to power the light curve of these Supernovae. This model could explain why in the ejecta of type I b Supernovae features of Fe and He appear first, and then those of intermèdiate mass elements.

iii)**C–O white dwarf accreting material at low efficiency** from the wind of a red giant companion (Branch and Nomoto, 1986). The clock for the explosion (again He–off center detonation) is given by the lifetime of the red giant which should have a mass in the range $4 - 7M_\odot$.

The last two scenarios may however present some difficulty if the amount of oxygen found in type Ib Supernovae, which is very uncertain, would exceed $0.5M_\odot$.

Summary: where and how the elements are formed

Here we summarize where and how the chemical elements relevant to our discussion are synthetized.

-^{12}C: quiescent He–burning in high and intermediate mass stars.

-^{13}C: quiescent and explosive H–burning (cold and warm CNO cycle, respectively), both in single and binary intermediate mass stars (Novae).

This element is generally "secondary", but there could be some primary production as a result of the third dredge–up in asymptotic giant branch stars operating in conjunction with envelope burning.

-^{14}N: quiescent H–burning (cold CNO cycle) in low and intermediate mass stars. It could also be produced as a primary element, in the same way as discussed for ^{13}C.

-^{15}N: explosive H–burning occurring in Novae and Supernovae. The evidence for a production of ^{15}N in explosive processes is that the ratio $^{15}N/^{14}N$ from cold CNO cycle is 100 times lower than the same ratio in the solar system abundances.

-^{16}O: He–burning in massive stars.

-^{17}O: mostly in cold CNO cycle in single intermediate mass stars ($M > 4M_{\odot}$) but also in explosive H–burning.

-^{18}O: from destruction of ^{14}N by the reaction $^{14}N(\alpha,\gamma)^{18}F(\beta^{+})^{18}O$ occurring in stellar regions suffering He–burning but also from quiescent and explosive H–burning.

-^{20}Ne and ^{24}Mg: C–burning in massive stars and also in C–deflagration in C–O white dwarfs. Mg can be produced also in quiescent and explosive Ne–burning.

-^{28}Si and ^{32}S: quiescent and explosive O–burning in massive stars and also during C–deflagration in C–O white dwarfs.

-^{40}Ca: Explosive O–and Si–burnings in massive stars and in C–deflagration in white dwarfs.

-^{56}Fe: C–deflagration in C–O white dwarfs and quiescent and explosive Si–burning in massive stars.

Before closing this section we would like to remind which are the main uncertainties still existing in stellar evolution and nucleosynthesis:

- the rate of the reaction $^{12}C(\alpha,\gamma)^{16}O$ which is now believed to be higher by a factor from 3 to 5 (Kettner et al., 1982; Caughlan et al. 1985), with respect to previous estimates (Fowler et al. 1975)
- the treatment of convection
- the nature of the explosion mechanism in massive stars (prompt or delayed explosion?, see Woosley, 1987, for a discussion)
- the range of stellar masses which indeed explode
-the velocity of the deflagration front in C–deflagration models of type I Supernovae.

Galactic Chemical Evolution

The basic parameters of a chemical evolution model are:
Initial Conditions
Stellar Birthrate Function
Stellar Yields
Besides those there are some supplementary parameters, which, nevertheless, can be as fundamental as the basic ones. These parameters are:
Infall of Primordial or Enriched Material
Radial Inflow
Galactic Winds

Initial Conditions. One can start the evolution of a galaxy with primordial gas or with gas where a certain level of enrichment has already been reached (prompt initial enrichment (PIE), or prompt initial pollution (PIP), if the enrichment level is rather low). Besides, one can assume that initially all the gas, out of which the galaxy will form, is already there at t=0, or that the gas will be slowly accreted in time. While the first case may be appropriate for spheroidal systems the second one seems to reproduce better the features of galactic disks.

Stellar Birthrate Function. Generally speaking the stellar birthrate should be a function of both time and stellar mass. However, in the absence of a detailed understanding of the star formation process and for simplicity reasons, the stellar birthrate is usually separated in to two independent functions:

$$B(m,t) \;=\; \psi(t)\,\varphi(m)$$

$\psi(t)$ is usually known as the star formation rate (SFR) and $\varphi(m)$ as the initial mass function (IMF).

I will briefly recall here the most common parametrizations of these two functions:

i) **SFR**

a) time constant SFR (Twarog and Wheeler, 1982;1987)

b) exponentially decreasing SFR (Tosi, 1988a)

c) proportional to some power (between 1 and 2) of the total surface gas density (Lacey and Fall, 1985). This formulation derives from Schmidt's (1959) original suggestion that the SFR should vary as a power of the volume density of gas. In principle, Schmidt's formulation of the SFR and this one are equivalent when the exponent is equal to 1, whatever the scale height of gas, but they are also equivalent for any value of the exponent as long as the scale height of gas is constant with galactocentric radii. The advantage of the formulation with the surface gas density lies in the fact that it does not require the computation of the scale height of the gas layer.

A slightly more complicated formula can be obtained by introducing a dependence also on the total surface mass density (Talbot and Arnett, 1975; Chiosi, 1980; Matteucci and Greggio, 1986; Dopita, 1985).

d) proportional to the surface density of molecular hydrogen only (Rana and Wilkinson (1986), where also a dependence on the metal content of gas is assumed).

e) Bimodal SFR: high mass mode of star formation exponentially decreasing and low mass mode constant (Larson, 1986; Wyse and Silk, 1987).

f) Finally, a more physical approach based on the analysis of induced and spontaneous star formation (Franco and Shore, 1984; Shore et al., 1987, Matteucci et al., 1989)

ii) IMF

Generally the IMF is expressed as a power law . The exponent can be constant over the whole range of stellar masses (unique power law) or it can vary so that the IMF is divided in various segments. Usually the IMF is assumed to be constant in space and time.

However, several suggestions about a variable IMF have been proposed in recent years.

Güsten and Mezger (1983) suggested that, in order to explain the SFR inferred from the distribution of ionizing photons, as measured by the thermal radio emission, the lower mass limit in the IMF in spiral arms should be 2-3$M_{\odot}$, whereas in the interarm regions should increase with decreasing radius. They were able to reproduce the gradient of oxygen but they did not show predictions for the gradient of nitrogen, which is synthetized mainly in low and intermediate mass stars and therefore may be affected by this kind of assumption. This leaves the doubt that this bimodal IMF may not satisfy all the observational constraints.

Larson (1986), on the basis of the claim by Scalo (1986) of a double- peaked IMF (but see also D'Antona and Mazzitelli, 1983,1986) , suggested that the SFR may be bimodal, as discussed previously. In this framework, Larson could reproduce the missing mass in the solar neighbourhood, in the form of white dwarfs, as claimed by Bahcall (1984), but this conclusion has been challenged by some recent work by Kuijken and Gilmore (1989). However, his model overestimates the solar abundance of oxygen unless masses greater than 15$M_{\odot}$ are not contributing to galactic enrichment: an unrealistic hypothesis after the occurrence of SN1987a, the progenitor of which should be a star of $\simeq 20M_{\odot}$.

Following the same line of Larson, Wyse and Silk(1987) proposed a slightly different bimodal SFR.

It is worth noting that a recent study on fragmentation processes (Ferrini et al., 1989) has shown that the existence of a bimodality in the star formation process lasting for the whole galactic lifetime would require very peculiar physical conditions.

Finally, Terlevich and Melnick (1983) have suggested the possibility of variations of the IMF with the metal content of gas, but the way of parametrizing this variation is subject to uncertainties.

In conclusion, it is not clear at present whether and how the IMF should vary and how this variability should be parametrized. Therefore, the assumption of an IMF constant in space and time is still to be preferred in galactic chemical evolution models.

Stellar Yields

The amount of processed and unprocessed material restored to the ISM by stars at their death, depends on some fundamental parameters, namely:

-M_α: the mass of the He–core (where H is turned into Helium)

-M_{CO}: the mass of the C–O core (where He is turned into heavier elements)

-M_r: the remnant mass

-$(M_{ej})_i$: the mass ejected by the star in the form of the element i.

These quantities, which are the result of stellar evolution and nucleosynthesis, should be specified for any value of the initial stellar mass and, possibly, for different values of the initial stellar chemical composition.

Infall Rate. The possibility of the existence of infall, as a natural consequence of galaxy formation from extended halos, was first introduced as a solution to the G-dwarf problem (Larson, 1972; Lynden-Bell, 1975), and still represents its best solution.

The existence of infall could also be desirable to prevent gas consumption in spirals in times shorter than their ages.

From an observational point of view High Velocity Clouds (HVC) and Very High Velocity Clouds (VHVC) are interpreted as infall.

The infall rate in galactic chemical evolution models is usually parametrized in the following ways:

a) constant in space and time (Twarog and Wheeler, 1982,1987),

b) exponentially decreasing in time and constant in space (Tosi, 1988a),

c) exponentially decreasing in time and also variable in space (Chiosi and Matteucci, 1982; Clayton,1987; Matteucci and François, 1989).

Generally models assume the chemical composition of the infalling material to be primordial, with few exceptions (Tosi, 1988b; Matteucci and François, 1989).

Radial Inflow. Recently, Lacey and Fall (1985) and Tosi (1988a) have studied the effects of radial inflow on galactic chemical evolution. Their main conclusions are:

a) radial inflow with velocities not exceeding $2kmsec^{-1}$ improve the predicted abundance gradients in models with a SFR linearly dependent on the total surface gas density, whereas they do not substantially change the results for a quadratic dependence of the SFR on surface gas density, and in any case infall is required (Lacey and Fall, 1985).

b) By assuming that the mass flowing inward in the galactic disk is just a fraction of the infalling gas, Tosi suggests that this fraction should not exceed 50%.

Outflow: galactic winds. Supernova explosions can trigger galactic winds when the energy input from supernovae equates the binding energy of the gas.

Models with galactic winds have been proposed for ellipticals (Larson, 1974; Arimoto and Yoshii, 1987; Matteucci and Tornambè, 1987; Vader, 1986).

They represent a good explanation of the observed mass–metallicity relation in such objects, and can account for the observed almost solar iron abundance in the intracluster medium (Matteucci and Vettolani, 1988).

The possibility of galactic winds also in dwarf irregular galaxies (extragalactic HII regions) has been proposed by Matteucci and Chiosi (1983) and by Matteucci and Tosi (1985), in order to explain the observed spread in their chemical properties.

In fact, they suggested that different amounts of chemically enriched material lost from these objects, as a consequence of a different number of star burst events, could explain the observed spread in the distribution of N and O abundances.

The Basic Equations

Let us define M as the total mass of the system. M may increase as a result of accretion at a rate A(t) or even decrease as a result of mass outflow at a rate W(t), namely:

$$dM/dt \;=\; A(t) - W(t)$$

Be M_{gas} the total mass of gas. This will be consumed by forming stars at a rate $\psi(t)$ and by ejection of gas out of the system at a rate W(t), but also replenished by infall material at a rate A(t) and by the stellar ejecta at a rate E(t):

$$dM_{gas}/dt \;=\; -\psi(t) + E(t) + A(t) - W(t).$$

The abundance by mass of any elemental species X_i varies with time according to:

$$d(X_i M_{gas})/dt \;=\; -X_i(t)\psi(t) + E_i(t) + X_{inf}(t)A(t) - X_i(t)W(t),$$

where X_{inf} is the abundance of the infalling material which can be constant or variable in time.

The quantities E(t) and $E_i(t)$ are somewhat complicated since they depend on the number of stars dying at the time t. Any star of mass m, born at the time $(t - \tau_m)$, where τ_m is the stellar lifetime as a function of the mass, will contribute to the galactic enrichment at time t. The quantity E(t) can then be written as follows:

$$E(t) = \int_{m(t)}^{\infty} (m - M_r)\psi(t - \tau_m)\varphi(m)dm.$$

$E_i(t)$ is the same quantity but in the form of a specific element i:

$$E_i(t) = \int_{m(t)}^{\infty} ((m - M_r - mP_m)X_i(t - \tau_m) + mP_{im})\psi(t - \tau_m)\varphi(m)dm,$$

wheme P_m and P_{im} represent the fractions of material nuclearly processed and ejected by a star of mass m in the form of all elements and of the specific element i, respectively.

Under the assumption of instantaneous recycling approximation (IRA), e.g. stars less massive than $1M_\odot$ live forever and the others die instantaneously, one can write:

$$E(t) = \psi(t)R,$$

where R is defined as the "return fraction" and:

$$E_i(t) = RX_i\psi(t) + Y_i(1 - R)(1 - X_i)\psi(t),$$

where Y_i is known as the " yield" and is defined as:

$$Y_i = \int_{m(t)}^{\infty} mP_{im}/(1 - R).$$

In Fig.6 the quantity P_{im} as a function of the initial stellar mass is shown. He and heavy elements are distinguished as well as the different contributions from stellar wind and Supernovae of type II. This Figure is taken from Maeder (1985) and refers to old theoretical results, so it should be taken only as indicative.

In fact, this kind of diagram is a strong function of the treatment of mass loss and overshooting in stellar models, which influence the quantity M_α.

It is still controversial the effect of mass loss by stellar wind on M_α in massive stars. Originally, Chiosi et al.(1978; 1979) found a lower M_α with respect to conservative models as a consequence of mass loss. Later, Maeder(1981;1983) did not find any sensitive difference with respect to conservative models. At the same time, Bressan et al.(1981) including mass loss in the same way as Chiosi et al. together with strong overshooting, found that the net result, due to the predominance of overshooting, was M_α greater than in the conservative models. More recently, Maeder and Meynet (1987;1989) recomputed models with mass loss and moderate overshooting and found, as a net result, a moderate increase of M_α with respect to conservative models.

In any case, it is clear that the net effect of considering overshooting from convective cores in stellar models is to enlarge the mass of the He–core M_α, with

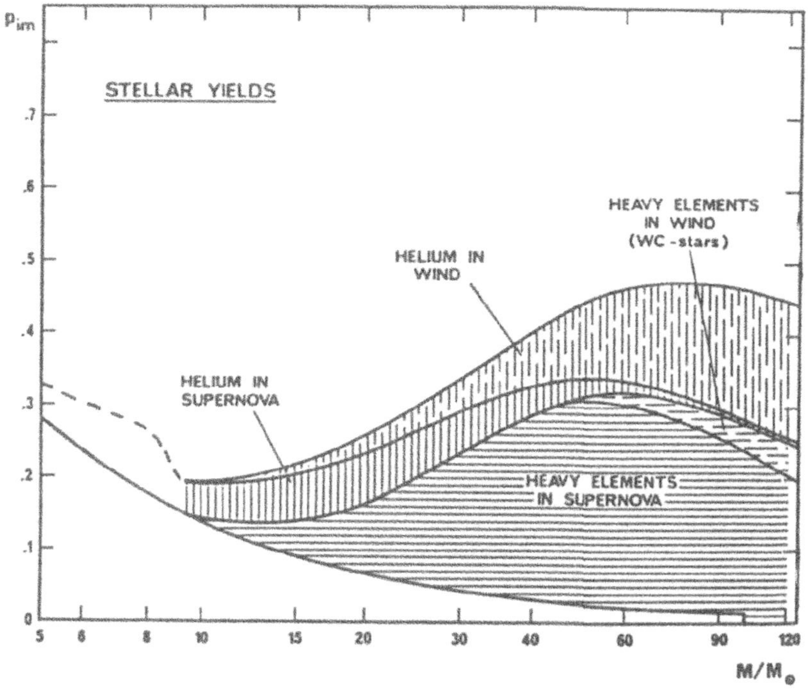

Fig.6. Representation of the stellar yields expressed in mass fractions. The Helium and the heavy elements ejected by stellar wind are distinguished from those ejected in the Supernova. Type I Supernova ejecta are not taken into account. The figure is from Maeder (1985).

respect to models with no overshooting. While an increase of the He–core mass in massive stars clearly leads to an increase in the production of metals, in low and intermediate mass stars the situation is more complicated. In fact, in these stars, most of the material inside the He–core is never ejected with the exception of the fraction which is dredged–up to the envelope. An increase of the mass of the He–core for the same initial mass leads to a smaller mass in the envelope, which will be ejected. Although there is a lack of complete nucleosynthetic models with overshooting for this mass range, the effect of overshooting on nucleosynthesis in intermediate mass stars can be summarized as follows (Greggio and Tosi, 1986; Serrano,1986): elements such as ^4He and ^{14}N (of secondary origin), brought to the surface during the second dredge–up, are enhanced by overshooting, whereas nuclear processing during the third dredge–up and envelope burning is decreased with consequent decrease of ^{14}N and ^{13}C of primary origin as well as He produced during envelope burning.

Generally, in galactic chemical evolution models, one accounts for these effects just by changing the $M_\alpha - M$ relation.

Chemical evolutionary models

The most simple model of galactic chemical evolution basically starts from pure gas (Z=0) and assumes:

a) instantaneous recycling approximation (I.R.A.),
b) gas well mixed at any time,
c) constant IMF in space and time and
d) constant "yield".

Deviations from the simple model give rise to:

i) Models with PIE or PIP (Truran and Cameron, 1971; Ostriker and Thuan, 1975; Gilmore, 1984; Cayrel, 1986).

ii) Models with infall (Tinsley, 1980; Chiosi, 1980; Diaz and Tosi, 1984; Lacey and Fall, 1985; Matteucci and Greggio, 1986; Clayton, 1987; Twarog and Wheeler, 1982,1987; Matteucci and François, 1989).

iii) Models with infall plus inflow (Mayor and Vigroux, 1981; Lacey and Fall, 1985; Tosi, 1988a).

iv) Models with outflows (Matteucci and Chiosi, 1983; Matteucci and Tosi 1985; Arimoto and Yoshii, 1986,1987; Matteucci and Tornambé, 1987; Gilmore and Wyse, 1986).

v) Models with metal enhanced star formation (MESF) (Talbot and Arnett, 1973).

vi) Models with variable IMF or variable yield (Güsten and Mezger, 1983; Larson, 1986; Edmunds and Pagel, 1984; Serrano and Peimbert, 1983).

It is worth noting that the majority of models assume I.R.A.. On the other hand, it has been demonstrated (Matteucci and Greggio, 1986; Matteucci, 1986; Matteucci and François, 1989) that I.R.A. is a very bad approximation in studying the evolution of those elements which are mainly produced in low and intermediate mass stars, such as Fe and N. Therefore, it is of fundamental importance to relax I.R.A. if one wants to predict the G–dwarf distribution as a function of iron abundance, the age- $|Fe/H|$ relation and the evolution of the ratios of the various heavy elements with respect to iron.

It is also worth noting that the majority of models do not reproduce all the observational constraints of the solar vicinity and the whole disk at the same time.

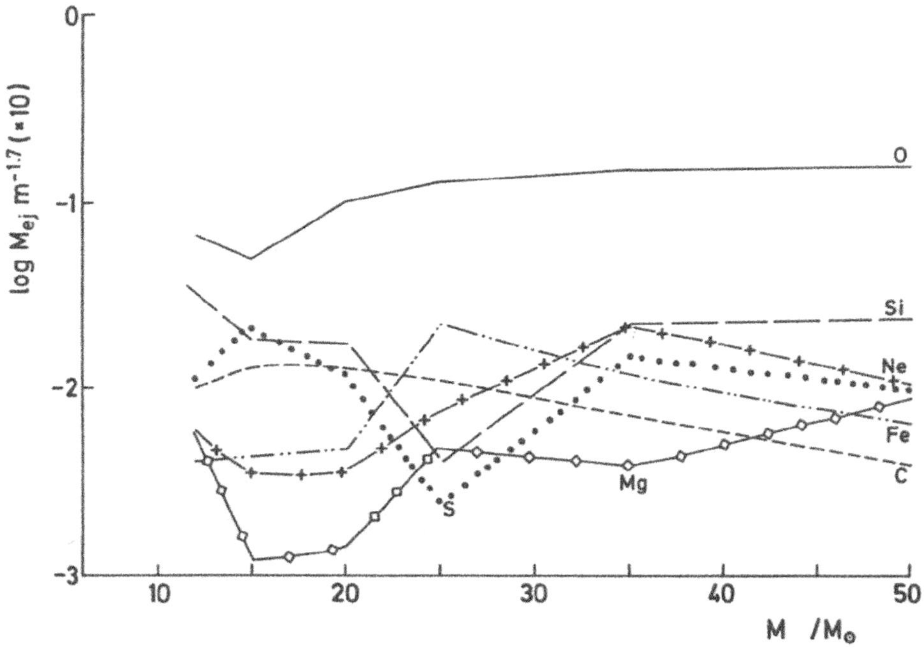

Fig.7. Nucleosynthesis in massive stars. The ejected material, in the form of several heavy elements, averaged over the IMF, is plotted as a function of the initial stellar mass. The nucleosynthesis prescriptions are from Woosley and Weaver (1986) and Woosley (private communication).

A Possible Scenario for the Evolution of Single Elements in the Solar Vicinity and in the Whole Disk of our Galaxy

Our Galaxy offers a unique opportunity to set constraints on theories of galaxy formation and evolution , since only in the Milky Way one can obtain the tridimensional spatial density distribution, stellar kinematics and stellar chemical abundances.

Eggen et al. (1962) found a correlation between the ultraviolet excess, $\delta(U-B)$, orbital eccentricity and vertical motions implying that halo stars had formed first during a free–fall time collapse and the residual gas had dissipated in a comparable time to form the disk.

More recently, Norris et al.(1985) discovered metal poor stars with low eccentricity and Zinn (1985) discovered a flattened population of globular clusters with intermediate metallicity. These objects seem to belong to a "thick disk" , the existence of which was already suggested by Gilmore and Reid (1983). The following schematic picture arises : the halo formed first, in a time scale which should not exceed 10^9years, as indicated by chemical considerations, especially from the constant overabundance of oxygen and $\alpha - elements$ with respect to iron until $|Fe/H| = -1$. In fact, the iron abundance of -1 dex corresponds also to the border between halo and disk (thick and thin) stars, as suggested by their kinematical properties.

The simplest explanation for the behaviour of oxygen and α– elements with respect to iron is that, assuming the IMF to be constant in space and time, halo stars have been enriched only by the products of massive stars, wheras disk stars contain also the contamination from stars of smaller mass (and longer life), in particular from type I Supernovae which greatly contribute to iron enrichment.

From calculation of chemical evolution (Matteucci and Greggio, 1986; Matteucci and François, 1989; Wyse and Gilmore, 1989) the time scale corresponding to a relative iron abundance of -1 dex is found to lie in the range of several times 10^8 up to 10^9 years.

Matteucci and Greggio(1986) and Matteucci (1987,1988) computed detailed evolution of several single elements (^4He, ^{12}C, ^{13}C, ^{14}N, ^{16}O, ^{20}Ne, ^{24}Mg, ^{28}Si, ^{32}S and ^{56}Fe) in the solar vicinity, by taking into account the most recent ideas on type I and II SN progenitors together with the most recent nucleosynthesis prescriptions (Fig.7).

In this model the disk is gradually forming by infall of primordial material with a time scale of $\simeq 3 - 4Gyr$ at the solar neighbourhood, assumed to be at 10 Kpc from the Galactic centre. The SFR is assumed to be a function of the surface gas density as well as of the total surface mass density. The IMF is constant in space and time and defined in the mass range 0.1-80$M_\odot$ (the prescriptions are from Scalo, 1986).

In Fig.8 the predicted ratios, relative to the sun, of several heavy elements with respect to iron as functions of the relative abundance of iron are shown. As one can easily see , elements such as O, Ne, Mg, Si and S show an almost constant overabundance with respect to iron for $|Fe/H| \leq -1.0$, followed by a decline down to the solar value for iron abundances greater than this value, in very good agreement with observations (see Table I).

This behaviour can be easily explained, as already mentioned before, as the consequence of the different roles played by type I and II Supernovae in galactic enrichment. In order to understand this better we show in Fig. 9 the temporal behaviour of type II and I Supernovae. As one can see, the maximum of the type II Supernova rate is reached before than the maximum in the type Ia Supernova rate, for any reasonable assumed scenario for type Ia Supernova progenitors.

Another interesting point is that each element shows a different level of over-abundance with respect to iron, and this is mostly due to the different percentages at which each element is produced in type I and type II SNe, respectively.

The predictions for C are roughly in agreement with a solar $|C/Fe|$ in the whole iron abundance range. The small bump around $|Fe/H| = -1.0$ should not be real: it is instead due to the assumption of not fully consistent nucleosynthesis calculations in the ranges of massive and low plus intermediate mass stars, respectively. It is easy to understand the behaviour of C , since C and Fe are produced in almost the same percentages by the same stellar mass ranges, although C is produced in single stars and Fe in binary systems evolving to type I Supernovae.

On the other hand, the situation for N is not so clear, and better agreement with the observations could be achieved if N produced in massive stars had a primary origin, contrary to what is predicted by standard nucleosynthesis (Matteucci, 1986). Perhaps this could be achieved if some extra mixing could take place in massive stars.

More recently, Matteucci and François (1989) have computed a model where the halo and disk phases are treated separately. The Galactic halo is treated as an homogeneous sphere of gas evolving as a one-zone model. A fast collapse is simulated so that the situation is the same as in a closed model. The halo phase is considered over when $|Fe/H| = -1.0$. At this point one should decide whether the disk will form out of primordial gas coming from outside the halo, or out of the already enriched residual gas from the halo collapse. The relative iron abundance of the residual gas, in this latter case, is exactly to $|Fe/H| = -1.0$, due to the assumption of complete instantaneous mixing of gas.

Before proceeding further it is worth noting that the results of Fig.8 do not change at all when halo and disk are considered separately, nor even when enriched infall is considered in the process of disk formation, as long as the abundances in the infalling material are constant in space and time.

The disk is assumed to form gradually by infalling material, with a time scale of formation which is a function of galactocentric distance. In particular, the innermost Galactic regions are assumed to form more rapidly than the outermost ones, in agreement with the results of Larson (1976).

In the case of the disk forming out of residual enriched halo gas a difficulty arises, namely the model predicts too many stars with $|Fe/H| \leq -1.0$ ($\simeq 20\%$) in the solar neighbourhood with respect to what is observed (only $\simeq 3\%$), as shown in Fig.1b. This is due to the fact that, in the simple picture adopted, where gas is lost only when the process of halo formation is over, all the available gas is involved in

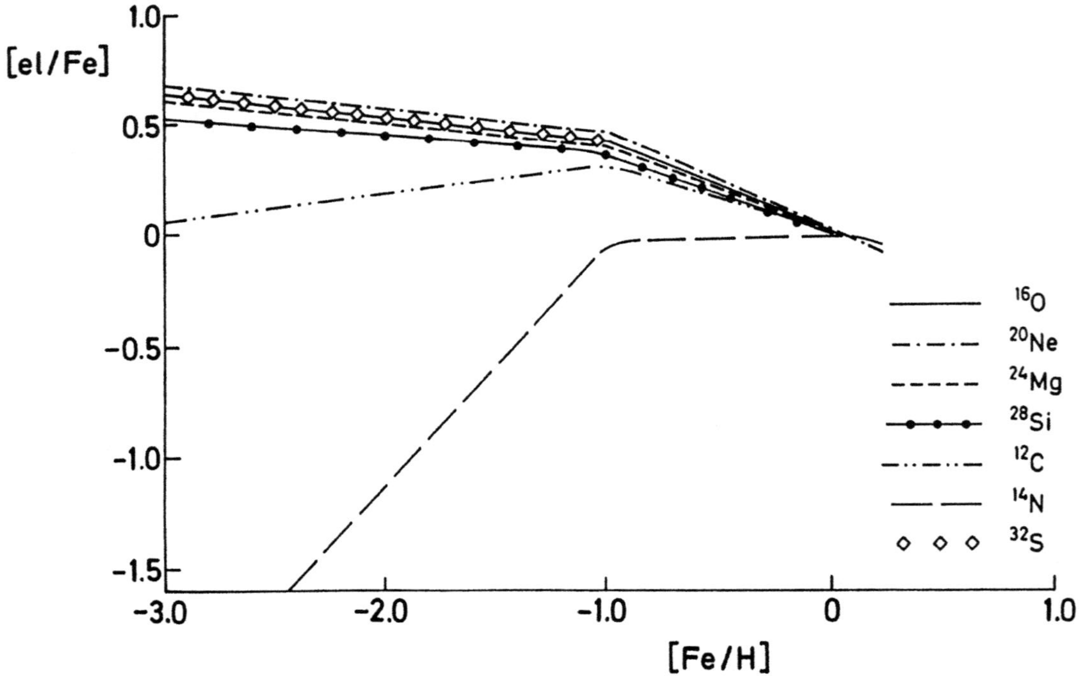

Fig.8. Elemental ratios of several heavy elements with respect to iron, relative to the sun, as functions of the relative iron abundances, as predicted by the model with k=1.1 in the star formation law.

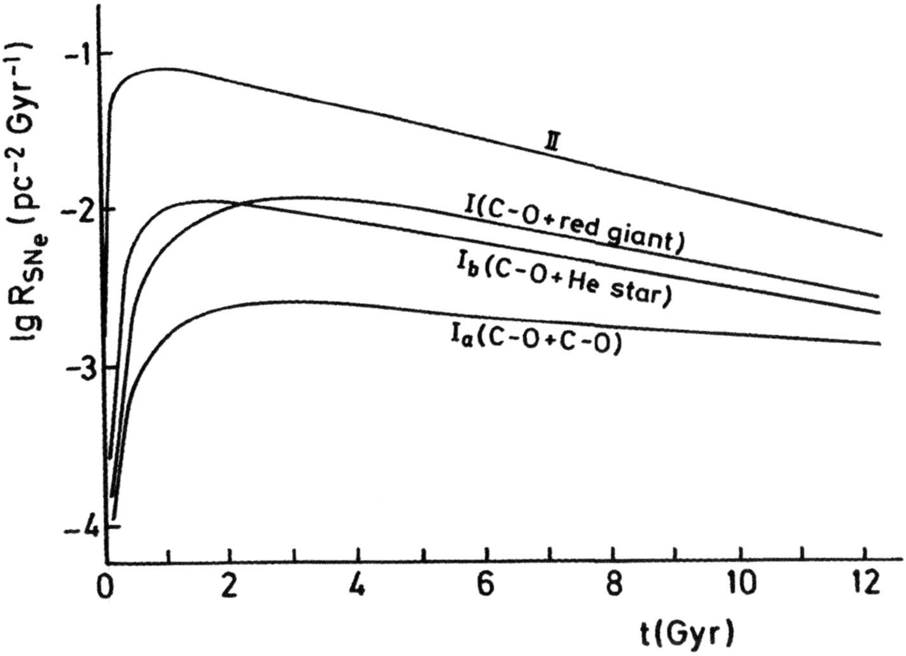

Fig.9–Predicted Supernova rates in the solar neighbourhood as functions of time. The labels on each curve indicate the Supernova type, and in the case of type I also the adopted model for progenitors is shown.

forming stars, whereas in a more realistic picture of halo formation gas loss should occur since the beginning. Another way of solving this problem could be to relax the assumption of a constant IMF.

In the paper of Matteucci and François (1989) several cases are presented, namely primordial and enriched infall and different star formation rates, in an attempt to find a model which reproduces, at the same time, the elemental ratio evolution in the solar neighbourhood as well as gas distribution, SFR distribution and abundance gradients along the Galactic disk. The most important results of Matteucci and François can be summarized as follows:

a) the evolution of the relative (to the sun) ratios between several elements (C, N, O, Ne, Mg, Si, S) and iron, as functions of the relative iron abundance can be reproduced by different choices of the SFR parameters as long as iron is assumed to be produced both in type I and II Supernovae (the former produce more) and that $|Fe/H| = -1.0$ is reached in a time scale between 0.5 Gyr and 1.5 Gyr.

b) The G–dwarf iron distribution is better reproduced by a model assuming primordial infall than by a model assuming enriched infall from the halo. In fact, in this second case the predicted number of halo stars is too large when compared with Pagel (1989) data. Parameters influencing this distribution are the exponent of the gas density in the star formation law and the time scale of formation of the disk in the solar vicinity (here $3Gyr$).

c) The age–iron relationship in the solar neighbourhood is better reproduced by models where the star formation rate is also a function of the total surface mass density.

d) Chemical abundance gradients along the Galactic disk are well reproduced by a model with a SFR law depending on both surface gas density and total surface mass density (see Table V where the results of the best model of Matteucci and François, 1989, are shown), and with a time scale of disk formation assumed to vary with galactocentric distance, in the sense that the innermost regions must have formed in a shorter time than the outermost ones, which could actually be still forming now.

The different gradients found for the different elements are reflecting the different nucleosynthetic histories of each element. In particular, they are reflecting the lifetimes of the element progenitors and the nature (primary or secondary) of each element.

Elements mostly secondary, such as N and ^{13}C which are also produced in long living stars, show the steepest gradients, whereas primary elements produced mostly in massive stars such as S, Ne and O have shallower gradients. Some small differences are present also among the gradients of these latter elements, being the gradient of S and Ne shallower than that of oxygen.This fact can be interpreted as the consequence of the bigger range of oxygen producers as compared with the producers of S and Ne. In fact, as shown in Fig.7, oxygen production increases with the stellar mass up to $50M_{\odot}$, whereas S and Ne production is more concentra-

Table V : Predicted and Observed Gradients in the Galaxy

	Model	Observations
$\frac{\Delta \log(N/H)}{\Delta R}$ (dex Kpc^{-1})	–0.08	$-0.09 \pm 0.015^{1)}$
$\frac{\Delta \log(O/H)}{\Delta R}$ (dex Kpc^{-1})	–0.058	$-0.07 \pm 0.015^{1)}$
$\frac{\Delta \log(Ne/H)}{\Delta R}$ (dex Kpc^{-1})	–0.041	—
$\frac{\Delta \log(S/H)}{\Delta R}$ (dex Kpc^{-1})	–0.043	$-0.01 \pm 0.02^{1)}$
$\frac{\Delta \log(Fe/H)}{\Delta R}$ (dex Kpc^{-1})	–0.043	-0.05–$0.01^{2),3)}$

1) Shaver et al. (1983)
2) Janes (1979)
3) Panagia and Tosi (1981)

ted around some specific stellar masses. Moreover, oxygen production in type I Supernovae, although negligible relative to its production in massive stars, is less negligible than the corresponding one for S and Ne.

Finally, primary elements produced in long living stars, such as Fe, behave as secondary ones.

In the light of this we can interpret the results on gradients of isotopic ratios shown in Table II. The increase of the $^{13}C/^{12}C$ ratio towards the Galactic center indicates that ^{13}C is mostly a secondary element whereas ^{12}C is a primary element. The same kind of explanation can be applied to the other isotopic ratios.

Therefore, observations of gradients of abundances of elements with respect to hydrogen and of elemental ratios can put constraints on stellar evolution and nucleosynthesis.

From the previous discussion it is also clear how important is to take into account the stellar lifetimes when one follows the evolution of single elements and that I.R.A. is not even a good approximation for elements mainly produced in massive stars.

e) It is generally very difficult to reproduce, at the same time, gas distribution and abundance gradients in models not assuming the current observed gas distribution as an input parameter. On the other hand, the observed gas distribution is very uncertain. In Fig.4 are reported the observed data together with the predictions of Matteucci and François(1989). If the peak at 5Kpc is real, radial inflow may be required.

Some longstanding problems and future prospects

- The relative enrichment of He with respect to heavy elements, $\frac{\Delta Y}{\Delta Z}$, is still an open problem.

Peimbert (1986) suggested a value of 3.5$\pm$0.7, based on observations of Galactic and extragalactic HII regions.

Pagel et al. (1986) claimed an even greater value, 5.7$\pm$2.7, from extragalactic HII regions, although they attributed it to contamination from Wolf-Rayet stellar winds.

Old models of Galactic chemical evolution adopting standard nucleosynthesis and mass loss in massive stars, as suggested by Chiosi et al.(1978,1979), predicted a value for that ratio of 2–3 (Serrano and Peimbert, 1981; Matteucci and Chiosi, 1982).

More recent models adopting the stellar evolution results of Maeder (1981;1983) predicted a value of $\simeq 1$ (Matteucci and Greggio, 1986; Matteucci and François, 1989).

If overshooting from convective cores were considered this ratio would be further lowered (Serrano, 1986).

A possible solution to the discrepancy between observations and theoretical results has been proposed by Schild and Maeder (1985), who suggested the existence of an upper limit for type II SN explosion of $\simeq 50M_\odot$. Stars above this limit should never explode but all the stellar mantle should implode in a black hole.

However, very recently Maeder (private communication), by adopting a different algorithm for mass loss in massive stars leading to a greater loss of material, found that a lot of ^4He is lost during the wind phase and therefore subtracted to further processing inside the stars. The net production of metals would therefore be decreased in favour of He production, leading to a $\frac{\Delta Y}{\Delta Z}$ of the order of 2.

- A great deal of discussion has been devoted recently to the possibility of a missing mass in the solar vicinity and to its nature.

Bahcall (1984) derived, from data on F and K giants, a total surface mass density for the solar neighbourhood of $\simeq 67M_\odot pc^{-2}$. Since $\simeq 35M_\odot pc^{-2}$ is in the form of stars and $5 - 10M_\odot pc^{-2}$ in the form of gas (molecular plus neutral hydrogen), the implication is that $\simeq 30M_\odot pc^{-2}$ is in the form of unseen matter.

Very recently, Kujiken and Gilmore (1989) from a similar analysis found that there is no need to claim unseen mass in the solar vicinity. In fact, they derived a total surface mass density for the solar neighbourhood of only $46\pm 9M_\odot pc^{-2}$.

To conclude we would like to outline some future prospects in the field of Galactic chemical evolution:

a) Models with halo and disk coupled evolution are a necessary step to compute the chemical evolution of our Galaxy.

b) The evolution of single elements, especially C, N, O, Ne, Mg ,Al, Si Ca, Fe and s– and r– process elements should be calculated in detail by taking into account the different type of Supernovae and their nucleosynthetic products.

c) Detailed nucleosynthesis for the whole range of stellar masses of interest for Galactic enrichment by taking into account overshooting, mass loss and the new rate for the $^{12}C(\alpha,\gamma)^{16}O$ should be calculated.

An effort should be done also in computing these quantities for different initial stellar metal contents.

d) The instantaneous recycling approximation should be used only in very special cases.

Especially, should not be used if one's aim is to compute the detailed evolution of iron in order to reproduce the G–dwarf distribution, the age–iron relation and the evolution of the other elements with respect to iron. Alternatively, (see Pagel, 1989) one could transform the data for iron into data for oxygen and use the simple model with instantaneous recycling approximation.

e) Since it is now clear that the abundance ratios of the "cosmic abundances" have varied with time, one should be careful in not assuming their constancy when studying objects much less chemically enriched than the Sun, like for example in modelling the evolution of stars in Globular Clusters.

Acknowledgements:
I would like to thank Laura Greggio for patiently reading this manuscript.

References

Abia, C., Rebolo, R., Beckman, J., Crivellari, L.: 1988, *Astron. Astrophys.* **206**, 100

Appenzeller, I., Fricke, K.J.: 1972, *Astron. Astrophys.* **18**, 10

Appenzeller, I., Tscharnuter, W.: 1973, *Astron. Astrophys.* **25**, 125

Arimoto, N., Yoshii, Y: 1986, *Astron. Astrophys.* **164**, 260

Arimoto, N., Yoshii,Y.: 1987, *Astron. Astrophys.* **173**, 23

Arnett, W.D.: 1971, *Astrophys. J.* **166**, 153

Bahcall, J.N.: 1984 *Astrophys.J.* **276**, 169

Bhat, C.L., Houston, B.D., Issa, M.R., Mayer, C.J., Wolfendale, A.W.: 1984, in *Gas in the Interstellar Medium* ed. by P.M.Gondalekhar ,RAL

Barbuy, B., Spite, F., Spite, M.: 1985, *Astron. Astrophys.* **144**, 343

Becker, S.A., Iben, I. jr.: 1979, *Astrophys. J.* **232**, 831

Branch, D., Nomoto, K.: 1986, *Astron. Astrophys.* **164** , L13

Bressan, G.A., Bertelli, G., Chiosi, C.: 1981, *Astron. Astrophys* **150**, 33

Cameron, A.G.W.: 1982, in *Essays in Nuclear Astrophysics* eds. C. Barnes et al., Cambridge University Press, p.377
Cappellaro, E, Turatto, M.: 1988, *Astron. Astrophys.* **190**, 10
Carlbergh, R.G., Dawson, P.C., Hsu, T., VandenBergh, D.A.: 1985, *Astrophys. J.* **249**, 674
Caughlan, G.R., Fowler, W.A., Harris, M.J., Zimmerman, B.A.: 1985, *Atomic Data and Nuclear Data Tables* **32**, 197
Cayrel, R.: 1986, *Astron. Astrophys.* **168**, 81
Chiosi, C.: 1980, *Astron. Astrophys.* **83**, 206
Chiosi, C.: 1987, in *Nucleosynthesis and Stellar Evolution* 16th Saas Fee Course, eds. B. Hauck et al., Geneva Observatory, p.199
Chiosi, C., Nasi, E., Bertelli, G.: 1979, *Astron. Astrophys.* **74**, 62
Chiosi, C., Nasi, E., Sreenivasan, S.R.: 1978, *Astron. Astrophys.* **63**, 103
Chiosi, C., Matteucci, F.: 1982, in *Astron. Astrophys.* **105**, 140
Clayton, D.D.: 1987, *Astrophys. J.* **315**, 451
Crane, P., Hegyi, D.J.: 1988, *Astrophys. J.* **326**, L35
D'Antona, F., Mazzitelli, I.:1983, *Astron. Astrophys.* **127** , 149
D'Antona, F., Mazzitelli, I.:1986, *Astron. Astrophys.* **162** , 80
Diaz, A., Tosi, M.: 1984, *M.N.R.A.S.* **208**, 365
Dopita, M. : 1985, *Astrophys. J.* **295**, L5
Eggen, O.J., Lynden-Bell, D., Sandage, A.R.: 1962, *Astrophys. J.* **136**, 735
Edmunds, M.G.: 1989, *Rev. Mod. Astron.*, in press
Edmunds, M.G., Pagel, B.E.J.: 1984, in *Stellar Nucleosynthesis* eds. by C.Chiosi and A.Renzini, (Dordrecht:Reidel), p.341
Evans, I.N.: 1987, *Astrophys. J.* **309**, 544
Evans, R., van den Bergh, S., Mc Clure, R.: 1989, preprint
Ferrini, F., Palla, F., Penco, U.: 1989, *Astron. Astrophys.*, in press
Filippenko, A.V., Sargent, W.L.W: 1986, *Astron. J.* **91**, 691
Fowler, W.A., Caughlan, G.R., Zimmerman, B.A.: 1985, *Ann. Rev. Astron. Astrophys.* **13**, 113
Franco, J., Shore, S.N.: 1984, *Astrophys.J.* **285**, 813
François, P.: 1987, *Astron. Astrophys.* **176**, 294
François, P.: 1988, *Astron. Astrophys.* **195**, 226
Gallino, R., Busso,M., Picchio, G., Raiteri, C.M., Renzini, A.: 1988, *Astrophys. J. Lett.* **334**, L45
Gaskell, C.M., Cappellaro, E., Dinerstein, H.L., Garnett, D.R., Harkness, R.P., Wheeler, J.C.: 1986, *Astrophys. J.* **306**, L77
Gehren, T., Nissen, P.E., Kudritzki, R.P., Butler, K.: 1985, in *Production and Distribution of C,N,O Elements*, eds. by I.J. Danziger, F.Matteucci and K.Kjär, E.S.O. Publ. p.171
Gehren, T.: 1988, *Rev. Mod. Astron.* in press
Gilmore, G.: 1984, *M.N.R.A.S.* **207**, 223
Gilmore, G., Reid, I.N.: 1983, *M.N.R.A.S.* **202**, 1025
Gilmore, G., Wyse, R.F.G.: 1986, *Nature* **322**, 806

Gilroy, K.K., Sneden, C., Pilachowski, C.A., Cowan, J.J.: 1988, *Astrophys. J.* **329**, 298

Greggio, L., Tosi, M.: 1986, *Astron. Astrophys.* **156**, L1

Grenon, M.: 1987, *J. Astrophys. Astron.* **8**, 123

Gustafsson, B: 1987, in *Stellar Evolution and Dynamics in the Outer Halo of the Galaxy* eds. M.Azzopardi and F.Matteucci, E.S.O. Publ. p. 33

Güsten, R.: 1986, in Evolution of Galaxies eds. C.Chiosi and A.Renzini, (D.Reidel: Dordrecht), p.449

Güsten, R., Mezger, P.G.: 1983, *Vistas in Astronomy* **26**, 159

Hawking, I., Jura, M.: 1987, *Astrophys. J.* **317**, 926

Kettner, K.V., Becker, H.W., Buchmann, L., Görres, J., Kräwinkel, H., Rolf, C., Schmalbrock, P., Trautvetter, H.P., Vlieks, A.: 1982, *z. Physik* **A308**, 73

Koester, D. Reimers, D.: 1985, *Astron. Astrophys.* **153**, 260

Kujiken, K., Gilmore, G.: 1989, *M.N.R.A.S.* **239**, 605

Iben, I.Jr., Nomoto, K., Tornambè, A., Tutukov, A.: 1987, *Astrophys. J.* **317**, 717

Iben, I.jr., Renzini, A.: 1983, *Ann. Rev. Astron. Astrophys.* **21**, 271

Iben, I.Jr., Tutukov, A.: 1984, *Astrophys. J. Suppl.* **54**, 335

Iben, I.Jr., Tutukov, A.:1985, *Astrophys. J. Suppl.* **58**, 661

Janes, K.A.: 1979, *Astrophys.J.Suppl.* **39**, 154

Lacey, G., Fall, S.M.: 1985, *Astrophys. J.* **290**, 154

Lambert, D.L.: 1988, in *Cosmic Abundances of Matter* ed. C.J. Waddington, AIP Conference Proceed. 183, p.168

Larson, R.B.: 1972, *Nature Phys. Sci.* **236**, 7

Larson, R.B.: 1974, *M.N.R.A.S.* **166**, 585

Larson, R.B.: 1976, *M.N.R.A.S.* **176**, 31

Larson, R.B.: 1986, *M.N.R.A.S.* **218**, 409

Lester, D.F., Dinerstein, H.L., Werner, M.W.,Watson, D.M., Genzel, R.L.: 1983, *Astrophys. J.* **320**, 573

Li, TiPei, Riley, P.A., Wolfendale, A.W.: 1982, *J.Phys.G.* **8**, 1141

Luck, R.E., Bond, H.E.: 1983, *Astrophys. J.* **271**, L75

Luck, R.E., Bond, H.E.: 1985, *Astrophys. J.* **292**, 559

Lynden–Bell, D.: 1975, *Vistas in Astr.* **19**, 299

Maeder, A.: 1981, *Astron. Astrophys.* **102**, 401

Maeder, A.: 1983, *Astron. Astrophys.* **120**, 113

Maeder, A.: 1985, in *Nucleosynthesis and its Implications on Nuclear and Particle Physics* Fifth Moriond Astrophysical Meeting, eds. J. Audouze and N. Mathieu, (D.Reidel: Dordrecht), p.207

Maeder, A., Meynet, G.: 1987, *Astron.Astrophys.* **182**, 243

Maeder, A., Meynet, G.: 1989, *Astron. Astrophys.*, in press (MM89)

Magain, P.: 1988, *Astron. Astrophys.* **209**, 211

Matteucci, F.: 1986, *M.N.R.A.S.* **221**, 911

Matteucci, F.: 1987, in *Stellar Evolution and Dynamics in the Outer Halo of the Galaxy*, eds. by M.Azzopardi and F.Matteucci, E.S.O. Publ.p.609

Matteucci, F.: 1988, in *Origin and Distribution of the Elements*, ed. by G.J. Mathews, World Scientific, p.186
Matteucci, F.: 1989, in *Evolutionary Phenomena in Galaxies*, ed. J.Beckman, in press
Matteucci, F., Chiosi, C.: 1983, *Astron. Astrophys.* **123**, 121
Matteucci, F., Franco, J., François, P., Treyer, M.A.: 1989, *Rev. Mex. Astr. Astrophys.*, in press
Matteucci, F., François, P.: 1989, M.N.R.A.S., **239**, 885
Matteucci, F. Greggio, L.: 1986, *Astron. Astrophys.* **154**, 279
Matteucci, F., Tosi, M.: 1985, *M.N.R.A.S.* **217**, 391
Matteucci, F., Tornambè, A.: 1987 *Astron. Astrophys.* **185**, 51
Matteucci, F., Vettolani, G.: 1988, *Astron. Astrophys.* **202**, 21
Mayor, M., Vigroux, L.: 1981, *Astron.Astrophys.* **98**, 1
Meyer,J.P.: 1985, *Astrophys. J. Suppl.* **57**, 173
Miller, G.E., Scalo, J.H.: 1979, *Astrophys. J. Suppl.* **41**, 453
Nissen, P.E., Edvardsson, B., Gustafsson, B.: 1985, in *Production and Distribution of C,N,O Elements*, eds, by I.J.Danziger, F.Matteucci and K.Kjär, E.S.O., Publ. p.131
Nomoto, K., Thielemann, F.K., Yokoi, Y.: 1984, *Astrophys. J.* **286**, 644
Norris, J., Bessel, M.S., Pickles, A.J.: 1985, *Astrophys. J. Suppl.* **58**, 463
Ostriker, J.W., Thuan, T.X.: 1975, *Astrophys. J.* **202** , 353
Pagel, B.E.J.: 1989, in *Evolutionary Phenomena in Galaxies* ed. J.Beckman, in press
Pagel, B.E.J., Terlevich,R., Melnick, J.: 1986, *P.A.S.P.* **98**, 1005
Panagia, N.: 1987, in *High Energy Phenomena Around Collapsed Stars*, ed. by F.Pacini, (Dordrecht:Reidel), p. 33
Panagia, N. Tosi, M.: 1981, *Astron.Astrophys.* **96**, 306
Peimbert, M.: 1979, in *Les Elements et leurs Isotopes dans l'Universe* eds. Boury et al.
Peimbert, M.: 1986, *P.A.S.P.* **98**, 1057
Rana, N.C., Wilkinson, D.A.: 1986, *M.N.R.A.S.* **218**, 497
Renzini, A. Voli, M.: 1981, *Astron. Astrophys.* **94**, 175
Scalo, J.H.: 1986, *Fund. Cosmic Phys.* **II**, p.1
Schild, H. Maeder, A.: 1985, *Astron.Astrophys.* **143**, L7
Schmidt, M.: 1959, *Astrophys. J.* **129**, 243
Serrano, A.: 1986, *P.A.S.P.* **98**, 1066
Serrano, A., Peimbert, M.: 1983, *Rev.Mex.Astr.Astrofis.* **8**, 117
Serrano, A., Peimbert, M.: 1981, *Rev. Mex. Astr. Astrofis.* **5**, 109
Shaver, P.A., Mc Gee, R.X., Newton, L.M., Danks, A.C., Pottasch, S.R.: 1983, *M.N.R.A.S.* **204**, 53
Shore, S.N., Ferrini, F., Palla, F.: 1987, *Astrophys.J.* **316**, 663
Spite, M., Spite, F.: 1985, *Ann.Rev.Astron.Astrophys.* **23**, 225
Talbot, R.J.: 1980, *Astrophys. J.* **235**, 821
Talbot, R.J., Arnett, D.W.: 1973, *Astrophys.J.* **186**, 69

Talbot, R.J., Arnett, D.W.: 1975, *Astrophys. J.* **197**, 551

Tamman, G.A.: 1982, in *Supernovae: a Survey of Current Research* eds. by M.J. Rees and R.J.Stoneham, (Dordrecht:Reidel), p.371

Terlevich, R., Melnick, J.: 1983, *E.S.O. preprint* **n.264**

Thielemann, F.K., Arnett, D.W.: 1985, *Astrophys. J.* **295**, 604

Thielemann, F.K., Hashimoto, M., Nomoto, K.: 1989, *Astrophys. J.*, in press

Thielemann, F.K., Nomoto, K., Yokoi, K.,: 1986, *Astron. Astrophys.* **158**, 17

Tinsley, B.M.: 1980, *Fund. Cosmic Phys.* **5**, 287

Tornambè, A. Matteucci, F.: 1987, *Astrophys. J.* **318**, L25

Tosi, M.: 1988a, *Astron.Astrophys.* **197**, 33

Tosi, M.: 1988b, *Astron.Astrophys.* **197**, 47

Truran, J.W., Cameron, A.G.W.: 1971, *Astrophys. Space Sci.* **14**, 179

Truran, J.W.: 1985, in *Nucleosynthesis and its Implications on Nuclear and Particle Physics* Fifth Moriond Astrophysical Meeting, eds. J. Audouze and N. Mathieu, (D.Reidel: Dordrecht), p.87

Twarog, B.A.: 1980, *Astrophys. J.* **242**, 242

Twarog, B.A., Wheeler, J.C.: 1982, *Astrophys. J.* **261**, 636

Twarog, B.A., Wheeler, J.C.: 1987, *Astrophys. J.* **316**, 153

Vader, P.: 1986, *Astrophys. J.* **305**, 669

van den Bergh, S., Mc Clure, R.D., Evans, R.: 1987, *Astrophys.J.* **323**, 44

van den Bergh, S.: 1988, *Comments on Astrophysics* **Vol.XII** N.3, p.13

Vilchez, J.M., Pagel, B.E.J., Diaz, A.J., Terlevich, R., Edmunds, M.G. : 1988, preprint

Webbink, R.F.: 1984, *Astrophys. J.* **277**, 355

Weidemann, V: 1987, *Astron. Astrophys.* **188**, 74

Wheeler, J.C., Sneden, C., Truran, J.W.: 1989, preprint

Whelan, J.C., Iben, I.Jr.: 1973, *Astrophys. J.* **186**, 1007

Woosley, S.E.: 1987, in *Nucleosynthesis and Chemical Evolution* 16th Saas Fee Course, eds. B.Hauck et al. Geneva Observatory, p.1

Woosley, S.E., Axelrod, T.S., Weaver, T.A.: 1984, in it Stellar Nucleosynthesis eds. C.Chiosi and A.Renzini, (D.Reidel: Dordrecht) p.263

Woosley, S.E., Weaver, T.A.: 1986, *I.A.U. Coll. 89*, eds. D. Mihalas and K.A.Winkler, p.91

Wyse, R.F.G., Gilmore, G.: 1989, preprint

Wyse, R.F.G., Silk, J.: 1987, *Astrophys. J.* **313**, L11

Zinn, R.: 1985, *Astrophys. J.* **293**, 424

A BRIEF INTRODUCTION TO THE COSMIC RADIATION

MAURICE M. SHAPIRO*
Department of Physics and Astronomy
University of Maryland
College Park, MD, USA

ABSTRACT. Some prominent features of the Galactic Cosmic Rays (GCR) are summarized for the non-specialist. These include the composition of the "primaries" reaching the earth and the "primordial" or source composition; the energy spectra and their modulation in the heliosphere; the effects of magnetic fields, notably on the directional distribution; the propagation of the GCR in the interstellar medium (ISM), and what is inferred about abundances, especially the isotopic ones; lifetime of the GCR; acceleration of the particles and their injection into the ISM. Finally the significance of cosmic neutrinos, cosmic gamma rays, and antiprotons is briefly treated.

1. INTRODUCTION

The Galactic cosmic rays are mostly relativistic atomic nuclei that pervade the interstellar medium, with an energy density comparable to that of starlight. This much has been known for some years, but for several decades after their discovery in 1912, their essential nature was in dispute. And even today, their origin is not firmly established. Yet we have learned a great deal about this ubiquitous phenomenon--enough to fill many volumes of conference proceedings. Hence, a treatment of some salient features of the cosmic radiation in a couple of lectures--even qualitatively--must be sketchy at best. Some topics will scarcely be mentioned; the Bibliography may help to fill the voids.

In Sec. 2 the cosmic-ray discipline is related to the other contemporary astronomies. This is followed by an historical perspective on the first half-century of cosmic-ray research. The rest of the paper describes some important characteristics of the cosmic rays: Sec. 3, their composition as observed locally, i.e., the "arriving composition;" Sec. 4, energy spectra and solar modulation;

*Address for Correspondence: 205 Yoakum Parkway, #1720
Alexandria, VA 22304 USA

J. M. Greenberg and V. Pirronello (eds.), Chemistry in Space, 43–69.

Sec. 5, magnetic fields and isotropy; Sec. 6 propagation, pathlength, source composition; Sec. 7, injection and acceleration; Sec. 8, isotopic composition; Sec. 9, high-energy neutrinos and gamma rays; Sec. 10, other aspects of cosmic radiation.

2. HISTORICAL PERSPECTIVE

The discovery of cosmic rays by Hess in 1912 marked the advent of the first new astronomical discipline of the 20th century--although an appreciation of its astrophysical nature and impact was not to be realized for many years. Two decades later Jansky opened up the channel of radio astronomy, which was to acquire a special symbiotic relationship to the cosmic radiation. In the sixties the X-ray sky was revealed by Friedman, Giacconi, Rossi and their colleagues. Soon thereafter came gamma-ray astronomy, and today we are at the threshold of two more channels of cosmic information--neutrino astrophysics and gravitational wave astronomy. All of these contemporary astronomies are complementary to each other and to the classical discipline of optical astronomy

The mutability--and, indeed, the fallibility--of cherished physical theories is exemplified by the early history of the cosmic-ray discipline (yet, paradoxically, certain concepts, such as Fermi's mechanisms of acceleration, persist like the proverbial cat with nine lives.) Widely held views as to the identity of the penetrating "primaries" changed radically from decade to decade. First they were gamma rays; then, successively, electrons (and positrons), protons, heavier nuclei up to iron and beyond, culminating in the periodic table of the elements. The progression came full circle with the advent of high-energy gamma ray astronomy--albeit the relevant fluxes are a thousand times lower that those of the relativistic nuclei. Electrons are present among the primaries, but they only comprise about 10^{-2} of the nuclear component. To the radioastronomer, they are the most "visible" of the relativistic particles. High-energy cosmic neutrinos--tracers of cosmic-ray sources--are yet to be observed, but their detection seems imminent.

2.1 <u>The Development of Particle Astrophysics</u>. From observations in his manned balloon flights, Victor Hess concluded that a penetrating radiation of extra atmospheric origin was responsible for the response of his electroscope. He could hardly have imagined that his discovery portended the birth of two new sciences--elementary-particle physics and high-energy astrophysics. We are here concerned mainly with astrophysics; the field of elementary particles--in its experimental embodiment--has become the province of huge, ground-based accelerators.

In the decades of the twenties and thirties, particle physics at high energies (far beyond those then available in the laboratory) was the main concern of cosmic-ray scientists, both experimental and theoretical. Among the noteworthy discoveries were the positron, the process of pair production of electrons, electromagnetic cascades, muons, "stars", i.e. nuclear disintegrations, and extensive air showers. These phenomena were successfully incorporated into a coherent theory of secondary cosmic-ray production and propagation through the atmosphere, thanks to the work of Bethe, Heitler, Oppenheimer, Bhabha and their collaborators. Some remaining enigmas about the atmospheric secondaries, e.g., the origins of the soft and hard components (i.e., the electron-photon showers and the muons) were resolved in the late forties and early fifties through the discoveries of neutral and charged pions, repectively.

Meanwhile, the latitude effect, discovered by Clay and established by Compton, showed that the "primaries" striking the top of the atmosphere must be charged particles. These are predominantly positive as determined by the East-West affect, and at first they were thought to be positrons. In 1941 Schein, Jesse and Wollan disproved this assumption. Their findings were consistent with the primary particles being penetrating protons.

In 1948 Freier et al. a team of Minnesota and Rochester physicists, discovered that the arriving protons were accompanied by nuclei of elements from helium to iron. (In due course, Fowler was to find evidence for cosmic-ray nuclei much heavier than iron, in fact up to the actinides). Soon thereafter, V.L. Ginzburg and others (notably Pikelner, Shklovsky, and Hayakawa) interpreted the synchrotron radiation emanating from the Crab Nebula as due to a copious flux of relativistic electrons. The mechanism responsible for energizing electrons could be expected to accelerate protons and other heavy ions as well.

For the first time one could point to an object in the sky that generated cosmic rays, albeit the clear evidence was only for relativistic electrons. The idea of supernovae as cosmic-ray sources was attractive. Astronomical evidence showed that the fast electrons were moving in large-scale magnetic fields. Moreover, the theory of nucleosynthesis in stars, developed in the fifties by Burbidge et al. and by Cameron, predicted the production of heavy elements in supernova explosions, and the cosmic rays were known to be rich in those elements. The source strength seemed to meet the requirements of cosmic-ray replenishment in the Galaxy. Fermi's stochastic mechanisms of acceleration, which encountered difficulties in interstellar space, might work more efficiently within the confines of supernova remnants. (Concern over adiabatic losses in the expanding shells came later.)

We have reviewed rather briefly certain landmarks in cosmic-ray history during its first half-century. We shall now elaborate on some salient features of the cosmic radiation, as they are understood today.

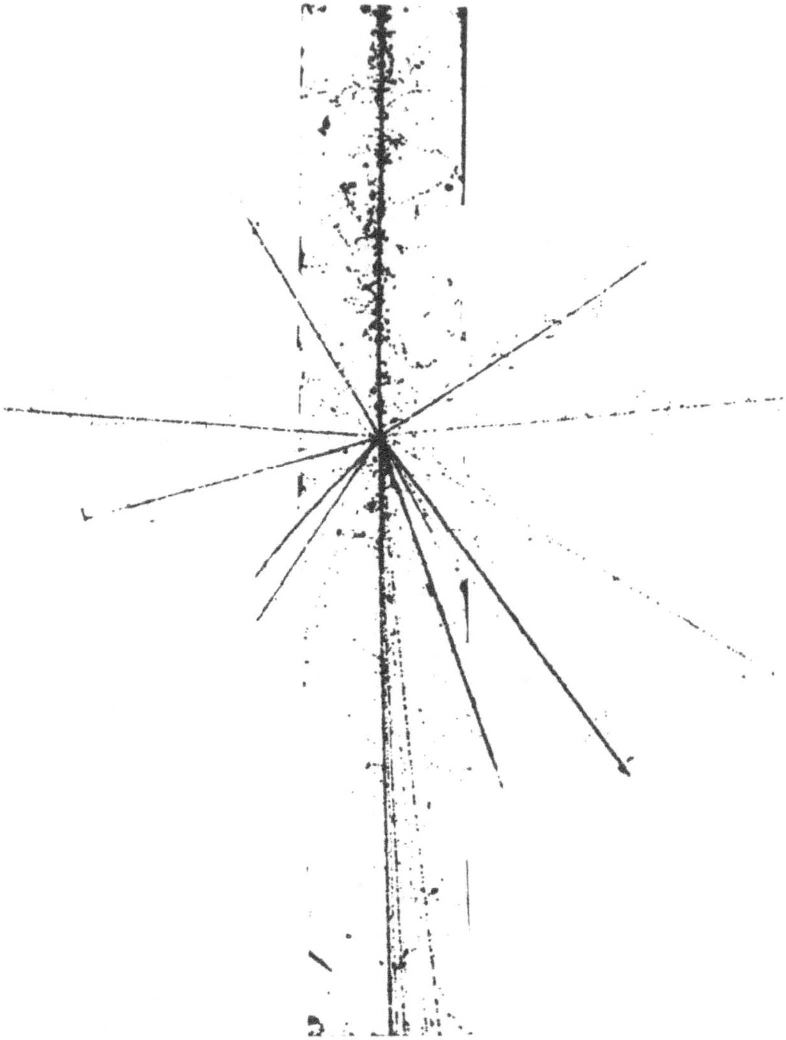

Figure 1. Breakup of a heavy primary cosmic-ray nucleus in collision with a nucleus in a photographic emulsion detector. The narrow-angle shower of particles proceeding beyond the star-like collision point are fragments of the incident cosmic-ray nucleus. The other, lateral tracks radiating from the collision point are fragments of the "complex-target" nucleus; these would be absent in a collision with a hydrogen nucleus in the ISM.

3. COMPOSITION OF THE COSMIC RAYS NEAR THE EARTH

The Galactic cosmic rays (GCR) arriving in our vicinity include a substantial admixture of secondaries generated en route by progenitor nuclei in collision with interstellar matter. These transformations of source particles into secondaries--e.g., nuclei, electrons, antiprotons, gamma rays--provide a veritable Rosetta Stone that helps us unravel the complexities of cosmic-ray propagation and origin. Figure 1 is a photomicrograph of an "event" observed in a photographic emulsion exposed near the top of the atmosphere. It portrays the breakup of a heavy cosmic-ray nucleus into lighter fragments. In this case the target was an atomic nucleus within the emulsion. The corresponding events in outer space would usually involve collisions with hydrogen or helium targets. By investigating the details of such interactions we have learned much of what is known about the many effects to which cosmic rays are subject in their tortuous paths from sources to earth.

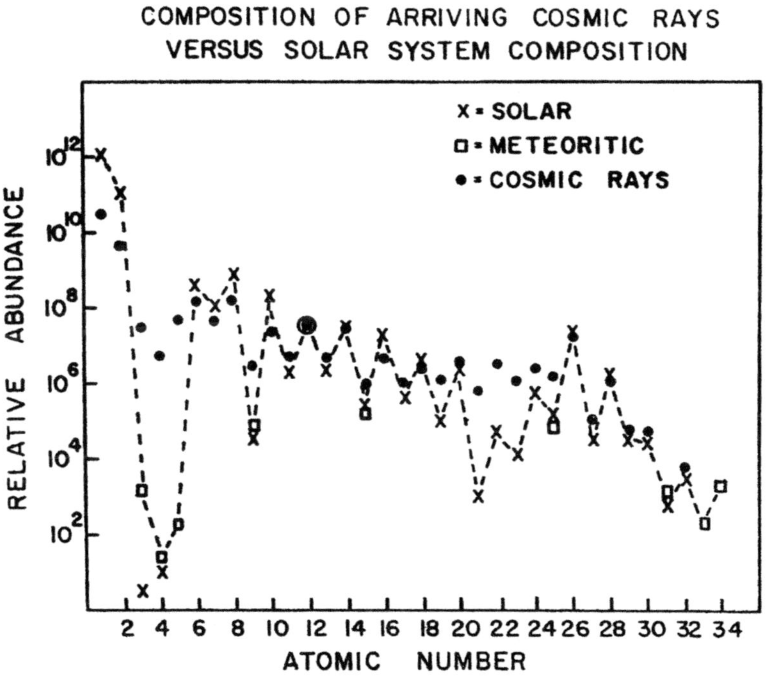

Figure 2. The relative abundances of elements in the cosmic rays and in the solar system. The abundance of solar hydrogen is normalized to 10^{12}, and the cosmic-ray abundances are normalized to the solar values at magnesium.

Figure 2 compares the composition of the arriving GCR with the (thermal) composition of solar system material. In the diagram solar abundances are plotted relative to 10^{12} hydrogen atoms, and the cosmic-ray composition has been normalized to the solar one at Mg. Looking first at the cosmic-ray values, represented by circles in the figure, we see that hydrogen and helium are more abundant than the principal heavier elements by some two orders of magnitude. In the solar system the disparity is greater, three orders of magnitude. Carbon and oxygen are distinctly more plentiful that the other principal nuclides--those of Li, Be, B, N, Ne, Mg, Si and Fe. Still lower in flux than Ne, Mg, Si are the odd-Z nuclei of F, Na, Al. The elements S, Ar, Ca, Ti, and Cr--all some 10 times lower in abundance than Mg or Fe--show enhanced abundances relative to their odd-numbered neighbors, phosphorus to vanadium. However, for all $Z > 8$, and especially for the sub-iron group, the odd-even ratios are less pronounced than they are in the solar abundances. This is consistent with the production of secondary nuclides by collision in the interstellar medium (ISM).

Comparing the cosmic-ray and solar compositions, it is noteworthy that, normalized to Mg (or to other "peak" elements like Si or Fe) H and He are relatively deficient in the cosmic radiation. To a lesser degree, O, N, and C are also underabundant relative to the solar values, although this seems like a small effect in the logarithmic scale of Figure 2. The fluxes span ten orders of magnitude, so many of the abundances look deceptively close.

The deepest "valley" in the solar distribution, due to the extreme scarcity of Li, Be, and B, contrasts sharply with their abundance in the arriving cosmic rays. The observational uncertainty -- and prolonged controversy--surrounding these light nuclei [do they truly come from space or are they solely atmospheric secondaries?] was disspelled by the definitive work of the NRL group (O'Dell et al. 1962). These three elements have been the most revealing breakup products, i.e., secondary nuclides, of cosmic-ray collisions in space, yielding a wealth of information on propagation, path lengths, and source composition. These topics will be treated further in Sec. 6 below.

Until the decade of the sixties the cosmic-ray composition had been investigated only for elements up to the iron group. It was known that in the solar system the abundances of the elements with $Z \geq 30$ comprises $< 10^{-6}$ that of the lighter ones. A similar relative scarcity characterizes the ultra-heavy nuclei (UHN) with $Z \geq 30$ in the cosmic rays. The latter were first detected in meteoritic crystals through ancient particle tracks developed by chemical etching (Fleischer et al. 1967). The formidable task of accumulating enough data for useful studies of contemporary UHN was tackled by P. Fowler et al. (1967). They deployed very-large-area stacks of photographic emulsions at balloon altitudes. Soon hybrid stacks, consisting of layers of plastics and emulsion, were used for detection and identification of UHN. Since the density of tracks in emulsions increases with Z^2, the tracks of UHN are very dense indeed, as shown in Figure 3.

On NASA's HEAO-3 (High-Energy Astronomy Observatory), a consortium of physicists from Washington University, Caltech, and the University of Minnesota improved the precision of charge resolution for the UHN (see Binns 1986 and references therein). Their apparatus incorporated ionization chambers for measuring dE/dx and a Cerenkov counter. Figure 4 is a histogram showing the "charge spectrum" of elements with $Z \gtrsim 50$. These abundances contain information about the special reactions of nucleosynthesis (Burbidge et al. 1957) that are considered to be responsible for most of the UHN, i.e., the s-process ("slow" addition of neutrons) and the r-process ("rapid" capture). The terms "slow" and "rapid" are relative to the rates of beta decay. Owing to the rarity of the heaviest (actinide) elements, the data on these have thus far been scanty.

4. ENERGY SPECTRA, SOLAR MODULATION

The differential intensities of arriving protons and helium nuclei at solar-quiet (Q) times and in active (A) periods are displayed in Figure 5. Over a wide interval of energies, starting at a few GeV/amu, the intensities of H, He, and heavier nuclides conform to a power law $E^{-\gamma}$, where $\gamma = 2.7 \pm 0.1$. At energies $\lesssim$1 GeV/amu during solar-active times the intensities are suppressed by solar modulation. Thus the cosmic-ray flux observed inside the "solar cavity" has an inverse correlation with solar activity in the 11-year cycle (see Figure 5). [Contrary behavior would be expected if the particles were mainly of solar origin.] As cosmic rays enter the heliosphere, their intensity is reduced by the action of the solar wind, especially at low energies and at times of solar maximum.

Various aspects of solar modulation are shown schematically in Figure 6. The solar wind moves out from the sun into the solar cavity and, owing to the sun's rotation, solar magnetic fields are drawn out into a spiral structure. The solar wind also transports irregular components of the magnetic field. The paths of cosmic rays entering the heliosphere are altered: while diffusing, the particles are also being convected outward. As they interact with the expanding magnetic fields, the charged particles are adiabatically decelerated. Hence, those observed at a given energy had posessed considerably higher energies before reaching the solar system.

5. MAGNETIC FIELDS AND ISOTROPY

The Galaxy is pervaded by plasma clouds and by large-scale magnetic fields aligned approximately along the spiral arms. The magnetic fields deflect and store the cosmic-ray ions so that, to a first approximation, the incidence of cosmic rays striking the earth's magnetosphere at energies > 10^{10} eV is nearly uniform in direction and

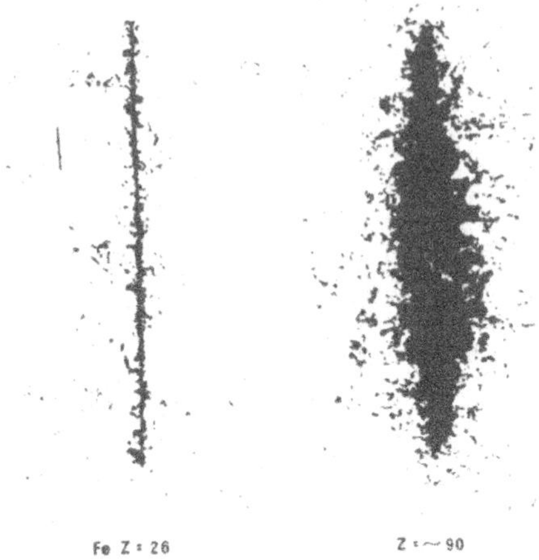

Figure 3. Comparison of tracks of an Fe nucleus and that of a very heavy primary. (Courtesy of P. H. Fowler).

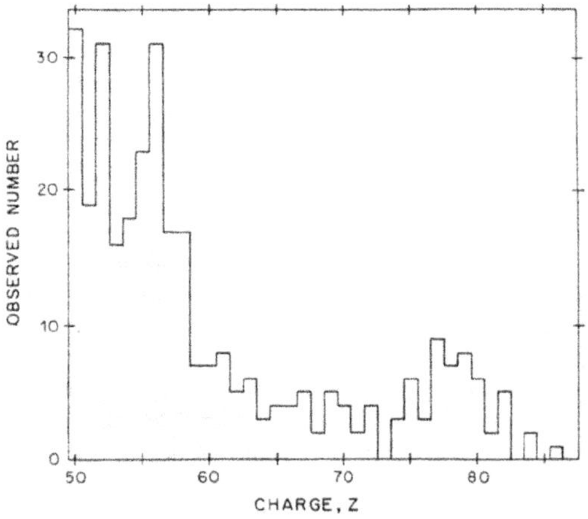

Figure 4. Histogram of nuclides in the region of atomic numbers between 50 and 90 (after Binns et al.)

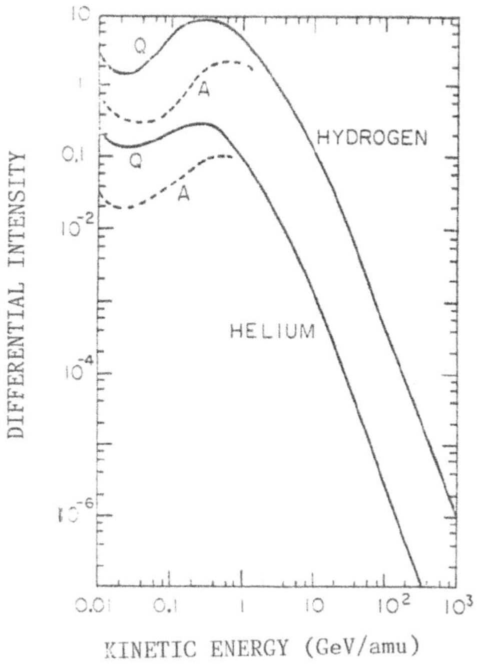

Figure 5. Energy spectra of cosmic-ray protons and helium nuclei at solar-quiet (Q) and solar-active (A) periods.

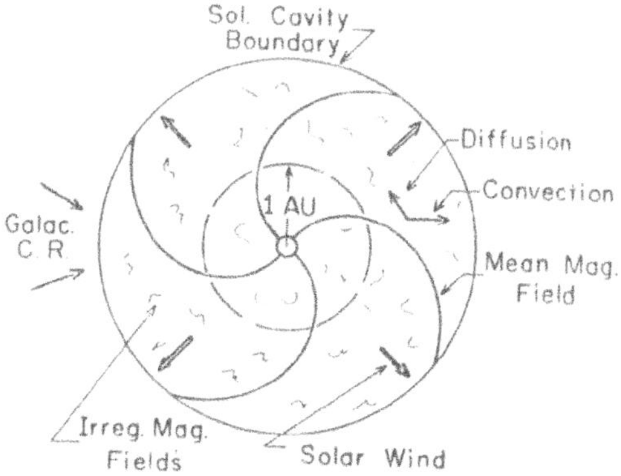

Figure 6. Aspects of solar modulation. The effects are described in Section 4.

time of arrival.

Actually, both anisotropies and time variations have been observed in the GCR. At energies near 10^{12} eV the magnitude of the anisotropy is $\simeq 10^{-4}$. At still higher energies the reported anisotropies are less well known, but they seem to increase with energy; the effects are of the order of a few per cent or less. At the highest energies, $> 10^{19}$ eV, larger anisotropies have been reported (e.g., a higher flux from the direction of the Virgo cluster), but the statistical weight of these observations is low. The study of time variations, on the other hand, has provided a rich harvest of information, e.g., on solar modulation (results based heavily on the work of Forbush and on the Simpson neutron monitors), and on cosmic-ray chronology over geologic time spans.

The substantial isotropy of the cosmic-rays--the fact that the particles have "forgotten" their initial directions--distinguishes charged-particle astronomy from the electromagnetic channels. Indeed, it was a long time before many classical optical astronomers recognized the cosmic radiation as a branch of astrophysics. It is worth noting parenthetically that another branch of particle astronomy--that of high-energy cosmic neutrinos, closely related to cosmic rays--does not suffer from loss of directionality. Since neutrinos, like photons, travel in straight lines, it follows that neutrino astronomy can "point back" to the locations of sources, as do the established channels in astronomy.

6. PROPAGATION, PATH LENGTH, AND SOURCE COMPOSITION

The propagation of the GCR in the Galaxy from the time of birth to their egress from the CR reservoir (by escape or energy loss) encompasses the history of the particles in space and time, once the constituent nuclei have been formed by element synthesis. Among the processes comprising "propagation" are diffusion, acceleration (see Sec. 7), fragmentation that generates secondary nuclei, including radioactive ones (see Sec. 8), and other secondaries (to be discussed in Sec. 9).

A prime task of propagation theory, however, is to deduce the composition of the cosmic rays at the "sources(s)," for the light that it can shed on CR origin. The theory is subject to the constraint that it must yield, finally, the observed, arriving composition. The latter is bound to be very different from the source composition, especially in its isotopic constitution.

The first comprehensive program of calculations on the transformation and propagation of CR nulcei in the interstellar medium (ISM) was carried out by the NRL group (Shapiro, et al, 1970a, b, and c). They applied the latest cross sections and took account of their dependence on energy, as well as the effects of ionization loss and geomagnetic and solar modulation. Clearly, a knowledge of nucleon-nucleus interactions is essential for these computations. Only a fraction of the needed cross sections, however, have been measured; most have been calculated by semi-empirical methods that depend on nuclear systematics. Silberberg and Tsao (1973) have used experimental cross sections for products from proton interactions with various nuclides to formulate semi-empirical formulae for cross sections. These relations have been widely utilized by CR observers: they are applicable to all products having a mass number $A \gtrsim 6$, and for targets ranging from lithium to uranium. For targets with atomic number $Z \gtrsim 30$, the relations are applicable from 100 MeV up to many GeV.

An important parameter for CR propagation is the mean path length of the particles. It tells us how much material the CR have penetrated during their space travels. The magnitude of this path length is connected with the regions and mechanisms of cosmic-ray diffusion and trapping; it was used, together with interstellar densities, to get a preliminary estimate of the mean confinement time of the nuclei. Also, a knowledge of the path length is needed for deducing the cosmic-ray source composition. The best source of information on path lengths is the abundance ratio of the so-called "light" nuclei to heavier ones arriving at the top of the atmosphere. Since the light nuclei are the breakup products of heavier ones that have collided with interstellar gas and dust, the relative flux of light nuclei reflects the amount of material traversed. An NRL investigation (O'Dell et al. 1961) on the relative flux of Li, Be, and B nuclei showed that the ratio for CR in the GeV range "light"/"medium" = 0.25 ± 0.02. (The "medium nuclides are C,N,O,F.) This result has been confirmed by subsequent work.

The simplest approach to deducing a mean path length from the foregoing data involves the so-called slab approximation, in which all the cosmic-ray nuclei are assumed to have the same path. By applying the appropriate diffusion equations, a mean path length $\lambda_e = 5 g/cm^2$ was inferred. This value has also been confirmed through studies of the isotopic composition of cosmic-ray helium; in "stripping" reactions ^4He is converted to ^3He. In fact, the distribution of path lengths does not conform to a delta-function (slab model). No unique value of λ_e could satisfy both (a) the observed production and survival of the group of light nuclides, and also (b) the production, from Fe, of the "sub-iron" set of nuclides having $17 \lesssim Z \lesssim 25$. Investigation of various distribution functions revealed that an exponential type of distribution can satisfy the abundance ratios in both regions (a) and (b).

The transformation of CR nuclei by collisions in the ISM can be described by a diffusion equation (sometimes called a transport equation) of the type published by Ginzburg and Syrovatskii (1964); see also Shapiro et al. 1972, and Silberberg et al. 1989. In simplified form, the solution to this equation can be written:

$$J_i(E) = (J_{i,1} + J_{i,2} + J_{i,3} + J_{i,4})\, f_m, \tag{1}$$

where J_i is the total flux of primary i-type nuclei; $J_{i,1}$ is the intensity of <u>primordial</u> i-type nuclei surviving a path x, and arriving at the top of the atmosphere with energy E; $J_{i,2}$ is the flux of i-type secondaries generated over a path x; $J_{i,3}$ and $J_{i,4}$ are the tertiary and quaternary products; f_m is the modulation coefficient that incorporates both solar and geomagnetic effects.

For a path-length distribution of the form e^{-x/λ_e}, where λ_e is the mean path, the first and second terms of Equation (1) are, respectively:

$$J_{i,1}(E) = \frac{1}{\lambda_e}\int_0^\infty f_{ion}(E,x)\, J_{i,o}\, e^{-x/\lambda_i}\, e^{-x/\lambda_e}\, dx \tag{2}$$

$$J_{i,2}(E) = \frac{1}{\lambda_e}\int_0^\infty e^{-x/\lambda_i}\, e^{-x/\lambda_e} \left\{ \sum_j \frac{P_{ji}}{\lambda_j} J_{j,o} \cdot \right.$$

$$\left. \int_0^x f_{ion}(E,x_i)\, f_{col}(E')\, f_{ion}(E',x_j)\, e^{-x_j\left(\frac{1}{\lambda_j} - \frac{1}{\lambda_i}\right)} dx_j \right\} dx \tag{3}$$

$J_{i,o}$ and $J_{j,o}$ are the intensities of i- and j-nuclei in the absence of interstellar matter and of modulation effects. f_{ion} is an ionization loss factor, and f_{col}, the collision energy loss factor. E' is the energy of a j-type nucleus just before it collides. In Equation (3) x_i is the path of the daughter nucleus and x_j that of the parent; $x_i + x_j = x$. The tertiary and quaternary intensities, $J_{i,3}$ and $J_{i,4}$ involve more complex expressions which have been used in the calculations. Actually, cross sections rather than mean free paths are usually measured. So it is convenient for purposes of calculation to replace the latter in Equns. (2) and (3) by suitable transformations (see Shapiro et al. 1972)

In order to deduce a source composition, one can solve the diffusion equations numerically: the fragmentation of the primordial nuclei, and the production and attenuation of the secondaries are calculated in successive increments of path length. First, however, one adopts a trial source composition resembling, say, the solar one, and also a path-length distribution. One then derives an arriving composition, and compares it to the observed one. Both the trial source composition and the distribution function of path lengths are modified to improve the fit with observations.

Results of this iterative procedure are displayed in Figure 7 showing the relative abundances of cosmic rays at the sources and near the earth. The arriving particles are subdivided into the surviving "original" nuclei and the secondaries generated en route. All abundances have been normalized to a value of 100 for the total flux of arriving carbon. For each element in the figure (with few exceptions), there are two adjacent ordinates. The diagonally hatched column gives the source abundance; the black-and-white column represents the overall abundance of the arriving nuclide. In the latter, the white portion gives the surviving flux of nuclei from the source, and the dark portion specifies the surviving secondary nuclei produced en route. In order that the abundances of the rarest elements be visible, the right side of the histogram was expanded in scale compared to the left side. We see that, of the principal arriving cosmic-ray elements, substantial fractions of C, O, Ne, Mg, and Si have survived. On the other hand, each of these is accompanied by an appreciable admixture of secondaries. Li, Be, and B are entirely secondary, as is flourine, and the total arriving nitrogen exceeds the primordial contribution. Owing to its large geometric cross section, cosmic-ray iron is sharply depleted, much of it being transformed into nuclides of the sub-iron group.

7. INJECTION AND ACCELERATION

How and where do the cosmic rays acquire their prodigious particle energies? In recent years some promising advances have been made in answering these questions. The favored mechanisms of acceleration can be traced to the concepts of magnetohydrodynamics (MHD) due to Hannes Alfven. Enrico Fermi built upon these ideas in his seminal theories of acceleration. Alfven emphasized that the acceleration of ions in space arises from variations in the magnetic fields, and is produced by "magnetic pumping". He argued that ions can be energized in the earth's magnetosphere, in the interplanetary medium, in the ISM, and even in extragalactic space as well. In the light of subsequent observations, his predictions seem well founded.

Fermi devised a scheme of stochastic acceleration in which charged particles gain energy by encounters with irregularities in moving magnetic fields. These could be clouds of plasma in the ISM; if a moving ion and a cloud are approaching each other, then the ion gains energy in the "collision" at the expense of the cloud. To be sure, if the ion is overtaking the cloud, it loses energy. However, on the average, head-on-collisions occur more often than overtaking ones, so the particle gains energy in the long run. A success of

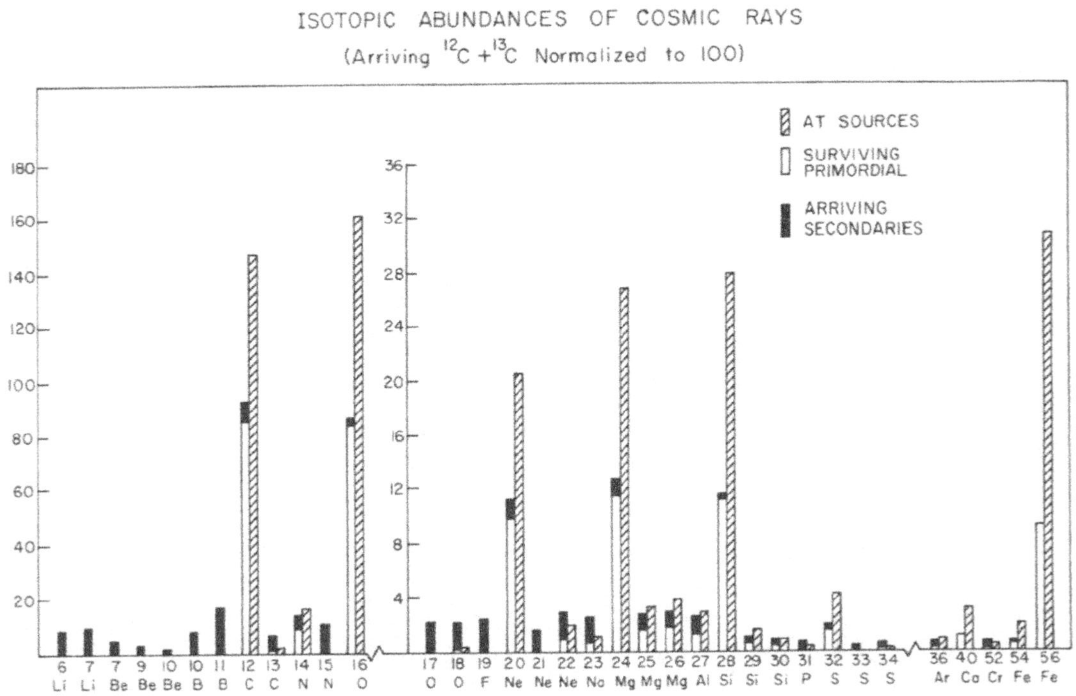

Figure 7. The relative abundances of nuclides at cosmic-ray sources and near the earth. In the arriving columns, the secondaries and the surviving source component are shown separately.

Fermi's so-called "second-order" theory was that it yielded a power-law energy spectrum, as is observed. Some difficulties with this early theory were that it generated a relatively slow, inefficient acceleration, and required what seemed to be a high threshold for injection--especially for the heavier particles having a high rate of ionization loss. In his later ("first order") theory, Fermi proposed that the spiral arms of the Galaxy might operate as magnetic bottles with a configuration like that of a plasma "mirror machine". Suppose that the magnetic field lines in a segment of a spiral arm converge toward the ends. If the magnetized plasma near the two ends of the "trap" approach each other, then the confined ion will gain energy. A special, analogous case of such efficient acceleration would be the relative approach of two contiguous clouds, with an ion between them gaining energy in successive collisions. This could happen efficiently within the confines of a supernova remnant, as realized by Ginzburg (1953).

Implicit in early theories of acceleration was the supposition that particles of modest energy (say, 1-10 MeV) are somehow injected into the regions where a further boost in energy occurs. The nature of the injection process has been elusive, but evidence has been accumulating that initially the thermal particles destined to become cosmic rays are pre-selected according to the first ionization potential (FIP) of the parent atoms. Generally, elements with low FIP are more abundant than those with higher FIP, as can be seen in Figure 8 (Silberberg et al. 1989). The data in this figure are based mainly on the satellite HEAO-3 experiment of the French-Danish collaboration (Lund, 1984).

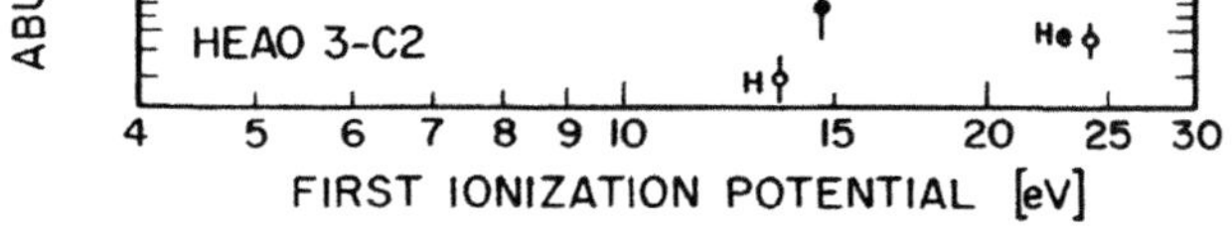

Figure 8. Comparison of the relative abundances of the galactic cosmic-ray source and the local galactic abundances, normalized at Mg. [Courtesy Silberberg et al. (1989)]

For many years, supernova explosions were regarded as the "machines" that could produce the GCR (see Shapiro 1962 and references therein). About 10^{53} ergs are released from the outburst within seconds after collapse of the stellar core, most of the energy going into neutrinos. Some 10^{51} ergs appear as kinetic energy of the ejected debris. In the precursor star, successive generations of nucleosynthesis have reached the stage of silicon-burning, with the formation of an iron core. As its exothermic energy sources are exhausted, the stellar core collapses to a neutron star, while the outer envelope explodes, giving rise to a supernova remnant, of which the Crab Nebula is the most famous example. If only one percent of the kinetic energy goes into high-energy particles, the source strength of supernovae suffices energetically to produce the observed CR intensity. (The frequency of these stellar explosions is estimated to be about twice per century). Current theories attribute the main acceleration of CR to their encounters with shockwaves in the ISM. The direct evidence that shocks occurring in outer space can boost particle energies comes from observations with space probes in the interplanetary medium. The shocks in the ISM are believed to be mainly those at the leading edge of expanding shells of supernovae. An important difference of this view from the older supernova theories of origin is that formerly the CR were thought to be boosted directly from freshly ejected material <u>within</u> the supernova remnant. Likewise, the acceleration itself was supposed to occur within the remnant. While modern theories still regard the supernova explosions as the main energy source, the ions that are ultimately promoted to CR energies reside in the ISM, and are accelerated by the shock fronts that they encounter (see, for example, Axford 1981). It will be noted that shock acceleration is still basically a Fermi mechanism, in that it is due to the ion's interaction with the magnetic inhomogeneities in the shock front.

We may now ask, is injection into the ISM really needed in order to start the accelerating process? There are experts (e.g., Eichler 1979) who believe that the "seed" particles can be boosted to CR energies directly from the thermal pool in the ISM. Others, however, such as Casse (1983) and Meyer (1985) have argued that the CR source composition precludes such direct acceleration. One reason is that the temperature in the ISM--notably in the hot phase that fills most of the Galactic volume--are not conducive to FIP selection. Accordingly, injection of, say, 1-10 MeV ions into the ISM from suitable sources is most probably required. Though many eruptive stars are likely to contribute some seed ions into the ISM, two particularly favorable classes of stars seem to be the dMe and dKe stars (Shapiro 1987). The possibility that these flare stars might provide seed particles has cropped up several times in recent years, but a detailed argument establishing energetic and compositional plausibility has only recently been advanced (Shapiro 1989). It appears that the powerful and frequent flares of the numerous dMe and dKe red dwarfs can populate the ISM with the seed particles required for the shock acceleration of the bulk of the cosmic rays.

8. ISOTOPES IN THE COSMIC-RAY COMPOSITION

The GCR nuclei arriving in the solar system are a mixture of stable and radioactive species. It has been demonstrated that the interactions occurring en route must have a profound effect on the relative abundances detected near the earth. As a result, the composition of the cosmic rays arriving at the top of the atmosphere is quite different from that just after production at the sources, not only in their elemental makeup, but particularly in their isotopic distribution. For example, such nuclides as ^{3}He, ^{7}Be, and ^{15}N are relatively abundant in the arriving cosmic rays, though absent or scarce in the general composition of matter.

In theories of origin and propagation of cosmic rays a central problem is this: what are the relative cosmic-ray abundances when the particles are just leaving the sources? This composition provides a starting point for calculating the production and attenuation of secondary nuclides. To do so requires, as we have seen, the solution of diffusion equations in which the path length distribution, the partial cross sections for fragmentation, and the particle energies are among the parameters.

The composition at the cosmic-ray sources has been calculated by adjusting the values of the source-abundances in the diffusion equation until the calculated arriving composition near the earth agreed with the observed composition of cosmic-ray nuclei at the top of the earth's atmosphere. The calculations require detailed knowledge of the many modes of cosmic-ray spallation and their respective cross sections as a function of energy.

8.1 Cross Sections for Fragmentation. In recent years, progress has been made both in Laboratory measurements and in calculations of break-up cross sections. Originally, the group at the Bernas Laboratory in Orsay applied mass spectrometry to this problem, and the NRL group devised semiempirical relationships based upon nuclear systematics and the available measured values. Meanwhile, relativistic beams of heavy ions were developed at Berkeley, Dubna and other accelerator laboratories, with energies exceeding those hitherto available by two orders of magnitude. In this way, it became possible to measure the production rates of many more secondaries. The predictive value of the semiempirical formulations could be tested and confirmed, and the formulae themselves could be further refined.

8.2 Calculation of Isotopic Composition. These developments enhanced our confidence in deducing the relative abundance of the cosmic ray elements at the sources. Once a source composition has been derived (by a best fit to the arriving elemental composition), the procedure can be turned around to estimate the arriving isotopic composition. Starting with their calculated elemental source composition, the NRL group computed the expected isotopic distribution of cosmic rays at the top of the earth's atmosphere (Tsao et al. 1973). The isotopic abundances of the various elements at the sources were taken to be the same as those in the solar system, according to Cameron. Some typical results are depicted in Figure 9 which compares some of the predicted cosmic-ray abundances with the corresponding nuclidic abundances in the general (thermal) composition found in the solar system.

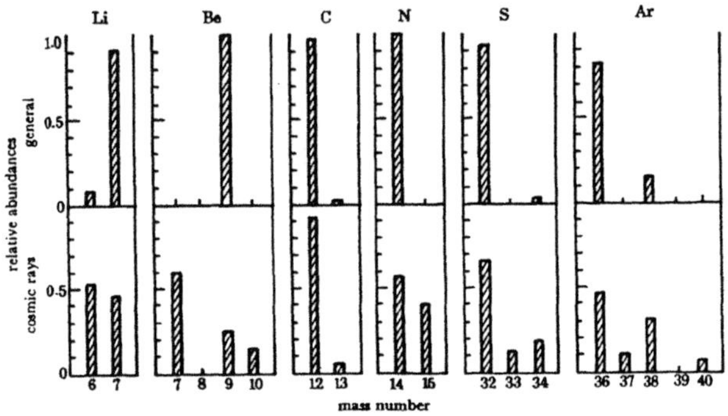

Figure 9. Comparison of the calculated isotopic abundances for several elements in the arriving cosmic rays with observed abundances in solar system material (Tsao et al, 1973). The columns are adjusted so that the total for each element is normalized to unity.

It can be seen that for most of the elements the two distributions are strikingly different. For example, in the general composition, 93 per cent of lithium consists of ^{7}Li; in the cosmic rays, about half of it is expected to be ^{6}Li. "Normal" beryllium is almost entirely ^{9}Be, while in the arriving cosmic rays, only about 30 per cent of this element should be ^{9}Be, about 60 per cent ^{7}Be, and $\lesssim$ 10 per cent ^{10}Be. (As discussed below, nuclei like ^{7}Be, that decay only by electron capture, are expected to survive in cosmic rays.) ^{15}N comprises only 4×10^{-3} of ordinary nitrogen, but it should constitute very nearly one half of the element in cosmic rays. For sulfur, argon, potassium, and calcium, the lightest isotopes predominate in the general abundances, while in the arriving cosmic rays, the heavier isotopes contribute about half of the total.

Measurements of relative isotopic abundances for many of the cosmic-ray elements are still in their infancy. Certain striking qualitative features, however, have already been confirmed: the predominance of ^{7}Be among the beryllium isotopes, and the comparable fluxes of ^{15}N and ^{14}N in the arriving cosmic rays.

Thus far, the measured and calculated isotopic compositions of the elements from lithium to oxygen agree rather well. This suggests that the NRL model is a reasonable representation of the isotopic composition at the sources. Any significant isotopic differences that may be discovered between predicted values for the local cosmic rays and observations, would show that the isotopic composition at the sources must differ from that elsewhere in nature. Such discrepancies would provide guidelines for modifying an initially adopted source model; they would tell us about special conditions at the sources, (or at least at <u>some</u> sources), and the processes of nucleosynthesis operating there. (See Sec. 8.6 below).

8.3 <u>The Helium Isotopes</u>. Among the well-established results on galactic cosmic-ray isotopes are those for helium at low energies. About ten per cent of helium nuclei at energies of ~ 100 to 500 MeV are ^3He. The flux of ^3He is virtually all secondary, for its relative abundance in the general composition is hundreds of times lower. If we inquire what mean path length of interstellar material is required to produce by spallation the observed fraction of ^3He, the answer (4 to 6 g/cm^2) agrees rather well with the mean path deduced from the production of Li, Be, and B.

Stripping of ^4He in collisions with interstellar gas is the main source of the ^3He, both indirectly (by stripping off a neutron) and directly, through proton stripping and the subsequent decay of tritium. It may be remarked that the sizeable cross sections for nucleon stripping of energetic ions make this process one of the more important agents of cosmic-ray transformations. Thus, the two nuclides next in abundance to ^4He, i.e., ^{16}O and ^{12}C, undergo stripping, and give rise to rather copious fluxes of ^{15}N and ^{11}B, respectively. These two nuclides, somewhat like ^3He, are generated directly--by proton stripping--and also indirectly, through the intermediate production of radioactive ^{15}O and ^{11}C by neutron stripping.

8.4 <u>Electron-Capture Isotopes and Cosmic-Ray Chronology</u>. Among the radioactive cosmic-ray nuclides are some neutron-deficient species that decay by capturing one of their K-electrons, rather than by emitting a positron. However, at high velocities in a tenuous medium, a K-capture nucleus is effectively stable since in cannot then capture an electron. A striking example is ^7Be which has a half-life T of 53 days at rest. At energies $\gtrsim$ 10 MeV it can survive for millions of years in interstellar space, and it is in fact the most abundant of the three Be isotopes arriving in the cosmic rays. If it is to avoid decay the ^7Be ion must, of course, be completely stripped of its orbital electrons--a condition satisfied by cosmic-ray nuclei that have been accelerated to relativistic energies: of these, K-capture nuclei can survive as long as they have not been unduly slowed down (to ~ 10 MeV/nucleon for light nuclei like ^7Be, and to ~ 300 MeV/nucleon for the Fe group).

Cosmic-ray ^7Be is a <u>secondary</u> fragment of heavier nuclei. Certain other K-capture nuclides are formed mainly during nucleosynthesis rather than by subsequent spallation. Among these are three--^{57}Co, ^{44}Ti, and ^{59}Ni-- which have been proposed as tracer-clocks for estimating the magnitude of the time t_a from nucleosynthesis until acceleration up to, say, ~ 300 MeV/nucleon. A fast K-capture atom which has not decayed in the time t_a will have been stripped of its electrons during the period of acceleration, and thereby avoid decay indefinitely while propagating through interstellar space. Thus ^{57}Co (T = 1 year) will be present in the arriving cosmic rays if these were accelerated within a year after nucleosynthesis, and will be essentially absent if acceleration requires more than a few years, for it will have decayed into ^{57}Fe. Since the present state of our knowledge by no means precludes time t_a >> 1 year, we may require chronometers of longer half life; two such nuclides are ^{44}Ti (with T = 47 years) and ^{59}Ni (T = 8 x 10^4 years). The approximate bounds that can be placed upon t_a by detecting the survival or decay of these tracers--and of ^{57}Co are shown in the following table:

K-Capture Nuclide	Time for Acceleration t_a (years) if nuclide is	
	Present	Absent
^{57}Co	$\lesssim 1$	> 1
^{44}Ti	$\lesssim 50$	> 50
^{59}Ni	$\lesssim 10^5$	> 10^5

Available data on the relative abundances of cobalt and nickel in the cosmic rays imply that ^{57}Co has decayed, hence that t_a > 1 year. Isotopic measurements of nickel may tell us which of two models is more tenable--acceleration of fresh supernova material or acceleration of old interstellar gas. In the former case, ^{59}Ni should be present in the arriving cosmic rays, in the latter case it should be absent. Similarly, various models of acceleration in supernova remnants predict different values of t_a which can be tested from the presence or absence of ^{44}Ti.

8.5 The Residence Time of Cosmic Rays. Among the long-lived radioactive tracers for measuring cosmic-ray lifetime in the Galaxy, ^{10}Be has been especially useful (Hayakawa et al. 1958; Peters 1963; Shapiro and Silberberg 1967, 1968, 1970). The ^{10}Be has a mean life of 2.5 x 10^6 years; it decays into ^{10}B.

In 1969, the NRL group reported (11th Conf. Budapest) based on a measured rate of Be/B at relativistic energies, an estimate of T < 10^7 y for the age of the CR nuclei. Later, measurements of the Be isotopes at low energies became feasible through satellite experiments.

The observed abundance of ^{10}Be must be compared with the calculated rate of beryllium production in space, from the heavier cosmic rays. If very little of the isotope is found in the arriving cosmic rays, this means that they have been traveling for a long time --long enough to allow the radioisotope to die out. If, on the other hand, most of the beryllium has survived, this would signify that the cosmic rays have lived for a time short compared to the "mean life" of the beryllium nuclei. The principle of the "nuclear tracer-clock" in determining the CR dwell time is illustrated in Figure 10. Limiting values of calculated relative abundances of the tracer are displayed for the extreme cases of total survival and complete decay of a long-lived radionuclide. The dashed curve refers to a hypothetical case in which the half-life is somewhat shorter than the escape time.

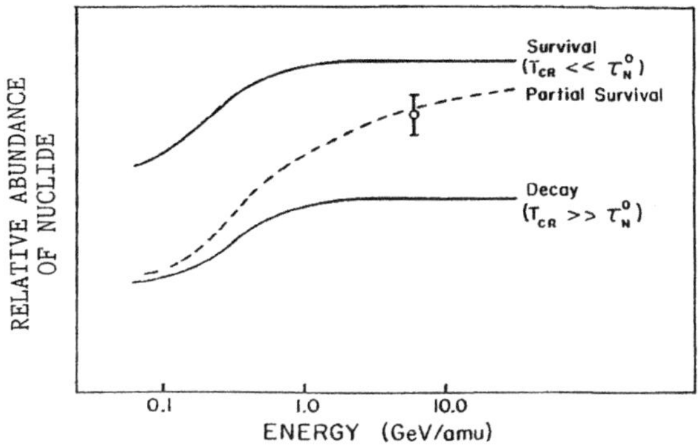

Figure 10. Tracer method for determining the galactic confinement time of cosmic rays. Limiting values of calculated relative abundances of the tracer are shown for the extreme possibilities of total survival and complete decay of a long-lived radionuclide. The dashed curve (and experimental point) illustrate an hypothetical case in which the half-life is somewhat shorter than the dwell time.

The Chicago group, in a satellite experiment, obtained sufficient, well-resolved isotopic data to estimate the cosmic-ray age (Simpson 1983). Taken together with subsequent work by Wiedenbeck (1983), a reasonable value of the residence time is ~ 10^7 years.

8.6 Isotopic "Anomalies" at the Source

Of the major isotopic differences thus far deduced between cosmic-ray source material and local (e.g., solar) matter, the abundance of ^{22}Ne in the sources is very striking: it has about 3.5 times the solar value. In addition, ^{25}Mg, ^{26}Mg, ^{29}Si and ^{30} Si appear to be "excessive" in the sources. These anomalies suggest that at least some of the CR sources may have had different conditions for nucleosynthesis (e.g., neutron-rich environments) than the sources responsible for the solar system composition.

As we acquire more reliable information about isotopic abundances in the cosmic rays, it should become possible to solve fundamental problems of their propagation, chronology, acceleration, and sites of origin. Consider, for example, the primordial nuclidic material prior to acceleration: is it young, newly born in nucleosynthesis, is it old (swept up from the interstellar medium), or does it consist of a mixture of these two? Assuming that it is fresh material, must it be produced in special processes of nucleosynthesis or can its composition be generated in the same processes that give rise to ordinary thermal matter? How much time is then required for acceleration? This time--and also the duration of magnetic trapping in the galaxy--can be estimated from measurements of radioactive tracers in the cosmic rays. The total time since nucleosynthesis can also be inferred from radioactive species, i.e. those of thorium and uranium.

9. HIGH-ENERGY NEUTRINOS AND GAMMA RAYS

Neutrinos are neutral particles with little or no mass which interact so weakly with matter that they can pass through the whole earth--and even through a star like the sun--without interaction. Relativistic electrons have been detected in remote regions of space through the radiations they emit, and especially the "synchrotron" radiation produced by electrons circulating in magnetic fields. But the more numerous cosmic-ray nuclei, mainly protons, have not been observed in sources at great distances from our solar system, owing to their tortuous paths in the ISM. The nascent field of neutrino astronomy can help solve this problem.

Neutrinos pervade all of space, along with photons and cosmic rays. However, according to the standard big-bang cosmology, the density of neutrinos is about a billion times higher than the average space density of atoms in the universe. That we have only begun to see extraterrestrial neutrinos poses a tantalizing challenge.

9.1 Neutrinos as Tracers of Cosmic-Ray Sources. Most of the neutrinos are exceedingly low in energy and extremely difficult to detect. Actually all neutrinos tend to elude observation. For the high-energy neutrinos that arise from cosmic-ray collisions with matter, however, the prospects of detection are brighter. It is these very energetic neutrinos that may reveal observationally the origin of cosmic-ray nuclei in exotic objects like pulsars and quasars and the cores of other active galaxies.

One may ask: could not radio astronomy, or x-ray or gamma-ray astronomy provide such information? Radio emission is produced by high-energy electrons in magnetic fields, and x-rays when such electrons collide with photons. They thus point back to the locations of sources of the cosmic-ray electrons. But the energy distribution of electrons becomes distorted by energy losses in magnetic fields, which actually prevent the acceleration of electrons to the higher energies. This is a severe shortcoming; to learn about the acceleration process, it is essential to know the energy spectrum. Furthermore, a study a cosmic-ray protons is required, since they vastly outnumber the electrons arriving at the earth.

Gamma-rays are produced as tertiary offspring of cosmic-ray nuclei colliding with matter; the ephemeral "neutral pions", (directly produced) disintegrate into gamma-rays. These could, in principle, tell us about the sites of acceleration of cosmic-ray protons and of heavier nuclei. However, gamma-rays are also generated by the Compton effect--electrons colliding with photons. Other gamma rays are produced by the "braking" of electrons giving so-called "Bremsstrahlung." These latter processes seem to dominate over those due to proton interactions in their production of gamma rays. Thus, in important respects, neutrinos provide a unique test for models of cosmic-ray sites and acceleration processes.

9.2 Detection of Cosmic Neutrinos. Neutrino astronomy is beset by the great difficulty in observing the elusive particles. Due to their very low cross sections for interaction, a large detector volume is required. This problem could be solved by the proposed DUMAND (Deep Underwater Muon and Neutrino Detection) experiment: an appreciable fraction of a cubic kilometer of ocean would be the basic detector, and the target mass as well. Neutrino-initiated showers and muons will be "seen" in this detector by photo-electric particle sensors, spaced in an array some tens of meters apart.

For over a decade intensive studies have been carried out to identify candidate celestial sources of neutrinos. These studies have focused principally on the high-energy domain (above 10^{11} eV). Much of the pioneering work was done by participants in Project DUMAND whose efforts have stimulated research on lower-energy cosmic neutrinos as well. With the detection of neutrinos from Supernova 1987a, neutrino astronomy has finally emerged as an observational science. This, in turn, has prompted the design of new instrumental arrays for cosmic high-energy neutrino and gamma-ray detection. The main neutrino background is due to cosmic-ray collisions in the earth's atmosphere. Some promising sources of cosmic neutrinos have been investigated. Among these are the expanding shells around young pulsars or neutron stars formed in a supernova collapse.

Outside of our galaxy, if a central massive black hole is the power source of a Seyfert galaxy like NGC 1068 or NGC 4151, these are also plausible candidate sources. NGC 1068 is the Seyfert closest to us (some 15 megaparsecs away). Its compact core pours out as much power as 200 billion suns. Neutrino astronomy gives promise of teaching us about acceleration mechanisms and other processes occurring, e.g., in pulsars, in compact binary systems and in active galactic nuclei.

9.3 Very-High-Energy and Ultra-High-Energy Gamma Rays. Cosmic gamma rays have come into their own as an astronomical discipline, and also as tracers of cosmic-ray interactions. At energies up to the GeV domain much has been learned from the SAS-2 and COS-B satellites (see, for example, Ramana Murthy and Wolfendale 1986 and Chupp 1976). Here we shall call attention to recent discoveries at ground-based observatories in the VHE (10^{11} - 10^{14} eV) and the UHE (> 10^{14} eV) regions. The former has depended on mirrors for collection of

Cerenkov light; the latter has employed extensive-air-shower (EAS) arrays.

There is a growing body of data on celestial sources that generate these UHE and VHE photons. Among these are the binary systems Cyg X-3, Her X-1, Vela X-1, and the Crab Nebula. The EAS apparently generated by UHE photons, have aroused controversy and even skepticism. Confusion has arisen partly because of the sporadic character of the signals and partly because of the anomalously high muon content in their cascades. Extragalactic sources have also been reported, e.g., Cen A, 3C 273, and M 31; the reports will require confirmation. The discovery by Samorski and Stamm (1983) of pulsed UHE gamma rays from Cyg X-3 excited great attention mainly because the nuclear primaries of the putative photons (up to $\simeq 10^{15}$ eV) would have to be about 20 times as energetic. Thus a "point source" of cosmic rays capable of boosting protons up to $\sim 10^{16}$ eV would be established. Hillas (1984) and Wdowczyk and Wolfendale (1983) emphasized the importance of such gamma-ray sources as possible sites of copious cosmic-ray production. Several large ground based arrays are now being enlarged or constructed with a view to clearing up the enigmas engendered by these UHE observations.

9.4 Antiprotons. A small but finite flux of antiprotons must be expected in the primaries detected above the atmosphere. These are secondaries of cosmic-ray collisions in the ISM. Some measurements of this intensity have yielded values higher than those predicted by propagation theory. One result, especially puzzling, was the considerable number of low-energy $\bar{p}$ reported by Buffington (1981). Within the past year, however, upper limits obtained for the antiproton flux at energies of a few hundred MeV have shown that the earlier results must have been in error. The new measurements (Salamon et al. 1988, Moats et al. 1988) agree with "standard" propagation theory.

10. OTHER ASPECTS OF THE COSMIC RADIATION

Constraints of time and space have led to omission of important topics in CR astrophysics. A few that come to mind are the Anomalous Component, the highest-energy CR, extragalactic CR, nucleosynthesis, components and properties of the ISM, the role of CR in the dynamics of the Galaxy, low-energy cosmic rays as catalytic agents in space chemistry, gamma-ray spectroscopy, instrumental advances in CR detection, etc. The interested reader is referred to the literature, some of it in the General Bibliography.

REFERENCES

Axford, W. I. (1981), in G. Setti, G. Spada, and A.W. Wolfendale (eds.), Origin of Cosmic Rays, D. Reidel Co., Dordrecht, p. 339 ff.
Binns, W. R. (1986), in M.M. Shapiro and J.P.Wefel (eds.), Genesis and Propagation of Cosmic Rays, Reidel Publ. Co., Dordrecht, p. 71 ff.
Buffington, A., Schindler, S.M. and Pennypacker, C.R. (1981), Ap. J. 248, 1179
Burbidge, E.M., Burbidge, G.R., Fowler, W.A., and Hoyle, F. (1957), Rev. Mod. Phys. 29., 547
Casse',M. (1983), in M.M. Shapiro (ed), Composition and Origin of Cosmic Rays, Reidel Publ. Co., Dordrecht, p. 193 ff.
Chupp, E. L. (1976), Gamma Ray Astronomy, Reidel Publ. Co., Dordrecht
Eichler, D. (1979), Ap. J. 229, 419
Fleischer, R.L., Price, P.B., and Walker, R.M. and Hubbard, E.L. (1967) Phys. Rev., 156, 353
Fowler, P.H., Adams, R.A., Cowen, V.G., and Kidd, J.M. (1967), Proc. Roy. Soc. A 301,39
Freier, P.S., Lofgren, E.J., Ney, E.P., Oppenheimer, F., Bradt, H. L., and Peters, B. (1948), Phys. Rev. 74, 213
Ginzburg, V.L. (1953) Uspekhi Fiz. Nauk 51, 343
Ginzburg, V.L. and Syrovatskii, S.I. (1964), Origin of Cosmic Rays, Pergammon Press, New York
Hayakawa, S., Ito, K., Terashima, Y. (1958), Prog. Theo. Phys., Kyoto, Suppl. 6, 1
Hillas, A.M.,(1984), Ann. Rev. Astron. and Astrophys. 22, 425
Lund, N. (1984), in M.M. Shapiro (ed.), Cosmic Radiation in Contemporary Astrophysics, Reidel Publ. Co., Dordrecht, p. 1, ff.
Meyer, J.P. (1985), Ap. J. Suppl., 57, 173
Moats, A., Bowen, T., Golden, R., Strittmatter, R. Stochaj, S., Ormes, J. and Lloyd-Evans, J. (1988), in M.M. Shapiro and J.P. Wefel (eds.), Kluwer Academic Publ., Dordrecht, p. 475 ff.
O'Dell, F., Shapiro, M.M., and Stiller, B.(1962), J. Phys. Soc. Japan, Supplem. A-III, 17, 23
Peters, B. (1963), Pontif. Acad. Sci. Scripta Varia 25, 1
Ramana Murthy, P.V. and Wolfendale, A.W. (1986), Gamma Ray Astronomy, Cambridge University Press
Salamon, M.H. et al., (1989), in M.M. Shapiro and J.P. Wefel (eds.), Cosmic Gamma Rays, Neutrinos, and Related Astrophysics, Kluwer Academic Publ., Dordrecht, p. 465 ff.
Samorski, M. and Stamm, W. (1983), Ap. J. 268, L17
Schein, M., Jesse, W.P., and Wollan, E.O. (1941), Phys. Rev. 59, 615
Shapiro, M.M., (1962) Science 135, 175
Shapiro, M.M., (1987), Proc. 20th Internat. Conf. on Cosmic Radiation (ICRC), Moscow, 2, 260
Shapiro, M.M. (1990), Proc. 21st ICRC, Adelaide, Australia, Paper OG 9. 1-7
Shapiro, M.M. and Silberberg, R. (1967), in B.P.S. Shen (ed.), High Energy Nuclear Reactions, W. A. Benjamin, Inc., Addison-Wesley Publ., Reading, Pa., p. 33 ff.
Shapiro, M.M. and Silberberg, R. (1970), Acta Phys. Hungarica 29, Suppl. 1, 485

Shapiro, M.M., Silberberg, R., and Tsao, C.H. (1970a), Acta Phys. Hungarica, 29, Suppl. 1, p. 463 ff.
_______ (1970b), ibid., p. 471 ff.
_______ (1970c), ibid., p. 479 ff.
_______ (1972), in F. Reines (ed.), Cosmology, Fusion and Other Matters, Univ. of Colorado Press, Boulder, Co., p. 124 ff.
Silberberg, R. and Tsao, C.H. (1973), Ap. J. Suppl. 25, 315 and 335
Silberberg, R., Tsao, C.H. and Letaw, J.R. (1989), in M.M. Shapiro and J.P. Wefel, Cosmic Gamma Rays, Neutrinos, and Related Astrophysics, Kluwer Acad. Publ., Dordrecht, p. 491 ff.
Simpson, J.A. (1983), in M.M. Shapiro (ed.), Composition and Origin of Cosmci Rays, Reidel Publ. Co., Dordrecht, p. 1, ff.
Tsao, C.H., Silberberg, R., and Shapiro, M.M. (1973), Conference Papers, 13th ICRC, Denver, University of Denver, 1, 107
Wdowczyk, J. and Wolfendale, A. W. (1983), Nature 305, 609
Wiedenbeck, M.E., (1983), in M. M. Shapiro (ed.), Origin and Composition of Cosmic Rays, Reidel Publ. Co., Dordrecht, p. 65 ff.

GENERAL BIBLIOGRAPHY

Proceedings of the International Conferences on Cosmic Radiation (ICRC), sponsored by the International Union of Pure and Applied Physics:
17th ICRC (1981), Paris, France
18th ICRC (1983), Bangalore, India
19th ICRC (1985), La Jolla, CA, USA
20th ICRC (1987), Moscow, USSR
21st ICRC (1990), Adelaide, Australia

Books

Creutz, E. (ed.) (1958, Nuclear Instrumentation, in Hdb. Physik (Encyclop. of Physics), 45/1 and 45/2 Springer-Verlag, Berlin
Fleischer, R.L., Price, P.B., and Walker, R.M. (1975), Nuclear Tracks in Solids, Univ. of California Press, Berkeley
Ginzburg, V.L. and Syrovatskii, S.I. (1964), Origin of Cosmic Rays, Pergammon Press, New York
Hayakawa, S. (1969) Cosmic Ray Physics, Wiley-Interscience, New York
Longair, M.S. (1981), High Energy Astrophysics, Cambridge University Press
Morrison, P. (1959)Hdbk. d. Physik, 46/1, Springer-Verlag, Berlin
Powell, C.F., Fowler, P.H., and Perkins, D.H. (1959), The Study of Elementary Particles by the Photographic Method, Pergammon Press, New York
Shapiro, M.M. (ed.) (1982), Composition and Origin of Cosmic Rays, Reidel Publ. Co., Dordrecht
Shapiro, M.M. (ed.) (1984), Cosmic Radiation in Contemporary Astrophysics, Reidel Publ. Co., Dordrecht
Shapiro, M.M. and Wefel, J.P. (eds.) (1986), Genesis and Propagation of Cosmic Rays, Reidel Publ. Co., Dordrecht

Shapiro, M.M. and Wefel, J.P. (1989) (eds.), Cosmic Gamma Rays, Neutrinos, and Related Astrophysics, Kluwer Academic Publishers, Dordrecht

Sitte, K. (ed.) (1963), Hdbk. d. Physik 46/2, Springer-Verlag, Berlin

Stecker, F.W. (1971), Cosmic Gamma Rays, NASA Publication SP-249

Wilson, J.G. and Wouthuysen, S.A. (eds.) (1958-67), Prog. Elementary Particle and Cosmic-Ray Physics, Vols. IV to IX, North-Holland Publishing Co., Amsterdam, and John Wiley & Sons, Inc., New York

Wolfendale, A. W. (1963), Cosmic Rays, Philosophical Library Inc., New York

Review Articles

Cesarsky, C.J. (1980), Confinement of Cosmic Rays in the Galaxy, Ann. Rev. Astron. Astroph., 18, 289

Fleischer, R.L., Price, P.B. and Walker, R.M., Ann. Rev. Nucl. Sci., 15,1.

Meyer, P. (1969), Cosmic Rays in the Galaxy, Ann. Rev. Astron. Astroph., 7, 1

Shapiro, M.M. (1958), in Creutz, E. (ed.), Nuclear Instrumentation, in Hdbk. d. Physik, Vol. 45, Springer-Verlag, Berlin

Shapiro, M.M. (1962), Supernovae as Cosmic-Ray Sources, Science, 135, 175

Shapiro, M.M. and Silberberg, R.,(1970), Heavy Cosmic-Ray Nuclei, in Ann. Rev. Nucl. Sci., 20, 323

Shapiro, M.M. and Silberberg, R. (1974), Phil. Trans. Roy. Soc. London, A 277, 319

Simpson, J.A. (1983), Ann. Rev. Nucl. Part. Sci. 33, 323

INTERSTELLAR CHEMISTRY

A. DALGARNO
Harvard-Smithsonian Center for Astrophysics
60 Garden Street
Cambridge, Massachusetts 02138
USA

ABSTRACT. A description is presented of the basic elements of the chemistry of interstellar clouds and the success of models of diffuse, translucent and dense clouds is assessed. The chemical differences between non-dissociative and dissociative shocks are reviewed. A brief discussion is given of the effects of intense ultraviolet radiation fields.

1. Introduction

Molecules have been discovered beyond the solar system in the atmospheres of stars, in emission, reflection and planetary nebulae, in diffuse, translucent and dense interstellar clouds, in the outflow gas from protostellar objects, evolving stars, novae and supernovae and beyond the Milky Way in Seyfert galaxies, starburst galaxies and merging galaxies. The diatomic molecule, carbon monoxide, is ubiquitous and provides a unique diagnostic probe of molecular clouds and the distribution of mass in galaxies. In the Milky Way galaxy a remarkable diversity of molecular species has been observed in dense interstellar clouds. These dense clouds are the sites of star formation. Table 1 is a list of the molecules that have been detected. It contains over seventy molecules and includes a rich variety of organic molecules. There is some evidence that still larger molecules containing more than twenty carbon atoms exist in diffuse regions of the interstellar medium.

Most molecules have been detected by the observation of emission lines in the radio and millimetre regions of the electromagnetic spectrum. For recent surveys see Jewell et al. (1989) and Turner (1989). Fig. 1 is a reproduction of the spectrum of the molecular cloud in Orion between 24.15 GHz and 242.5 GHz (Sutton et al. 1985), which demonstrates the presence of several species. In diffuse molecular clouds, which are of limited extent, the presence of molecules has been established also by observations of absorption lines in the ultraviolet, visible and infrared regions. Fig. 2 is a reproduction of a narrow region of the visible spectrum taken towards the star ζ Ophiuchi (Meyer and Jura 1984). The spectrum contains absorption lines P(J) and R(J) originating in the J=0, 1 and 2 rotational levels of the ground electronic state and ground vibrational level of the molecule CN. From the measured strengths of the absorption lines, the relative populations of the different rotational levels can be determined and an excitation temperature inferred. The value of the rotational excitation temperature is 2.8K. The level populations are controlled by the universal blackbody background radiation, the relict of the burst of light than in the now conventional big-bang cosmology (cf. Weinberg 1972) heralded the beginning of the Universe; its temperature is 2.8 K.

Molecules constitute powerful diagnostic probes of the characteristics of the physical conditions in which they exist. They have provided measures of the density, the

J. M. Greenberg and V. Pirronello (eds.), Chemistry in Space, 71–87.

temperature, the radiation field, the magnetic field and the cosmic ray ionization rate. My lectures are an introductory survey of the chemistry of molecular clouds.

TABLE 1. Interstellar Molecules

H_2	Hydrogen	CH	Methylidyne
CH	Methylidyne Ion	OH	Hydroxyl
C2	Carbon	CN	Cyanogen
CO	Carbon Monoxide	NO	Nitric Oxide
CS	Carbon Monosulphide	Sio	Silicon Monoxide
SO	Sulphur Monoxide	NS	Nitrogen Sulfide
SiS	Silicon Sulphide	PN	Phosphorus Nitride
HCl	Hydrogen Chloride	C_2H	Ethynyl
H_2O	Water	HNC	Hydrogen Isocyanide
HCN	Hydrogen Cyanide	HNC	Hydrogen Isocyanide
HCO	Formyl	HCO^+	Formyl ion
N_2H^+	Protonated Nitrogen	H_2S	Hydrogen Sulphide
HNO	Nitroxyl	OCS	Carbonyl Sulphide
SO_2	Sulphur Dioxide	HCS^+	Thioformyl ion
H_2CO	Formaldehyde	C_2S	Dicarbon Sulphide
NH_3	Ammonia	H_2CS	Thioformaldehyde
HNCO	Isocyanic Acid	HCNS	Isothiocyanic Acid
C_3N	Cyanoethynyl	HOCO	Protonated Carbon Dioxide
C_3O	Tricarbon Monoxide	C_3S	Tricarbon sulphide
C_2H_2	Acetylene	C_3H_2	Cyclopropenylidene
CH_2CO	Ketene	HCOOH	Formic Acid
HCOOH	Formic acid	HC_3N	Cyanoacetylene
C_4H	Butadinyl	NH_2CH	Cyanamide
CH_2NH	Methanimine	CH_3CH	Methyl Cyanide
CH_2CN	Cyanomethyl radical	CH_3SH	Methyl Mercaptan
C_5H	Pentynylidyne	CH_3C_2H	Methyl Acetylene
CH_3OH	Methyl Alcohol	CH_2CHCN	Vinyl Cyanide
HCC_2HO	Propynal	$HCOOCH_3$	Methyl Formate
C_6H	Hexatrinyl	HC_5N	Cyanodiacetylene
NH_2CHO	Formamide	CH_3C_4H	Methyl Diacetylene
CH_3NH_2	Methylamine	$(CH_3)_2O$	Dimethyl Ether
CH_3CHO	Acetaldehyde	CH_3CH_2CN	Ethyl Cyanide
CH_3C_3N	Methyl Cyanoacetylene	CH_3C_4H	Methyl Diacetylene
CH_3CH_2OH	Ethyl Alcohol	HC_7N	Cyanohexatriyne
HC_9N	Cyano-octatetra-yne	$HC_{11}N$	Cyano-decapenta-yne

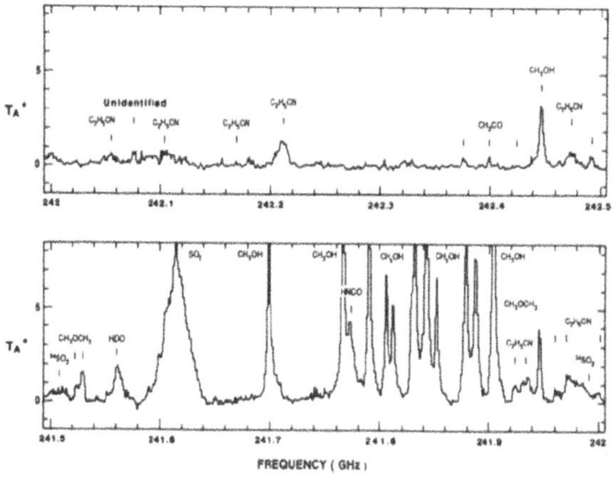

Figure 1. The spectrum of the molecular cloud Orion A from 241.5 GHz to 242.5 GHz. The identified emitting molecules are indicated (from Sutton et al. 1985).

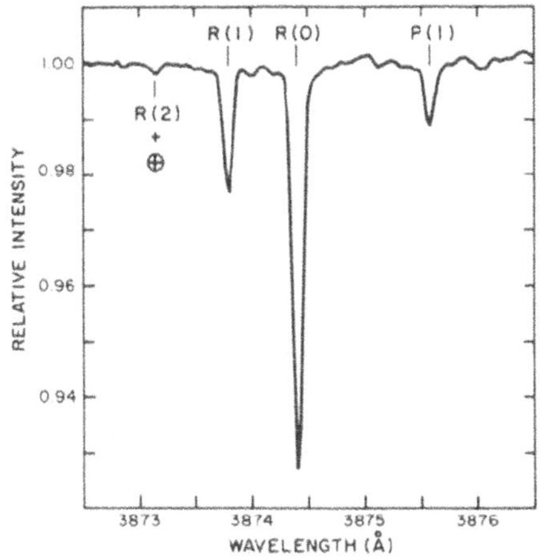

Figure 2. The 3874Å band of CN observed towards ζ Ophiuchi, showing absorption lines originating in the J=0, 1 and 2 rotational levels (from Meyer and Jura 1984).

2. Chemistry of molecular clouds

Molecules containing carbon, oxygen, nitrogen, sulphur, silicon, phosphorous and chlorine have been found in interstellar clouds but hydrogen and helium are by far the most abundant constituents and the chemistry of molecular clouds is essentially the chemistry of a gas composed of hydrogen and helium with a small admixture of heavy elements. The densities are low and only two-body collisions and collisions with grains occur. The interstellar gas is subjected to the radiation field produced by the assembly of stars and to energetic cosmic rays. The cosmic rays penetrate into the clouds and ionize and heat them. The presence of the molecular ions HCO^+, $HN_2{}^+$, HCS^+, $HOCO^+$ and $HCNH^+$ is a clear testimony of the importance of cosmic rays. In the gas there are solid refractory particles, dust grains, which absorb and scatter the interstellar radiation field and shield the molecules from its destructive effects and which provide a site for the formation of hydrogen molecules and most likely of the chemically saturated species H_2O, NH_3, CH_4, H_2S and PH_3.

The interstellar photons incident upon a cloud penetrate into the cloud material and the ultraviolet photons photoionize atoms and molecules

$$X + h\nu \rightarrow X^+ + e$$

and photodissociate molecules

$$YZ + h\nu \rightarrow Y + Z \ .$$

Surveys of photodissociation and photoionization of interstellar molecules have been presented by van Dishoeck (1987, 1988a). With increasing depth into the cloud, the photons are absorbed by the grains. Thus molecular clouds have envelopes in which photoionization and photodissociation processes dominate the chemistry and interiors in which processes driven by interstellar photons are unimportant.

For clouds near to external intense sources of ultraviolet radiation, processes initiated by the absorption of photons exert a major influence on the chemistry for substantial distances into the clouds. Molecular envelopes subjected to intense radiation have been called photodissociation regions (Tielens and Hollenbach 1985) or alternatively photochemical regions (van Dishoeck 1988b).

The chemistry of the envelope of a dense molecular cloud is similar to that of diffuse and translucent molecular clouds. Diffuse and translucent molecular clouds are distinct entities in the interstellar gas. Diffuse clouds have visual extinctions A_v of up to about 2 magnitudes. Translucent clouds are more extensive with A_v between 2 and 10 magnitudes. The visual extinction A_v is a measure of the column density of material. If N_H is the total hydrogen nuclei density, $N_H = 2N(H_2) + N(H) + N(H^+)$,

$$N_H \sim 1.9 \times 10^{21} \ A_v \ cm^{-2} \ .$$

The photoionization and photodissociation rates by the interstellar radiation field diminish exponentially with A_v or N_H.

The molecules H_2 and CO are special in that they can be self-shielding and can survive even in the absence of absorption by grains. They are both dissociated by processes initiated by discrete line absorption into excited electronic states. The absorption is followed for H_2 by radiative emission and for CO by predissociation into the vibrational continuum of the ground electronic state.

The photodestruction of H_2 is balanced by its formation which in the absence of an efficient gas phase mechanism is attributed to grain surface catalysis

$$H + H_g \rightarrow H_2 + g \ .$$

The abundances of H and H_2 in diffuse clouds have been derived from measurements of ultraviolet line absorption. Comparison of the formation and destruction rates suggests that an H_2 molecule is produced with an efficiency approaching unity in each collision of a hydrogen atom with a grain. In the interior of dense clouds where photodissociation of H_2 is negligible, the hydrogen exists almost entirely as H_2.

Except in shocked regions or regions subjected to intense sources of radiation, interstellar clouds are cold, with temperatures less than 100K and in dense interiors close to 10K. Neutral particle reactions often have activation barriers and proceed slowly at low temperatures. In contrast, ion-molecule reactions are usually rapid and indeed for heteronuclear molecules the rate coefficients of many reactions increase with decreasing temperature. Thus the chemistry is mainly an ion-molecule chemistry though specific neutral reactions are of critical importance. Thus the chemistry is mainly an ion-molecule chemistry though specific neutral reactions are of critical importance.

3. Dense Clouds

With the formation of H_2, the chemistry in the interiors of dense clouds is driven by the cosmic ray ionization of H_2 which produces the molecular ion H_2^+. Because H_2 is by far the major component, if anything can react with H_2 it will. Thus the H_2^+ ions are immediately converted to H_3^+ ions by the reaction

$$H_2^+ + H_2 \rightarrow H_3^+ + H$$

The H_3^+ ions do not react with H_2 and instead undergo proton transfer with any of a range of neutral species X,

$$H_3^+ + X \rightarrow H_2 + XH^+ \ .$$

X may be O, C, CO, S, Si, N_2 or SiO, though not N or O_2. Exothermic proton transfer reactions are rapid at the low temperatures between 10K and 50K that prevail in dense clouds (cf. Rowe 1988). Following proton transfer, more complex ions are built by hydrogen abstraction reaction sequences

$$X H_{n-1}^+ + H_2 \rightarrow XH_n^+ + H \ .$$

The sequence terminates when the ion no longer reacts with H_2. The terminal ion in the sequence is then neutralized by dissociative recombination

$$X H_n^+ + e \rightarrow XH_{n-1} + H \ .$$

$$\rightarrow XH_{n-2} + H_2 \ .$$

With the exception of H_3^+, dissociative recombination is rapid.

An example is the production of OH and H_2O which proceeds by the sequence

$$H_3^+ + O \rightarrow OH^+ + H_2$$

$$OH^+ + H_2 \rightarrow H_2O^+ + H$$

$$H_2O^+ + H_2 \rightarrow H_3O^+ + H$$

$$H_3O^+ + e \rightarrow H_2O + H$$

$$\rightarrow OH + H_2 \ .$$

A similar sequence leads to the formation of CH_3^+ from neutral carbon atoms. In addition, the carbon chemistry is modified by reactions with C^+ which is a product of the destruction of CO by He^+,

$$He^+ + CO \rightarrow C^+ + O + He$$

and which in molecular envelopes is produced by photoionization of C. The C^+ ion does not react with H_2. Unlike dissociative recombination, radiaative recombination is slow and the C^+ may participate in a radiative association reaction with H_2.

$$C^+ + H_2 \rightarrow CH_2^+ + h\nu \ .$$

Radiative association,

$$X^+ + Y \rightarrow XY^+ + h\nu$$

$$X + Y \rightarrow XY + h\nu \ ,$$

is a powerful mechanism for creating large systems. Thus

$$CH_3^+ + H_2O \rightarrow CH_3^+ \cdot H_2O + h\nu$$

followed by dissociative recombination

$$CH_3^+ \cdot H_2O + e \rightarrow CH_3OH + H$$

leads to methyl alcohol and

$$H_3O^+ + C_2H_4 \rightarrow CH_3CH_2OH_2^+ + h\nu$$

followed by

$$CH_3CH_2OH_2^+ + e \rightarrow CH_3CH_2OH + H$$

leads to ethyl alcohol.

Methane can be formed by

$$CH_3^+ + H_2 \rightarrow CH_5^+ \quad + h\nu$$

$$CH_5^+ + CO \rightarrow CH_4 \quad + HCO^+$$

and reactions of C^+ with CH_4 and its products can build the complex hydrocarbons by insertion reactions such as

$$C^+ + CH_4 \rightarrow C_2H_3^+ + H$$

to make acetylene through

$$C_2H_3^+ + e \rightarrow C_2H_2 + H \ .$$

The build-up may continue through

$$C^+ + C_2H_2 \rightarrow C_3H^+ + H \ .$$

Condensation reactions such as

$$CH_3^+ + CH_4 \quad \rightarrow C_3H_5^+ + H_2$$

also contribute to the formation of complex hydrocarbon ions. Fig. 3 (Sternberg, Dalgarno and Lepp 1987) illustrates the sequence of chemical reactions. The exchange reactions of the molecular ions with neutral species leads to molecules containing heavy elements. Thus the reaction with nitrogen

$$C_nH_n^+ + N \rightarrow C_nH_{n-1}N^+ + H$$

is followed by

$$C_nH_{n-1}\,N^+ + e \rightarrow C_nH_{n-2}N + H \ .$$

As an example of the nitrogen chemistry, the cyanomethyl radical CH_2CN is produced by

$$C_2H_4^+ + N \rightarrow C_2H_3N^+ + H$$

$$C_2H_3N^+ + e \rightarrow CH_2CN + H \ .$$

The final distribution of interstellar species is then determined by a small group of neutral particle exchange reactions which remain rapid at low temperatures such as

$$O + OH \rightarrow O_2 + H \ .$$

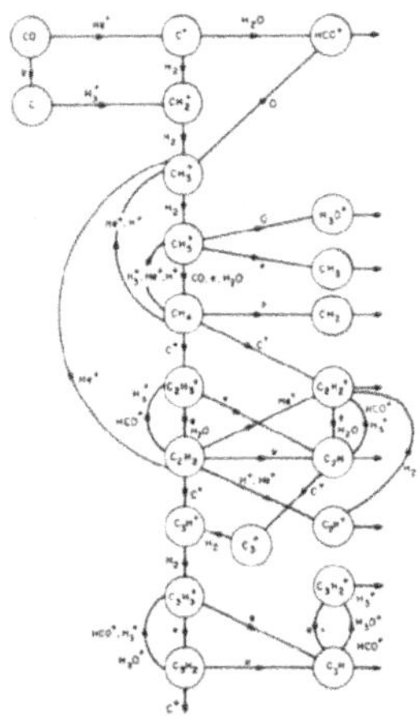

Figure 3. Schematic diagram of the principal elements in the ion-molecule hydrocarbon chemistry in dense clouds (from Sternberg, Dalgarno and Lepp 1987).

Recent studies of various aspects of interstellar chemistry have been carried out by Herbst et al. (1989), Langer and Graedel (1989), Gredel et al. (1989), Herbst et al. (1989) and Langer and Glassgold (1989), and a comprehensive discussion of the formation of complex molecules in dense interstellar clouds has been given by Herbst and Leung (1989). A valuable review has been presented by van Dishoeck (1988b).

Although interstellar photons do not penetrate into the interiors of dense clouds, photons are created there by cosmic rays (Prasad and Tarafdar, 1983; Sternberg, Dalgarno and Lepp 1987). The fast electrons ejected in cosmic ray ionization lose energy in exciting H_2 to electronic states that radiate in the ultraviolet. Fig. 4 illustrates the calculated spectrum (Gredel et al. 1989) between 80 nm and 180 nm. The destructive effects of the photons are amplified as more carbon atoms are added. Thus in Fig. 3, the abundance of acetylene C_2H_2 is reduced by photodissociation and its reduction diminishes the supply of C_2H_2 in the formation of C_3H_2, whose abundance is further decreased by its own photodissociation. The destructive effects are partially offset by the photodissociation of carbon monoxide

$$CO + h\nu \rightarrow C + O$$

which also occurs and which provides a source of neutral carbon atoms inside dense clouds (Gredel, Lepp and Dalgarno 1987). The neutral atoms then participate in reaction sequences that reform the complex hydrocarbon molecules (Gredel et al. 1989).

The steady-state ratios are considerably smaller than the measured ratios. In many clouds the C/CO ratio is of the order of 0.1 (cf. Frerking et al. 1989) whereas the predicted values do not exceed 0.006 (Gredel et al. 1989). The steady-state fractional abundances of complex molecules are also much below the measured values. As an example, the C_3H_2/H_2 abundance ratio found for a range of molecule clouds (Cox, Walmsley and Güsten 1989) lies between $10^{-9.5}$ and $10^{-8.4}$ whereas the predicted values are about 10^{-11}.

The low ratios occur because most of the available carbon is utimately locked up in carbon monoxide. A possible resolution is to postulate that the abundance of gaseous carbon exceeds that of gaseous oxygen so that there is a plentiful supply of carbon available after the carbon monoxide has formed. The solar and galactic HII region abundances of carbon are half those of oxygen but differential depletions in dense interstellar gas may enhance the C/O ratio (cf. Blake et al. 1987).

There is considerable evidence that large molecules containing in excess of twenty carbon atoms exist in the diffuse interstellar medium (cf. Puget and Leger 1989). If they also exist in dense clouds, the chemistry will be modified. The large molecules LM will convert the C^+ ions to neutral C atoms by charge transfer

$$C^+ + LM \rightarrow C + LM^+ \ .$$

Additionally the LM will scavenge the free electrons

$$LM + e \rightarrow LM^- + h\nu$$

(Omont 1986) and C atoms will be formed by mutual neutralization

$$C^+ + LM^- \rightarrow C + LM \ .$$

The large molecules limit the conversion of C^+ to CO and enhance the formation of complex molecules.

The presence of large molecules produces a qualitative change in the chemistry. Dissociative recombination

$$XY^+ + e \rightarrow X + Y$$

may be replaced as the main ion-removal process by mutual neutralization

$$XY^+ + LM^- \rightarrow XY + LM \ .$$

(Lepp and Dalgarno 1988).

The presence of large molecules enhances the abundance of C and of the complex interstellar molecules but not enough to explain the observations unless the large molecules are themselves a direct source of interstellar molecules through chemical reactions with oxygen and other species (Omont 1986, Duley and Williams 1986, Lepp et al. 1988, Brown et al. 1989), Flower et al. 1989) or by cosmic ray secondary electron and photon impact. Even so, there is a problem of supply as the large molecules are diminished.

Another possiblity for explaining the high C/CO ratios and the high abundances of complex molecules is to invoke time-dependence. The time scale for forming CO in a cloud in which the carbon is initially neutral C atoms is longer than that for forming the complex molecules. At times of order 10^5 years for example, the C_3H_2/H_2 ratio is 10^{-7}, larger than observed (cf. Herbst and Leung 1989, Gredel et al. 1989).

Not all molecular abundances can be reproduced by time-dependent models, in which the heavy elements are initially atomic. The ion-molecule scheme for the formation of NH_3 probably fails because the possible initiating reactions

$$N^+ + H_2 \rightarrow NH^+ + H$$

$$N + H_3^+ \rightarrow NH^+ + H_2$$

are endothermic.

A potential source of NH_3 is provided by grains. There seems little doubt that grain surfaces contain chemically saturated species and infrared absorption studies have established the presence of solid NH_3, CO, H_2O, and recently CO_2 (d'Hendecourt and Jourdain de Muizon 1989).

The mechanisms by which these molecules can be released into the gas phase is less certain. Desorption by cosmic ray impacts has been considered by Leger, Jura and Omont (1985) and the explosive ejection in grain-grain collisions by d'Hendecourt, Allamandola and Greenberg (1985). The grains are not only sources but they are surely sinks for the heavy atoms and molecules. Their surfaces are cold and condensation should occur on every collision. In the absence of a returning mechanism, the clouds would be devoid after a time of only a few 10^5 years of all molecules except H_2, andthe molecular ions H_3^+, H_2D^+ and HeH^+.

Time-dependent models in which grains are both sources and sinks have been investigated by d'Hendecourt, Allamandola and Greenberg (1985). They did not study complex molecules but it appears that the abundances would diminish quickly after about 10^5 years, along with the C/CO ratio.

A more complicated scenario seems to be required in which the chemistry is influenced by events associated with star formation and high velocity outflows which not only modify the environment but are themselves sources of molecules. For recent discussions of the chemistry in circumstellar shells, see Mamon, Glassgold and Huggins (1988), in protostellar winds see Rawlings, Williams and Canto (1988) and in nova outbursts see Rawlings (1988). The interactions of the outflowing gas with grains will remove grain mantles and deposit molecules into the gas phase. Observations of large abundances of NH_3 (Walmsley et al. 1987) H_2O (Jacq et al. 1988) and methanol (Wilson et al. 1989) in a region of active star formation, Orion-KL, are consistent with evaporation from heated grain mantles. Different chemical compositions in different regions inside molecular clouds also suggest varying depletions, presumably associated with star-forming events.

A detailed model of the molecular cloud Barnard 5 has been constructed by Charnley et al (1988, 1989) in which ionized hydrogen and helium mix with ablated molecular material and modify it by chemical reactions.

I turn now to what should be a simpler chemistry, that of diffuse and translucent molecular clouds.

4. Diffuse and Translucent Cloud Models

Diffuse and translucent clouds have no internal energy sources and their physical environment can be more reliably characterized than can that of dense clouds. They provide a critical assessment of the chemical processes that have been suggested. The chemistry of diffuse clouds is dominated by processes initiated by photoabsorption, gradually changing to a chemistry like that of dense clouds as the clouds become more extensive.Recent steady-state models have been constructed by van Dishoeck and Black (1986, 1988, 1989) Viala (1986), Viala, Roueff and Abgrall (1988) and Nercessian, Benayoun and Viala (1988).

Only diatomic molecules have been detected in diffuse and translucent clouds. Because of the efficiency of photodissocation it is difficult to build polyatomc molecules though the models do predict high abundances of some triatomic species. In diffuse clouds the time-scales for molecular formation and destruction are short and steady-state models may suffice (see however Wagenblast and Hartquist 1988, 1989).

The chemistry is qualitatively similar to dense clouds. With increasing depth into the clouds, the atomic hydrogen is converted into molecular form and for a cloud with a visual extinction of unity, the column densities of atomic and molecular hydrogen are comparable. As the conversion of atomic hydrogen to molecular hydrogen is occurring the C^+ ions produced by photoionization are participating in sequences that lead to CH, CH^+, CN, C_2, CS and CO. Carbon monoxide has the largest dissociation energy of the diatomic molecules and the chemistry places almost all the carbon in the form of CO for A_V greater than unity.

Some reactions with atomic hydrogen are important, particularly the charge transfer process

$$H^+ + O \rightarrow O^+ + H$$

which is followed by

$$O^+ + H_2 \rightarrow OH^+ + H \ ,$$

leading to the production of OH. The OH is removed by photodissociation

$$OH + h\nu \rightarrow O + H \ ,$$

and by reactions with C^+,

$$C^+ + OH \rightarrow CO + H^+$$

$$\rightarrow CO^+ + H \ .$$

Because the formation of OH is driven by cosmic ray ionization of H and H_2, its abundance provides a measure of the cosmic ray ionization rate. The inferred value is about $5\times10^{-16}\ s^{-1}$ (van Dishoeck and Black 1986).

A comparison of observations towards the star ζ Ophiuchi and one of the models of van Dishoeck and Black (1986, 1988) (see also Drdla, Knapp and van Dishoeck 1989) is presented in Table 2. The agreement for CH is obtained by postulating a rate coefficient of 7×10^{-16} cm^3 s^{-1} for

$$C^+ + H_2 \rightarrow CH_2^+ + h\nu \ .$$

The agreement for OH is obtained by an appropriate (and plausible) cosmic ray ionization rate. The molecule HD is formed by the sequence

$$H^+ + D \rightarrow H + D^+$$

$$D^+ + H_2 \rightarrow HD + H^+$$

so it also reflects the cosmic ray flux. Given the cosmic ray flux from the OH data, the [D/H] abundance ratio can be derived from the abundance of HD. The agreement for C_2 is supportive of the chemical model.

TABLE 2. Calculated and observed molecular abundances seen towards ζ Ophiuchi.*

	OH		H_2O		H_2O^+	
Observed	4.8±0.5	(13)	< 2.2	(13)	1.5	(13)
Model	4.5	(13)	5.5	(11)	1.4	(11)
	HD		CH		CH^+	
Observed	2.1±0.2	(14)	2.5±0.1	(13)	2.9±0.1	(13)
Model	2.9	(14)	2.1	(13)	3.2	(10)
	C_2		CO		NH	
Observed	1.5±0.2	(13)	2.0±0.3	(15)	< 7.5	(12)
Model	1.7	(13)	1.9	(14)	3.8	(11)
	CN		HCL			
Observed	2.5±0.1	(12)	<4.5	(11)		
Model	1.2	(12)	3.0	(11)		

*column densities in units of cm-2

The discrepancies for CN and CO can be resolved by postulating a less intense radiation field at the dissociating wavelengths since

$$CN + h\nu \rightarrow C + N$$

$$CO + h\nu \rightarrow C + O$$

are major destruction processes.

The carbon-containing molecule C_2, CH, C and CO have been detected in the more extensive translucent clouds. Models have been constructed by van Dishoeck and Black (1988, 1989) and Nercessian, Beneyoun and Viala (1988). Similar discrepancies occur for CO and CN as in diffuse cloud models which again may be partly remedied by postulating lower radiation fields.

There are also discrepancies in the relative abundances of C^+ and C obtained from observation and from the models. Table 3, reproduced from Dalgarno (1988), is a comparison for four clouds. The discrepancies could be resolved if large molecules LM were present at a fractional abundance of a few times 10^{-7} (Lepp et al. 1988), the C^+ being neutralized by reactions with LM and LM^-. These large molecules might then be a source of diatomic molecules.

TABLE 3. Observed and model abundances of C and C^+ towards four stars

	ζ Per				ζ Oph			
	Observed		Model		Observed		Model	
C	3.3±0.4	(15)	3.3	(15)	3.2±0.6	(15)	3.1	(15)
C^+	3.0+11.0	(17)	6.7	(17)	0.93±0.45	(17)	4.5	(17)
	χ Per				o Oph			
	Observed		Model		Observed		Model	
C	1-10	(15)	3.4	(15)	3.3-6.4	(15)	3.6	(15)
C^+	0.3-4	(17)	8.7	(17)	0.1-2.0	(17)	6.1	(17)

5. Shocked Regions

When a shock propagates in an interstellar cloud, the energy of the directed motion of the shock is randomized in a thin shock front and converted to thermal energy. The initial temperature of the shocked gas is determined by the shock velocity. In slow shocks with velocities less than about 25-50 km s^{-1} propagating in the interstellar magnetic field, the temperature is not so high that molecules are dissociated by collisions and the chemistry of the heated gas is controlled by reactions with H_2 and H, both exothermic and endothermic,

$$XH_{n-1} + H_2 \rightleftarrows XH_n + H \ .$$

The chemistry of a warm hydrogen-rich atmosphere tends to convert oxygen into water, nitrogen into ammonia and sulphur into hydrogen sulphide. How far the conversion proceeds depends upon the energetics of the abstraction reactions, on the H/H_2 ratio and on the initial temperature. Possible chemical diagnostic probes of shocked gas are high abundances of CH^+ (Elitzur and Watson 1980) and of molecules containing sulphur (Oppenheimer and Dalgarno 1980, Hartquist, Pineau des Forêts, Roueff and Flower 1986). A recent observational study of the chemistry of a shocked region associated with the supernova remnant IC443 has been presented by Ziurys, Snell and Dickinson (1989).

Despite the additional flexibility provided by postulating that a region of shocked gas lies along the line of sight towards ζ Ophiuchi, satisfactory chemical models have not yet been achieved (Draine 1986, Pineau des Forêts et al. 1986).

The chemical consequences can be substantial. For example, HCO^+, produced by

$$H_3{}^+ + CO \rightarrow HCO^+ + H_2 \ ,$$

is a major molecular ion in cold insterstellar gas. But if a gas with a high H_2/H ratio is shocked, the oxygen sequence

$$O + H_2 \rightarrow OH + H$$

$$OH + H_2 \rightarrow H_2O + H$$

carries through to H_2O and HCO^+ is converted to H_3O^+ by

$$HCO^+ + H_2O \rightarrow H_3O^+ + CO \ .$$

The sulphur chemistry is also significantly affected. The reactions

$$S + H_2 \rightarrow SH + H$$

$$SH + H_2 \rightarrow H_2S + H$$

may enhance the abundance of hydrogen sulphide which is difficult to produce in a cold gas (Hartquist, Oppenheimer and Dalgarno 1980).

Despite the additional flexibility provided by postulating a region of shocked gas, a satisfactory explanation of the CH^+ abundances has not yet been given (Draine 1986, Pineau des Forêts et al. 1986). The models do not allow for any shock-processing of the grains. Its inclusion may yield a resolution of the discrepancies (Brown et al. 1989).

If the shock velocities exceed 50 km s^{-1}, the molecules in the gas are destroyed (Hollenbach and McKee 1980). Following the shock, the ionized gas cools and recombines and molecules may form. The cooler gas in which the molecules may form is subjected to precursor radiation emitted by the downstream hot shocked gas.

The chemistry differs from that of a heated molecular gas because most of the material is in the form of ions and atoms. Eventually by gas phase reactions such as

$$H + e \rightarrow H^- + h\nu$$

$$H + H^- \rightarrow H_2 + e$$

molecular hydrogen is formed. Then

$$O + H_2 \rightarrow OH + H$$

produces OH and reactions with OH lead to other molecules such as CO, SiO, SO and NO. Molecular ions H_2^+, H_3^+, OH^+, SO^+ and CH^+ are produced in some abundance in regions where the shocked gas has a temperature of several thousand degrees and might be detectable by emission for vibrationally excited v=1 levels (Neufeld and Dalgarno 1989).

6. Molecular Clouds in Strong Radiation Fields

Warm gas can also be produced in edge regions of clouds subjected to intense ultraviolet radiation fields (Tielens and Hollenbach 1985, Sternberg and Dalgarno 1989). In the warm region, the chemistry is similar to that of shocked regions but modified by rapid photoionization and photodissociation. With increasing depth, the gas becomes cooler and photoabsorption becomes less important. The locations of the H/H_2 and C^+/CO transition regions change with the radiation field intensity. The C/CO ratio is high because of photodissociation of CO and of O_2. Photodissociation of CO is a major source of C atoms and the absence of O_2 reduces the loss of C by

$$C + O_2 \rightarrow CO + O \ .$$

The calculations suggest that a high C/CO ratio will be produced whenever the radiation field can penetrate. Genzel et al. (1988) have attributed the high abundance of C, and of C^+ in M82 to ultraviolet photons penetrating into a cloud consisting of dense clumps of gas embedded in a diffuse inter-clump medium. These clumpy clouds will have an interesting temperature, chemical structure and dynamical structure (cf. Falgarone and Puget 1985, Chieze and Pineau des Forêts 1987, 1989, Boissé 1989).

7. Acknowledgment

This work was supported by the National Science Foundation, Division of Astronomical Sciences, under Grant AST 86-17675.

8. References

Blake, G.A., Sutton, E. C., Masson, C.R. and Phillips, T.G. 1987 Ap. J.315, 621.
Brown, P.D., Duley, W.W., Jones, A. P. and Williams, D. A. 1989 M.N.R.A.S. in press.
Boissé, P. 1989 Astr. Ap. in press.
Charnley, S. B., Dyson, J.E., Hartquist, T.W. and Williams, D.A. 1988 M.N.R.A.S. 231, 269.
Charnley, S. B., Dyson, J.E., Hartquist, T.W. and Williams, D.A. 1988 M.N.R.A.S. in press.
Chièze, J. P. and Pineau des Forêts, G. 1987 Astr. Ap. 183, 98.
Chièze, J. P. and Pineau des Forêts, G. 1989, Astr. Ap. 221, 89.
Cox, P., Walmsley, C. M. and Giuster, R. 1989 Astr. Ap. 209, 382.
Dalgarno, A. 1988 Astro. Lett. Comm. 26, 153.
d'Hendecourt, L.B., Allamandola, L.J. and Greenberg, J.M. 1985 Astr. Ap. 152, 130.
d'Hendecourt, L. B. and Jourdain de Muizon, M. 1989 Astr. Ap. in press.
Draine, B.T. 1986 Ap. J. 310, 408.
Drdla, K., Knapp, G.R. and van Dishoeck, E.F. 1989 Ap. J. in press.
Duley, W.W. and Williams d. A. 1986 M.N.R.A.S. 219, 859.
Elitzur, M. and Watson, W. D. 1980 Ap. J. 236, 182.
Falgarone, E. and Puget, J. L. 1985 Astr. Ap. 142, 157.
Flower, D. R., Heck, L., Pineau des Forêts, G. and Millar, T. J. 1989, M.N.R.A.S. in press.
Frerking, M.A., Keene, J., Blake, G.A. and Phillips, T. G. 1989 ap. J. 344, 311.
Genzel, R., Harris, A.I., Jaffe, D. T. and Stutzki, J. 1989 Ap. J 332, 1049.
Hartquist, T. W., Oppenheimer, M. and Dalgarno, A. 1980 Ap. J. 236, 182.
Hollenbach, D. and McKee, C. F. 1980 Ap. J. Lett. 241, L47.
Jacq, T. Jewell, P.R., Henkel C., Walmsley, C.M. and Baudry, A. 1988 Astr. Ap. 199, L5.
Jannuzi, B. T., Black, J. H., Lada, c. J. and van Dishoeck, E.F. 1988 Ap. J. 332, 995.
Leger, A., Jura, M. and Omont, A. 1985 Astr. Ap. 144, 147.
Lepp, S. and Dalgarno, A. 1988 Ap. J. 324, 553.
Lepp, S., Dalgarno, A., van Dishoeck, E.F. and Black, J.H. 1988 Ap. J. 329, 418.
Mamon, G.A., Glassgold, A.E. and Huggins, P.J. Ap. J. 328, 797.
Meyer, D. M. and Jura, M. 1985 Ap. J. 297, 101.
Nercessian, E., Benayoun, J. J. and Viala, Y. P. 1988 Astr. Ap. 195, 245.
Neufeld, D.A. and Dalgarno, A. 1989 Ap. J. 340, 869.
Omont, A. 1986 Astr. Ap. 186, 184.
Pineau des Forêts, G., Flower, D. R., Hartquist, T. W. and Dalgarno, A. 1986 M.N.R.A.S. 220, 801.
Pineau des Forêts, G., Roueff, E. and Flower, D.R. 1986 M.N.R.S. 223, 743.
Puget, J. L. and Léger, A. 1989 Ann. Rev. Astr. Ap. in press.
Rawlings, J.M.C., Williams D.A. and Canto, J. 1988 M.N.R.A.S. 230, 695.
Rawlings, J.M.C. 198 M.N.R.A.S. 232, 507.
Rowe, B. R. 1988 in T. J. Millar and D. A. Williams (eds.) Rate Coefficients in Astrochemistry, Kluwer, Dordrecht, p. 135.
Sternberg, A. and Dalgarno, A. 1989 Ap.J. 338, 197.
Sternberg, A., Dalgarno, A. and Lepp, S. 1987 Ap. J. 320, 676.
Sutton, E.C., Blake, G. A., Masson C. R. and Phillips, T. G. 1985 Ap. J. Suppl. 58, 341.

Tielens, A. G. G. M. and Hollenbach, D. 1985 Ap. J. 291, 722.
Turner, B. E. 1989 Ap. J. Suppl. 70, 539.
van Dishoeck, E. F. 1987 in M. S. Vardya and S. P. Tarafdar (eds.), Astrochemistry IAU Symp. 120, Reidel, Dordrecht, p. 51.
van Dishoeck, E. F. 1988 in T. J. Millar and D. A. Williams, (eds.), Rate Coefficients in Astrochemistry, Kluwer, Dordrecht, p. 49.
van Dishoeck, E. F. 1988a in W. B. Burton and R. D. Wolstencroft (eds.), Kluwer, Dordrecht, p. 117.
van Dishoeck, E. F. and Black, J. H. 1986 Ap. J. Suppl 62, 109.
van Dishoeck, E. F. and Black, J. H. 1989 Ap. J. 340, 273.
van Dishoeck, E. F. and Black, J. H. 1988 Ap. J. 334, 771
Viala, Y. P. 1986 Astr. Ap. Suppl. 64, 391.
Viala, Y. P., Roueff, E. and Åbgrall, H. 1988 Astr. Ap. 190, 215.
Wagenblast, R. and Hartquist, T.W. 1988 M.N.R.A.S. 230, 363.
Wagenblast, R. and Hartquist, T.W. 1988 M.N.R.A.S. 237, 1019.
Wilson, T. L., Johnston, K. J., Henkel, C. and Menten, K.M. 1989, Astr. Ap. 214, 321.
Ziurys, L. M., Snell, R. L. and Dickman, R. L. 1989 Ap. J. in press.

THE MOLECULAR COMPOSITION OF DENSE INTERSTELLAR CLOUDS

WILLIAM M. IRVINE
Five College Radio Astronomy Observatory and
Department of Physics and Astronomy
University of Massachusetts
Amherst, Massachusetts 01003, USA

1. Introduction

The dense interstellar clouds in our Milky Way Galaxy and other galaxies are the most massive molecular objects known in the Universe. Their composition and chemical evolution are, in consequence, fundamental aspects of nature whose investigation is basic to space chemistry. Because the physical conditions in these clouds are quite extreme by the standards of the terrestrial laboratory, they provide a chemical environment which is quite different from that normally encountered on Earth.

In fact, the physics and chemistry of interstellar clouds are intimately related. The chemical composition will evolve in response to changes in the physical environment associated, for example, with star formation; likewise, the thermodynamics of cloud evolution is closely coupled to the chemical state of the medium through the presence of trace molecular constituents which act as cooling (or heating) agents or which influence the degree of ionization and hence coupling to magnetic fields. Moreover, the molecular constituents provide probes of the physical conditions to outside observers, since the strength of molecular transitions depends critically on such parameters as local kinetic temperature, density, and kinematics. It is also probable that preserved interstellar molecular material present in primitive solar system objects such as carbonaceous chondrites and comets may provide a link to the interstellar cloud from which the sun and planets formed, and thus provide critical information about the conditions and processes associated with this formation (the subject of subsequent lectures by J. Lewis).

It should also not be forgotten that the low temperature and density in interstellar molecular clouds provides a laboratory for basic molecular physics. That is, the radio and infrared telescope can be used to study basic molecular parameters for species which are unstable by terrestrial standards and hence very difficult to study on Earth. Thus, a number of radicals, positive ions, and even closed-shell molecular species which are very reactive have been observed

J. M. Greenberg and V. Pirronello (eds.), Chemistry in Space, 89–121.

astronomically prior to being studied in the laboratory, and some of these have even today only been observed in space. Likewise, some chemical processes such as radiative association are extremely difficult to observe in the laboratory, but are thought to play crucial roles in the synthesis of observed interstellar molecules, and deductions about the rates of such processes can be made from the astronomical data.

In this chapter we shall be primarily concerned with the gas phase chemistry, both because the composition of the gas phase can be determined with much less ambiguity than is possible for the solid interstellar grains (dust particles), and because the chemistry of the dust will be considered in a subsequent chapter by J.M. Greenberg. An effort to investigate the distribution of the volatile elements between the gas and solid phases will, however, be made in Section 7 below. We shall consider only dense molecular clouds, because the diffuse and translucent interstellar clouds will be the subject of lectures by A. Dalgarno. Other recent reviews on the subject of this chapter have been given by Irvine et al. (1985, 1987), Irvine and Knacke (1989), Guélin (1985), Turner and Ziurys (1988), Winnewisser and Herbst (1987), Duley and Williams (1984), and Friberg and Hjalmarson (1989).

We begin in Section 2 by reviewing molecular spectra, particularly the rotational spectra which are critical for the investigation of cold clouds and which have also to this point provided most of the chemical information even on hotter regions. Sections 3 and 4 provide an inventory of known interstellar molecules and a review of the physical state of dense interstellar clouds, respectively. Section 5 considers problems associated with accurately determining abundances of molecular species in these environments, and both similarities and differences in the chemical composition of quiescent and of hotter molecular clouds. Section 6 discusses isotopic fractionation, particularly for hydrogen. Section 7 considers the chemical reservoirs for the most abundant elements, with an emphasis on volatile species. Lastly, Section 8 presents an example of the identification of a new interstellar molecule on the basis of astronomical, quantum chemical, and finally laboratory, data.

2. Rotational Spectra of Molecules

The following brief review of molecular spectra is oriented to astronomers, not quantum chemists or molecular physicists. We shall draw heavily on the more rigorous monograph by Kroto (1975). The reader may also wish to consult the classic texts by Herzberg (1950, 1945, 1966) and by Gordy and Cook (1984). We shall frequently use the term "molecule" to include not only closed-shell species but also radicals (species with unpaired electrons and hence with non-zero net electronic angular momentum) and ions (all molecular ions observed to date in the interstellar medium are positive).

2.1 BASIC EQUATIONS

A molecule is a many-body system consisting of both nuclei (n) and electrons (e). The energy levels of this system are in principle obtained by solving the appropriate Schrödinger equation, which in the non-relativistic approximation takes the form

$$H \Psi = E \Psi \tag{1}$$

where the Hamiltonian operator H includes the sum of a kinetic energy term

$$T = \sum_n \frac{\mathbf{P}_n^2}{2m_n} + \sum_e \frac{\mathbf{P}_e^2}{2m_e} \tag{2}$$

and a potential energy term

$$V = V_{nn} + V_{ee} + V_{ne} \ . \tag{3}$$

In these expressions the vectors $\mathbf{P}_n$ and $\mathbf{P}_e$ represent the momenta of the nuclei and of the electrons, respectively, m_n and m_e are the corresponding masses, and V_{nn}, V_{ee}, and V_{ne} are the nucleus-nucleus, electron-electron, and nucleus-electron potential energy terms, which are functions of the generalized coordinates $\mathbf{q}_n$ and $\mathbf{q}_e$ and are near-Coulombic in form. Note that Equation 1 neglects the effects of spin, which in fact plays an important role in molecular rotational spectra for radicals ("fine structure") and for all molecules containing nuclei with non-zero spin (hyperfine structure). In Equation 2, the momemtum operators must of course be expressed in the Schrödinger equation as differential operators through

$$\mathbf{P} = -i\hbar\nabla \ , \tag{4}$$

so that the Hamiltonian becomes

$$H = -\sum_n \frac{\nabla_n^2}{2m_n} - \sum_e \frac{\nabla_e^2}{2m_e} + V_{nn} + V_{ne} + V_{ee} \tag{5}$$

where $\nabla^2 = \nabla \cdot \nabla$ is the Laplacian operator and the momenta are expressed in units of $\hbar = h/2\pi$, where h is Planck's constant.

Equations 1-5 are of course extremely complex. In most cases the problem may be simplified by the so-called "Born-Oppenheimer approximation," which separates the electronic and the nuclear motion. Let us write Equation 5 as

$$H = H_e(\mathbf{q}_e,\mathbf{q}_n) + H_n(\mathbf{q}_n) \ , \tag{6}$$

where

$$H_e = -\frac{1}{2m_e} \sum_e \nabla_e^2 + V_{ee} + V_{ne} + V_{nn} \ , \tag{7}$$

$$H_n = - \sum_n \frac{1}{2m_n} \nabla_n^2 \quad , \tag{8}$$

and we express the wave function Ψ as the product

$$\Psi(q_e, q_n) = \psi_e(q_e, q_n) \; \psi_n(q_n) \quad . \tag{9}$$

If now cross terms such as $\nabla_n^2 \; \Psi_e$ can be neglected, so that the electronic wave function does not change rapidly with nuclear configuration, the Schrödinger equation separates into a portion describing the electronic motion

$$H_e(q_e, q_n) \; \psi_e(q_e, q_n) = E_e(q_n) \; \psi_e(q_e, q_n) \tag{10}$$

and a part describing the nuclear motion

$$[H_n(q_n) + E_e(q_n)] \; \psi_n(q_n) = E_{ne} \; \psi_n(q_n) \quad , \tag{11}$$

where E_e (q_n) can be considered as a potential which governs the nuclear motion, while the q_n can be taken as fixed parameters in solving for the electronic wave function. Physically, the Born-Oppenheimer approximation corresponds to assuming that the electronic distribution comes into instant equilibrium with the position of the nuclei, reflecting the fact that the electrons are much lighter particles and thus will move much more rapidly. Born and Oppenheimer applied perturbation theory to the problem with $(m_e/m_n)^{1/4}$ as the expansion parameter. They showed that in general for small displacements of the nuclei the electronic energy E_e, the vibrational energy E_v, and rotational energy E_r of the molecule are related such that

$$E_r \sim (m_e/m_n)^{1/2} \; E_r \sim m_e/m_n \; E_e \quad . \tag{12}$$

Thus, typical transitions between electronic states for molecules tend to occur in the visible and ultraviolet and to correspond to temperature differences of thousands of Kelvin, transitions between vibrational states within a given electronic state tend to produce infrared photons with energies corresponding to hundreds of K, and the vibrational states in turn are subdivided into rotational levels, separated by energies corresponding to radio (in the broad sense) wavelength photons or temperatures of tens of K. Within this approximation the nuclear wave functions are independent of the electronic coordinates for a particular electronic state. If the electronic state is degenerate or nearly so, these approximations break down and a more general treatment is needed.

2.2 RIGID ROTORS

As we shall be considering cold clouds, we shall not deal further with the classification of electronic states for a molecule except to note that each electronic eigenfunction and hence each electronic state corresponds to a possible symmetry species of the point group of the molecule in question (cf. Herzberg, 1966). This leaves us with Equation

11 to describe the nuclear motion, which is itself of course quite difficult to handle. For example, the potential function E_e is usually approximated by an empirical expression such as the Morse function (cf. Townes and Schawlow, 1975). A solution then corresponds to the coupled vibrational and rotational motions of the molecule. Since even low-lying vibrational states of heavy interstellar molecules are typically excited only in the vicinity of regions of massive star formation (see discussion in this volume by R. Genzel), we shall further simplify the problem of molecular spectra by taking the nuclei to be at fixed distances from one another (the so-called "rigid rotor approximation"). We shall thus deal with the pure rotational spectra of the interstellar molecular constituents. Note that by neglecting nuclear relative motion we also are ignoring the effects of centrifugal distortion, which are easily measurable for interstellar molecules; as well as effects such as internal rotation and inversion, which are important for several interstellar molecules which contain methyl (CH_3) groups and for ammonia, for which the flexible nature of the molecule gives rise to the well-studied inversion transitions at wavelengths near 1 cm; cf. Ho and Townes (1983).

For pure rotation the relative nuclear coordinates are fixed and the only relevant contribution to the Hamiltonian comes from that portion of H_n corresponding to rigid rotation,

$$T_r = (1/2)\ (I_A\ \omega_A{}^2 + I_B\ \omega_B{}^2 + I_C\ \omega_C{}^2), \tag{13}$$

where I_A, I_B, and I_C are the principal moments of inertia of the molecule and the ω's are the corresponding angular velocities. It is customary to define the rotational constants $A = (2I_A)^{-1}$, $B = (2I_B)^{-1}$, and $C = (2I_C)^{-1}$ (note that strictly speaking the rotational constants depend on the vibrational state of the molecule). By convention the principal axes are chosen such that $A \geq B \geq C$. Because the eigenvalues of angular momentum operators are well-studied in quantum mechanics, it is convenient to express the rotational kinetic energy operator corresponding to Equation 13 as

$$H_r = AJ_A{}^2 + BJ_B{}^2 + CJ_C{}^2\ , \tag{14}$$

where the J's are the components of molecular angular momentum. The Schrödinger equation to be solved for the rigid rotor is now

$$H_r\ \psi_n = E_r\ \psi_n\ . \tag{15}$$

We note in passing that we have skipped over the subtleties associated with the difference between a *laboratory-fixed* coordinate system, in which basic physical laws are valid, and a *molecule-fixed* coordinate system, such as that defining the principal axes for a given molecule. Suffice it to say that the three components of the total angular momemtum J of a rotating molecule relative to Cartesian axes fixed in the laboratory can be related at any given moment in time to the principal axes fixed in a molecule to determine the "molecular-fixed components of J," and that there exist functions which are simultaneously the eigenfunctions of J^2 and of single components of both the space-fixed and molecule-fixed angular momenta (cf. Kroto, 1975).

The eigenvalues of Equation 15 now depend on the symmetry of the molecule being considered, and hence on the form taken by the Hamiltonian, Equation 14. The simplest situation corresponds to the so-called spherical top molecules for which A = B = C. Equation 14 then reduces to $H_r = BJ^2$. Since a well-known result of quantum mechanics is that the eigenvalues of the operator J^2 are J(J+1), we find for the energy levels of a spherical top molecule E(J) = BJ(J+1). A molecule with this degree of symmetry cannot, however, possess a permanent electric dipole moment; as a result, there are no allowed electric dipole transitions and no pure rotational spectrum. This applies to astrochemically significant molecules such as CH_4, in the interstellar medium and planetary atmospheres, and G_eH_4, which has been reported in the atmospheres of the giant planets. In fact, centrifugal distortion effects can induce a small dipole moment, and corresponding transitions have been seen in the laboratory for CH_4, but have not been confirmed in the interstellar medium.

2.2.1 *Linear Molecules.* For linear molecules it can be shown that rotation about the molecular axis can be treated as part of the electronic energy, so that one can write

$$H_r = B\ (J_B{}^2 + J_C{}^2) = BJ^2\ , \tag{16}$$

which again has the eigenvalues

$$E_r(J) = BJ(J+1)\ . \tag{17}$$

(Recall that our convention of measuring angular momemtum in units of $\hbar$ results in the rotational constants having the units of energy.) Each energy level specified by the angular momentum quantum number J is 2J + 1-fold degenerate in the "magnetic" quantum number M (corresponding to a given component of the angular momentum in the laboratory-fixed coordinate system). It is frequently convenient to express the rotational constant B and hence the energy of the Jth level in either frequency units such as MHz, or in temperature units, recalling that Boltzmann's constant k = 20.836 GHz per Kelvin. Thus, for carbon monoxide for which B = 2.77 K, one would not expect a significant population in the level corresponding to J = 4 (E = 55 K) in a cloud with kinetic temperature 10 K, at least in the absence of a significant radiation field.

It can be shown that the selection rules for a linear molecule are $\Delta J = \pm 1$, corresponding to the transition frequencies

$$\nu(J \rightarrow J-1) = E(J) - E(J-1) = 2BJ \tag{18}$$

(for emission lines). Thus, ignoring centrifugal distortion, all transition frequencies are harmonics of the fundamental $J = 1 - 0$ transition, a relation which has been invaluable in efforts to identify observed interstellar lines which have not been previously assigned in the laboratory (a recent example is C_4Si, cf. Ohishi et al., 1989).

2.2.2. *Symmetric Tops.* A molecule is called a symmetric top or symmetric rotor if two moments of inertia are equal but different from the third. This is the case if the molecule has an axis of 3- or more fold symmetry: that is, if rotation about some axis by an angle $2\pi/n$, where $n \geq 3$, leaves the distribution of atoms in space unchanged. Astronomical examples include molecules with a linear "backbone" terminated by a methyl group, such as CH_3CN and CH_3C_2H, and the pyramidal molecule ammonia (NH_3). Such a molecule is called a prolate top if $A > B = C$, and an oblate top if $A = B > C$. For the prolate case, we see from Equation 14 that the Hamiltonian becomes

$$\begin{aligned} H_r &= AJ_A^{\,2} + B(J_B^{\,2} + J_C^{\,2}) \ , \\ &= BJ^2 + (A - B)J_A^{\,2} \ . \end{aligned} \qquad (19)$$

Since only the square of the total angular momentum and one component of that angular momentum (J_A) enter the Hamiltonian, and simultaneous eigenfunctions exist of J^2 and J_A, the eigenvalues of Equation 19 may be found immediately as

$$E(J_K) = BJ\,(J+1) + (A-B)\,K^2 \ , \qquad (20)$$

where K, the eigenvalue of J_A, corresponds to the component of the angular momentum along the molecular axis (see Figure 1). The energy levels are normally labelled J_K, as illustrated in Equation 20. The energy is independent of the sign of K, so that all levels for which $K > 0$ are doubly degenerate. The statistical weight of each level has an additional factor $2J + 1$, corresponding to the possible orientations of J in the laboratory coordinate system (degeneracy in the "magnetic" quantum number M). The statistical weight for symmetric tops is typically further complicated by the existence of identical nuclei in equivalent positions in the molecule (cf. Townes and Schawlow, 1975).

The selection rules for symmetric tops can be shown to be

$$\Delta J = 0, \pm 1 \qquad \Delta K = 0 \qquad + \leftrightarrow - \ , \qquad (21)$$

where the last selection rule applies to the symmetry or anti-symmetry of the wave function under inversion about the origin, an effect which we shall not discuss further but which is critical for the radio frequency spectrum of NH_3. Note that transitions among "K ladders" are forbidden ($\Delta K = 0$), which turns out to be very important in analyzing emission spectra of symmetric tops to probe physical conditions in

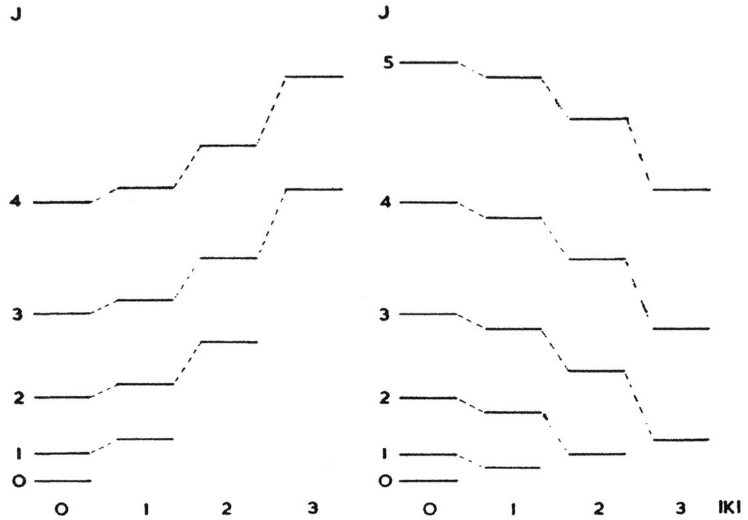

Figure 1. Energy levels for a symmetric top molecule labelled with the quantum numbers for the total angular momentum (J) and the projection of the angular momentum along the molecular axis (K). Prolate top on left, oblate on right. For a linear molecule K≡0 and the energy levels correspond to the leftmost "ladder" (from Kroto, 1975).

interstellar clouds. For K = 0, the energy levels are identical to those for a linear molecule, and in fact if Equation 20 were strictly accurate, application of Equation 21 would lead to the conclusion that the spectrum was identical to that of a linear molecule irrespective of the value of K. In practice, however, centrifugal effects (which depend on the value of K) lead to a splitting of the transition $J \rightarrow J-1$ into J K-components corresponding to $J_K \rightarrow (J-1)_K$. Such a set of components has the interesting property that the frequencies are close together, hence eliminating differences in instrumental sensitivity and allowing accurate measurements of relative intensity, while at the same time corresponding to quite different energies above the ground state, thus providing a very convenient and accurate probe of physical conditions (e.g., Askne et al., 1984).

For an oblate symmetric top an analogous discussion leads to the eigenvalues

$$E(J_K) = BJ\ (J+1) + (C-B)\ K^2 \tag{22}$$

where now the coefficient of K^2 is negative. This is the case for NH_3, while CH_3CN and CH_3C_2H are prolate tops.

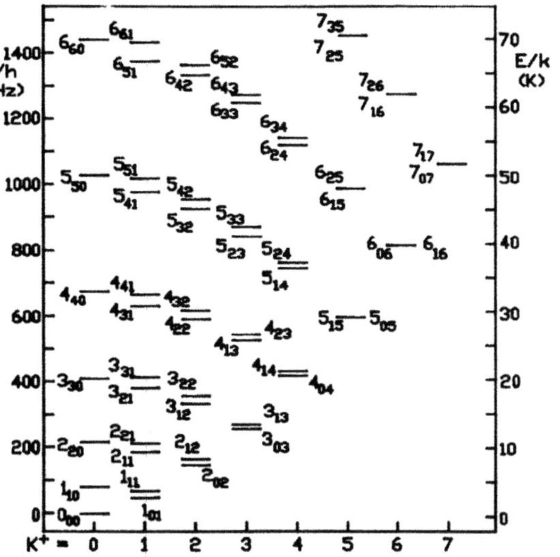

Figure 2. Energy level diagram for the asymmetric top C_3H_2, a small oblate, planar ring. Note the splitting of the levels for a given J and K^+, as the two-fold degeneracy for $K \neq 0$ present for symmetric tops is removed in the asymmetric case (adapted from Vrtilek et al., 1987).

2.2.3. *Asymmetric Tops.* If all three rotation constants are unequal ($A \neq B \neq C$), the molecule is referred to as an asymmetric top. The two-fold degeneracy that exists for $K > 0$ for symmetric tops is now lifted, and each value of the total angular momentum J is now associated with 2J + 1 distinct energy levels which are normally labeled with two subscripts; e.g., $J_{K^-K^+}$, where K^- and K^+ are the values of K that would result if the molecule could be smoothly deformed into either a prolate (K^-) or oblate (K^+) top. It is helpful to remember that, as energy increases for a given value of J, K^- increases and K^+ decreases (e.g., as a function of increasing energy for J = 2 the five possible levels are labeled 2_{02}, 2_{12}, 2_{11}, 2_{21}, and 2_{20}). The rotational spectrum can now become quite complicated, depending on the symmetry of the molecule and the orientation of the electric dipole moment (Townes and Schawlow, 1975); see Figure 2 for the case of C_3H_2.

In interpreting astronomical spectra, it is important to remember that many interstellar molecules have electronic angular momenta which has been ignored up to this point in our discussion, but which will create important and distinctive spectral signatures. For example, the "spin doubling" or "fine structure" produced by the unpaired electron spin of the radicals C_3N and C_4H played a critical role in the identification of these species from the astronomical data, since they had not at that point been observed in the laboratory. In some cases the splitting so introduced can be comparable to the difference between rotational energy levels, particularly for triplet (spin one) molecules like SO and C_2S.

It is also important to emphasize that non-zero nuclear spin will give rise to hyperfine structure of the rotational spectrum which, in many cases, is easily resolved with the heterodyne techniques employed by radio astronomers. The different intrinsic strength of hyperfine components of a rotational transition may give information as to their optical depth, and hence on the column density and abundance of the species in question (e.g., Irvine et al., 1987).

3. Chemical Constituents of Interstellar Clouds

The richness of the spectra with which molecular astrophysicists currently have to deal, given the sensitivity of present-day telescope and receiver systems, is illustrated in Figure 3. Essentially every upward spike corresponds to a real emission feature in the spectrum of the Orion molecular cloud, and almost all of these features have been identified. Although the spectrum illustrated is of course very compressed and the Orion core is one of the two or three richest molecular sources in the Galaxy (from our observational platform on the Earth), it is nonetheless clear that enormous progress has been made in

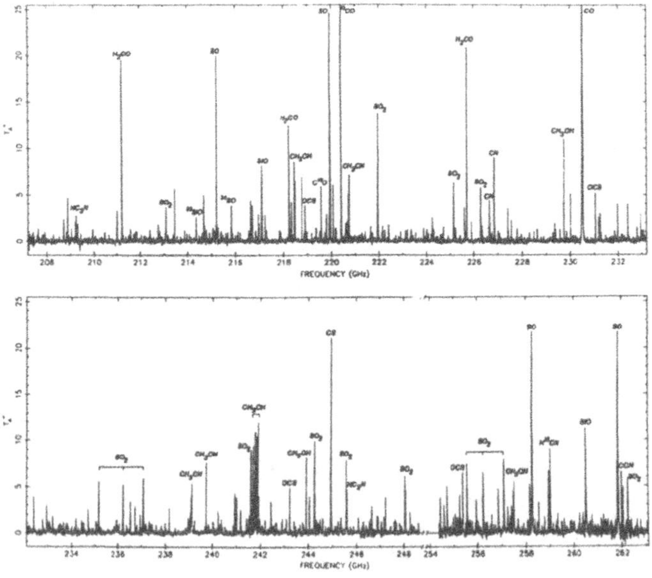

Figure 3. Emission spectrum of gas in the Orion molecular cloud in the 1mm wavelength region. More than 800 resolved spectral features are present and have been assigned to some 29 molecular species (Blake et al., 1987.)

this field since the detection of the first interstellar polyatomic molecules in 1968. In fact, in some sources and some frequency regions considerable care must be exercised in assigning a particular feature, because of the possibility of frequency coincidences among transitions from different molecular species. This becomes a particular problem when trying to assign unidentified spectral lines to new molecular carriers; in a source such as Orion, several lines of consistent intensity must be observed before a new identification can be confirmed.

Although it is virtually certain to be out of date by the time this book is published, Table 1 contains an inventory of the currently known molecular species identified in interstellar clouds and the extended envelopes expelled by evolved stars (photospheric lines are excluded). Although a few assignments are tentative (in most cases where only a single feature has been observed), more than ninety molecular species have been identified, including molecules with up to thirteen atoms and molecular weights up to 147 (which is about twice that of the simplest amino acid, glycine). Even without quantitative abundances, the list in Table 1 provides immediate evidence for the very non-equilibrium character of interstellar chemistry. This is evident firstly from the large number of reactive radicals and ions that have been identified, species which are very short-lived under most laboratory conditions and which in many cases were observed astronomically before being studied in the laboratory. The second feature illustrating the predominance of chemical kinetics over thermodynamics is the large number of highly unsaturated molecular species in Table 1, even though the cosmic abundance of hydrogen is three to four orders of magnitude greater than that of carbon, oxygen, or nitrogen. Of course not all the molecules listed in Table 1 are present in all types of sources, as will be discussed below.

Many of the molecules which have been detected in interstellar and circumstellar sources have been observed not only in the principal isotopic species, but in one or more rare isotopic variants. Because of the extremely high frequency resolution possible using heterodyne methods, isotopic line shifts are easily measurable. Such observations provide critical information on the evolution of nucleosynthesis in the Galaxy, variations in nuclear processing of material as a function of distance from the Galactic Center, the abundance of molecular constituents (from a comparison of the intensity of lines of different isotopic species), and detailed chemical processes (cf. Section 6 below). Nuclei which have been detected which are less abundant than the principal isotope for a given chemical element are ^{2}H (D), ^{13}C, ^{15}N, ^{17}O, ^{18}O, ^{29}Si, ^{30}Si, ^{33}S, ^{34}S, and ^{37}Cl.

4. Physical Conditions in Molecular Clouds

Interstellar molecular material in the Galaxy exists over a range of physical conditions, and the classification of regions is somewhat subjective. Nonetheless it might be expected that the chemical composition of clouds would vary as a function of physical parameters such as kinetic temperature and density, as well as potentially of other

TABLE 1 Identified Interstellar Molecules[a]

Simple Hydrides, Oxides, Sulfides, Halides, and Related Molecules				
H_2	CO	NH_3	CS	$NaC\ell$*
$HC\ell$	SiO	SiH_4*	SiS	$A\ell C\ell$*
H_2O	SO_2	CC	H_2S	$KC\ell$*
	OCS	CH_4*	PN	$A\ell F$*
	HNO ?			
Nitriles, Acetylene Derivatives, and Related Molecules				
CCC	HCN	H_3CNC	$H_3C{-}CH_2{-}CN$	$H_2C{=}CH_2$*
$CCCCC$	$HC{\equiv}C{-}CN$	H_3CCN	$H_2C{=}CH{-}CN$	$HC{\equiv}CH$
CCO ?	$H(C{\equiv}C)_2{-}CN$	$H_3C{-}C{\equiv}C{-}CN$	HNC	
$CCCO$	$H(C{\equiv}C)_3{-}CN$	$H_3C{-}C{\equiv}CH$	$HN{=}C{=}O$	
$CCCS$	$H(C{\equiv}C)_4{-}CN$	$H_3C{-}(C{\equiv}C)_2{-}H$	$HN{=}C{=}S$	
C_4Si	$H(C{\equiv}C)_5{-}CN$	$H_3C{-}(C{\equiv}C)_2{-}CN$?		
$HC{\equiv}CCHO$				
Aldehydes, Alcohols, Ethers, Ketones, Amides, and Related Molecules				
$H_2C{=}O$	H_3COH	$HO{-}CH{=}O$	H_2CNH	
$H_2C{=}S$	$H_3C{-}CH_2{-}OH$	$H_3C{-}O{-}CH{=}O$	H_3CNH_2	
$H_3C{-}CH{=}O$	H_3CSH	$H_3C{-}O{-}CH_3$	H_2NCN	
$NH_2{-}CH{=}O$	$(CH_3)_2CO$?	$H_2C{=}C{=}O$		
Cyclic Molecules				
C_3H_2	SiC_2*	$c{-}C_3H$		
Ions				
CH^+	HCO^+	$HCNH^+$	H_3O^+ ?	
HN_2^+	$HOCO^+$	SO^+ ?	HOC^+ ?	
	HCS^+		H_2D^+ ?	
Radicals				
OH	C_3H	CN	HCO	C_2S
CH	C_4H	C_3N	NO	NS
C_2H	C_5H	H_2CCN	SO	SiC*
CH_2 ?	C_6H			

(a) Irvine (1989), plus recent detections by Bernath et al. (1989), Hinkle et al. (1988), Hollis et al. (1989), Lacy et al. (1989), Cernicharo et al. (1989), and Ohishi et al. (1989).
(*) Indicates detection only in the envelopes around evolved stars.
(?) Claimed but not yet confirmed.

variables including evolutionary state, elemental depletions onto the solid grains, flux of energetic particles and photons, special events such as the passage of shock waves or the influx of material from stellar outflows, etc. Since the detailed quantitative chemical analysis of molecular clouds is currently limited to a very few examples, it follows that the dependence of chemical composition on all of the factors mentioned above remains highly uncertain. It nonetheless is useful to define certain classes of interstellar clouds, and to compare available data within such classes to theoretical models of the chemistry. We shall omit from the present discussion the situation in circumstellar envelopes, the subject of the lectures by A. Omont in this volume.

The least dense molecular environments in the interstellar medium are the diffuse and translucent clouds, which probably include the outer envelopes of denser cloud structures. Operationally, as discussed by A. Dalgarno in this volume, these regions are defined in terms of the extinction at visual wavelengths (A_V) suffered by light from a background star passing through such a cloud. Since the extinction is due to the interstellar grains, and the gas and grains are normally assumed to be well-mixed, this amounts to a definition in terms of column density rather than volume density of material. The significance of the definition is that there is sufficient penetration of ultraviolet photons to ensure that photochemistry plays a significant chemical role. The molecular hydrogen density and kinetic temperature in diffuse/translucent clouds are thought to be $\lesssim 10^3$ cm^{-3} and $\lesssim 100$ K, respectively.

We shall concentrate our attention on the denser clouds, for which $A_V > 5$. The nearest such regions are the so-called dark or dust clouds, apparent on optical photographs as patches within which background stars are obscured (at distances greater than several hundred pc these clouds are not readily detected in this manner). These dark clouds have overall masses of a few hundred $M_\odot$ (solar masses), densities in the range 10^3 to 10^5 cm^{-3}, and temperatures of order 10-20 K. They frequently contain one or more "cores" with densities toward the upper portion and temperatures toward the lower portion of the ranges indicated, and masses of order a few $M_\odot$. Two examples that have been relatively well-studied chemically are Taurus Molecular Cloud 1 (TMC-1) and Lynds 134N (L134N, also known as L183). Portions of such clouds are extremely quiescent, having linewidths close to that expected from thermal broadening alone (linewidths in more massive clouds are typically many times the thermal value, reflecting the doppler effect of mass motions within the clouds). These regions lack embedded high luminosity stars or protostars, although lower mass stars may be present. The energy for driving the chemistry thus should come primarily from cosmic ray ionization, just the situation envisioned in most computational models of interstellar cloud chemistry.

The so-called giant molecular clouds are extremely massive objects which are often the sites of on-going star formation, including both the massive O and B stars which are prolific sources of ionizing radiation, as well as less massive solar-type stars. These clouds or cloud complexes are regions of great heterogeneity in temperature, density, ionization state, and kinematics. Overall masses may approach $10^6 M_\odot$, although dense cores such as those observed in Orion may be tens to hundreds of $M_\odot$. Densities range from presumably diffuse cloud values up to 10^9 or higher in the vicinity of star formation. Likewise, temperatures span the range from tens of K to hundreds of K or higher (see the discussion of star forming regions in the chapter in this volume by Genzel). There is a growing body of evidence suggesting that these giant clouds have an exceedingly clumpy structure, even in the innermost regions. As a result, for example, ultraviolet radiation may penetrate the interiors of clouds to a much greater degree than had been realized until quite recently. The chemistry, in consequence, will illustrate features typical of photodissociation regions such as the translucent clouds, as well as characteristics of much denser regions (cf. chapter by A. Dalgarno).

The closest such giant molecular cloud is that in Orion. A central core of current massive star formation in the Orion molecular cloud is referred to as Orion A or Orion KL (for Kleinmann-Low). The molecular material towards Orion KL includes several subregions which have distinctive chemistries as well as kinematics and spatial distributions. The more quiescent gas is distributed in an extended ridge of material that is characterized by kinetic temperatures of order 30-60 K and molecular hydrogen densities reaching at least 10^5 cm^{-3} (with lower densities towards the periphery). The most intrinsically luminous embedded source (protostar?) in the core is IRc 2, which is believed to power the energetic mass motions which are observed. Near the position of IRc 2 there is a sharp velocity gradient, suggesting the division of the extended ridge material into northern and southern components with slightly different mean velocities. Near the interface of these two clouds is material which it is clear has been heated and set into mass motion by effects associated with the formation of massive stars. Various subregions are referred to as the "plateau", the "hot core", and the "compact ridge", with the plateau itself being sometimes subdivided into a disc or "doughnut", and a roughly orthogonal bipolar outflow. The physical details, while fascinating from the point of view of star formation and the cycling of material between the interstellar medium and stars themselves, need not concern us here (see, for example, Irvine et al., 1987 and references therein). We shall instead concentrate below on the differences between the quiescent material, as exemplified in the spatially extended ridge, and the material in the vicinity of active formation of massive stars.

One of the most massive giant molecular cloud in the Galaxy is thought to be that close to the Galactic Center and known as Sgr B2. Although extremely distant relative to the Orion cloud, the large column density of material has meant that Sgr B2 is one of the richest sources of molecular emission observed in the Galaxy. In fact, many interstellar molecules were first detected in Sgr B2, and some remain detected only in this source. Even at its large distance of approximately 10 kpc, it is clear that this cloud is extremely complex, with different regions containing embedded young stars or protostars as well as zones of strikingly different chemistry (e.g., Goldsmith et al., 1987).

5. Determination of Chemical Abundances

5.1 SELECTION EFFECTS

The study of astrochemistry through high-frequency radio astronomical observations has significant selection effects. Thus, allowed rotational transitions are only observable for molecules with permanent electric dipole moments. This excludes several potentially abundant species, including homonuclear molecules such as N_2, symmetric hydrocarbons such as CH_4 and HCCH, and other symmetric molecules including CO_2. Estimates of the abundance of such molecules can sometimes be made from that of a related, observable species, for example the protonated ion (e.g., N_2H^+ and HCO_2^+; see Herbst et al., 1977; Linke, Guélin, and Langer, 1983; Minh, Irvine, and Ziurys, 1988). It is also possible in principle to search for species such as the hydrocarbons by absorption at the wavelengths of vibrational transitions against background (or embedded) luminous infrared sources. The limited sensitivity obtained with existing systems and the dearth of appropriate infrared sources has, however, limited the usefulness of this method. Nonetheless, detection of acetylene (HCCH) has recently been reported by this approach, and limits have been obtained on the abundance of CH_4 (Knacke et al., 1985; Lacy et al., 1989).

A second, critically important selection effect is the sequestering of material in the solid grains, where unambiguous identification of molecular components becomes much more difficult. Although the grains are thought to contain only about 1% by mass of the material within molecular clouds, the fraction of elements heavier than helium incorporated into grains in these regions may be very much higher (Section 7 below).

Moreover, astronomers continue to be plagued by the age-old problem of absorption due to the Earth's atmosphere. This is particularly critical in the present context at submillimeter wavelengths, as well as in more limited frequency ranges where constituents such as molecular oxygen effectively prevent ground-based measurements. The submillimeter region is important as the spectral range which includes rotational transitions of the light hydrides, which are critical components of the chemical mixture in molecular clouds. Some measurements are possible from very high sites such as Mauna Kea, where two new submillimeter telescopes recently began operation (the Caltech Submillimeter Observatory and the James Clerk Maxwell Telescope). A better "site" is furnished by the Kuiper Airborne Observatory which can fly at altitudes above the troposphere, but even here there remains significant residual atmospheric absorption. It is clear that satellite facilities are needed, and a first step in this direction will be the Submillimeter Wave Astronomy Satellite in NASA's Small Explorer Program, a joint endeavor of a team including the Smithsonian Institution, the University of Massachusetts, the University of California at Berkeley, and the

University of Cologne in West Germany. It should be emphasized that such observations are necessary even for H_2O, which has been widely observed by radio astronomers, but only at transitions which appear to be anomalously excited and hence difficult to interpret in terms of abundances, or through the isotopic variant HDO.

Another procedure that has been used to avoid telluric absorption is to search for isotopic variants such as $^{16}O^{18}O$. This approach for molecular oxygen has yielded only upper limits, since the transitions are weak both because of the isotopic dilution and because they are magnetic dipole rather than electric dipole transitions (e.g., Goldsmith et al., 1985). Another imaginative investigation sought O_2 in a galaxy with a sufficiently large red shift that the transition in question occurred at a readily observable (lower) frequency, but again only an upper limit on the O_2/CO ratio was obtained (Goldsmith and Young, 1989).

A more subtle selection effect is that imposed by the energy level structure of a given molecule. As discussed in Section 2, linear molecules have a relatively small number of rotational levels that are populated at the densities and temperatures characteristic of interstellar clouds, and hence any given transition is much more readily detectable than for more complex species, for which an equal number of molecules will be distributed over a much larger number of states (in the terminology of molecular physics, the partition function is much larger for more complex species).

An additional selection effect is imposed by the lack of laboratory data for many interesting or potentially interesting interstellar molecules. As has been pointed out above, many interstellar chemical constituents are extremely unstable in the laboratory, so that molecular parameters and rotational spectra were not available prior to the detection of corresponding lines with the radio telescope. The identification and determination of abundance for such species requires an intimate collaboration between observational astronomers, quantum chemists, and laboratory molecular spectroscopists. For stable but more complex species, there is also a lack of data. This is particularly the case for molecules with low vapor pressure; thus, of the amino acids only glycine has a measured and assigned rotational spectrum. Interestingly enough, microwave spectra have recently been assigned to the nucleic acid bases uracil, adenine, thymine, and cytosine (e.g., Brown et al, 1989a,b,c).

5.2 PROCEDURE FOR DETERMINING ABUNDANCES

Now let us turn to some of the problems more directly associated with the measurement techniques. It is evident, first of all, that it is extremely difficult to determine *in situ* molecular abundances, although approaches have been used to drive the volume density n of molecular hydrogen and the relative fraction of the species of interest from an analysis of the collisional excitation and observed line strength; this requires significant assumptions about the geometry of the source and the details of radiative transfer within the source, as well as the likelihood of a correlation between errors in the density and abundance so derived (e.g., Stenholm, 1983). It is more straightforward, although inherently less informative, to measure abundance in terms of column density ratios, so that the fractional abundance with respect to hydrogen of species X is defined by

$$f(X) = N(X)/N(H_2), \tag{23}$$

where N is the number of molecules per square centimeter along the line of sight. Adoption of this procedure naturally means that the results refer to some sort of average along the line of sight through the cloud.

Even measuring fractional abundance by comparing column densities is more difficult than it might seem *a priori*. Since molecular hydrogen is by far the most abundant constituent in dense interstellar clouds, it is natural to define abundances with respect to that of H_2. This homonuclear molecule has, however, no electric or magnetic dipole transitions between the ro-vibrational levels of the ground electronic state. Electric quadrupole transitions may occur and have been observed from vibrationally excited, "shock-heated" gas in the vicinity of star formation regions in molecular clouds, planetary nebulae, and even some external galaxies (Shull and Beckwith, 1982); likewise, H_2 has been observed in absorption at ultraviolet wavelengths when hot stars are observed through diffuse interstellar clouds. In quiescent molecular clouds the fundamental pure rotational electric quadrupole transition might in principle be observed at 28 microns with detectors above the atmosphere, or H_2 might be observed in absorption against bright infrared sources in or behind molecular clouds, but at present there are no such direct measurements. Instead, the H_2 column density is normally estimated from observations of carbon monoxide, or in fact a rarer isotope such as ^{13}CO or $C^{18}O$ since CO itself is typically saturated (optically thick) in dense interstellar clouds. The column density of CO is related to that of H_2 through the visual extinction by dust, on the basis of measurements of A_V, $N(H_2)$, and CO in diffuse or translucent clouds and the assumptions of a constant hydrogen gas/dust ratio and consistent optical properties of the dust within dense clouds. Other relevant data is supplied by comparing CO emission and measurements of thermal emission from dust and by observations of galactic gamma-ray emission (e.g., Schwartz et al, 1989; Dickman, 1988; Bloemen, 1988; see also Irvine et al., 1985, and Guélin, 1987, who give relevant references and also consider other instrumental problems related to telescope beam size and efficiency). A related problem is that of density variations ("clumping") within the antenna beam (that is, within the area on the sky sampled by the telescope). It is clear that measurements will refer to some sort of average over this area, as well as along the line of sight. This is probably more of a problem in giant molecular clouds, where there is an increasing amount of evidence that such clumping occurs (e.g., Stutzki et al., 1988; Taylor and Dickman, 1989), than in the nearby dark clouds, where linear spatial resolution is much higher because the clouds are much closer to the solar system.

The conversion of an observed spectral line intensity (typically measured in temperature units by radio astronomers) to a molecular column density raises in principle questions of radiative transfer through the source, unless the lines are known to be optically thin (unsaturated). An indication of whether this is the case may be obtained if data on more than one isotopic variant is available, or if fine or hyperfine components with different intrinsic line strengths can be observed; such procedures may, of course, require knowledge of additional factors, such as the isotope ratio. Details of relevant procedures are discussed by Irvine et al. (1987) and Winnewisser et al. (1979). Even if the column density in the states corresponding to a particular transition can be determined, there remains the question of determining the population in other energy levels that are likely to be populated under the relevant physical conditions in the cloud. In the case of thermodynamic equilibrium, this is accomplished through determination of the cloud temperature; in fact, such equilibrium does not typically apply in the interstellar medium, and there is in fact no reason why a single rotational temperature will characterize the population distribution over all relevant energy levels for a given molecule. It is thus extremely useful in specifying an abundance quantitatively to measure more than one transition, hopefully spanning a range in energies above the ground state.

There are other pitfalls in determining relative molecular abundances through comparison of the column density of a given species with that of CO (and hence ultimately with that of H_2), since observable emission will only be produced from molecules in excited states and the excitation in turn involves competition between processes such a collisions (primarily with H_2) and spontaneous emission (dependent upon the molecular electric dipole moment). As a result, a molecule with a particularly small dipole moment such as CO will be excited to emit over a broader range of densities and hence over a more extensive cloud region than will a higher dipole moment molecule such as HCN. There will thus be a systematic tendency to underestimate the relative abundance of species other than CO, which is the common tracer of molecular gas. How significant this effect will be depends on the extent of low density envelopes around molecular clouds, a subject concerning which there is still considerable uncertainty.

Given all of the above caveats, let us now proceed to consider the relative molecular abundances which have been measured for dense interstellar clouds. The most accurate relative abundances for a particular astronomical object will be obtained through uniformly calibrated observations with a single telescope, over a frequency range throughout which the beam size does not vary greatly. Such systematic spectral surveys at radio frequencies are very limited, having been carried out only for the Orion KL region, Sgr B2, and the cold cloud TMC-1 (as well as the envelope of the evolved carbon star, IRC+10216). Moreover, in each case the spectral coverage is limited, and such an approach cannot address the relative abundance of molecular species whose relevant transitions are widely separated in frequency. The most detailed analyses of the data have been published by Johansson et al. (1984), Blake et al. (1987), and Cummins et al. (1986), with subsequent tabulations by Irvine et al. (1987) and Guélin (1985). The general character of the results may be partly summarized in terms of the following characteristics.

5.3 ABUNDANCES IN QUIESCENT CLOUDS

Molecular hydrogen, the only constituent that seems clearly to be formed on grain surfaces, is by far the dominant component in dense clouds. For quiescent cloud material there is a general uniformity in abundance for a number of relatively simple molecular species over a range of perhaps a factor of 100 in $n(H_2)$ and a factor of several in temperature. This uniformity, and in fact the abundances themselves, are in reasonable agreement with gas phase models of ion-molecule chemistry, within which the ultimate energy source is cosmic ray ionization of H_2 and only exothermic reactions with negligible activation energies are permitted (e.g., calculations by Herbst and Leung, 1989; Langer and Graedel, 1989; Millar et al., 1987; and earlier references therein). With an abundance by number about 10^{-4} that of H_2, CO is the predominant trace constituent in the gas phase. This relative abundance of CO is difficult to determine accurately, for reasons discussed above, but there is no strong evidence for significant variations. The distribution of carbon and other volatile elements among possible chemical repositories is discussed in Section 7 below.

A number of relatively simple molecules have abundances in the range of 10^{-4}–10^{-3} times that of CO, including HCN and its isomer HNC, formaldehyde (H_2CO), the cyclic molecule C_3H_2, the radicals CN and C_2H, and inorganic molecules such as ammonia (NH_3) and presumably OH and CH. Only slightly less abundant are species such as HCO^+, CS, SO, and CH_3C_2H (see the abundances tables in Irvine et al., 1987).

There are, however, some fundamental questions concerning the gas phase chemical models. First of all, it may be quite easily shown that the "freeze-out" time for molecules in the gas phase to stick to the solid grains is short compared to anticipated cloud lifetimes. The very existence of a significant gas phase chemistry thus becomes a problem. Clearly there must be mechanisms for recycling molecules from the grains back to the gas phase, and various processes have been suggested (see the chapter in this volume by Greenberg, references in Irvine et al., 1987, recent models such as that of Charnley et al., 1988, and the discussion of the gas-grain interaction in Tielens and Allamandola, 1987). A second problem is that the observed abundances, particularly of complex organic species, appear to match the calculations much better at model ages shorter than the probable (but poorly known) physical cloud ages than they do after steady state has been reached. The problem here is that in an evolving cloud, at least if the physical parameters such as temperature and density remain fixed, the bulk of the carbon will eventually become locked up in CO if the cosmic abundance ratio $C/O < 1$ applies. Some method for recycling carbon from carbon monoxide back into more chemically active forms seems called for. Observationally, the problem is manifested also by the unexpectedly high abundances of atomic carbon observed in the interior of dense clouds (see discussion by Genzel in this volume). It should be emphasized, however, that even the relatively sophisticated chemical models employed at present, which utilize up to 2000 reactions and perhaps 300 molecular species, are necessarily vastly oversimplified relative to the actual physical and chemical conditions. Thus, they typically consider the chemistry to evolve under given physical conditions, rather than considering simultaneously the physical and chemical evolution of a (collapsing?) molecular cloud [an exception is the work of Prasad et al. (1987) which, however, consequently considers a somewhat limited chemical model]. Likewise, the models are developed for homogeneous regions and an insignificant radiation field, whereas (as discussed above) real clouds are quite inhomogeneous, with significant clumping in the more massive clouds which may allow significant penetration of ultraviolet photons. Internal generation of photons may also play a role in cloud chemistry (Gredel et al., 1987), as, potentially, would the presence of very large organic molecules such as the putative polycyclic aromatic hydrocarbons (PAH's; Lepp and Dalgarno, 1988).

All in all, it is perhaps surprising how well the gas phase models can reproduce the observed abundances in quiescent clouds, given the problems discussed in the previous paragraph. Of course the lack of much relevant laboratory data leads to the presence of many adjustable parameters in the models. An example concerns the almost entirely unmeasured branching ratios for the products of dissociative electron recombination of polyatomic positive ions, thought to be the ultimate step in the production of many neutral interstellar molecules. Useful articles are contained in Millar and Williams (1988). Specific aspects of the chemical models can sometimes be tested by measuring the abundance of closely related species; e.g., isotopic variants (cf. Section 6 below; Wootten, 1987), isomers such as HCN/HNC and CH_3CN/CH_3NC (DeFrees, 1987; Cernicharo et al., 1988), and molecules with a putative common precursor (C_3O/H_2C_3O, Irvine et al., 1988; CH_2CN/CH_3CN, Turner et al., 1989).

5.4 COMPARISON OF QUIESCENT CLOUDS WITH WARMER CLOUDS

In the vicinity of massive young stars or protostars a variety of processes may take place which may alter the chemistry with respect to that observed in cold, quiescent regions. Shock waves may be generated, mass outflows may result in the exchange of material between reservoirs of different chemical composition, and radiative heating may become important; as a result, reactions with significant activation energies or even endothermic reactions may take place, resulting in a rather different chemical composition, at least for trace constituents. There is a variety of evidence that this indeed occurs (Blake et al., 1987; Johansson et al., 1984). A review of the theory incorporating some of these effects is given by Herbst (1988); several other relevant articles are also given in Millar and Williams (1988). Let us consider some specific examples.

The "hot core" in the Orion KL region appears to consist of dense cloudlets that are heated by the nearby infrared source IRc2. Striking abundance differences with respect to the surrounding quiescent material are observed, including order of magnitude or more enhancements in the abundance of such species as HDO, NH_3, HCN, CH_3CH_2CN, CH_2CHCN, and to a lesser extent CH_3CN and HC_3N. It has been speculated that such effects may involve a heterogeneous chemistry including both the solid phase and the gas phase (e.g., Blake et al., 1987). Thus, relatively saturated species such as CH_3CH_2CN, CH_2CHCN, water, and ammonia may be easily formed on grain surfaces because of the high mobility for hydrogen atoms on the grains (Jones and Williams, 1984). Water ice is in fact observed as a grain component in dense clouds from its near infrared absorption band at 3.1 microns. NH_3 is another plausible constituent of icy grain mantles, and there have been suggestions that HCN may also be a grain component in comets. The relatively high abundance of HDO in the gas phase suggests that previously frozen interstellar water which retained a high deuterium fractionation upon its incorporation into grain mantles has been evaporated and has not had time to equilibrate at the relatively high gas temperatures present in this source.

Another molecular species whose abundance seems to be significantly enhanced (perhaps by up to three orders of magnitude) in warm, dense regions is methanol (CH_3OH; e.g., Wilson et al., 1989). Methanol is another possible constituent of icy grain mantles in dense clouds, on the basis of infrared spectra. It might also be produced in the gas phase from reactions of H_2O and CH_3^+, after evaporation of H_2O from grain mantles.

Even more violent grain disruption may be taking place in the gas associated with the Orion "plateau" source, also in the immediate vicinity of the luminous object IRc2. In this case, orders of magnitude abundance enhancements relative to the quiescent cloud are observed for several sulfur- and silicon-containing species, including SiO, H_2S, SO_2, SO, and OCS, as well as somewhat smaller enhancements in HDO, HCN, and HC_3N. The suggestion is that sulfur and silicon ordinarily incorporated into the grains has here been released to the gas phase. In both the plateau and the hot core regions there is a significant deficit with respect to the adjacent quiescent gas for chemically reactive radicals and ions such as CN, C_2H, and HCO+. That is, the composition in these regions is closer to that expected under chemical equilibrium for the given physical conditions, as would be expected for the higher densities and temperatures observed.

Two abundance ratios for which correlations with kinetic temperature in molecular clouds have been reported are that of the isomers HCN and HNC and the abundance of SiO relative to HCN. Although the energy difference between HNC and HCN is sufficient that at chemical equilibrium in a cloud at 10 K there should be barely a single molecule of HNC present, in fact the energy barrier to rearrangement from HNC to HCN is sufficiently high that HNC formed in such regions is stable. The abundance ratio HNC/HCN is in fact observed to be greater than 1 in cold clouds, which can be understood from kinetic considerations (cf. Section IV in Irvine et al., 1985). As the temperature increases, however, the relative abundance of HNC diminishes significantly so that at kinetic temperatures of order 200 K, HCN is more than two orders of magnitude more abundant (Goldsmith et al., 1986; Irvine et al., 1987). It is unclear precisely which processes are responsible for such dramatic effects, although an enhanced production of HCN due to neutral reactions has been suggested. In the case of SiO, extremely low upper limits on the abundance have been obtained in cold, dark clouds, but this species has been detected in warmer regions and is significantly enhanced in the vicinity of active star formation as well as in the presumably shocked gas associated with the supernova remnant IC443. A linear relationship between the log of the abundance ratio SiO/HCN and the inverse of the kinetic temperature has been inferred from the data by Ziurys et al. (1989), who propose that the formation of SiO involves a chemically specific process that contains an activation barrier of order 90 K.

5.5 CHEMICAL VARIATIONS AMONG AND WITHIN PHYSICALLY SIMILAR CLOUDS

Whereas chemical differences between regions with significantly different energy inputs are understandable, significant abundance variations between clouds with very similar kinetic temperatures and densities are more puzzling. Just this phenomenon is observed, however, between the cold, quiescent, dark clouds TMC-1 and L134N. In each case estimates of the kinetic temperature are of order 10 K, and core regions appear to have molecular hydrogen densities of a few times 10^4 cm^{-3}. In spite of these similarities, many highly unsaturated molecules are vastly more abundant in TMC-1, including the cyanopolyynes [$H(C{\equiv}C)_nCN$, n=0-5], related methylated species, hydrocarbon radicals (C_nH, n=1-6), the sulfur-containing carbon chains (C_nS, n=1,2,3), and other unsaturated species such as the small ring C_3H_2 and the molecules C_3N and C_3O. In contrast, several oxygen-containing or fully saturated molecules appear to be more abundant in L134N than in TMC-1; included in these categories are SO, SO_2, NO, HCOOH (formic acid), NH_3, and H_2S (e.g., Minh et al., 1989). There are also several oxygen-containing organic molecules which have comparable abundances in both these clouds; e.g., CH_3OH, CH_3CHO, OCS, and H_2CO. Surely these differences provide an interesting probe of chemical processes, if they could be understood. Their origin is certainly not clear at the present time, although it might be due to differences in the gas phase carbon-to-oxygen ratio between these two clouds (perhaps caused by differential depletion of water with respect to organics onto the grains) or possibly to different evolutionary stages in the chemistry. A model along the latter lines has been developed by Millar and Freeman (1984), although more data is available since that paper was published, and a detailed reanalysis is warranted.

Some of the above differences between cold, quiescent clouds may be mirrored on a smaller scale within individual clouds. A particularly well-studied case is L134N, where Swade (1987, 1989) has demonstrated the existence of gradients in the relative abundance of some carbon-rich relative to oxygen-rich species. Likewise, a gradient in the relative abundance of ammonia with respect to the cyanopolyynes seems to be established in TMC-1 (e.g., Olano et al., 1988).

6. Isotopic Fractionation

One of the most distinctive signatures of the chemistry of dense interstellar clouds is the very large isotopic fractionation observed for many molecular species, particularly those containing hydrogen. Such fractionation can provide valuable clues to chemical processes that the material in question has undergone. An interesting example may be provided by the isotopic anomalies observed in certain meteorites for a number of chemical elements. Fractionation is particularly pronounced for hydrogen, and the large deuterium enhancements observed in certain components of carbonaceous chondrites have been cited as evidence for the preservation of interstellar molecular material in these "primitive" objects (for example, Zinner, 1988).

From observations of the sun, the atmospheres of the giant planets, and the diffuse interstellar medium it is deduced that the overall (cosmic) abundance ratio of deuterium to hydrogen is $D/H \cong 0.8\text{-}2 \times 10^{-5}$. Values three orders of magnitude larger than this are found for many hydrogen-containing interstellar molecules in cold clouds, and values almost four orders of magnitude larger (that is, of order 0.1) have been found for C_3H_2 and NH_3, both molecules which contain multiple hydrogens (see Table and References in Irvine and Knacke, 1989). Large deuterium fractionation has been observed for molecules as complex as CH_3OH and HC_5N.

The mechanism of such fractionation in dense molecular clouds is reasonably well understood. It is basically the result of gas phase ion-molecule reactions which transfer deuterium from the principal reservoir (HD) to reactive ions which play key roles in the production of heavier species. In particular,

$$H_3^+ + HD \rightarrow H_2D^+ + H_2 + \Delta E_1 \qquad (\Delta E_1 \sim 230\ K) \tag{24}$$

and

$$CH_3^+ + HD \rightarrow CH_2D^+ + H_2 + \Delta E_2 \qquad (\Delta E_2 \sim 370\ K) \tag{25}$$

proceed predominantly from left to right at the temperatures of cold clouds, whereas reactions in both directions occur at approximately equal rates at room temperature. Equation 24 is particularly important in the cold, quiescent clouds, while reaction 25 and other processes become important at the somewhat warmer temperatures appropriate to the gas in regions of star formation (a recent theoretical discussion is given by Millar et al., 1989). Observational verification for the models is available in part from the temperature-dependence of deuterium enhancement for certain species (cf. Wootten, 1987).

Isotopically heavy carbon has also been observed in carbonaceous chondrites, with the suggestion that it may have an interstellar or circumstellar origin (Anders, 1988; Kerridge and Chang, 1985). Values of the $^{12}C/^{13}C$ ratio lower than the terrestrial one have been well established in the envelopes expelled by evolved stars, where they are an expected result of nucleosynthesis in the stellar interiors (Olofsson, 1988). The problem is complicated, however, since how "heavy" the carbon in the present interstellar medium is, or was 4.5 billion years ago, is not entirely clear from existing data. Observations of CO in the nearest dense molecular clouds show a $^{12}C/^{13}C$ ratio of 75 ± 8 (the terrestrial value is 89), while recent optical studies of CH^+ in the solar neighborhood find $^{12}C/^{13}C = 43 \pm 4$ (Hawkins and Meyer, 1989). Although processes such as isotopic selective photodissociation coupled with variations in optical depth can produce carbon fractionation in molecular clouds, there are no obvious processes to produce such effects in the high-temperature regions where CH^+ is thought to be produced. A more extended discussion and references are given in Irvine and Knacke (1989).

7. Chemical Reservoirs for the Volatile Elements

Questions of how the cosmic complement of each element is partitioned between the gas and the solid phases in dense interstellar clouds, and within each phase among the chemical constituents, are obviously of major importance to astrochemistry. In this Section we shall consider this problem for several of the volatile elements. For most refractory elements such as metals, visible and ultraviolet wavelength studies of diffuse clouds show that even in these regions of relatively large ultraviolet flux the bulk of this material is missing from the gas phase and hence sequestered into the solid grains. A somewhat more complete discussion of the subject of this Section, including background references, is given by Irvine and Knacke (1989).

A fundamental difficulty in these investigations concerns the determination of the true cosmic abundance of the elements. Although this question will not be pursued further here, it should be borne in mind that significant complications do arise from the possibility of differences between solar and interstellar values, gradients within and among galaxies, inaccuracies in models of the solar atmosphere, errors in the values of oscillator strengths for weak lines, and difficulties arising from the incorporation of material into the interstellar grains.

Consider first the case of carbon. After molecular hydrogen, the most abundant detected gas phase constituent of dense molecular clouds is carbon monoxide. Even determining the fraction of the total carbon abundance represented by CO presents significant problems, however (cf. van Dishoeck and Black, 1987). A basic difficulty, mentioned in Section 5 above, concerns measuring the total mass of material along the observational line of sight, which is necessary in order to derive the fractional abundance of CO. The difficulty arises in determining the column density of H_2, which is to date unobservable in quiescent molecular clouds. Somewhat roundabout arguments suggest that typically in dense clouds about 10% of the total carbon is present as CO. Two other oxides of carbon, CO_2 and C_3O, have been observed in molecular clouds, but neither is a major carbon repository at least under typical conditions (estimates of CO_2 are made from that of its protonated ion HCO_2^+). Observational data on the fraction of carbon in hydrocarbons is primarily in the form of upper limits, although a recent detection of acetylene has been reported. All told, the data indicate that for both CH_4 and HCCH the abundance is less than 10^{-2} that of CO. Other possible reservoirs of carbon include both neutral and ionized atomic carbon (CI and II). A surprisingly high abundance of CI has been observed in dense clouds, but it still appears that $C/CO \lesssim 0.1$. The very limited data on CII suggests that it is less abundant than CI.

A summation of the observed values and upper limits for various gas phase carbon-containing constituents leads one to the conclusion that the bulk of this element in dense interstellar clouds must be present on or in the solid grains. This conclusion is supported by various observations of infrared absorption and emission features and the 2200 Å "bump" in the interstellar extinction curve, which have been attributed to organic material. Quantitative characterization of such material is difficult, but various carbonaceous species seem to be present, depending in part on the local physical conditions. Suggestions include graphite, amorphous carbon, polycyclic aromatic hydrocarbons, and volatile species such as solid CO and CH_3OH.

In contrast with carbon, most nitrogen may be present in molecular clouds in the gas phase. Although N_2 has no permanent electric dipole moment, its abundance may be estimated from that of the protonated ion N_2H^+. A difficulty with this approach, that the N_2H^+ lines are frequently saturated in these regions, has been circumvented by observations of the rare ^{15}N isotope in one giant molecular cloud (Linke et al., 1983). The results in this case are consistent with most nitrogen existing in the form of molecular nitrogen. Particularly at early stages of chemical evolution, however, a significant amount of nitrogen may exist in atomic form; unfortunately, there are no observations to confirm this prediction. Both NH_3 and HCN are reasonably abundant in dense clouds, and both show enhanced abundances in warmer regions which may be due to evaporation from icy grains or processes taking place on grain surfaces. In neither case, however, is the quantity sufficient to include most of the cosmic abundance of nitrogen. Nitrogen-containing chemical groups such as CN, ammonia hydrate, and NH_4^+ have been suggested as possible carriers of infrared absorption features attributed to grains in dense clouds. Nonetheless, it seems most likely that gas phase N_2 is the major repository for nitrogen, although further observations to confirm this are extremely desirable.

The situation is more confused for oxygen. The chemical state of gas phase oxygen depends critically on the carbon-to-oxygen ratio. Even though the overall cosmic ratio of C to O is of order 1/2, scenarios exist in which the gas phase ratio in dense clouds might be significantly greater than this. If cloud regions existed for which $C/O > 1$, then the major repository of oxygen is predicted to be carbon monoxide, which would correspond to about 6% of the cosmic complement of oxygen. For the probably more realistic case that $C/O < 1$, significant fractions of the available oxygen in dense clouds may be in the form of O_2, atomic oxygen (OI), and H_2O. Unfortunately, it is extremely difficult to determine the abundance of these constituents from ground-based observations (cf. Section 5.1). To date only upper limits for O_2 are available, indicating that $O_2/CO < 1$ in the sources observed. The H_2O abundance should be definitively established once appropriate equipment is available above the atmosphere. Some transitions have been observed from the ground or from the Kuiper Airborne Observatory, but these lines show evidence of maser action or at least large optical depths. Three transitions of HDO (deuterated water) have also been observed in molecular clouds, but it is difficult to convert these measurements to a total water abundance without knowing the degree of isotopic fractionation. Although there are problems in the assignment due to blending with nearby lines, Jacq et al. (1988) have argued from observations of a rather highly excited line of $H_2{}^{18}O$ that the abundance of water in several regions of active star formation (like the Orion "hot core") is of order 10^{-5}, and thus less than that of CO. Atomic oxygen may be observed in hot, dense clouds via its fine structure transition at 63 microns. In the presumably shocked gas in the core of Orion KL the abundance of this species appears to be similar to that of carbon monoxide, but it is unfortunately difficult to observe this transition in colder clouds. Oxygen is certainly present in both the refractory and the volatile components of the interstellar grains, in the former case in the silicates which are identified via features at 9.8 and 18 microns and in the latter case via absorption features due to solid H_2O and CO. It is not clear, however, that these grain sources contain more than some 30% of the cosmic oxygen abundance. Overall, more than half of the total oxygen complement is not well accounted for.

Sulfur has been observed in several interstellar molecules, in both reduced and oxidized states. Nonetheless, comparison with gas phase models suggests that in typical cold clouds only of order 0.1% of the sulfur remains in the gas phase.

Only one phosphorus-containing molecule has been observed in dense clouds, PN. Moreover, this species has been securely identified only in warmer cloud regions. The distribution of phosphorus between the gas and solid phases in dense clouds thus remains quite uncertain.

Sensitive searches in cold, quiescent clouds have failed to detect silicon-containing molecules in the gas phase. Abundances of SiO, for example, are certainly much larger in warm and shocked regions, such as the supernova remnant IC443. In general most silicon appears to be in silicate grains (cf. Ziurys et al., 1989).

8. Identifying a New Interstellar Constituent, CH_2CN

We have emphasized above that many of the molecular constituents of dense interstellar clouds were observed astronomically before being studied in the laboratory, and that the presence of many of these "unstable" species provides evidence for the very non-equilibrium character of interstellar chemistry. It is natural to ask how assignments are made for such molecules, and we illustrate below a recent example. The critical interplay between observational astronomers, quantum chemists, and laboratory molecular spectroscopists is obvious. Another "classic" example where the chain of deductive logic was even freer from laboratory constraints concerned the radicals C_4H and C_3N (Guélin and Thaddeus, 1977; Guélin et al., 1978; Friberg et al., 1980; Guélin et al., 1982; Gottlieb et al., 1983).

The story to be told here begins with the observation by Thaddeus et al. (1985) that two unidentified emission lines detected in a 3mm spectral survey of the giant molecular cloud Sgr B2 at the Bell Laboratories were harmonically related to each other and to a third line observed in the different source TMC-1. Observed lines in Sgr B2 are sufficiently broad that very accurate frequency assignments are not possible, but the lines appeared to fall at 20.12, 80.48, and 100.60

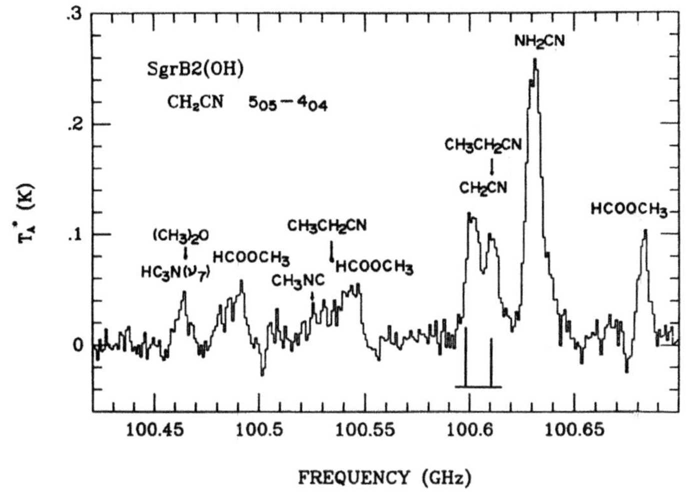

Figure 4. Spectrum in the region of 100 GHz of the dense molecular cloud Sgr B2 showing the 5_{05}-4_{04} transition of the CH_2CN radical; note splitting of the line as a result of "fine structure" (predicted intensity pattern shown below lines). Observations with the 14m antenna of the Five College Radio Astronomy Observatory, operated by the University of Massachusetts (cf. Irvine et al., 1988 and Saito et al., 1988).

GHz. As discussed in Section 2 above, the strong implication is that, if the observed lines do indeed have a common carrier, this carrier is linear or nearly linear. Thaddeus et al. suggested an assignment to ethynyl methylene (HCCCH), an isomer of the recently discovered small ring, cyclopropenylidene (C_3H_2).

Subsequent observations with higher signal-to-noise showed that the lines in Sgr B2 appeared to be doublets (see Figure 4) and that no emission was observed at the intermediate frequency of 90.54 GHz. This established that the lowest frequency rotational transition was at 20.12 GHz, corresponding to a rotational constant of about 10.06 GHz. Moreover, the fine structure suggested that we were dealing with a radical (spin-rotation doubling). Additional observations of TMC-1 showed that there was indeed a line present at 40.24 GHz, twice the frequency of the lowest rotational transition (Irvine et al., 1989). Emission was also observed in TMC-1 at 80.48 and at 100.60 GHz, supporting the idea that the lines observed in TMC-1 and in Sgr B2 had a common carrier.

More detailed observations of TMC-1, carried out as part of a systematic survey of the 20 GHz spectral region, then showed that there were at least ten unidentified emission features in the interval from 20115 to 20140 MHz. In view of the paucity of lines observed at other frequencies in this source, it seemed very likely that most or all of these features were related to each other and thus to the harmonically related lines at higher frequencies. The multitude of lines suggested that the species in question was exhibiting hyperfine as well as fine structure, and, moreover, that one contributor to the hyperfine patterns might be nitrogen (since the ^{14}N nucleus has spin-one, a more complex pattern will be observed than that for a spin-one-half species). Quantum chemical calculations suggested that about the right rotation constant would be obtained for the molecule HCCNH (D. DeFrees, private comm.). Before further astronomical observations were carried out, however, transitions at higher but harmonically related frequencies were observed in the laboratory during a search for rotational transitions of the HCCN radical, and both the laboratory and the astronomical lines were assigned to the cyanomethyl radical CH_2CN (Saito et al., 1988). This species is the heaviest non-linear radical observed to date in the interstellar medium. Interestingly enough, its abundance in TMC-1 is significantly greater than that of the related, fully-saturated molecule CH_3CN (Irvine et al., 1988). It appears that this can be understood on the basis of gas phase reactions.

This review was supported in part by NSF grant AST 88-15406 and NASA grant NAGW-436. The author is grateful to Jan Whitaker for editorial assistance and preparation of the manuscript.

9. Bibliography

Anders, E. (1988) "Noble Gases, Carbon, and Nitrogen", in J.F. Kerridge and M.S. Matthews (eds.), Meteorites and the Early Solar System, U. Arizona Press, Tucson, 927-955.

Askne, J., Höglund, B., Hjalmarson, Å., and Irvine, W.M. (1984) "Methyl Acetylene as a Temperature Probe in Molecular Clouds", Ast. Ap. 130, 311-318.

Bernath, P.F., Hinkle, K.H., and Keady, J.J. (1989) "Detection of C_5 in the Circumstellar Shell of IRC+10216", Science 244, 562-564.

Blake, G.A., Sutton, E.C., Masson, C.R., and Phillips, T.G. (1987) "Molecular Abundances in OMC-1: The Chemical Composition of Interstellar Molecular Clouds and the Influence of Massive Star Formation", Ap. J. 315, 621-645.

Bloemen, H. (1988) "Gamma-ray Insight into the CO-H_2 Calibration Problem", in R.L. Dickman, R.L. Snell, and J.S. Young (eds.), Molecular Clouds in the Milky Way and External Galaxies, Springer-Verlag, Berlin, 71-75.

Brown, R.D., Godfrey, P.D., McNaughton, D., and Pierlot, A. (1989a) "Microwave Spectrum of the Major Gas-phase Tautomer of Thymine", J. Chem. Soc., Chem. Comm., 37-38.

Brown, R.D., Godfrey, P.D., McNaughton, D., and Pierlot, A. (1989b) "A Study of the Major Gas-phase Tautomer of Adenine by Microwave Spectroscopy", Chem. Phys. Lett. 156, 61-63.

Brown, R.D., Godfrey, P.D., McNaughton, D., and Pierlot, A. (1989c) "Tautomers of Cytosine by Microwave Spectroscopy", J. Amer. Chem. Soc. 111, 2308-2310.

Cernicharo, J., Kahane, C., Guélin, M., and Gomez-Gonzalez, J. (1988) "Tentative Detection of CH_3NC Towards Sgr B2", Ast. Ap. 189, L1-L2.
Cernicharo, J., Gottlieb, C.A., Guélin, M., Thaddeus, P., and Vrtilek, J.M. (1989) "Astronomical and Laboratory of the SiC Radical", Ap. J. (Lett.) 341, L25-L28.
Charnley, S.B., Dyson, J.E., Hartquist, T.W., and Williams, D.A. (1988) "Chemical Limit Cycles for Models of a Region of Low-mass Star Formation", M.N.R.A.S. 235, 1257-1271.
Cummins, S.E., Linke, R.A., and Thaddeus, P. (1986) "A Survey of the Millimeter-wave Spectrum of Sagittarius B2", Ap. J. Suppl. 60, 819-878.
DeFrees, D.J. (1987) "Theoretical Studies of Interstellar Isomers", in M.S. Vardya and S.P. Tarafdar (eds.), Astrochemistry, D. Reidel, Dordrecht, 203-204.
Dickman, R.L. (1988) "H_2 Masses in the Interstellar Medium: Current Techniques and Selected Problem Areas", in R.L. Dickman, R.L. Snell, and J.S. Young (eds.), Molecular Clouds in the Milky Way and External Galaxies, Springer-Verlag, Berlin, 55-70.
Duley, W.W. and Williams, D.A. (1984) Interstellar Chemistry, Academic Press, London.
Friberg, P., Hjalmarson, Å., Irvine, W.M., and Guélin, M. (1980) "Interstellar C_3N: Detection in Taurus Dark Clouds", Ap. J. (Lett.) 241, L99-L103.
Friberg, P. and Hjalmarson, Å. (1989) "Molecular Clouds in the Milky Way" in T.W. Hartquist (ed.), Molecular Astrophysics, Cambridge U. Press, in press.
Goldsmith, P.F., Snell, R.L., Erickson, N.R., Dickman, R.L., Schloerb, F.P., and Irvine, W.M. (1985) "Search for Molecular Oxygen in Dense Interstellar Clouds", Ap. J. 289, 613-617.
Goldsmith, P.F., Irvine, W.M., Hjalmarson, Å., and Elldér, J. (1986) "Variations in the HCN/HNC Abundance Ratio in the Orion Molecular Cloud", Ap. J. 310, 383-391.
Goldsmith, P.F., Snell, R.L., Hasegawa, T., and Ukita, N. (1987), "Small-scale Structure and Chemical Differentiation in the Central Region of the Sagittarius B2 Molecular Cloud", Ap. J. 314, 525-534.
Goldsmith, P.F. and Young, J.S. (1989) "Search for Molecular Oxygen in VII Zw 31", Ap. J. 341, 718-721.
Gordy, W. and Cook, R.L. (1984) Microwave Molecular Spectra, J. Wiley, New York.
Gottlieb, C.A., Gottlieb, E.W., Thaddeus, P., and Kawamura, H. (1983) Laboratory Detection of the C_3N and C_4H Free Radicals", Ap. J. 275, 916-921.
Gredel, R., Lepp, S., and Dalgarno, A. (1987) "The C/CO Ratio in Dense Interstellar Clouds", Ap. J. (Lett.) 323, L137-L140.
Guélin, M. and Thaddeus, P. (1977) "Tentative Detection of the C_3N Radical", Ap. J. (Lett.) 212, L81-L86.
Guélin, M., Green, S., and Thaddeus, P. (1978) "Detection of the C_4H Radical toward IRC+10216", Ap. J. (Lett.) 224, L27-L30.
Guélin, M., Friberg, P., and Mezaoui, A. (1982) "Astronomical Study of the C_3N and C_4H Radicals: Hyperfine Interactions and Rho-type Doubling", Ast. Ap. 109, 23-31.

Guélin, M. (1985) "Chemical Composition and Molecular Abundances of Molecular Clouds" in G.H.F. Diercksen, W.F. Huebner, and P.W. Langhoff (eds.), Molecular Astrophysics, D. Reidel, Dordrecht, 23-43.
Guélin, M. (1987) "Radio and Millimeter Observations of Less Complex Molecules", in M.S. Vardya and S.P. Tarafdar (eds.), Astrochemistry, D. Reidel, Dordrecht, 171-181.
Hawkins, I. and Meyer, D.M. (1989) "Probing the Possibility of A $^{12}C/^{13}C$ Abundance Gradient from Observations of Interstellar CH^+", Ap. J. 338, 888-901.
Herbst, E., Green, S., Thaddeus, P., and Klemperer, W. (1977) "Indirect Observation of Unobservable Interstellar Molecules", Ap. J. 215, 503-510.
Herbst, E. (1988) "Dense Interstellar Cloud Chemistry", in T.J. Millar and D.A. Williams (eds.), Rate Coefficients in Astrochemistry, Kluwer, Dordrecht, 239-262.
Herbst, E. and Leung, C.M. (1989) "Gas-phase Production of Complex Hydrocarbons, Cyanopolyynes, and Related Compounds in Dense Interstellar Clouds", Ap. J. Suppl. 69, 271-300.
Herzberg, G. (1950) Molecular Spectra and Molecular Structure. I. Spectra of Diatomic Molecules, Van Nostrand Reinhold, New York.
Herzberg, G. (1945) Molecular Spectra and Molecular Structure. II. Infrared and Raman Spectra of Polyatomic Molecules, Van Nostrand Reinhold, New York.
Herzberg, G. (1966) Molecular Spectra and Molecular Structure. III. Electronic Spectra and Electronic Structure of Polyatomic Molecules, Van Nostrand Reinhold, New York.
Hinkle, K.W., Keady, J.J., and Bernath, P.F. (1988) "Detection of C_3 in the Circumstellar Shell of IRC+10216", Science 241, 1319-1322.
Ho, P.T.P. and Townes, C.H. (1983) "Interstellar Ammonia", Ann. Rev. Ast. Ap. 21, 239-270.
Hollis, J.M., Jewell, P.R., and Lovas, F.J. (1989) "A Search for Methylene in the Orion Nebula", Ap. J. (Lett.), submitted.
Irvine, W.M., Schloerb, F.P., Hjalmarson, Å., and Herbst, E. (1985) "The Chemical State of Dense Interstellar Clouds: An Overview", in D.C. Black and M.S. Matthews (eds.), Protostars and Planets II, U. Arizona Press, Tucson, 579-620.
Irvine, W.M., Goldsmith, P.F., and Hjalmarson, Å. (1987) "Chemical Abundances in Molecular Clouds", in D.J. Hollenbach and H.A. Thronson (eds.), Interstellar Processes, D. Reidel, Dordrecht, 561-609.
Irvine, W.M., Friberg, P., Hjalmarson, Å., Ishikawa, S., Kaifu, N., Kawaguchi, K., Madden, S.C., Matthews, H.E., Ohishi, M., Saito, S., Suzuki, H., Thaddeus, P., Turner, B.E., Yamamoto, S., and Ziurys, L.M. (1988) "Identification of the Interstellar Cyanomethyl Radical (CH_2CN) in the Molecular Clouds TMC-1 and Sagittarius B2", Ap. J. (Lett.) 334, L107-L111.
Irvine, W.M., Brown, R.D., Cragg, D.M., Friberg, P., Godfrey, P.D., Kaifu, K., Matthews, H.E., Ohishi, M., Suzuki, H., and Takeo, H. (1988) "A New Interstellar Polyatomic Molecule: Detection of Propynal in the Cold Cloud TMC-1", Ap. J. (Lett.) 335, L89-L93.
Irvine, W.M., Madden, S.C., Ziurys, L.M., Friberg, P., Hjalmarson, Å., Matthews, H.E., and Turner, B. (1989) "Observations of the CH_2/CN

1_{01}-0_{00} and 4_{04}-3_{03} Transitions", in A. Webster (ed.), Submillimeter and Millimeter Astronomy, Kluwer, Dordrecht, in press.
Irvine, W.M. and Knacke, R.F. (1989) "The Chemistry of Interstellar Gas and Grains", in S.K. Atreya, J.B. Pollack, and M.S. Matthews (eds.), Origin and Evolution of Planetary and Satellite Atmospheres, U. Arizona Press, Tucson, 3-34.
Jacq, T., Jewell, P.R., Henkel, C., Walmsley, C.M., and Baudry, A. (1988) "$H_2{}^{18}O$ in Hot Dense Molecular Cloud Cores", Ast. Ap. 199, L5-L8.
Johansson, L.E.B., Andersson, C., Elldér, J., Friberg, P., Hjalmarson, Å., Höglund, B., Irvine, W.M., Olofsson, H., and Rydbeck, G. (1984) "Spectral Scan of Orion A and IRC+10216 from 72 to 91 GHz", Ast. Ap. 130, 227-256.
Jones, A.P. and Williams, D.A. (1984) "The 3 μm Ice Band in Taurus: Implications for Interstellar Chemistry", M.N.R.A.S. 209, 955-960.
Kerridge, J.F. and Chang S. (1985) "Survival of Interstellar Matter in Meteorites: Evidence from Carbonaceous Material", in D.C. Black and M.S. Matthews (eds.), Protostars and Planets II, U. Arizona Press, Tucson, p. 738-754.
Knacke, R.F., Geballe, T.R., Noll, K.S., and Tokunaga, A.T. (1985) "Search for Interstellar Methane", Ap. J. (Lett.) 298, L67-L69.
Kroto, H.W. (1975) Molecular Rotation Spectra, J. Wiley, London.
Lacy, J.H., Evans, N.J., Achtermann, J.M., Bruce, D.E., Arens, J.F., and Carr, J.S. (1989) "Discovery of Interstellar Acetylene", Ap. J. (Lett.) 342, L43-L46.
Langer, W.D. and Graedel, T.E. (1989) "Ion-molecular Chemistry of Dense Interstellar Clouds: Nitrogen-, Oxygen-, and Carbon-bearing Molecule Abundances and Isotopic Ratios", Ap. J. Suppl. 69, 241-269.
Lepp, S. and Dalgarno, A. (1988) "Polycyclic Aromatic Hydrocarbons in Interstellar Chemistry", Ap. J. 324, 553-556.
Linke, R.A., Guélin, M., and Langer, W.D. (1983) "Detection of $H^{15}NN^+$ and HNN^{15+} in Interstellar Clouds", Ap. J. (Lett.) 271, L85-L88.
Lis, D. (1989) "The Sagittarius B2 Molecular Cloud - An Extreme Case of a Galactic Center Giant Molecular Cloud", Ph.D. Dissertation, Univ. of Massachusetts.
Millar, T.J. and Freeman, A. (1984) "Chemical Modelling of Molecular Sources. II. L183", M.N.R.A.S. 207, 425-432.
Millar, T.J., Leung, C.M., and Herbst, E. (1987) "How Abundant Are Complex Interstellar Molecules?", Ast. Ap. 183, 109-117.
Millar, T.J. and Williams, D.A. (1988) Rate Coefficients in Astrochemistry, Kluwer, Dordrecht.
Millar, T.J., Bennett, A., and Herbst, E. (1989) "Deuterium Fractionation in Dense Interstellar Clouds", Ap. J. 340, 906-920.
Minh, Y.C., Irvine, W.M., and Ziurys, L.M. (1988) "Observations of Interstellar $HOCO^+$: Abundance Enhancements Toward the Galactic Center", Ap. J. 334, 175-181.
Minh, Y.C., Irvine, W.M., and Ziurys, L.M. (1989) "Detection of Interstellar Hydrogen Sulfide in Cold, Dark Clouds", Ap. J. (Lett.), submitted; also this volume.
Ohishi, M., Kaifu, N., Kawaguchi, K., Murakami, A., Saito, S., Yamamoto, S., Ishikawa, S., Fujita, Y., Shiratori, Y., and Irvine, W.M. (1989)

"Detection of a New Circumstellar Carbon-chain Molecule, C_4Si", Ap. J. (Lett.), in press.

Olano, C.A., Walmsley, C.M., and Wilson, T.L. (1988) "The Relative Distribution of HN_3, HC_7N and C_4H in the Taurus Molecular Cloud 1 (TMC1)", Ast. Ap. 196, 194-200.

Olofsson, H. (1988) "Radio and Infrared Observations of Circumstellar Envelopes", Space Sci. Revs. 47, 145-174.

Prasad, S.S., Tarafdar, S.P., Villere, K.R., and Huntress, W.T. (1987) "Chemical Evolution of Molecular Clouds", in D.J. Hollenbach and H.A. Thronson, Jr. (eds.)., Interstellar Processes, D. Reidel, Dordrecht, 631-666.

Saito, S., Yamamoto, S., Irvine, W.M., Ziurys, L.M., Suzuki, H., Ohishi, M., and Kaifu, N. (1988) "Laboratory Detection of a New Interstellar Free Radical CH_2CN ($^2B_1$)", Ap. J. (Lett.) 334, L113-L116.

Schwartz, P.R., Snell, R.L., and Schloerb, F.P. (1989) "1300 Micron Continuum and $C^{18}O$ Line Mapping of Giant Molecular Cloud Cores. II. W3, NGC 2264, NGC 6334I, Rho Ophiuchi and S140", Ap. J. 336, 519-525.

Shull, J.M. and Beckwith, S. (1982) "Interstellar Molecular Hydrogen", Ann. Rev. Ast. Ap. 20, 163-190.

Stenholm, L.G. (1983) "Structure of Molecular Clouds", Ast. Ap. 117, 41-45.

Stutzki, J., Stacey, G.J., Genzel, R., Harris, A.I., Jaffee, D.T., and Lugten, J.B. (1988) "Submillimeter and Far-Infrared Line Observations of M17 SW: A Clumpy Molecular Cloud Penetrated by Ultraviolet Radiation", Ap. J. 332, 379-399.

Swade, D.A. (1987) "The Physics and Chemistry of Dark Molecular Clouds: a Detailed Study of L134N", Ph.D. Dissertation, University of Massachusetts.

Swade, D.A. (1989) "Radio Wavelength Observations of the L134N Molecular Core", Ap. J. Suppl., in press.

Taylor, D.K. and Dickman, R.L. (1989) "A Reassessment of the Double Isotope Ratio [^{13}CO]/[$C^{18}O$] in Molecular Clouds", Ap. J. 341, 293-298.

Thaddeus, P., Vrtilek, J.M., and Gottlieb, C.A. (1985) "Laboratory and Astronomical Identification of Cyclopropenylidene, C_3H_2", Ap. J. (Lett.) 299, L63-L66.

Tielens, A.G.G.M. and Allamandola, L.J. (1987) "Composition, Structure, and Chemistry of Interstellar Dust", in D.J. Hollenbach and H.A. Thronson, Jr. (eds.), Interstellar Processes, D. Reidel, Dordrecht, 397-469.

Townes, C.H. and Schawlow, A.L. (1975) Microwave Spectroscopy, Dover, New York.

Turner, B.E. and Ziurys, L.M. (1988) "Interstellar Molecules and Astrochemistry" in G.L. Verschuur and K.I. Kellermann (eds.), Galactic and Extragalactic Radio Astronomy, Springer-Verlag, Berlin, 200-254.

Turner, B.E., Friberg, P., Irvine, W.M., Saito, S., and Yamamoto, S. (1989) "Interstellar Cyanomethane", in preparation.

van Dishoeck, E.F. and Black, J.H. (1987) "The Abundance of Interstellar CO", in G.E. Morfill and M. Scholer (eds.), Physical Processes in Interstellar Clouds, D. Reidel, Dordrecht, 241-274.

Vrtilek, J.M., Gottlieb, C.A., and Thaddeus, P. (1987) "Laboratory and

Astronomical Spectroscopy of C_3H_2, the First Interstellar Organic Ring", Ap. J. 314, 716-725.
Wilson, T.L., Johnston, K.J., Henkel, C., and Menten, K.M. (1989) "The Distribution of Hot Thermal Methanol in Orion-KL", Ast. Ap. 214, 321-326.
Winnewisser, G., Churchwell, E., and Walmsley, C.M. (1979) "Astrophysics of Interstellar Molecules", in G. Chantry (ed.), Modern Aspects of Microwave Spectroscopy, Academic Press, 313-499.
Winnewisser, G. and Herbst, E. (1987) "Organic Molecules in Space", Topics in Current Chemistry 139, 121-172.
Wootten, A. (1987), "Deuterated Molecules in Interstellar Clouds", in M.S. Vardya and S.P. Tarafdar (eds.), Astrochemistry, D. Reidel, Dordrecht, 311-320.
Zinner, E. (1988) "Interstellar Cloud Material in Meteorites", in J.F. Kerridge and M.S. Matthews (eds.), Meteorites and the Early Solar System, U. Arizona Press, Tucson, 956-983.
Ziurys, L.M., Friberg, P., and Irvine, W.M. (1989) "Interstellar SiO as a Tracer of High Temperature Chemistry", Ap. J., August 1.

MOLECULAR CLOUDS IN REGIONS OF MASSIVE STAR FORMATION

Reinhard Genzel
Max-Planck Institut für extraterrestrische Physik
Garching, FRG

I. INTRODUCTION

At a distance of 5 pc from a massive O star ($>10^6$ $L_\odot$) the flux of UV photons is about 1000 times larger than in the average radiation field of the solar neighborhood ($G_0 \approx 1.6 \times 10^{-3}$ erg s^{-1} cm^{-2}). Furthermore, massive stars also have supersonic winds with velocities up to several thousand km/s carrying between 0.1 and 1% of the star's luminosity in form of kinetic energy. At the end of its lifetime an O star explodes as a supernova, ejecting 10^{51} erg of energy into its surroundings. Since the lifetime of the massive stars is relatively short ($\leq 10^7$ years), the parental cloud is within 5 pc or so of the star during a significant fraction of the star's lifetime.

Radiation and winds from nearby OB stars have profound effects on the physical conditions, structure and chemistry of dense clouds. Discussion of observations and theoretical models of dense clouds affected by nearby massive star formation and the physical processes occuring there is the purpose of this review, with emphasis on the infrared, submillimeter and millimeter wavelength bands. UV radiation with energies >13.6 eV creates HII regions where the gas including hydrogen is ionized and all molecules are destroyed. Somewhat softer UV photons between 6 and 13.6 eV create regions where hydrogen is neutral, but low ionization atoms are ionized and molecules are photodissociated. These photodissociation (or photo-chemical) regions (PDR's) have gas temperatures of 100 to 1000 K and emit intense infrared (1 to 300μm), submillimeter (300 to 1000μm) and millimeter line and continuum radiation. Observations and theoretical models of PDR's will be discussed in detail in section III. Supersonic winds and mass outflows result in shocks that raise gas temperatures to between a few 10^3 K (in C-shocks) and 10^5 K (in J-shocks). Again the result is intense infrared and submillimeter line emission and a chemistry that is quite different from cold clouds. Shocks will be discussed in section IV. The effect of a supernova exploding within a cloud is briefly adressed in V.

J. M. Greenberg and V. Pirronello (eds.), Chemistry in Space, 123–170.

The most important global application of the effects of massive star formation on the interstellar medium as a whole is in star burst galaxies. The average flux of UV photons in the central 700 pc of the infrared luminous galaxy M82 ($3x10^{10}$ $L_{\odot}$) is about the same as that 5 pc from an O star. The entire interstellar medium of star burst galaxies can thus be expected to resemble star forming cloud cores rather than cold clouds within our Galaxy. The issue of star burst galaxies, the properties of their interstellar media and the consequences for star formation will be discussed in section VI. The very fact that photodissociation regions, shocks etc. emit intense infrared, submillimeter and millimeter lines from molecular and atomic gas in many stages of excitation gives us a sensitive tool for determining physical conditions and excitation mechanisms. This is the topic of DISCOURSE I, followed by a discussion of the derivation of cloud column densities and masses in DISCOURSE II at the end of this paper.

We begin with a brief overview of the observational phenomena in the case of the nearest cloud with massive star formation, the Orion Molecular Cloud (for a more detailed review see Genzel and Stutzki 1989).

II. AN EXAMPLE OF A STAR FORMING CLOUD: ORION

1.Global Structure of the Cloud

Fig.1 (left) gives an overview of the Orion molecular cloud complex (adapted from Maddalena et al. 1986). The complex is made up of several sub-complexes (Orion A and B clouds etc.) and has a size of about 120 pc (15° at a distance of 450 pc). This massive interstellar cloud is mostly made up of molecular hydrogen molecules, a conclusion which is not based on direct measurement of H_2 itself (which does not have "allowed" electric dipole transitions), but on extensive observations of dust and trace molecules (CO, NH_3, H_2CO, CS etc.). Its mass is estimated to be a few 10^5 $M_{\odot}$ from observations of CO 1→0 line emission (Maddalena et al. 1986). This corresponds to a mean hydrogen density of 100 cm^{-3} or more.

In this cloud massive OB stars ($M>10M_{\odot}$) have been forming for the past 10^7 y or so. Four distinct OB associations can be distinguished from visible studies (Blauuw 1964). The three older ones are marked by hatched circles in Fig.1. The youngest is associated with the Orion nebula. In addition to the associations which are in front of or near the cloud, the most recent star forming activity (within the last 10^6 y) has occured within the cloud itself. Fig.1(right) shows an image of the 12 to 100μm continuum emission that was obtained with the IRAS satellite (Beichman 1988). The emission comes from dust grains that are heated by the short-wavelength radiation from embedded stars to temperatures between 30 to 70 K and then reemit in the far-infrared. The IRAS image marks the sites of the most recently formed massive stars and gives estimates of their luminosities. It is clear that star formation in the Orion cloud is taking place at a high rate.

The most prominent and luminous of these star forming regions is

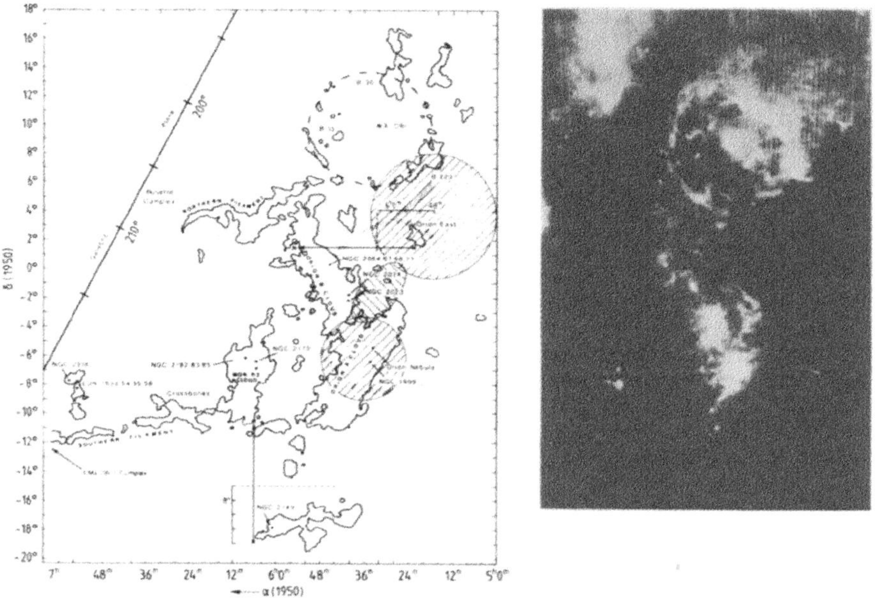

Fig. 1: Large scale distribution of clouds and star formation in Orion. Left: Outlines of the molecular cloud complexes in Orion and Monoceros, from CO mapping by Maddalena et al.(1986), along with the standard designations of clouds and individual star froming regions. The hatched circles are the Orion Ia,Ib and Ic OB associations (NW to SE). The Orion Id association is located at the Orion nebula. The shaded arc is Barnard's loop. Right: Composite 12 to 120μm IRAS image of the same region (Beichman 1988).

the Orion nebula itself. It contains the Trapezium OB association ($L \approx 10^5 L_\odot$) as well as a remarkable concentration of lower mass stars surrounding it. About 0.1pc behind the HII region that is ionized by the Trapezium stars is the Orion-KL region, containing a luminous ($10^5 L_\odot$) cluster of compact infrared sources near the center of the densest core of the molecular cloud. The size of this core is about 0.1pc, its mass 200 $M_\odot$ and hydrogen volume densities range between 10^5 and a few 10^7 cm^{-3}.

2.Infrared and Submillimeter Line Emission from Warm Gas

Fig.2 shows that in addition to the intense infrared continuum this region emits also very bright infrared line radiation in a large number of atomic, ionic and molecular transitions. They have all been detected for the first time within the last decade as the result of dramatic progress in infrared instrumentation (see Genzel and Stacey 1985 for a review). Hydrogen and ionic line emission from the Orion

HII region in the visible range has of course been studied for a long time. The infrared fine structure line emission for ions with ionization potentials >13.6 eV also comes from the HII region.

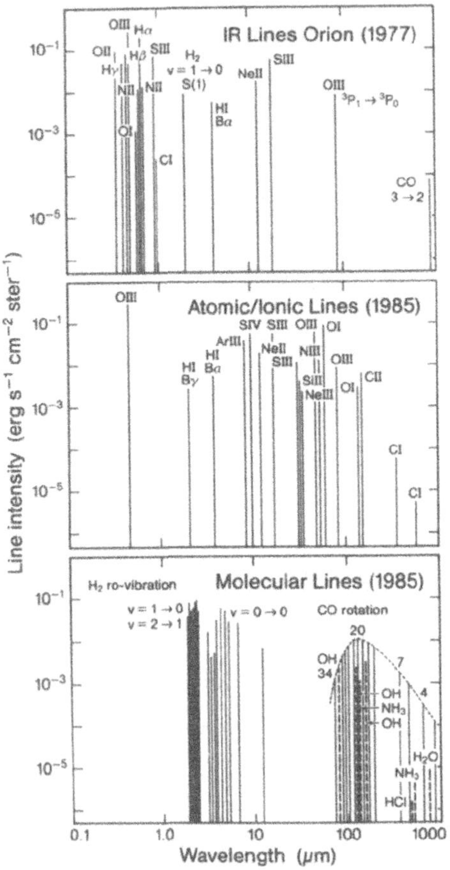

Fig. 2: Line Emission in the central 0.2pc of the Orion nebula. The three panels show visible, infrared atomic and infrared molecular lines.

Quite remarkable is the intense far-infrared and submillimeter emission from species with ionization potentials <13.6 eV (OI,CI,CII). They indicate the presence of a large amount of partially ionized or atomic gas along the line of sight through the Orion Trapezium/Orion-KL region.

Detailed observational and theoretical studies during the past 10 years show that the atomic and molecular infrared line emission shown in Fig.2 can plausibly be accounted for by two main processes that are direct results of the star forming activity: first, photodissociation of the dense quiescent molecular cloud by the intense far-ultraviolet radiation coming from the Trapezium OB cluster and second, shock excitation of molecular gas by mass outflow from the center of the

Orion-KL cluster. These two processes will be the subjects of the more detailed discussion in the next two sections.

Fig.3 shows the first velocity resolved spectra of the 63μm [OI] line which is the brightest of all the infrared lines (Crawford et al. 1986). Toward the Trapezium the emission appears to come from a barely resolved ($\Delta v \approx 5$ km/s) line of Planck brightness temperature 150 to 200 K. Toward the Orion-KL region ("H_2 peak 1") there is an additional component of velocity width $\approx$30 to 50 km/s. These data show that the atomic gas has different components and comes from gas which has greater motions than the quiescent molecular cloud ($\Delta v_{cloud} \approx 5$ km/s) and from gas with a temperature of a few hundred K. That temperature is also significantly higher than the dust temperature or the gas kinetic temperature of the quiescent molecular cloud near Orion-KL (70 to 100 K). The atomic gas also must be dense, as the hydrogen density required to populate the upper level of the [OI] ground state is about 10^5 cm^{-3} (see DISCOURSE I).

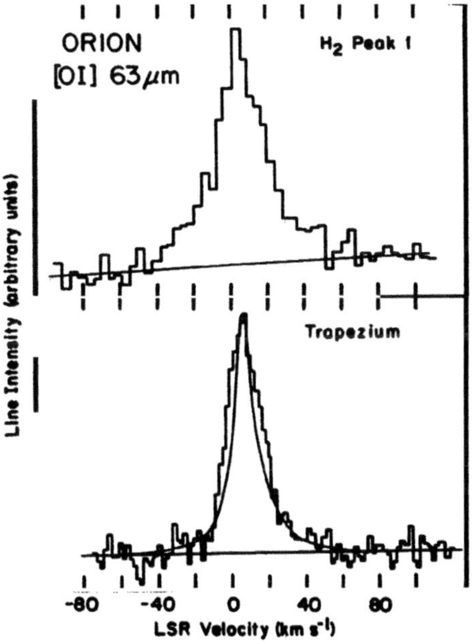

Fig. 3: Profiles of [OI] 63μm line emission toward the H_2 emission peak 1 in Orion-KL and the toward the Trapezium. Velocity resolution is about 15 km s^{-1}. The peak Planck brightness temperature is about 160 K (from Crawford et al. 1986).

Returning to Fig.2, the lowest panel shows that there are also many molecular emission lines. First, there is intense quadrupole line emission from vibrationally and rotationally excited levels of molecular hydrogen. The v=1 level of H_2 is 6500 K above ground. The detection of near-infrared H_2 line emission thus provides evidence for a component of hot (a few thousand K) molecular gas. High spectral resolution measurements show that the H_2 lines, like [OI], come from dynamically active gas with a velocity range of several tens of km/s.

Second, there is far-infrared and submillimeter/mm rotational line emission from CO, OH, NH_3 and H_2O which also require high temperatures

(100 to 1000 K) and high densities (10^5 to 10^8 cm^{-3}) for their excitation. The gas kinetic temperature of the bulk of the molecular gas close to the newly formed massive stars is between 50 and 200 K, significantly greater than in the rest of the Orion cloud (20 to 50 K) or in dark clouds (10 K).

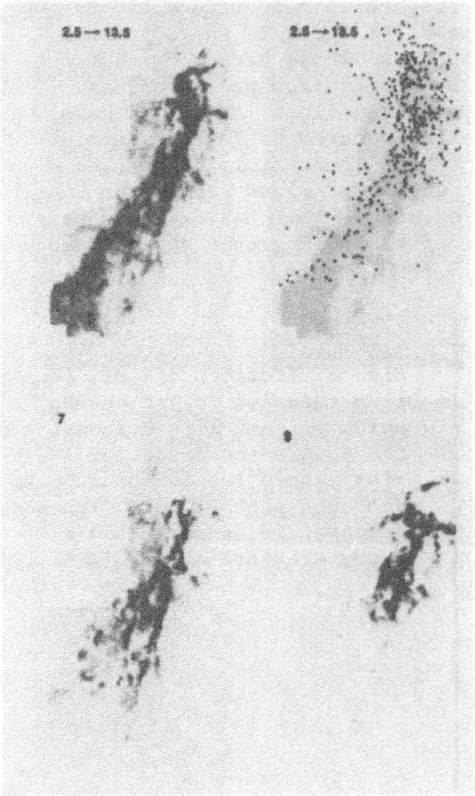

Fig. 4: Large scale structure and star forming activity in Orion A. Top left: Velocity integrated ^{13}CO 1→0 map of Orion A (90" resolution, Bally et al. 1987). Top right: The same overlaid with the distribution of Hα emission stars. Bottom left and right: ^{13}CO channel maps at 6 and 9 km s^{-1} LSR.

3.Clumpiness

While these findings already suggest profound changes of the physical conditions of molecular cloud material near newly formed stars, there is also a dramatic impact of star formation on the spatial structure and dynamics of molecular clouds. Fig.4 shows radiographs of the ^{13}CO 1→0 line emission in the Orion A sub-complex of the cloud from the survey by Bally et al.(1987). The ^{13}CO line is much less optically thick than the "standard" tracer ^{12}CO and gives thus a better picture of the distribution of column density. The data in Fig.4 very clearly show that the molecular cloud is highly filamentary or clumpy, with many bubbles and cavities. The clumps have volume densities of 10^4 hydrogen molecules per cm^3 or more and a volume filling factor less than 10%. The cloud has an overall V-shape with its apex pointing toward the Orion OB associations. Bally et al.

conclude that this structure is evidence for the strong dynamical effects of the OB stars on the cloud structure. It is likely that not only the OB stars but also the lower mass stars contribute to this process. These lower mass stars are present throughout the cloud (Fig.4 top right). In this scenario the outflows and winds of the young stars may stir up the cloud, thereby maintaining or contributing to the cloud's "turbulence" that prevents large scale collapse (Norman and Silk 1981). The same process will also lead to compression of density condensations by external pressure, thus resulting in further (massive?) star formation.

III. PHOTODISSOCIATION REGIONS

1.Theory

Far-UV radiation (912 to 2000 Å) longward of the Lyman edge of hydrogen dissociates most molecules and ionizes a number of abundant atoms, such as carbon, silicon, sulfur and iron. Wherever neutral gas clouds are exposed to far-UV radiation there are thus zones where most of the gas is in atomic or partially ionized form. Such regions are common in the interstellar medium. Examples include interfaces between HII regions and surrounding molecular clouds, neutral shells around planetary nebulae, interstellar gas in the star forming nuclei of galaxies and, of course, the diffuse interstellar medium in our own Galaxy. Theoretical models of the heat balance and chemistry of clouds illuminated by far-UV radiation have been presented by a number of authors (Walmsley 1975, Langer 1976a, Gerola and Glassgold 1978, deJong, Dalgarno and Boland 1980). Tielens and Hollenbach (1985), Hollenbach, Tielens and Takahashi (1989), Sternberg and Dalgarno (1989) and Burton et al.(1989) treat the cases of high cloud densities and high UV intensities that are particularly interesting for the present discussion. The basic size scale of the dissociation region is given by the penetration depth of far-UV radiation. The most efficient broad-band absorber in this wavelength range is dust, and the thickness of a photodissociation region is given by the condition (Werner 1970) that the UV dust absorption optical depth be a few, or

$$d_{PDR} = \tau_{abs} \approx \text{a few, corresponding to a hydrogen column density of}$$

$$N_0(H) \approx 6\times10^{21}\ \mathrm{cm}^{-2} \qquad (1).$$

The chemical structure in such a cloud interface that is common to all models is displayed in Fig.5. At $A_V \approx 1$ to 2 from the surface there is a sharp transition from atomic to molecular hydrogen. This is a direct consequence of the fact that photodissociation of H_2 requires UV absorption from the $X^1\Sigma_g$ ground electronic state to the $B^1\Sigma_u$ and $C^1\Pi_u$ excited states, followed by a transition into the unbound

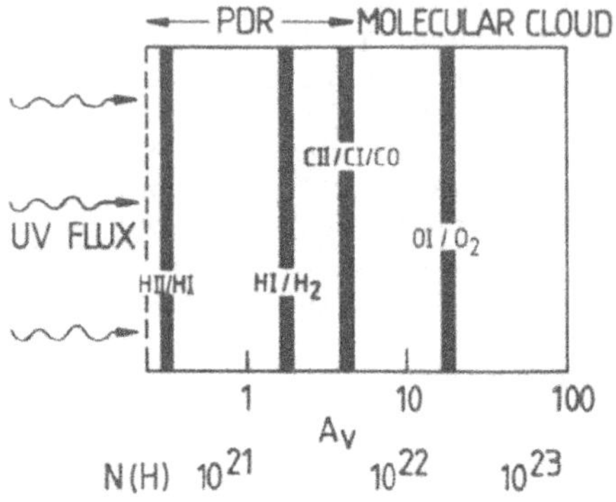

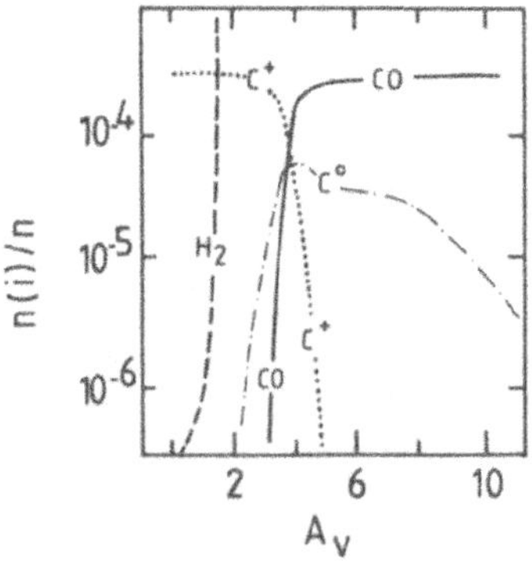

Fig. 5: Schematic of a PDR (top), and chemical abundances of the major species (bottom) in the model of Tielens and Hollenbach (1985a).

continuum of the ground state. There is no direct continuum dissociation process. The UV transitions of H_2 (the Lyman and Werner bands) rapidly become optically thick and self-shielding sets in (cf. Jura 1974). In this outer layer almost all of the gaseous carbon and oxygen is in form of C^+ and O^0. The transition between singly ionized carbon and CO occurs significantly deeper into the cloud ($A_V \approx 5$), with a layer of atomic carbon arising between $A_V \approx$ 3 and 8. In Fig.5 it is assumed that all carbon is in CO as soon as CO can form. Oxygen thus remains atomic deep into the cloud ($A_V \approx 20$) and OI may be the most important carrier of oxygen in interstellar clouds (see Dalgarno, this volume). In addition to these most abundant species, the photodissociation zone contains many less abundant ions, such as S^+, Mg^+, Si^+ and Fe^+ whose abundances depend heavily on depletion.

Photoelectric heating and UV pumping of H_2 are likely the most important gas heating mechanisms in dense PDR's. These mechanisms are summarized in Fig.6. The first mechanism involves the absorption of a UV photon by a dust grain, followed by ejection of an electron which then collisionally heats the gas (Draine 1978, deJong 1977). Per incoming 10 eV photon, the energy per electron available for heating is about 1 eV. The remaining 9 eV go into overcoming the binding energy of the electron to the grain (the "work function") and the grain's positive electrostatic potential. The potential energy which has to be overcome increases with increasing UV energy density (higher positive charge) and decreases with increasing density (higher rate of recombination). With a "yield" of conversion of incoming photons into photoelectrons of about 0.1 the photoelectric mechanism can plausibly convert about 1% of the UV photons into gas heating. The numerical values are, of course, grain-model dependent and also depend on carbon abundance and grain size distribution. Of great interest for future study is the importance of small dust grains and large molecules (PAH's) for this mechanism.

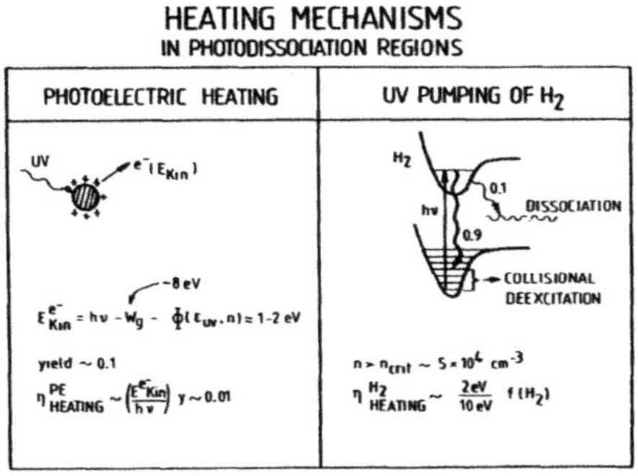

Fig. 6: Heating mechanisms in PDR's (see text).

UV pumping of H_2 molecules can convert a similar fraction of the UV if the hydrogen density is sufficiently high (Sternberg 1988). In this case, electronic excitation of H_2 molecules in the Lyman and Werner bands at 1000 Å leads to subsequent radiative decay into a vibrationally excited level of the electronic ground state (2 to 3 eV) in 9 out of 10 excitations (Fig.6). At densities greater than the critical density for collisional deexcitation of the vibrational states ($n(H_2)_{crit} \approx 5x10^4$ cm^{-3}, DISCOURSE I), that energy can again be converted into gas heating. In the UV transitions of H_2, the conversion efficiency of a single absorbed photon into kinetic energy can thus be as high as 30%. The total efficiency of converting far-UV photons into heating depends on the fraction of incident radiation absorbed by H_2 molecules throughout the cloud. This fraction depends on the dust to gas ratio and on the ratio of gas density to intensity of the UV field and is usually $\leq$10% (Sternberg 1988, Sternberg and Dalgarno 1989).

Gas cooling of the photodissociation zone is primarily by the fine structure lines of [OI] (63μm), [CII] (158μm) and [SiII] (35μm). For very dense gas the 26 and 35μm fine structure lines of [FeII] and the forbidden lines of [OI] and [CI] in the visible may also contribute. Of lesser importance are the H_2 ro-vibrational and rotational lines and the submm fine structure lines of [CI] largely because of the small column densities in the upper state and because of their small A-coefficients.

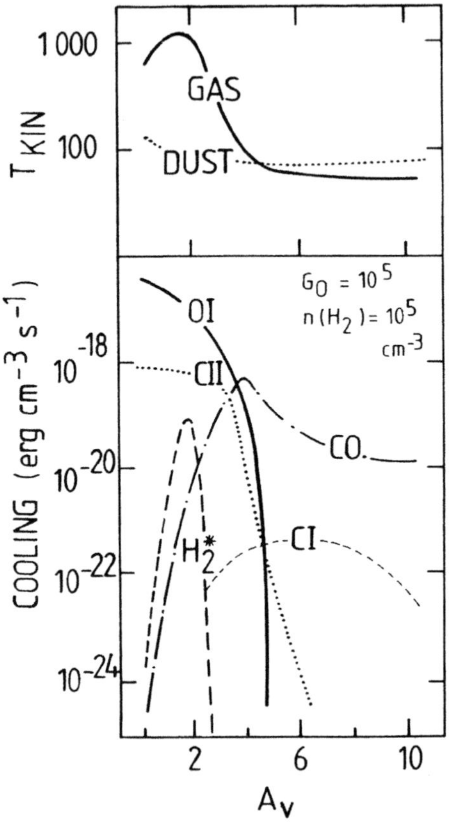

Fig. 7: Distribution of gas/dust temperatures (top) and cooling lines (bottom) through a PDR, for the specific case of the "standard model" of Tielens and Hollenbach (1985 a).

The major coolants and the resulting thermal structure is depicted in Fig.7 as a function of depth from the cloud surface, for the "standard" model of Tielens and Hollenbach (1985a,b: $G_{UV} \approx 10^5$ times the UV flux G_0 of the interstellar radiation field defined on page 1, $n(H_2) \approx 2x10^5$ cm^{-3}, "cosmic" abundances). The standard model is representative for dense photodissociation regions within a few tenths of a pc of a massive O star, such as the HII region/cloud interfaces in Orion and M17 or the circum-nuclear disk in the Galactic Center (Genzel et al. 1985). The gas temperature in the PDR ranges between about 1000 K at the surface to about 100 K at the C^+/CO transition.

In the following we discuss how these theoretical ideas compare to observations.

2. Observations of Photodissociation Regions.

The Orion Photodissociation Region

Fig.8 gives a direct comparison of Tielens and Hollenbach's (1985a) standard model with the observed infrared and submillimeter line and continuum emission toward the Trapezium region in Orion. Observations and theory agree fairly well. It seems quite convincing that the infrared measurements are accounted for by a PDR model for a dense molecular cloud at a distance of about 0.1 pc from the Trapezium OB association. About 0.6% of the impinging UV radiation is converted into line emission. Although the standard model is not a fit to the Orion data, the good agreement with the data should, of course, not come as a complete surprise, as the theoretical work was in part motivated by the observations of the intense infrared line emission.

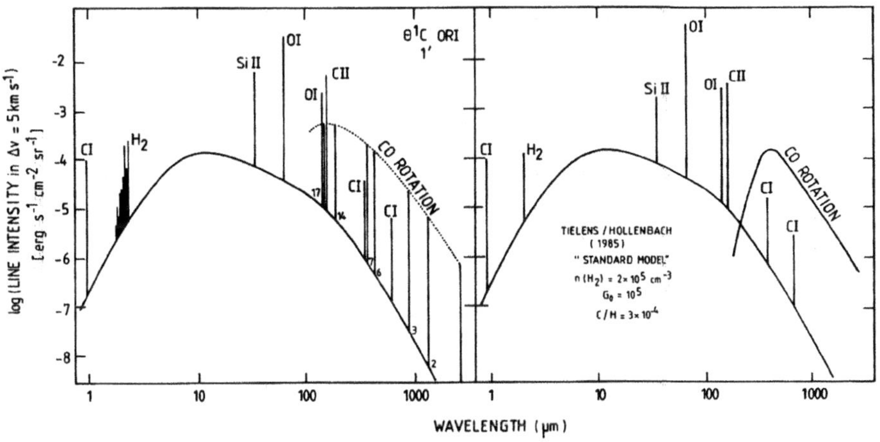

Fig. 8: Line emission from the Orion A PDR. The left diagram shows the observations, the right diagram shows the line emission predicted by the "standard model" of Tielens and Hollenbach (1985 a).

The standard model does not predict enough [SiII] emission which may indicate that the assumed gas phase abundance of silicon was too low. More serious is the failure of the model to account for the observed submm and far-infrared CO line emission. Again this is not too surprising, as the models discussed above predict very little CO at temperatures >200 K (Fig.6) that is required to explain the Orion CO data (Stutzki et al. 1989, Stacey et al. 1990b). The model does, however, predict fairly intense (50 to 100 K) ^{12}CO 1→0 emission originating just behind the transition region from C^+ to CO. The mid-and high-J CO emission in Fig.8 comes from warm, quiescent gas and not the high-velocity shock excited gas associated with Orion-KL (see section IV.2). Observations of the spatial distribution of the warm, quiescent CO in Orion and several other galactic star formation regions show a good correlation of bright (brightness temperatures 50 to 100 K) 7→6 emission and 158μm [CII] emission (Stutzki et al. 1988, Schmid-Burgk et al. 1989, Jaffe et al. 1989,1990), further supporting the inference of an origin of the warm CO in or close to photodissociation regions (Harris et al. 1987). Burton et al. (1989) propose that very dense ($n(H_2) \geq 10^6$ cm^{-3}) PDR's can account for the measurements. Heating efficiency and H_2/CO self/mutual shielding may then be sufficiently increased to keep a significant amount of CO at high temperature. However, while the model of a very dense PDR may plausibly fit the Orion observations, it is questionable for regions like NGC2023. The v=2→1 to v=1→0 S(1) H_2 line ratio in that source indicates pure fluorescent emission which requires densities less than a few 10^4 cm^{-3}. A similar and possibly related problem is the lack of current models to account for the CO abundance in diffuse clouds (van Dishoeck and Black 1986, 1989). In either case important uncertainties of the models lie in the adopted photodissociation rate of CO and the effects of CO shielding by itself and by H_2.

Global Characteristics and Importance of PDR's

From a historic perspective there are four different types of observations probing photodissociation regions. These are observations of neutral carbon and oxygen lines in the visible, radio observations of carbon recombination lines, far-UV measurements of C^+ resonance lines and finally, the observations of the infrared and submillimeter lines discussed here (see Genzel, Harris and Stutzki 1989 for references to the original work). One important aspect of these studies is that the radio and optical measurements typically refer only to "special" regions in the immediate neighborhood of O stars, such as the Orion Nebula or M17. The UV measurements, on the other hand, probe diffuse clouds that have been considered to be more typical of the global photodissociated interstellar medium in our Galaxy. The 158μm [CII] fine structure line is probably the major cooling line in the diffuse medium (Dalgarno and McCray 1972). Before extensive observations were possible, it was therefore commonly expected that on large scales in our Galaxy or globally in external galaxies the [CII] emission is dominated by diffuse gas associated with HI clouds or the HI interclump medium.

The observations give a totally different result (Crawford et al.1985, Stacey et al.1990). **The [CII] emission dominating the emission of galaxies with active star formation comes from moderately dense gas** ($n(H_2) \geq$ a few 10^3 cm^{-3}) **and is located in the UV illuminated surfaces of molecular clouds**. The dense photodissociated interstellar medium makes up between 5 and 30% of the mass of the interstellar medium. To demonstrate the first conclusion we show in Fig.9 a plot of [CII] line intensity as a function of HI column density (inferred from 21cm observations), for the nuclear regions of more than a dozen

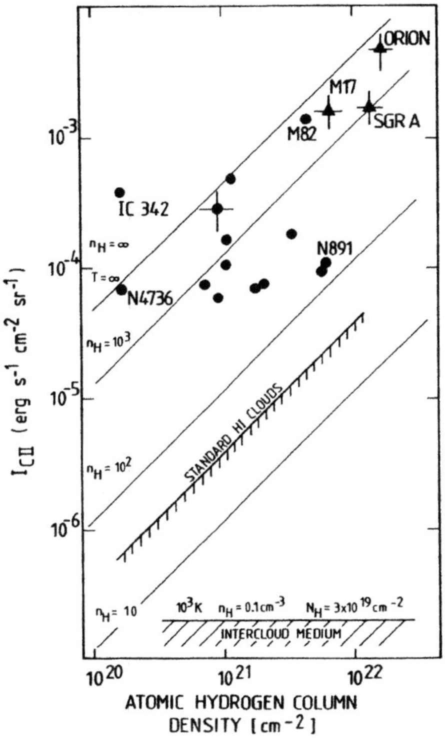

Fig. 9: Integrated 158μm [CII] intensity (1' beam) as a function of atomic hydrogen column density, derived from 21cm HI measurements. Black dots mark the gas rich galaxies investigated by Stacey et al.(1990). Triangles represent three representative, bright galactic HII regions. Lines of slope 45° mark different HI volume densities. The locations of HI clouds ($n_H \approx 30$ cm^{-3}, $T_{kin} \approx 100$ K) and of the HI intercloud medium are also indicated.

galaxies, together with three representative galactic sources (Crawford et al. 1985, Stacey et al. 1990). Thin lines of 45° angle mark different densities. The interpretation of this diagram is obvious. The observed [CII] emission cannot come from atomic hydrogen clouds or HI intercloud gas, as the atomic medium does not have large enough column and volume densities. The reverse is not true. A sizable fraction of the atomic column density in some galaxies may in fact come from dense PDR's (Eckart et al. 1989, Stacey et al. 1990). In about 10 sources, including the nuclear regions of the galaxies M82 and IC342, hydrogen densities of 10^3 to 10^4 cm^{-3} can also be inferred more directly from the ratio of the 63μm [OI] to the [CII] line.

A corresponding plot of [CII] intensity as a function of molecular column density (as derived from CO data, see DISCOURSE II) indicates that the molecular interstellar medium can in fact account for the [CII] data, as long as 5 to 30% of the molecular column density and mass is in C^+ regions (Crawford et al.1985). This is shown in more detail for four representative sources in Table I, along with estimates of the abundances of the most important species. The interpretation of a physical association between PDR's and molecular clouds is further strengthened by studies of the spatial distribution in a number of regions. Stacey et al.(1985) and Shibai et al.(1989) have shown that the scale height of the [CII] emission in our Galaxy is very similar to that of CO (≈0.6°).

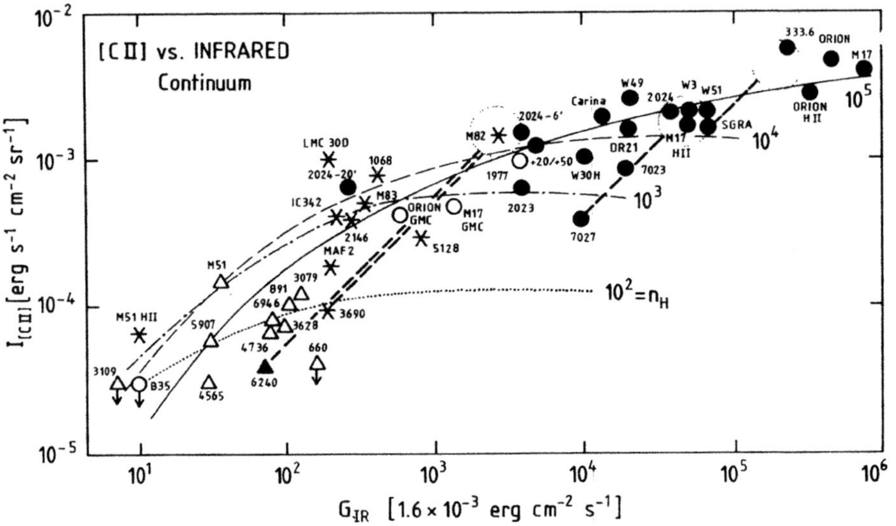

Fig. 10: [CII] intensity as a function of energy density of the radiation field. The data are represented by black dots in Galactic star forming regions, by asterisks for star burst galaxies, by open circles for Galactic clouds without very intense star formation and by triangles for other spiral galaxies (Stacey et al. 1990). The ultra-luminous galaxy NGC6240 is marked by a filled triangle. The radiation field's intensity G_{IR} in units of the local interstellar field is derived from the total infrared luminosity. Models of the [CII] intensity as a function of UV flux (assume $G_{IR} \approx G_{UV}$) and volume density are taken from Hollenbach, Tielens and Takahashi (1989). If the far-infrared continuum and line emission does not fill the beam, the data point has to be moved to the upper right along a 45° slope. This correction is indicated for three sources by dashed lines and dotted thin circles.

Fig.10, taken from Stacey et al.(1990), is a plot of the [CII] intensity as a function of the intensity (or energy density) of the radiation field, derived from the flux of the far-infrared continuum. Also plotted are the model predictions of Hollenbach et al.(1989) for different cloud densities with the assumption that the far-infrared luminosity is identical to the far-UV luminosity. It aslso assumes that all of the source's luminosity is concentrated in the center of a sphere of radius beam diameter divided by two. A more realistic assumption is perhaps that the UV luminosity is a fraction of the far-IR luminosity. In that case, the curves have to be shifted to the right by the reverse of that factor. The data in Fig.10, sampling about 40 galactic and extragalactic sources with widely different characteristics, make a fairly convincing case that the [CII] emission in gas rich galaxies can be interpreted as the result of the UV radiation from "naked" and embedded OB stars whose luminosity is inferred from the dust continuum radiation in the far-infrared. The [CII] luminosity in galaxies is thus a measure of the area (or filling factor) of the cloud surfaces exposed to intense UV radiation as well as of the intensity G of the UV radiation itself. In the theoretical models, the non-linear dependence of the [CII] intensity on the UV flux G and the "saturation" of the [CII] intensity at high G is a combination of decreasing heating efficiency (greater positive charge of the grains), increased fraction of cooling in [OI] and the effect of dust absorption on the radiation on the available column density in the PDR.

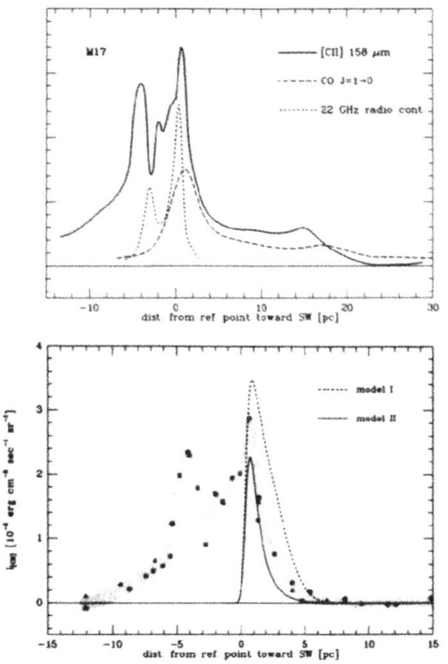

Fig. 11: [CII] distribution in M17 (from Stutzki et al.1988). The top diagram shows the [CII] line flux (1' beam) along a 1° long cut (p.a. ≈60°) through the M17 star forming region, along with the distributions of CO and radio continuum. The bottom graph shows two simple models of the [CII] distribution near the HII region in terms of a clumpy cloud penetrated by UV radiation from the M17 OB cluster.

Photodissociation Regions are Clumpy

Fig.11 is a comparison of the spatial distributions of 158μm [CII] emission, the 2.6mm ^{12}CO 1→0 emission and of the 1.2cm radio continuum flux along a north-east to south-west cut through the HII region M17 (adapted from Stutzki et al.1988). The OB cluster (10^6 $L_{\odot}$) is located at the base position of the cut. The interface between the ionized zone and the molecular cloud M17SW is nearly parallel to the line of sight, so that the observations are sampled approximately along a radius through the center of the OB cluster. As expected in the picture of photodissociation regions discussed so far, the [CII] emission peaks in two lobes just outside of the two radio continuum maxima. There are two unexpected results, however. First, there appears to be weak [CII] emission distributed throughout the M17SW molecular cloud (scale size 20pc or more). Stutzki et al. propose that this extended line emission is created by the external radiation field and/or by distributed B stars that are embedded in the cloud. It should be noted that recent ballon borne observations by Matsuhara et al.(1989) have essentially confirmed the observations of Stutzki et al.(1988) shown in Fig.11, but indicate that the intensity of the extended component is about a factor of 2 to 3 weaker than inferred from the spatially chopped KAO observations of Stutzki et al.(1988).

Second, the [CII] emission falls off rather slowly (scale size 1 to 2 pc) radially outward from the HII region interface and into the cloud. This result is particularly surprising. The H_2 column density at the point where the [CII] emission has fallen to about 1/e of the peak and measured radially outward from the OB cluster, corresponds to $A_V \approx 100$ or more (A_V and H_2 computed from CO column density, see DISCOURSE II). This thickness is at least an order of magnitude greater than expected for a uniform cloud (see section III.1). Stutzki et al.(1988) interpret the <u>slow dropoff as the result of the deep penetration of the UV radiation in a clumpy molecular cloud</u>. Fig.11 (bottom) gives two different simple models. The [CII] intensity as a function of distance into the cloud is computed from the models of Tielens and Hollenbach (1985a) and Hollenbach et al.(1989) for the specific case of the M17 UV energy density and a cloud structure, where 10 to 30% of the volume is filled by absorbing clumps. The model also requires that the UV radiation passes the volume between clumps essentially without absorption (A_V(interclump)< a few). The two curves are for two different cases of the UV flux distribution (see Stutzki et al.1988). The agreement with the data is quite good. Support for the clumpy cloud model also comes from high resolution $C^{18}O$ $J=2\rightarrow1$ mapping of the interface region with the IRAM 30m-telescope which shows the clumpiness of the molecular gas directly (clump sizes 0.5pc, Stutzki and Güsten 1989). Moreover, M17 is not the only case, another one being the circum-nuclear disk surrounding SGR A West (Genzel et al.1985). Giard et al.(1990) report evidence for a large penetration depth of UV radiation in a dark cloud. The deep penetration of UV radiation in M17, SGR A etc. raises an important problem for the medium between clumps. The requirement that it passes UV radiation without difficulty (τ(dust)$\leq$ a few), while the clumps are highly optically thick (τ(dust)$\approx$100 as derived from $C^{18}O$) means that the clump to interclump column density ratio would have to be about 10^2 if the gas to dust ratio is the same within and in between clumps. How could such a large clump to interclump contrast be maintained?

Puget (priv.comm.) proposes that instead the interclump dust absorption per nucleon is reduced, possibly because small dust grains have been destroyed.

The Origin of Submillimeter Emission of Atomic Carbon

The first observations of the 609μm $^3P_1 \rightarrow {^3P_0}$ [CI] fine structure line immediately showed that atomic carbon is widespread throughout molecular clouds, with line profiles and intensities very similar to ^{13}CO J=1→0 (Phillips et al.1980, Phillips and Huggins 1981). Phillips and coworkers conclude that the column density of atomic carbon is at least 10% of that of CO (see also Keene et al. 1985 and Table I). Zmuidzinas et al.(1986, 1988) and Genzel et al. (1988) come to similar conclusions from observations of the 370μm $^3P_2 \rightarrow {^3P_1}$ transition and find that the kinetic temperature in the [CI] emission region is between 30 and 80 K. These latter authors also infer from the [CI] line brightness temperatures that the [CI] lines have optical depths between 0.5 and 2. Figure 12 shows a fairly deep spectrum of the [CI] $^3P_2 \rightarrow {^3P_1}$ toward the [CI] emission peak of M17. From the lack of detection of the [^{13}CI] hyperfines of the $^3P_2 \rightarrow {^3P_1}$ transition, one can set an upper limit of $\tau([^{12}\mathrm{CI}]) \approx 5$ (Stutzki et al.1989). While [CI] emission is naturally expected to occur in PDR's, as discussed in section III.1, the large inferred column densities and the large spatial extent could not be explained by photodissociation region models at the time of the discovery of the [CI] emission. Furthermore, Keene et al.(1985) found that the [CI] emission in M17 peaks further into the cloud than CO, at an equivalent mean visual extinction of 100 magnitudes from the M17 OB cluster.

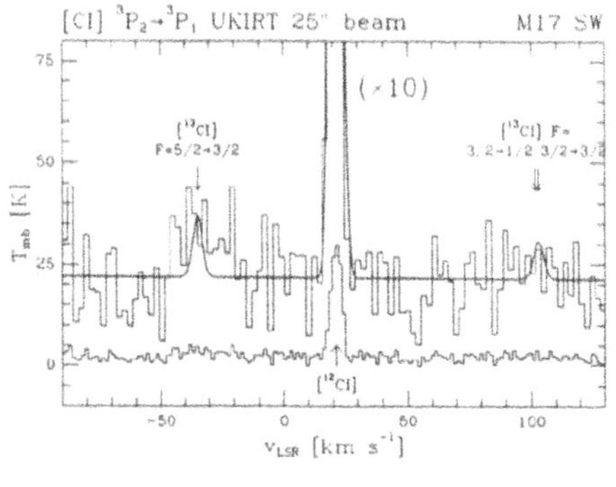

Fig. 12: $^3P_2 \rightarrow {^3P_1}$ 370μm [CI] profile toward the M17 interface region (Stutzki et al. 1989). The position and maximum strengths of the hyperfine components of [^{13}CO] are indicated.

A variety of mechanisms other than stellar UV radiation has been considered to account for the [CI] observations. Several authors have shown that in molecular clouds with time-dependent chemistry, a relatively large atomic carbon abundance exists if an initially atomic cloud has not evolved to an age greater than 10^6 y. At that time the conversion of CI into CO is not yet complete (Langer 1976b, Iglesias

1977, Gerola and Glassgold 1978, Graedel, Langer and Frerking 1982, Leung, Herbst and Huebner 1984, Tarafdar et al. 1985). Turbulence and shocks may return some of the carbon into atomic form (Boland and deJong 1982, Williams and Hartquist 1984). A weak UV field is created throughout the cloud through cosmic-ray induced, electronic excitation of H_2 (Prasad and Tarafdar 1983, Sternberg, Dalgarno and Lepp 1987) that may be intense enough to account for at least a fraction of the atomic carbon column density (Prasad and Tarafdar 1983, Gredel, Lepp and Dalgarno 1987). Finally, if the carbon to oxygen abundance ratio is greater than unity, an excess amount of CI is left after formation of CO (Langer et al. 1984).

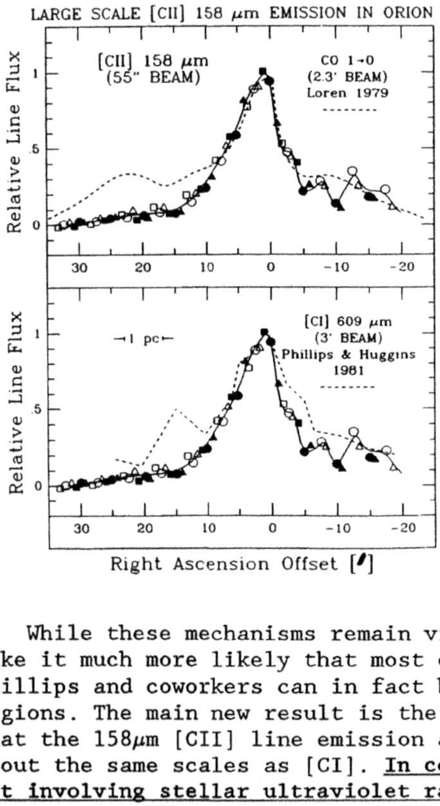

Fig. 13: East-west cut of [CII] (dots, triangles and rectangles connected with a thin continuous curve, from Stacey et al. 1989b), 610μm [CI] (dashed line in bottom inset, from Phillips and Huggins 1981) and CO (dashed line in top inset, from Loren 1979) through the Orion A ridge.

While these mechanisms remain viable, the most recent data now make it much more likely that most of the [CI] observations of Phillips and coworkers can in fact be explained by photodissociation regions. The main new result is the general finding, discussed above, that the 158μm [CII] line emission also appears to be extended on about the same scales as [CI]. **In contrast to [CI], chemical models not involving stellar ultraviolet radiation cannot produce observable amounts of [CII] radiation** (e.g.Gredel, Lepp and Dalgarno 1987). On the other hand, the most recent PDR models with large cloud densities and UV intensities that appear to account satisfactorily for the [CII]/[OI] data for clumpy clouds, probably can also explain the [CI] measurements in the same sources.

In the case of M17, high resolution mapping (25") of the 370μm [CI] line by Genzel et al. (1988) shows that there is a ridge of intense atomic carbon emission coincident with the column density peak of the molecular cloud and the south-western [CII] lobe of Fig.11. There is an excellent correlation between [CI] and $C^{18}O$ J=2→1 fluxes which can qualitatively be well accounted for with the clumpy cloud model by Stutzki et al.(1988). Orion is another case, where Phillips and Huggins (1981) had found very extended [CI] emission. Fig.13 shows their data along an east-west cut through the Orion A ridge, along with recent [CII] observations of the same region by Stacey et al.(1989b), and the CO 1→0 distribution from Loren (1979). Again there is [CII] emission over about the same scale as [CI], indicating the presence of an extended photodissociation region. The standard model of Tielens and Hollenbach (1985a) fails to explain the observed [CI] flux in Orion by about a factor of 2. By decreasing the dust to gas ratio by about that factor, Tielens and Hollenbach (1985b) are able to fit the [CI] observations.

It is quite possible that [CI] sources not associated with intense UV fields may be understood in terms of photodissociation by the interstellar UV field (see discussion in Frerking et al.1989). Phillips and Huggins (1981) and Frerking et al.(1989), for instance, have detected weak [CI] emission in the dark clouds L134N and ρ Oph. Beichman et al.(1983) found [CI] emission from the Egg Nebula (AFGL 2688), a circumstellar envelope around a carbon-rich late-type star. There are no nearby strong UV sources (other than the interstellar radiation field) in either case. Phillips and Huggins (1981) report a Rayleigh-Jeans brightness temperature of about 1 K in L134N. The models of Hollenbach, Tielens and Takahashi (1989) predict a Rayleigh-Jeans brightness temperature of about 2 K for a single PDR exposed to the interstellar radiation field (G_o=1, $n(H_2)$= 10^2 to 10^4 cm^{-3} and Δv=2 km s^{-1}), consistent with Phillips and Huggins' data. Frerking et al.(1989) compare the measured CI column densities in diffuse/translucent clouds and in ρ Oph with the photodissociation models of van Dishoeck and Black (1989). The column densities in diffuse clouds (Jenkins and Shaya 1979) are consistent with models of $G_{UV}\approx1$, but the ρ Oph measurements require $G\approx10$.

IV. SHOCKS IN DENSE INTERSTELLAR CLOUDS

Interstellar shock waves are generated by supersonic mass motions, such as fast moving clouds, outflows from young stars, stellar winds and supernovae. The kinetic energy of the supersonic motions is converted to thermal energy. In the process, shocks compress and accelerate gas and dust. Shocks radiate primarily in lines from the cooling gas behind the front where the flow motion is converted into random thermal motions. With the exception of high velocity shocks (v_s>50 km s^{-1}) in dense ($n(H_2)$>10^6 cm^{-3}) clouds, dust emission is small compared to line emission from hot gas. Shocks in dense molecular clouds often radiate mostly in the infrared and submillimeter lines of molecules, atoms and ions. High velocity, gas dynamic (J-shocks) dissociate most molecules and ionize atoms.

Magneto-hydrodynamic (C-shocks) create large column densities of moderately warm molecular gas.

1. Theory

J-Shocks

Shock waves in dense interstellar clouds have been recently reviewed in some detail by Shull and Draine (1987) and Hollenbach (1989) who also refer to the many detailed calculations of the last 10 years or so. We summarize here the main results of these theoretical studies. In cold gas (sound speed <1 km s^{-1}) with no or weak magnetic fields, a fast (>50 km s^{-1}) pressure disturbance creates a J-shock, where temperature, density and flow velocity suffer a virtually discontinuous jump across the shock front. Figure 14(top) gives a schematic of the change of the variables across the front. Conservation of mass and momentum (and energy for an adiabatic shock) yield relations between pre-shock variables and downstream variables. These Rankine-Hugoniot relations are, for instance, discussed in Spitzer's (1978) book. Immediately behind a fast shock in a molecular cloud the temperature is high,

$$T_s \approx 1.5\times10^5 \, (v_{shock}/100 \text{ km/s})^2 \quad [K] \qquad (2).$$

Consequently the gas dissociates and ionizes. It radiates in the UV and visible wavelength bands in resonance, semi-forbidden and forbidden lines of hydrogen, helium and the ions of oxygen, carbon, sulfur and so forth. Further downstream ($N_H \geq 10^{16}$ cm^{-2}), the hydrogen ionizing photons are absorbed, creating a 10^4 K temperature plateau. Hydrogen recombination lines and [NeII] fine structure lines originate there. Further downstream again ($N_H \approx 10^{20}$ cm^{-2}), the gas recombines and cools rapidly. Molecular formation sets in at temperatures of a few hundred K. As a result of the high temperatures, neutral-neutral chemical reactions proceed rapidly. Carbon is efficiently converted to CO. The remaining oxygen is converted to OH and H_2O. In dense gas ($n(H) \geq 5\times10^4$ cm^{-3}), the H_2 formation energy is converted into gas heating and creates another temperature plateau at $\approx$400 K (Neufeld and Dalgarno 1989a). The calculations show that much of the infrared emission comes from this plateau and from the temperature region $<10^4$ K where the gas is about 100 times larger than in the pre-shock gas. Typically 1 to 10% of the shock's energy emerges in infrared lines. Hollenbach and McKee (1989) and Neufeld and Dalgarno (1989a,b) have calculated the chemical and thermal structure and cooling of dissociative shocks with speeds between 40 and 150 km s^{-1} at pre-shock densities between 10^3 and 10^6 cm^{-3}. At medium densities ($\approx10^4$ cm^{-3}) and velocities, the dominant coolants are [OI] 63μm, [OI] 6300 Å and [CI] 9849 Å, followed by H_2 and CO rotational and ro-vibrational line emission. At densities of 10^6 cm^{-3} or greater the rotational line emission of H_2O and OH dominates and grain cooling becomes important. The [FeII] lines at 1.3/1.6μm, along with vibrational transitions of H_2, are the most prominent emission lines in the near-infrared range. Infrared hydrogen recombination lines of atomic hydrogen is also detectable from J-shocks, but may easily be confused with HII regions along the line of sight. Depending on the gas phase abundance of silicon, [SiII] 35μm emission may also be easily detectable. Submillimeter emission of highly excited rotational states of SiO, HCN, CN, SO and NO may be characteristic tracers of J-shocks in dense clouds (Neufeld and Dalgarno 1989b).

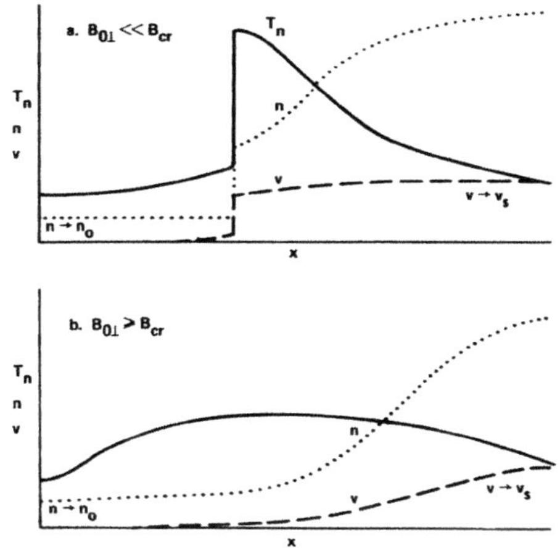

Fig. 14: Shock structure in J-and C-shocks (from Hollenbach 1989, see text)

C-Shocks

A different type of shock occurs if shock velocities are less than about 40 km s^{-1}, if there is a moderately strong magnetic field and if the ionization fraction in the cloud is moderately low ([ions]/[hydrogen]$\leq 10^{-6}$). In the presence of a magnetic field disturbances propagate at the magnetosonic, or Alfven speed v_A, given by

$$v_A = \{B^2 / (4\pi\rho)\}^{1/2} = 22 \, (B/1\text{mG})(n/10^4\text{cm}^{-3})^{-1/2} \, [\text{km s}^{-1}] \quad (3),$$

where ρ and n are the mass and number density of the gas. Shocks are possible for $v_A < v_{shock}$, otherwise the pressure disturbance is "communicated" and damped by Alfven-waves. In partially ionized gas, the ions react rapidly to changes in the magnetic field and then communicate those more slowly to the neutrals by ion-neutral collisions. Because of this difference in reaction speed, it is possible to transmit damped Alfven waves in the ion "fluid" at the ion magnetosonic speed, $v_A(\text{ion}) = (\rho/\rho(\text{ion}))^{1/2} v_A >> v_A$. For small ion fraction one may then have the situation $v_A < v_{shock} < v_A(\text{ion})$. In that case magnetic field and ion density $\rho(\text{ion})$ must vary continuously through the shock front. If the neutrals (because of their interaction with the ions) also vary continuously, the shock is called a "C-shock (for continuous, Draine 1980). This situation is depicted in the lower part of Fig.14. The shock sends a message to the upstream gas via the ions and magnetic field. The ions begin to compress and accelerate, so that they drift relative to the neutrals and heat and accelerate them. The neutral flow will then remain continuous as well if the heated neutral gas can rapidly radiate its thermal energy away. In practice this happens if molecules are not dissociated (T< a few 10^3 K) and the efficient cooling of H_2, CO, OH and H_2O is available. C-shocks radiate nearly the entire energy of the shock in many molecular and atomic infrared lines (Draine, Roberge and Dalgarno 1983). [OI] 63μm emission dominates the cooling for slow (10 km s^{-1}) shocks. Rotational and ro-vibrational line emission of H_2 is the main coolant for higher velocities ($10<v_{shock}<50$ km s^{-1}), but moderate pre-shock densities. For densities of 10^6 cm^{-3} or larger, H_2O and OH rotational emission become more and more important.

J-shocks vs. C-shocks vs. PDR's

How can J- and C-shocks be observationally distinguished and how can shocks be distinguished from photodissociation regions? In comparison to J-shocks, C-shocks have essentially no infrared emission from ionized species ([NeII], [FeII], [SiII], hydrogen recombination lines). Furthermore, C-shocks have much larger column densities of moderately high temperature (10^3 K) molecular gas, resulting in a relatively high ratio of far-infrared CO to near-infrared H_2 cooling. Another potential way of distinguishing J- from C-shocks is from the line profiles. In J-shocks, the cooling gas has been fully accelerated and thus emits a narrow line displaced from the velocity of the quiescent cloud (by the shock velocity). In C-shocks, there is a smooth, broad emission profile centered on the velocity of the quiescent cloud, as most of the emission occurs when the gas is just being accelerated.

Perhaps the most definite indicator of a photodissociation region vs. an (ionizing) J-shock are the absolute and relative intensities of the [CII] 158μm and [OI] 63μm fine structure lines. In shocks one finds weak [CII] emission (or no [CII] emission in C-shocks), but strong [OI] emission (I([OI])/I([CII])>>10), while the reverse tends to be true in PDR's. Another commonly used method is the ratio of the H_2 v=2→1 to v=1→0 S(1) line fluxes. If the ratio is small ($\leq$0.1), the emission is ascribed to shocks. If the ratio is $\approx$0.5, it is ascribed to a PDR. This is not generally a good indicator, however, as a small ratio merely indicates thermalized emission at a temperature not much exceeding $\approx 10^3$ K. A small v=2→1/1→0 ratio can also occur in a dense ($n(H_2) > 5\times10^4$ cm^{-3}) photodissociation region (Sternberg 1988, Sternberg and Dalgarno 1989).

2. Observations

The best investigated region is the shock in the Orion-KL star forming region. Since the first detection of H_2 ro-vibrational emission by Gautier et al. (1976), many infrared lines have been definitely shown to come from shock excited gas. Fig. 15 is a composite of the observed line emission. The shocked zone has a size of about one arcminute (0.15 pc) and is centered on the Orion-KL infrared cluster. The shock is created as a (clumpy) molecular outflow emanating from the young stars in the cluster (IRc2, BN) strikes the surrounding molecular cloud. Chernoff et al. (1982) and Draine and Roberge (1982) have successfully modeled most of the emission as coming from a 40 km s^{-1} C-shock, with a magnetic field of about 1 mG and a pre-shock hydrogen density of 10^5 to 10^6 cm^{-3}. The momentum flux required to drive the shock ($M\ v_{shock} \approx 10^{30}$ g cm/s^2) is quite consistent with other estimates of the mass outflow from IRc2. It is apparent from Fig. 15 that the qualitative features of the shocked gas in Orion-KL in fact closely match the theoretical expectations for a C-shock, as they were described in the last section. H_2 and CO infrared line emission are the dominant coolants ($L(H_2+CO) \approx 500\ L_\odot$). The H_2 emission comes from thermalized gas at a temperature near 2000 K. There is little ionization.

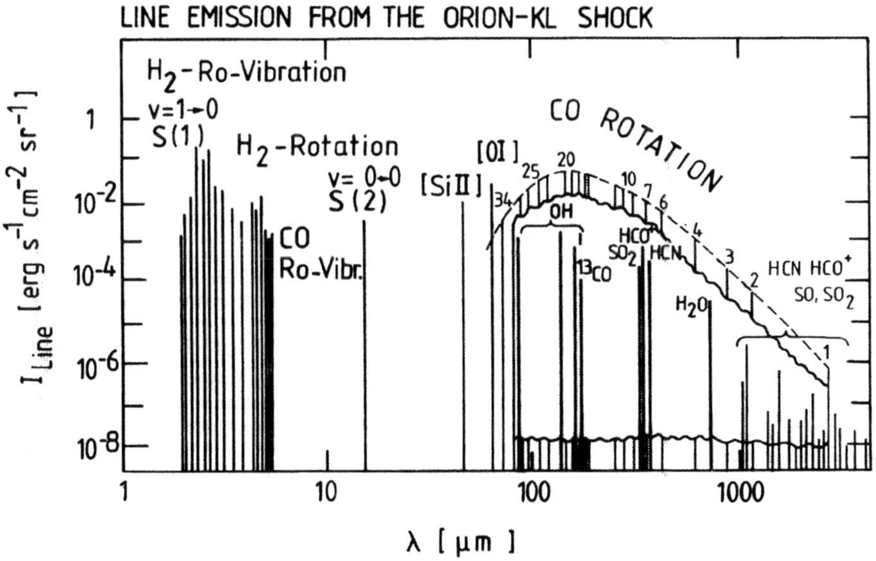

Fig. 15: Line Emission from the Orion-KL shock (dereddened and for a ≈30" beam).

While much of the infrared line emission in Orion-KL can be fit quite plausibly by a single C-shock, there remain several questions. The C-shock models cannot explain the strong [SiII] (and probably also the [OI]) emission and an additional J-shock (perhaps the "wind-shock, Hollenbach 1989) may be required. Brand et al.(1988) have recently argued that the H_2 lines can also be better explained by a J-shock. Finally the simple "cloud" C-shock models can also not account for the high velocity wings in the H_2 lines (e.g.Geballe et al. 1986). These high velocity wings may be created in collisions of high velocity cloudlets in the wind.

Chemistry

Fig.15 shows intense millimeter line emission from many sulfur- and oxygen-rich molecules. Fig.16 is a comparison of the resulting abundances derived for the shock (the "plateau") relative to those in the quiescent cloud (the "ridge"). This figure is adapted from the paper by Blake et al.(1987) who present an analysis of the Caltech 200 GHz line survey of the Orion region. The result is fairly clear and quite consistent with high temperature chemistry. Saturated molecules (HCN, H_2S) or molecules which are preferentially formed at high temperatures (SiO) are one to two orders of magnitude more abundant than in the quiescent cloud where ion-molecule chemistry dominates (see Irvine, this volume). Melnick et al.(1989) derive an OH/H_2 abundance of $2x10^{-5}$ from observations of far-infrared OH lines from the shock, again about 2 orders of magnitude greater than in the quiescent cloud and consistent with the theoretical predictions. One important remaining test is the observation of infrared H_2O lines which ought to be very bright as well ($\geq 10^{-1}$ erg s^{-1} cm^{-2} sr^{-1}: Neufeld and Melnick 1987).

V. SUPERNOVAE IN DENSE INTERSTELLAR CLOUDS

If a supernova explodes inside a dense molecular cloud, the resulting remnant will enter its radiative phase at an early time. A significant fraction of its 10^{51} erg kinetic energy will be radiated in form of soft X-rays which in turn will ionize, dissociate and heat the surrounding molecular cloud. If the cloud is dense and has a large column density, much of the cooling radiation emerges in the infrared. Since the X-rays can penetrate quite far into the molecular cloud, a significant fraction of the cooling may be in the 2μm lines of H_2. Draine and Woods (1989) have recently calculated the chemical and thermal structure, cooling and time evolution of a supernova exploding in a cloud of density $\approx 10^3$ cm^{-3}. They find that $9x10^{49}$ erg are emitted in form of H_2 line photons, including $6x10^{48}$ erg in the v=1→0 S(1) line. They also find a large abundance of H_3^+ ions. Ro-vibrational H_3^+ emission may in fact be an excellent telltale for this mechansim (Draine, priv. comm.).

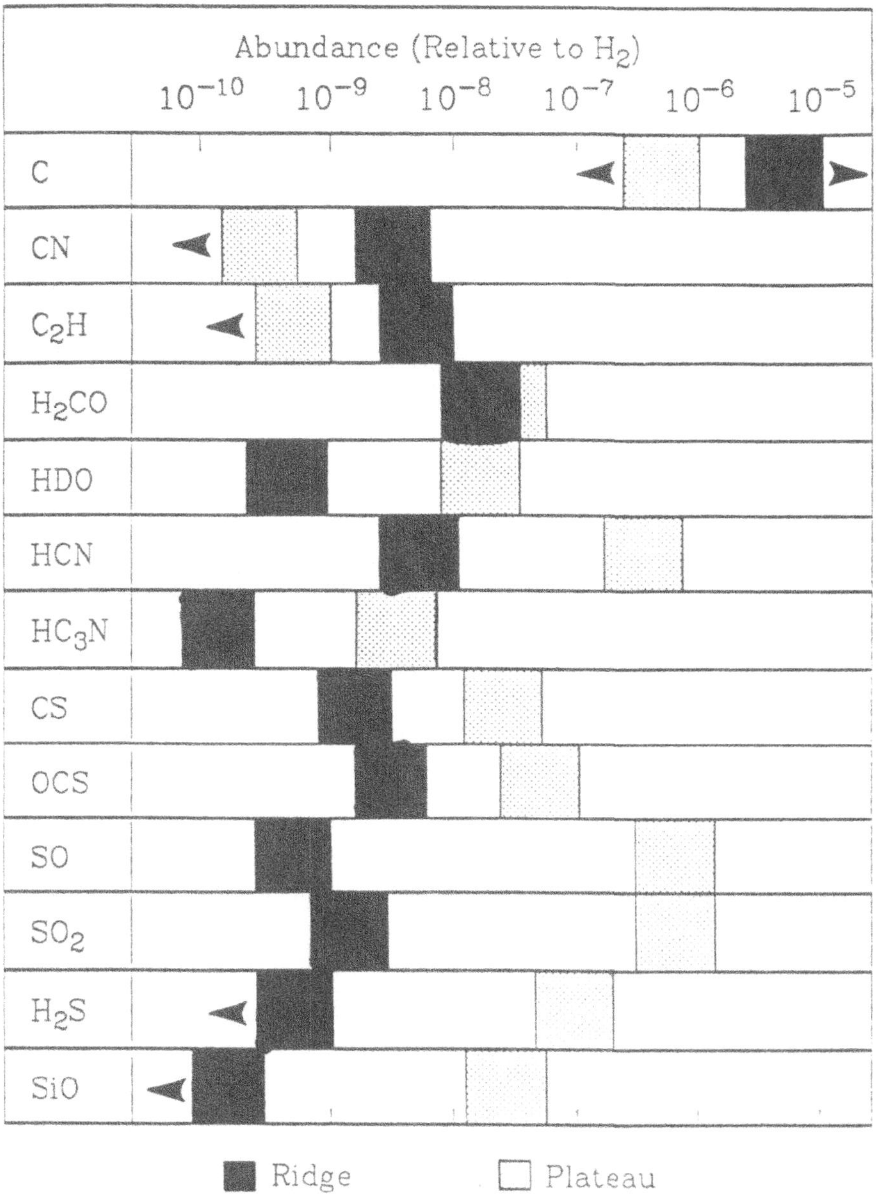

Fig. 16: Molecular abundances in the shocked ("plateau") and quiescent ("ridge") gas components of Orion-KL, derived from the 200 GHz CALTECH spectral survey (adapted from Blake et al. 1987).

Oliva and Moorwood (1989) find observational evidence for strong and extensive 1.3μm and 1.6μm [FeII] emission, as well as 2μm H_2 emission from supernova remnants. Draine (priv.comm.) and Moorwood (1989) propose that a large number of supernova remnants is the explanation of the intense H_2 line emission seen in galaxies like NGC 6240 (Rieke et al. 1985).

VI. CONSEQUENCES FOR THE INTERSTELLAR MEDIA OF STAR BURST GALAXIES

We will show in this section that the interstellar media in the nuclei of star burst galaxies are much more like the gas in star forming clouds than in the average disk clouds of our Galaxy. For the latter the results from several CO surveys have indicated typical kinetic temperatures of 10 K and volume averaged hydrogen densities of 200 cm^{-3} (e.g. Scoville and Sanders 1987). Local densities may be 5 to 10 times higher.

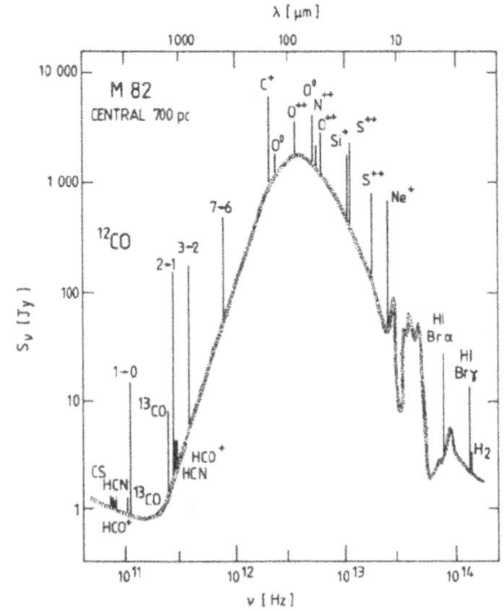

Fig. 17: Composite line and continuum spectrum of the central 45" (700 pc) of the star burst galaxy M82.

Fig.17 is a composite 1 to 3000 μm spectrum of the central 700 pc of the irregular galaxy M82. The characteristics of that spectrum are remarkably similar to that of the central few pc of our own Galaxy (Genzel et al. 1985), or of a luminous Galactic star forming region. The far-infrared luminosity is about $3x10^{10}$ $L_\odot$ and is probably due to a star forming activity that is 100 times greater than in the corresponding volume in our own Galaxy (Rieke et al. 1980). The spectrum also shows intense infrared line emission. The luminosity of the [CII] and [OI] lines alone is about $2x10^8$ $L_\odot$, or about 0.7% of the total luminosity. The ratio of [OI] to [CII] fluxes indicates hydrogen densities of about 10^4 cm^{-3} and gas temperatures of about 200 K (Lugten et al. 1986). The ionized gas has a density of about 100 cm^{-3} and is in approximate pressure equilibrium with the atomic gas (Houck et al. 1984, Duffy et al. 1987, Lugten et al. 1986). The molecular interstellar medium of M82 is highly excited. The CO emission indicates gas temperatures of about 40 K (Olofsson and Rydbeck 1984, Lo et al. 1987, Carlstrom 1989). Harris et al.(1990) and Wild et al. (1990) infer hydrogen densities of 10^4 to 10^6 cm^{-3} from the ratio of 3→2 to 1→0 HCN rotational lines. Harris et al.(1990) report the detection of intense CO 7→6 line emission which indicates the presence of a circum-nuclear "supercloud" complex (mass a few 10^7 $M_\odot$), with gas temperatures near 100 K. 3 to 12μm emission dust features in Fig.17 come from hot (1000 K), transiently heated small dust particles throughout the nucleus of M82.

These findings are in very good agreement with the star burst model first presented by Rieke et al.(1980) and clearly demonstrate the effects of the high star formation rate on the interstellar medium. About 20% (Table I) of the interstellar gas mass is directly affected by photoionization and UV heating (the corresponding number in our own Galaxy is 1 to 2%). The large energy density in the radiation field, and in cloud motions and cosmic rays due to stellar winds and supernovae explosions has heated the overall neutral interstellar medium to between 40 and 200 K. The interstellar gas is also compressed to densities at least an order of magnitude greater than in our own Galaxy. X-ray measurements show that there is a coronal medium with scale heights of several hundred parsec that probably indicates supernova-induced mass outflow.

One might speculate what consequences on present and future star formation this unusual state of the interstellar medium in M82 will have. First, the high temperatures and densities probably favor the formation of rather more massive stars, as was already inferred by Rieke et al. Second, the neutral interstellar clouds may eventually be dispersed and completely disrupted, thus preventing further star formation and bringing the star burst to an end.

While M82 is quite remarkable, it is by no means unique and may in fact be typical for other star burst galaxies. In support of this conclusion, we show in Fig.18 the [CII] flux as a function of ^{12}CO 1→0 flux in a number of Galactic and extragalactic sources (Stacey et al. 1990). The top diagram shows Galactic star forming regions and star burst galaxies. There is a pretty good linear correlation between CO and [CII] lines over two order of magnitude in line flux. The most likely interpretation is that in both types of sources the [CII] and CO lines sample warm gas in photodissociation regions at the surfaces of molecular clouds (Crawford et al 1985). Following Stacey et al. (1990) and Wolfire et al.(1990), the interpretation of the linear relationship is that the ratio of [CII] to CO brightness in each (photodissociation) region is about the same and that the beam filling factor (of PDR's) varies by about two orders of magnitude between the weakest (M51 HII) and strongest (M42) region. The existence of a linear correlation between [CII] and CO also makes it highly questionable whether CO can be used as a reliable tracer of molecular mass in galaxies with a high rate of star formation (Crawford et al.1985, Stacey et al. 1990). This issue is discussed in more detail in DISCOURSE II.

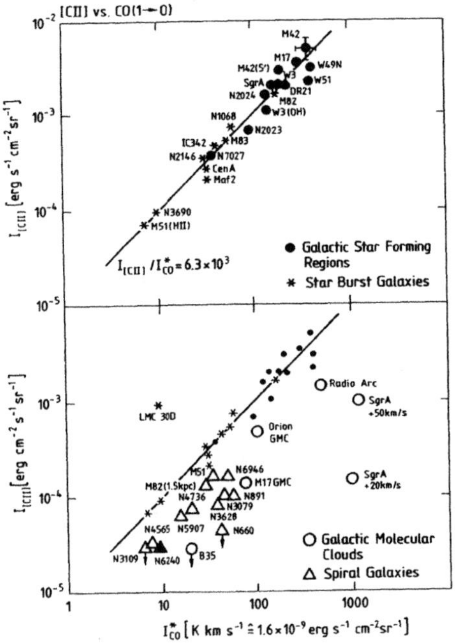

Fig. 18: [CII] vs. ^{12}CO 1→0 line fluxes in Galactic star formation regions (top, filled circles), Galactic clouds with lower star forming activity (bottom, open circles), star burst galaxies (top, asterisks) and normal spiral galaxies (bottom, open triangles). The ultra-luminous galaxy NGC 6240 is indicated by a filled triangle (from Stacey et al. 1990).

The bottom part of Fig.18 shows Galactic clouds and external galaxies with a lower star formation activity. They all fall below the line marking the correlation in star forming regions. The likely interpretation of this fact is that in these objects the gas temperatures and excitation of the interstellar medium is much lower (Stacey et al.1990). The [CII] line is thus correspondingly weaker. The CO line intensity does not drop by the same factor, as it samples now colder gas which does not emit detectable [CII] emission. In contrast, the low metallicity star forming cloud 30 Doradus in the LMC is significantly above the curve (Storey et al. 1990). A probable explanation of this fact is that the relative mass fraction of photodissociation regions is greater in low metallicity environments (Maloney et al. 1989).

DISCOURSE I

Excitation and Radiative Transport of Infrared, Submillimeter and Millimeter Spectral Lines

1) Atomic Fine Structure Lines

Atoms and ions with 1,2,4 and 5 p electrons in their outermost shell have fine structure in their electronic ground state as a result of the magnetic coupling between orbital angular momentum and spin. The splitting is typically 0.01 to 0.1 eV, resulting in transitions in the 100 to 10 μm region. They are magnetic dipole transitions with transition probabilities typically $\alpha^2 = (1/137)^2$, or 4 orders of magnitude smaller than those of electric dipole transitions. The upper fine structure levels are thus relatively easily excited in dense interstellar clouds by collisions with hydrogen atoms and molecules (in neutral clouds), or protons and electrons (in ionized or partially ionized regions). Since line optical depths are proportional to the above transition probability, infrared fine structure lines are usually optically thin ($\tau \approx 10^{-2}$ to a few).

As an example, let us consider the ground state of singly ionized carbon (^{2}P). The upper (J=3/2) and lower (J=1/2) fine structure levels are split by 78meV (157.7409 μm) with an Einstein A coefficient between them of $A_{ul} = 2.4 \times 10^{-6}$ s^{-1}. In the optically thin limit the line flux (erg s^{-1} cm^{-2}) is given by

$$F_{line} = (h\nu/4\pi)\ A_{ul} \iiint n_u\ dl\ d\Omega \qquad \text{(D1)}.$$

Here ν is the frequency, n_u is the number density (cm^{-3}) of ions in the upper state and the integration is over the beam (or source) solid angle and the line of sight. The population in the upper state is given by detailed balance between upward collisions (rate per ion per sec C_{lu}) and the sum of downward collisions (C_{ul}) and spontaneous emission (A_{ul})

$$n_u\ (\ A_{ul} + C_{ul}\ ) = n_l\ C_{lu} \qquad \text{(D2)}.$$

The up and down collisional rates at kinetic temperature T are related through detailled balance by

$$C_{lu}/g_u = C_{ul}/g_l\ \exp\ (-h\nu/kT) \qquad \text{(D3)},$$

the fractional population in the upper state is given by

$$\frac{n_u}{n_u+n_l} = \frac{g_u/g_l \exp(-h\nu/kT)}{1 + g_u/g_l \exp(-h\nu/kT) + A_{ul}/C_{ul}} \qquad (D4).$$

In equations (D3) and (D4) g_u and g_l are the statistical weights of the upper and lower levels. The collision rates C can be written as a product of the space density of collision partners, n (cm^{-3}), and a collissional rate coefficient γ ($cm^3\ s^{-1}$). There are obviously two regimes. Whenever n is smaller than $n_{crit}= A_{ul}/\gamma_{ul}$, the line intensity is proportional to emission measure ($\iiint n_u n\ dl\ d\Omega$). This is the subthermal regime; every collision leads to photon emission. For $n>n_{crit}$ the collisions are fast enough to thermally populate the upper level, and the photon emission rate is limited by the Einstein A coefficient. In that case, the line flux is proportional to column density in the upper state ($\iiint n_u\ dl\ d\Omega$). The critical density of the 158μm [CII] transition, for example, is about $4 \times 10^3\ cm^{-3}$ for excitation by atomic or molecular hydrogen. Note that the 1→0 transition of CO has a similar critical density, thus suggesting that these two transitions probe atomic and molecular abundances in similar regions. Table DI lists the parameters of the infrared fine structure lines of the most common atoms and ions. Most, but not all of these lines have now been observed in interstellar space.

Fine structure lines as probes of physical conditions in clouds

In addition to giving important handles on luminosities and spatial distribution of abundant atoms and ions, infrared fine structure lines give valuable information on the density, temperature and ionization structure of dense clouds with little or no dependence on dust extinction that hampers observations in the visible.

For instance, if an atom or ion has a triplet ground state (O^{++}, S^{++}, N^+, C^0, O^0) there are two transitions of the same species that have different critical densities (Table DI). In the density range where the longer wavelength transition is already thermalized, but the shorter wavelength transition is not, one can obtain a fairly reliable estimate of the density n from the line intensity ratio (cf. Watson and Storey 1980, Simpson 1975, Watson 1984, Pipher et al. 1984, Lester et al. 1987, Stacey 1989). Inspection of Table DI shows that the critical densities of infrared fine structure lines are well matched to the volume densities typically occuring in interstellar clouds. These lines will thus often be strong and thermalized. The Table also shows that one can "tune" to the desired density range by choosing line pairs located in an appropriate wavelength band ($n_{crit} \propto \lambda^3$). If in addition the kinetic temperature T is less than $h\nu$, the line ratio is also sensitive to temperature. A third line is then required to simultaneously determine density and temperature. Watson (1984) describes, for instance, how the double ratio of [OI] 63μm/[OI] 145μm and [OI] 63μm/[CII] 158μm can be used to determine the physical conditions in neutral, atomic clouds (see also Genzel, Harris and Stutzki 1989).

Another important application is the determination of the effective temperature of the UV radiation field in an HII region, giving important clues on the spectral type(s) of the exciting star(s). Here one takes advantage of the temperature dependence of the relative fluxes of ionic lines sampling different ionization stages (Lacy et al. 1980, Watson and Storey 1980). Briefly, in a moderately dense ($n_e \approx 10^4$ cm^{-3}) HII region excited by an O6.5 star ($T_{eff} \approx 40{,}000$ K) the strongest fine structure lines are 10.4μm [SIV], 18.7μm [SIII] and 51.8μm [OIII]. In a low density ($n_e \approx 10^2$ cm^{-3}) HII region surrounding a B0 star, on the other hand, the 33.5μm [SIII] line is strongest, followed by 18.7 μm [SIII], 121.9μm [NII] and 12.8μm [NeII]. The line ratios can change by several orders of magnitude from one case to the other. Detailed model calculations spanning a wide range of HII region parameters have been presented by Rubin (1985). Once the ionization structure is determined, the measurements can also be used to derive elemental abundances in HII regions (Lester et al. 1987, Herter 1989).

2) Molecular Lines

Molecular lines as probes of physical conditions in clouds

The infrared and submillimeter range contains the rotational and ro-vibrational lines of molecules. With the exception of homonuclear molecules such as H_2, O_2 and N_2 most molecules have a nonzero permanent electric dipole moment. Transition probabilities for electric dipole transitions are, therefore, large ($A \approx 0.1$ to 1 s^{-1} at 100μm) and radiative transport/optical depth effects usually are important. The optical depth for transitions at different wavelengths scales as

$$\tau \propto \mu^2 N_u (\exp\{-h\nu/kT\} - 1) \quad \begin{cases} h\nu \ll kT \quad \tau_{radio} \approx \mu^2_{rot}(h\nu/kT) & \text{mm/submm} \\ h\nu \geq kT \quad \tau_{FIR} \approx \mu^2_{rot} & \text{FIR rotation} \\ \text{IR ro-vibr.} \quad \tau_{IR} \approx \mu^2_{vib} & \end{cases} \qquad \text{(D5).}$$

N_u is the column density of molecules in the upper state, μ_{rot} is the transition moment for rotational transitions which is within a scale factor of order unity identical to the permanent electric dipole moment (typically 1 to 3 Debye, with the notable exception of CO: 0.1 Debye). μ_{vib} is the transition moment for ro-vibrational transitions and is of the order of 0.1 Debye. The result of equation (D5) is that far-infrared (100μm) rotational transitions, should they exist, have typically 100 times larger optical depths than either mm/submm transitions or mid-infrared (10 μm) ro-vibrational transitions. Far-infrared transitions from low-lying rotational levels thus couple efficiently to the far-infrared radiation field (molecules such as OH, CH, H_2O, NH_3). Infrared radiative pumping has been shown to be important, for instance, for a number of molecular lines in the Orion-KL "hot core", a region of dense molecular gas close to the luminous infrared source IRc2 (Genzel and Stutzki 1989 and references therein).

The concept of critical density described above for fine structure lines, of course, also approximately applies to molecular lines. Since the Einstein A coefficient depends on the transition moment squared and wavelength cubed, one obtains the following relationship

$$n_{crit} = \frac{A_{ul}}{\gamma_{ul}} = \frac{64\,\pi^3\,\nu^3(\mu_1)^2}{3\,h\,c^3\,\gamma_{ul}} = 2x10^9\,\lambda_{100}{}^{-3}\mu_1{}^2 \qquad (D6)$$

In this equation λ_{100} is the wavelength in units of 100μm, μ_1 is the transition moment in units of Debye and we chose γ_{ul}= 5x10^{-11} cm^3 s^{-1}. It is apparent that the density that is sampled by a given transition depends on dipole moment of the molecule and the wavelength of the transition. For simple linear molecules with rotational states of quantum number J, $\lambda \propto J$ and $n_{crit} \propto J^3$. For instance, the critical density is 4x10^3 cm^{-3} for the 2.6mm 1→0 transition of ^{12}CO (μ=0.12 Debye), 1.6x10^5 for the 371μm 7→6 transition and 1.6x10^7 cm^{-3} for the 77μm 34→33 transition. This sharp increase of n_{crit} with J allows the determination of cloud densities from measurements of different rotational (or ro-vibrational) lines, if there is one temperature component determining the emission at different J, if the excitation is collisional and if one or several transitions are subthermally populated. The same measurements will naturally also give an estimate of the cloud temperature, as the rotational quantum number where the strongest line emission for a molecule of rotation constant B (Hz) scales with rotation temperature T_{rot} as $J_{max} \approx (kT_{rot}/hB)^{1/2}$.

Fig.D1 gives an overview of molecular lines as probes of density and temperature (see also the discussion by Scoville 1984). For molecular hydrogen densities between a few 100 and 10^6 cm^{-3} and temperatures between 10 and 100 K, mm and submm transitions of CO or other heavy top molecules are ideally suited. Note that the same part of the parameter space can also been investigated with CO ro-vibrational lines in absorption against hot continuum sources. Submm and far-infrared transitions of heavy top molecules and far-infrared transitions of light hydrides sample very dense (10^7 to 10^9 cm^{-3}) and warm (a few 10^2 K) gas. Finally, molecular masers (OH,H_2O, SiO) and near-infrared emission lines from CO probe hot gas (10^3 K) with stellar densities (10^{10} to 10^{12} cm^{-3}). The H_2 molecule is an exception, as it doesn't have a dipole moment and even its 2μm ro-vibrational levels are thermalized at moderately low densities

($\approx 10^5$ cm^{-3}). Infrared line emission in the quadrupole transitions of H_2 can thus be observed from hot (10^3 K) gas of moderately low density (Shull and Beckwith 1982).

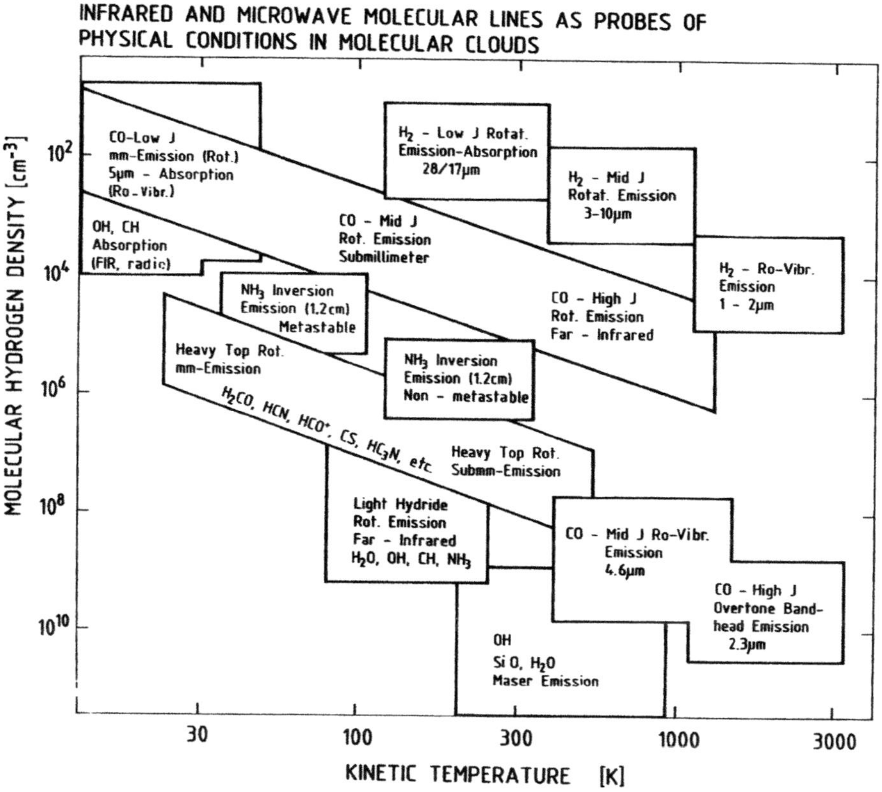

Fig. D1: Molecular lines as probes of physical conditions in molecular clouds.

Line Trapping

The discussion up to now has not taken into account two important aspects for infrared line formation of molecules. The first is that in most cases collisional transitions, and often radiative transitions, connect not just two, but several energy levels. A quantitative analysis of the population requires evaluation of rate equations involving all possible radiative and collisional channels. The second aspect missing so far is the inclusion of optical depth and trapping.

In order to qualitatively understand the effect of optical depth, let us consider the simple case of a cloud (volume V) emitting line photons in a thermalized transition. If the cloud is optically thin, a photon emitted anywhere within the volume of the cloud will escape without being reabsorbed. In this case the cloud's total line emission per sec is proportional to the number of molecules in the cloud and the spontaneous emission rate, $I_{thin} \propto n_u V A (s^{-1})$. n_u is again the volume density of molecules in the upper state of the transition under consideration, and A the Einstein coefficient. If the optical depth of the cloud is greater than one as the result of larger molecular volume or column densities or smaller velocity/frequency gradients through the cloud, the probability for a once emitted photon to escape without reabsorption is going to be less than one. Roughly speaking, the escape probability β will scale inversely proportional to the optical depth τ, $\beta \approx 1/\tau$. Once a photon is reabsorbed in our simple model, it will be "destroyed" in a collisional deexcitation and not contribute to the net photon emission. The total line emission rate in the optically thick case is then $I_{thick} \propto n_u V A\beta$. So the effect of "line trapping" in essence scales the spontaneous emission rate by β,

$$A \rightarrow A\ \beta(\tau) \approx A/\tau \qquad \text{(D7)}.$$

Another effect of line trapping is that it drives a level toward thermalization that would be subthermal in the optically thin case. This can be immediately seen from the fact that for a fixed collision rate C, the ratio of effective spontaneous emission rate to collision rate $A\beta/C = n_{crit}\beta/n$ is smaller for the optically thick ($\beta<1$) than for the optically thin ($\beta=1$) case. The generalization of the equation for line emission from a collisionally excited two level system (equations D 1 and 4) for the optically thick case thus is

$$F_{line} = h\nu/4\pi\ A\beta(\tau)N_{mol} \frac{g_u/g_l \exp(-h\nu/kT)}{1+g_u/g_l \exp(-h\nu/kT)+n_{crit}\beta(\tau)/n} \qquad \text{(D8)}.$$

N_{mol} is the total column density of molecules (or atoms, as equation D8 applies to fine structure lines as well). For optically thick, thermalized emission ($n_{crit}\beta/n << 1$) the line flux scales with $A\beta$, as discussed above. In the case of subthermal, optically thick emission ($n_{crit}\beta/n > 1$ and $\beta<1$), however, the line flux keeps growing linearly with N_{mol} since the β's in the numerator and denominator cancel each other. So an optically thick, but subthermal line behaves much like an optically thin and subthermal line. The difference is that the line trapping results in a greater population in the upper state and a higher excitation temperature T_{ex} ($n_u/n_l = g_u/g_l \exp(-h\nu/kT_{ex})$.

Line profiles

Line profiles contain important physical information as well. Let us first consider the simple case of a homogeneous cloud with a Gaussian profile of FWHM width Δv_{local} representing the thermal line width or microturbulence. The line profile as a function of velocity shift from line center v can be written as

$$I_{line}(v) = I_0(T_{ex}) \{ 1 - \exp(-\tau(v)) \} \qquad (D9),$$

where $\tau(v) = \tau_0 \exp(-\{v/\Delta v_{local}\}^2)$.

Increasing the line center optical depth τ_0 to greater than 1 than changes the profile from Gaussian to flat-topped with a width that depends logarithmically on τ_0 (c.f. Phillips et al. 1979). A line center optical depth of 10, for instance, results in a FWHM line width 1.96 times as large as in the optically thin case, τ_0=100 results in 2.7 and 1000 in 3.24 times the width .

The simple single Gaussian model often is not a good description of molecular cloud kinematics. The clouds are highly clumped (sections II.2 and III.2), with large scale motions significantly larger than the local motions. This will tend to decrease the dependence of line width on optical depth by approximately the factor $\Delta v_{cloud}/\Delta v_{local}$ where Δv_{cloud} is the FWHM width of the large scale motions. Martin, Sanders and Hills (1984) have calculated models of the line profiles in clumpy clouds in the simplified case that large scale motions (macroturbulence) and local line profile (microturbulence) can both be described by Gaussians. They have shown that such models can plausibly fit the ratios of line intensities and widths of various CO isotopic in the warm molecular cloud core M17.

DISCOURSE II

ESTIMATES OF CLOUD MASSES AND COLUMN DENSITIES

The main fraction of the mass of dense interstellar clouds is in form of molecular hydrogen. Atomic hydrogen is the dominant species in diffuse and translucent clouds, but contributes only a small fraction to the mass of dense molecular clouds. While atomic hydrogen can be easily detected via its 21cm hyperfine transition, molecular hydrogen can currently be observed directly only in special circumstances, such as through its infrared ro-vibrational emission in shocks and photodissociation regions (see sections III and IV), or via UV-absorption in diffuse clouds. In most cases it is thus necessary to determine the mass of a cloud with less direct indicators, such as the submillimeter/millimeter dust emission, through line emission of CO, the most abundant molecule after H_2, and through MeV γ-ray emission. We now briefly discuss these methods in turn.

1. Submillimeter dust emission

About one percent of the mass of interstellar matter is in form of dust grains (see Greenberg, this volume, for a more detailed discussion). The thermal emission of interstellar dust may be used to determine the dust optical depth and then infer a gas column density. Hildebrand (1983) has given a detailed discussion; the main results are summarized here. The flux density of dust emission at temperature T_d at frequency ν (or wavelength λ) is given by

$$S_\nu[\mathrm{Jy}] = 2h\nu^3/c^2 \{ \exp(h\nu/kT_d) - 1 \}^{-1} \{ 1 - \exp(-\tau_d(\nu)) \} \quad \text{(D10)},$$

where $\tau_d(\nu)$ is the dust optical depth at ν. Equation (D10) can be much simplified in the optically thin, Rayleigh-Jeans ($h\nu/kT_d \ll 1$) limit which is typically appropriate for the dust emission from dense clouds at wavelength longward of 100μm

$$S_\nu = 2kT_d\nu^2/c^2 \, \tau_d(\nu) \quad \text{(D11)}.$$

In this limit the flux density is only weakly (i.e. linearly) dependent on temperature and directly proportional to dust optical depth. Measurement of dust optical depth in the submillimeter or millimeter range also has the advantage that it likely is a measure of both warm and cold dust (Mezger et al. 1986). The optical depth depends on the gas to dust ratio and the dust extinction efficiency at ν. The dust extinction efficiency in the submillimeter range is unfortunately only poorly known (see Draine 1989 for a discussion). Whitcomb et al.(1981) have related the dust emission at 400μm to A_v in the reflection nebula NGC7023, giving a absolute scaling from τ(400μm) to A_v. It is also typically assumed that the extinction efficiency in the submillimeter scales with $\lambda^{-\beta}$ with $\beta \approx 1.5$ to 2 (see Hildebrand 1983, Draine 1989, Cox and Mezger 1989 for critical discussions of the uncertainties). A_v can then be connected to the nuclei of hydrogen atoms $N(H+2H_2)$ by using the calibration of Bohlin, Savage and Drake (1978: $N(H+2H_2)=1.9\times10^{21} A_v$) from UV measurements in diffuse clouds, or by assuming a gas to dust ratio (the canonical gas to dust mass ratio of 100 is equivalent to the Bohlin et al. calibration). The resulting "best" relationship near 400 μm then is (Hildebrand 1983, Cox and Mezger 1989, Draine 1989)

$$N(H+2H_2) = 1.2\times10^{25} \, Z/Z_\odot \, \tau_d(\lambda) \, (\lambda/400\mu\mathrm{m})^\beta \quad [\mathrm{cm}^{-2}] \quad \text{(D12)}.$$

Here $Z/Z_\odot$ is the abundance of heavy elements in units of the solar abundances.

2. CO Line Emission

The CO molecule has transitions in the UV (electronic transitions), at 2.3/4.6μm (ro-vibrational) and between 50μm and 2.6mm (rotational). It is also the most abundant molecule after H_2. Table DII is a list of determinations of the CO abundance (and of the fractional abundances of the isotopes of CO) from a variety of different methods (adapted in part from van Dishoeck and Black 1987). The determination of CO column densities in dense clouds is made through infrared, submillimeter and millimeter observations. In these clouds it is probably a safe assumption that the CO/H_2 fractional abundance is close to 10^{-4} (about 30% of carbon in CO). The most common method of determining CO column densities uses measurements of the low-J ($1\rightarrow0, 2\rightarrow1$) transitions in emission. ^{12}CO transitions in dense molecular clouds are very optically thick ($\tau \approx 100$).

Tracing mass with CO isotopes

One way of avoiding this problem is to observe the less abundant (optically thinner) ^{13}CO and $C^{18}O$ lines and multiply by the appropriate abundance ratio. If only one line is measured, an extrapolation of the column density in the measured levels to all other levels is necessary. Often this is done by assuming LTE population at some temperature. The critical density of the excited states increases rapidly with J ($n_{crit} \approx J^3$), so that typically only the lowest three or four levels are populated. Excitation/radiative transport calculations show that the integrated intensities of the $2\rightarrow1$ transitions of ^{13}CO (if optically thin) or $C^{18}O$ are a good measure of column density and more or less independent of temperature over a reasonably wide parameter range (a few $10^3 < n(H_2) <$ a few 10^6 cm^{-3} and $15 < T < 80$ K). With $[C^{18}O]/[H_2] = 1.7 \times 10^{-7}$ (Frerking et al. 1982) one then finds (Stutzki et al. 1990)

$$N(H_2) = 4 \times 10^{21}\ I(C^{18}O\ 2\rightarrow1) \quad [cm^{-2}]/[K\ km\ s^{-1}] \qquad (D13).$$

Similar relationships hold for optically thin ^{13}CO or $C^{17}O$ lines; one has to just choose the appropriate abundance (Table DII). The $1\rightarrow0$ transitions can also be used, although they are more sensitive to temperature; in that case multiply the coefficient on the right by a factor of four.

Tracing mass with ^{12}CO $1\rightarrow0$

In many cases the isotopic lines are too weak for measurement. A column density is then inferred from the optically thick ^{12}CO line(s) by assuming either that the integrated line flux is proportional to the filling factor of clouds of a constant flux, or by applying the virial theorem (e.g. Sanders and Scoville 1987). This method is widely used for external galaxies, for instance. In the following, the physical basis and possible pitfalls of the "virial" method is discussed.

Following van Dishoeck and Black (1987), let us consider an ensemble of clouds (or clumps) moving around each other or around the center of a galaxy with an overall velocity width Δv_{source} (FWHM). Assume that each cloud has an intrinsic velocity width Δv_{cloud} that is determined by a virial equilibrium (kinetic energy stabilizes cloud against gravitational collapse). Assume further that at each velocity the (beam) area filling factor of clouds $\Phi(v)$ is smaller than 1 (no shadowing). Each cloud emits optically thick ^{12}CO 1→0 lines with an excitation temperature T_{ex}, so that the integrated line flux [K km s^{-1}] is given by

$$I(^{12}CO) = 1/\Omega_b \iiint (T^*_{ex} - T^*_{BG}) (1 - \exp\{-\tau(v)\})\, dv\, d\Omega \quad (D14).$$

Ω_b is the solid angle of the beam. The integration is over the solid angle of the beam and the profile. T^*_{ex} and T^*_{BG} are the Rayleigh-Jeans tempertures corresponding to the excitation and 3K background temperatures,

$$T^* = h\nu/k \{ \exp(h\nu/kT) - 1 \}^{-1} \quad (D15).$$

For T_{ex}>>3 K and for τ>>1, equation (D15) becomes

$$I(^{12}CO) = T^*_{ex}\, \Phi(v)\, \Delta v_{source} = T^*_{ex}\, \Phi\, \Delta v_{cloud} \quad (D16),$$

where $\Phi = \Phi(v)\, \Delta v_{source}/\Delta v_{cloud}$ is the filling factor of clouds irrespective of velocity. The assumption of virialization of each cloud of radius R, mass M(R) and mean (volume averaged) density $\langle n(H_2)\rangle$ means

$$G\, M(R)/ (3R) = \sigma^2(\text{cloud}) + kT_{kin}/\langle m\rangle$$
$$\approx \sigma^2(\text{cloud}) = (\Delta v_{cloud}/2.35)^2 \quad (D17).$$

In this equation G is the gravitational constant. We have assumed that the intrinsic cloud motions are significantly larger than thermal. The beam averaged H_2 column density can be expressed as

$$\langle N(H_2)\rangle = \Phi\, [N(H_2)]_{cloud} = \Phi\, 2R\, \langle n(H_2)\rangle \quad (D18).$$

With $M(R) = 4\pi/3\ R^3 \langle n(H_2)\rangle \langle m\rangle$ ($\langle m\rangle = 4.3\times10^{-24}$ g is the mean mass per hydrogen molecule, including heavier elements), equations (D16), (D17) and (D18) can be combined to give

$$X_{CO} = \langle N(H_2)\rangle / I(^{12}CO) = 3\times10^{20}\ (8\ K/T^*_{ex})\ (\langle n(H_2)\rangle/200)^{1/2} \quad (D19).$$

The values for T^*_{ex} and $\langle n(H_2)\rangle$ chosen in equation (D19) are those quoted by Sanders and Scoville (1987) to be representative for giant molecular clouds in the Galaxy. The conversion factor derived in equation (D19) is the same within a factor of 2 as the ones obtained for galactic disk clouds by several other methods (isotopes, empirical correlation, γ-rays, Table DII). This forms the basis of the common assumption that the above relationship holds universally.

Given equation (D19) and the evidence, presented in sections II, III and VI, of much higher excitation temperatures of the ^{12}CO emitting gas near star forming regions and in the nuclei of star burst nuclei than in the average disk gas, this assumption has to be questioned. Yet the fact that the mean densities also strongly increase in the star forming clouds may "save" the relationship as a first order estimate, albeit with increased uncertainty. For instance, the integrated ^{12}CO 1→0 flux in the +20/+40 km/s clouds near SgrA is about 1600 K km/s. From $X = 3\times10^{20}$ one infers an H_2 column density of 4.8×10^{23} cm^{-2}, remarkably close to the estimates from submillimeter dust continuum and $C^{18}O$ 2→1 flux (3 to 6×10^{23}). This agreement must be fortuitous, as the ^{12}CO line brightness temperature is 47 K (6 times higher than $T^*_{ex} = 8$ K) and since the ^{12}CO line width is 40% broader than ^{13}CO and $C^{18}O$, indicating that the line width is not solely determined by bulk cloud motion. To understand the result, one has to assume that the mean density in these clouds is 70 higher than the 200 cm^{-3} assumed above. This is in fact quite reasonable in the environment of the Galactic Center. This lucky coincidence does not always work, however. In M82, for example, mass estimates via the standard conversion factor come out 3 to 5 times larger than from other lines of evidence, consistent with the greater kinetic temperature of the gas ($\approx$40 K). In the core of M17 it is the other way around. This uncertainty has to be kept in mind when applying the relationship to external galaxies.

Tracing mass with MeV γ-rays

High energy γ-rays (30 to 1000 MeV) in interstellar clouds are produced when cosmic rays collide with hydrogen nuclei. The π-mesons produced in this collision decay into γ's. If the cosmic ray density inside clouds is constant, the γ-ray flux thus measures the number of hydrogen nuclei in a given volume (see Bloemen 1989 for a review). Lebrun et al.(1983), Bloemen et al (1984) and Strong et al.(1988) have used the COS-B galactic plane survey to determine masses and the conversion rate between ^{12}CO flux and H_2 column density. The average conversion factor for the Galactic disk is listed in Table D II. The conversion factor in the Galactic Center may be 10 times lower (Blitz et al. 1985), supporting our conclusions above. The interpretation of the Galactic Center data is still under discussion, however.

Acknowledgements I am grateful to A.Harris and J.Stutzki for a critical reading of the manuscript.

TABLE D I

INFRARED FINE STRUCTURE LINES

LINES FROM H I/H_2 REGIONS [d)]

SPECIES	EXCITATION POTENTIAL [eV]	IONIZATION POTENTIAL [eV]	TRANSITION	λ [μm]	A [s^{-1}]	n_{crit} [cm^{-3}][a)]
C	-	11.26	^3P, J=1→0	609.1354	7.9(-8)	4.7(2) 1(4) for H_2[a)]
			2→1	370.415	2.7(-7)	1.2(3)
C^+	11.26	24.38	^2P, J=3/2→1/2	157.7409	2.4(-6)	2.8(3) 5(3) for H_2 50 for electrons
O	-	13.62	^3P, J=1→2	63.18372	8.95(-5)	$4.7(5)T_{300}^{-1/2}$ (b) $7(5)T_{300}^{-1/2}$ for H_2
			0→1	145.52547	1.7(-5)	$9.5(4)T_{300}^{-1/2}$ $\geq 1(5)T_{300}^{-1/2}$ for H_2[a)]
Si	-	8.15	^3P, J=1→0	129.68173	8.25(-6)	2.4(4)
			2→1	68.473	4.2(-5)	8.4(4)
Si^+	8.15	16.35	^2P, J=3/2→1/2	34.816	2.1(-4)	3.4(5)
S	-	10.36	^3P, J=1→2	25.249	1.4(-3)	1.9(6)
			0→1	56.309	3.0(-4)	7.2(5), but H_2?[a)]
Fe	-	7.87	^5D, J=3→4	24.04	2.5(-3)	3.1(6)
			2→3	34.71	1.6(-3)	3(6)
Fe^+	7.87	16.18	^6D, J=7/2→9/2	25.99	2.1(-3)	2.2(6)
			5/2→7/2	35.352	1.6(-3)	3.3(6)

a) if not otherwise specified critical density for collisions with atomic hydrogen: $n_{crit}=A_{ul}/\gamma_{ul}$; collisional rates with H_2 may be different, see Monteiro, T. and Flower, D. 1987, MNRAS 228, 101, for CI and OI excitation.

b) kinetic temperature in units of 300 K.

LINES FROM H II REGIONS [d)]

SPECIES	EXCITATION POTENTIAL [eV]	IONIZATION POTENTIAL [eV]	TRANSITION	λ [μm]	A [s^{-1}]	n_{crit} [cm^{-3}][a)]
O^{++}	35.12	54.93	^3P, J=1→0	88.356	2.6(-5)	5.1(2)
			2→1	51.815	9.8(-5)	3.6(3)
O^{+++}	54.93	77.40	^2P, J=3/2→1/2	25.87	5.2(-4)	1.0(4)
N^{+}	14.53	29.60	^3P, J=1→0	203.5±0.8	2.1(-6)	48
			2→1	121.89806	7.5(-6)	3.1(2)
N^{++}	29.60	47.45	^2P, J=3/2→1/2	57.317	4.8(-5)	3(3)
Ne^{+}	21.56	40.96	^2P, J=1/2→3/2	12.813	8.6(-3)	5.4(5)
Ne^{++}	40.96	63.45	^3P, J=1→2	15.55	5.99(-3)	2.9(5)
			0→1	36.02	1.2(-3)	4.2(4)
Ne^{++++}	97.11	126.21	^3P, J=1→0	24.28	1.3(-3)	5.4(4)
			2→1	14.33	4.6(-3)	3.8(5)
Ar^{+}	15.76	27.63	^2P, J=1/2→3/2	6.99	5.3(-3)	1.9(5)
Ar^{++}	27.63	40.74	^3P, J=1→2	8.991	3.1(-2)	3.1(5)
			0→1	21.84	5.2(-3)	3.5(4)
S^{++}	23.33	34.83	^3P, J=1→0	33.482	4.7(-4)	2.0(3)
			2→1	18.710	2.1(-3)	1.7(4)
S^{+++}	34.83	47.30	^2P, J=3/2→1/2	10.51	7.7(-3)	5.6(4)

c) critical density for collisions with electrons/protons.

d) for additional fine structure lines see Schmid-Burgk (1982, Landolt-Börnstein, VI, 2c, p.115).

TABLE D II

Measurements of the CO Abundance in Interstellar Clouds a)

Observation	Type of Cloud	H_2 Column Density [cm^{-2}]	Derived Abundance	Reference
UV Absorption toward πSco	diffuse	2(19)	CO/H_2=5(-8)	Federman et al. (1980)
UV Absorption toward ζOph	diffuse	4(20)	CO/H_2=5(-6) $^{12}CO/^{13}CO$=55±11	Wannier et al. (1982)
Near-IR (2μm) Absorption toward NGC 2024	dense molecular cloud	<1(23)	CO/H_2>8(-5)	Black and Willner (1984)
Near-IR (2.3 and 4.6μm) Absorption toward BN	dense molecular cloud	1(23)	$^{12}CO/^{13}CO$=96±5 $^{12}CO/C^{18}O$=426±150	Scoville et al. (1983)
Near-IR (2μm) H_2 and FAR-IR (100μm) CO Emission in Orion-KL	shocked, dense molecular cloud	3(21)	CO/H_2=1.2(-4) $^{12}CO/^{13}CO$≥32	Watson et al. (1985) Genzel et al. (1988)
Millimeter CO Emission and Dust Extinction	dense molecular cloud (Galactic disk)	>1(22)	CO/H_2=8(-5) $C^{18}O/H_2$=1.7(-7) $^{13}CO/H_2$=2.5(-6) $^{12}CO/^{13}CO$=60±8 $^{12}CO/C^{18}O$=500 $^{12}CO/C^{17}O$=3.2	Frerking et al. (1982) Dickman (1978) Wannier (1980) and refs

References

Bally, J., Langer, W.D., Stark,A.A. and Wilson, R.W. 1987, Ap.J. (Lett.) 312, L45

Beichman,C.,Keene,J.,Phillips,T.,Huggins,P.,Wootten,A.,Masson,C. and Frerking,M. 1983, Ap.J. 273, 633

Beichman, C. 1988, Ap.Lett.Comm. 27, 67

Blaauw, A. 1964, Ann.Rev.Astr.Ap. 2, 213

Black, J.H. and Willner, S.P. 1984, Ap.J. 279, 673

Blake, G.A., Sutton, E.C., Masson, C.R. and Phillips, T.G. 1987, Ap.J.

Observation	Type of Cloud	H_2 Column Density [cm^{-2}]	Derived Abundance	Reference
Millimeter CO Emission	dense molecular cloud (Galactic Center)	>1(23)	$^{12}CO/^{13}CO \sim 25$ $^{12}CO/C^{18}O \sim 250$ $C^{18}O/C^{17}O \sim 3.3$	Wannier (1980) and refs
Millimeter CO Emission – Virial Analysis	dense molecular clouds (Galactic disk)	>a few(22)	$\frac{N(H_2)}{I(^{12}CO\ 1\to 0)} = 1.8\to 4.5(20)$	Sanders et al. (1985) Scoville and Sanders (1987) Frerking et al. (1982)
γ-Rays and CO Millimeter Emission	dense molecular clouds (disk)	>a few(22)	$\frac{N(H_2)}{I(^{12}CO)} = 2.4(20)$	Bloemen et al. (1984) Strong et al. (1988)
	dense molecular clouds (Galactic Center)		$\frac{N(H_2)}{I(^{12}CO)} \approx 3(19)$	Blitz et al. (1985)

a) After van Dishoeck and Black (1987); for comparison, the "solar system" abundances are: $^{12}C/H = 3.3 \times 10^{-4}$, $^{12}C/^{13}C = 90$, $^{16}O/^{18}O = 500$, $^{18}O/^{17}O = 5.5$

315, 621
Blitz, L., Bloemen, J.B.G.M., Hermsen, W. and Bania, T.M. 1985, Astr.Ap. 143, 267
Bloemen, J.B.G.M. et al. 1984, Astr.Ap. 139, 37
Bloemen, J.B.G.M. 1989, Ann.Rev.Astr.Ap. 27, 469
Bohlin, R.C., Savage, B.D. and Drake, J.F. 1978, Ap.J. 224, 132

TABLE I
ABUNDANCES AND MASS FRACTIONS IN PDR's[a]

SOURCE	MOL. CLOUD M17 SW[c]	HII INTERFACE M17[c]	GAL. CENTER SGRA WEST CND[d]	STAR BURST GALAXY M82[e]
Spatial scale or distance from central UV source(s)	≥2 pc	~0.6 pc	≥2 pc	500 pc
$G_{IR}[1.7\times10^{-3}$erg s^{-1} $cm^{-2}]$	≤1(3)	2(5)	7(4)	3(3)
$N_o(H + 2\ H_2))^{b)}$ $[cm^{-2}]$	1.5(23)	3(23)	4(22)	2(22)
n^{PDR} (H, H_2) $[cm^{-3}]$	2(4)	3(4)	1→2(5)	1(4)
T_{kin} (PDR) [K]	200	300	350	200
$N(C^+)/N_o$	≤3(-6)	7(-6)	3(-5)	1(-4)
$N(CO{:}T{>}100\ K)/N_o$	1(-6)	≥3(-6)	6(-5)	–
$N(CI)/N_o$	8(-6)	1(-5)	–	–
$N(HI)/N_o$	≥4(-2)	–	~1	0.1
$N(O^\circ)/N_o$	–	1.5(-5)	7(-5)	2(-4)
$N(Si^+)/N_o$	–	5(-7)	2(-5)	2(-5)
Mass fraction in PDR	≤2%	≥5%	10→20%	20%

(a) ~1′ beam, dens/temp from line extinction except in M17 SW where adopted.
(b) Total column density of hydrogen nuclei derived from far-infrared/submm dust continuum and CO ($CO/H_2{:}8\times10^{-5}$).
(c) from Keene+85, Zmuidzinas+86,88, Harris+87, Genzel+88, Stutzki+88, Haas+89.
(d) from Genzel+85, Harris+85, Graf+88, Herter+89.
(e) from Watson+84, Crawford+85, Lugten+86, Lord+89a.

Boland,W. and de Jong,T. 1982, Ap.J. 268, 110
Brand, P.W.J.L., Moorhouse, A., Burton, M.G., Geballe, T.R., Bird, M. and Wade, R. 1988, Ap.J.(Lett), 334, L103
Burton,M., Hollenbach,D. and Tielens,A.A.W.G. 1989, in "IR Spectroscopy in Astronomy" , see Kessler and Glasse
Carlstrom, J. 1989, Ph.D.Thesis, Univ.Calif. Berkeley
Chernoff, D.F., Hollenbach, D. and McKee, C.F. 1982, Ap.J.(Lett.) 259, L97
Cox, P. and Mezger, P.G. 1989, Astr.Ap.Review, in press
Crawford,M.,Genzel,R.,Townes,C. and Watson,D. 1985,Ap.J. 291,755
Crawford,M.,Lugten,J.,Fitelson,W.,Genzel,R. and Melnick,G. 1986, Ap.J.(Lett.) 303, L57
Dalgarno,A. and McCray,R. 1972, Ann.Rev.Astr.Ap. 10,375
de Jong,T.,Dalgarno,A. and Boland,W. 1980, Astr.Ap. 91, 68
deJong 1977, Astr.Ap. 55, 137
Dickman, R.L. 1978, Ap.J.(Suppl.) 37, 407
Draine, B.T. 1978, Ap.J.(Suppl.) 36, 595
Draine, B.T. 1980, Ap.J. 241, 1021
Draine, B.T. and Roberge, W.G. 1982, Ap.J.(Lett.) 259, L91
Draine, B.T., Roberge, W.G. and Dalgarno, A. 1983, Ap.J. 264, 485
Draine, B.T. 1989, in "IR Spectroscopy in Astronomy", see Kessler and Glasse
Draine, B.T. and Woods, D.T. 1989, ibid.
Duffy, P.B., Erickson, E.F., Haas, M.R. and Houck, J.R. 1987, Ap.J. 315, 68
Eckart,A.,Downes,D.,Genzel,R.,Harris,A.,Jaffe,D. and Wild,W. 1989, Ap.J., in press
Federman, S.R., Glassgold, A.E., Jenkins, E.B. and Shaya, E.J. 1980, Ap.J.242, 545
Frerking, M.A., Langer, W.D. and Wilson, R.W. 1982, Ap.J. 262, 590
Frerking, M.A., Keene,J., Blake,G.A. and Phillips,T.G. 1989, Ap.J., in press
Gautier, T., Fink, U., Treffers, R. and Larson, H. 1976, Ap.J.(Lett.) 207, L129
Geballe, T.R., Persson, S.E., Simon, T., Lonsdale, C.J. and McGregoe, P.J. 1986, Ap.J. 302, 500
Genzel, R., Watson, D.M. ,Crawford,M.K. and Townes,C.H. 1985,Ap.J. 297,766
Genzel, R. and Stacey, G.J. 1985, Mittg.A.G. 63, 215
Genzel, R., Harris,A.I., Jaffe,D.T. and Stutzki,J. 1988, Ap.J. 332,1049
Genzel, R., Harris, A.I. and Stutzki,J. 1989, in "IR Spectroscopy in Astronomy, see Kessler and Glasse
Genzel, R. and Stutzki, J. 1989, Ann.Rev.Astr.Ap. 27, 41
Gerola,H. and Glasssgold,A. 1978, Ap.J.(Suppl.) 37, 1
Giard et al. 1990, Astr.Ap. in press
Graedel,T., Langer,W. and Frerking,M. 1982, Ap.J.Suppl. 48, 321
Graf,P.,Herter,T.,Gull,G. and Houck,J. 1988, Ap.J. 330, 803
Gredel,R.,Lepp,S. and Dalgarno,A. 1987, Ap.J.(Lett.) 323, L137
Haas,M., Meixner, M., Tielens, A.A. and Erickson, E.F. 1989, in prep.
Harris, A.I., Jaffe, D.T. , Silber, M. and Genzel, R. 1985,

Ap.J.(Lett.) 294,L93
Harris, A.I., Stutzki, J., Genzel, R., Lugten, J.B., Stacey, G.J. and Jaffe,D. 1987, Ap.J.(Lett.) 322,L49
Harris, A.I., Wild, W., Stutzki, J., Jaffe, D.T., Jackson, J.M., Eckart, A., Lugten, J.B. and Genzel, R. 1989, Ap.J. submitted
Herter, T. 1989, in "IR Spectroscopy in Astronomy", see Kessler and Glasse
Herter, T., Gull, G., Megeath, S., Rowlands, N. and Houck,J. 1989, Ap.J. in press
Hildebrand, R.H. 1983, Quart.J.Roy.Astr.Soc. 24, 267
Houck, J.R., Shure, M.A., Gulle, G.E. and Herter, T., 1984, Ap.J.(Lett.) 287, L11
Hollenbach,D. 1989, in "IR Spectroscopy in Astronomy" see Kessler and Glasse,
Hollenbach, D. and McKee, C.F. 1989, Ap.J. in press
Hollenbach, D., Tielens, A.A.W.G. and Takahashi, T. 1989, in prep.
Jaffe, D.T., Genzel, R., Harris, A.I., Lugten, J.B., Stacey, G.J. and Stutzki, J. 1989, Ap.J. 344, 265
Jaffe, D.T., Genzel, R., Harris, A.I., Howe, J.E., Stacey. G.J. and Stutzki, J. 1990, Ap.J. in press
Jenkins,E. and Shaya,E. 1979, Ap.J. 231, 55
Jura,M. 1974, Ap.J. 191, 375
Keene, J., Blake,G.A., Phillips, T.G., Huggins, P. and Beichman,C. 1985, Ap.J. 299,967
Kessler, M. and Glasse, A.(eds.), "Infrared Spectroscopy in Astronomy", ESA-SP series, Reidel in press
Lacy, J.H., Townes, C.H., Geballe, T.R. and Hollenbach, D. 1980, Ap.J. 241, 132
Langer, W. 1976a, Ap.J. 206, 699
Langer, W. 1976b, Ap.J. 210, 328
Langer, W.,Graedel, T., Frerking, M.A. and Armentrout,P. 1984, Ap.J. 277, 581
Lester, D.F., Dinerstein, H.L., Werner, M.W., Watson, D.M., Genzel, R. and Storey, J.W.V. 1987, Ap.J. 320, 573
Leung,C., Herbst,E. and Huebner,W. 1984, Ap.J.Suppl. 56, 231
Lo, K.Y., Cheung, K.W., Masson, C.R., Phillips, T.G., Scott, S.L. and Woody, D.P. 1987, Ap.J. 312, 574
Lord, S., Haas, M., Duffy, P. and Erickson, E. 1989a, in prep.
Loren,R. 1979, Ap.J.(Lett.) 234, L207
Lugten, J.B., Watson, D.M., Crawford, M. and Genzel,R. 1986b, Ap.J.(Lett.) 311,L51
Maddalena, R.J., Morris, M., Moscovitz, J. and Thaddeus, P. 1986, Ap.J. 303, 375
Maloney, P., van Dishoeck, E. and Black, J. 1989, Ap.J. in press
Martin, H.M., Sanders, D.B. and Hills, R.E. 1984, MNRAS 208, 35
Matsuhara,H. et al. 1989, Ap.J.(Lett.), in press
Melnick, G., Stacey, G.J, Genzel, R., Lugten, J.B. and Poglitsch, A. 1989, Ap.J. in press
Mezger, P.G., Chini, R., Kreysa, E. and Gemuend, H.P. 1986, Astr.Ap. 160, 324

Moorwood, A. 1989, in "IR Spectroscopy in Astronomy", see Kessler and Glasse
Neufeld, D.A. and Melnick, G.J. 1987, Ap.J. 322, 266
Neufeld, D.A. and Dalgarno, A. 1989a, Ap.J. 340, 869
Neufeld, D.A. and Dalgarno, A. 1989b, Ap.J. 344, 251
Norman, C. and Silk, J. 1981, Ap.J. 238, 158
Oliva, E. and Moorwood, A. 1989, Astr.Ap. in press
Olofsson, H. and Rydbeck, G. 1984, Astr.Ap. 136, 17
Phillips, T.G, Huggins, P.J., Wannier, P.G. and Scoville, N.Z. 1979, Ap.J. 231, 720
Phillips, T.G., Huggins, P.G., Kuiper, T. and Miller,R. 1980, Ap.J.(Lett.) 238, L103
Phillips, T.G. and Huggins, P.G. 1981, Ap.J. 251,533
Pipher, J.L., Helfer, H.L., Heter, T., Briotaa, D.A., Houck, J.R., Willner, S.P. and Jones, B. 1984, Ap.J. 285, 174
Prasad,S. and Tarafdar,S. 1983, Ap.J. 267, 603
Rieke, G.H., Lebofsky, M.J., Thompson, R.J., Low, F.J. and Tokunaga, A.T. 1980, Ap.J. 238, 24
Rieke, G.H., Cutri, R.M., Nlack, J.H., Kailey, W.F., McAlary, C.W., Lebofsky, M.J. and Elston, R. 1985, Ap.J. 290, 116
Rubin, R.W. 1985, Ap.J.(Suppl.) 57, 349
Schmid-Burgk, J. 1982, Landolt-Boernstein, VI,2c, p115
Schmid-Burgk, J. et al. 1989, Astr.Ap. in press
Sanders, D.B., Scoville, N.Z. and Solomon, P.M. 1985, Ap.J. 276, 182
Scoville, N.Z., Kleinmann, S.G., Hall, D.N.B. and Ridgway, S.T. 1983, Ap.J. 275, 201
Scoville, N.Z. 1984, in "Galactic and Extragalactic Infrared Spectroscopy", eds. M.Kessler and J.Phillips, AP-SP series 108, Reidel (Dordrecht), p167
Scoville, N.Z. and Sanders, D.B., in "Interstellar Processes", eds. D.Hollenbach and H.Thronson, Reidel(Dordrecht), p21
Simpson, J. 1975, Astr.Ap. 39, 43
Shibai,H. et al. 1989, preprint
Shull, M. and Beckwith, S. 1982, Ann.Rev.Astr.Ap. 19, 163
Shull, M. and Draine, B.T. 1987, in "Interstellar Processes", eds. D.Hollenbach and H.Thronson, Reidel (Dordrecht)
Spitzer, L. 1978, "Physical Processes in the Interstellar Medium", Wiley (New York) Stacey,G.,Viscuso,P.,Fuller,C. and Kurtz,N. 1985, Ap.J. 289,803
Stacey, G.J. 1989, in "IR Spectroscopy in Astronomy", see Kessler and Glasse
Stacey, G.J., Genzel, R., Lugten, J.B. and Townes, C.H. 1990, Ap.J. submitted
Stacey, G.J., Geis, N., Genzel, R., Harris, A.I., Jaffe, D.T., Poglitsch, A. and Stutzki, J. 1990b, Ap.J. to be submitted
Sternberg,A., Dalgarno,A. and Lepp,S. 1987, Ap.J. 320, 676
Sternberg,A. 1988, Ap.J. 332, 400
Sternberg, A. and Dalgarno, A. 1989, Ap.J. 338, 197
Storey, J.W.V. et al. 1990, in prep.
Strong, A.W. et al. 1988, Astr.Ap. 207, 1

Stutzki, J., Stacey, G.J, Genzel, R., Harris, A.I., Jaffe, D.T. and Lugten, J.B. 1988, Ap.J. 332,379
Stutzki,J. and Güsten,R. 1989, Ap.J. submitted
Stutzki,J. et al. 1989, in "Submillimetre and Millimetre Astronmy", ed. A.Webster, Kluwer, in press
Stutzki, J., Genzel, R. and Harris, A.I. 1990, in prep.
Tarafdar,S.,Prasad,S.,Huntress,W.,Villere,K. and Black,D. 1985, Ap.J. 289, 220
Tielens, A.A.W.G. and Hollenbach, D. 1985a, Ap.J.291,722
Tielens, A.A.W.G. and Hollenbach, D. 1985b , Ap.J.291,747
van Dishoeck,E. and Black,J. 1986, Ap.J.(Suppl.) 62, 109
van Dishoeck, E. and Black, J. 1987, in "Physical Processes in Interstellar Clouds", eds. G.Morfill and M.Scholer, NATO ASI 210, p241
van Dishoeck,E. and Black,J. 1989 , Ap.J. in press
Walmsley, M. 1975 in"HII Regions and Related Topics", eds. T.Wilson and D.Downes, Springer(New York), p17
Wannier, P.G. 1980, Ann.Rev.Astr.Ap. 18, 399
Wannier, P.G., Penzias, A.A. and Jenkins, E.B. 1982, AP.J. 254, 100
Watson,D. 1984,in "Galactic and Extragalactic IR Spectroscopy", eds. M.Kessler and J.Phillips, AP-SP series 108, Reidel (Dordrecht), p195
Watson, D.M. , Genzel, R., Townes,C.H. and Storey, J.W.V. 1985, Ap.J. 298, 316
Werner,M. 1970, Ap.J.(Lett.) 6,L81
Whitcomb, S.E., Gatley, I., Hildebrand, R.H., Keene, J., Sellgren K. and Werner, M.W. 1981, Ap.J. 246, 416
Wild, W. et al. 1990, in prep.
Williams, D. and Hartquist,T. 1984, MNRAS 210, 141
Wolfire, M., Hollenbach, D. and Tielens, A.A.W.G. 1990, Ap.J. in press
Wootten, A., Phillips, T.G. , Beichman, C. and Frerking, M.A. 1982, Ap.J.(Lett.) 256,L5
Zmuidzinas,J.,Betz,A. and Goldhaber,D. 1986, Ap.J.(Lett.) 307,L75
Zmuidzinas,J.,Betz,A.,Boreiko,R. and Goldhaber,D. 1988, Ap.J. 335,774

CIRCUMSTELLAR CHEMISTRY

Alain OMONT
Observatoire de Grenoble, Université Joseph Fourier
et
Institut d'Astrophysique de Paris, CNRS
98bis, boulevard Arago, F–75014 Paris

1. INTRODUCTION

Massive circumstellar envelopes (CSEs) ejected by the red giants at the very end of their evolution are among the most interesting objects for cosmic chemistry. Temperature, density and ultraviolet shielding, similar to the molecular interstellar clouds warrant the gas is mostly molecular, with dust grains. With respect to interstellar chemistry, CSEs have many rich peculiarities : the physical conditions span a very broad range of temperature and density, with a relatively simple structure and time evolution. The rapidity of the latter and the sharp photodissociation transition favor a very active photochemistry in the outer layers. However, two points make circumstellar chemistry nearly unique : a large proportion of CSEs are carbon–rich with the obvious advantage for synthesizing complex organic molecules (and the chance of benefiting of the presence of a prominent source, IRC+10216, exceptionally close to the Sun) ; CSEs are the most obvious location of formation of cosmic dust grains, with fascinating, albeit complex, processes of dust formation and interaction with the gas, and also the possibility that interstellar and solar system solid particles keep a memory of the circumstellar origin of some of their components.

CSEs are also by themselves important astronomical objects for the interstellar and galactic evolution. They represent a very brief, but crucial, phase, which probably terminates the evolution of every low and intermediate mass star, at the transition between red giants and planetary nebulae. The star of most of these molecular CSEs is on the late asymptotic giant branch (AGB) of red giants (although some of them are post–AGB stars or red supergiants). Its core has then properties very close to those of a white dwarf. The high temperature induces a powerful burning of hydrogen in a shell around the core with convective transport of energy. CSEs have thus very high luminosities ($\sim 10^4$ $L_\odot$ for AGB CSEs) and most of them are long period variables of different kinds (irregular, semi–regular, Miras, infrared) with typical period, a few 10^2 days. At time intervals $\sim 10^4$–10^5 years, the temperature of the helium shell around the core becomes high enough to briefly ignite the reaction $3\,^4He \longrightarrow {}^{12}C$ ("thermal pulses"). A powerful convection (dredge–up) brings to the surface of the star different products of the nucleosynthesis : ^{12}C, s–isotopes and products of CNO H–burning (^{14}N, ^{13}C, etc). Thus, the photosphere of the star ($T_{eff} \leq 3000$ K) and the CSE itself present a variety of chemical compositions dominated by the value of the C/O ratio : O–rich (M) stars for C/O < 1, carbon stars for C/O > 1 and S stars for C/O ~ 1.

Many details of this brief and crucial phase of evolution, which is dominated by the large mass–loss rate often called superwind, remain obscure. In particular the conditions of

J. M. Greenberg and V. Pirronello (eds.), Chemistry in Space, 171–197.

the transition from O– to C–richness are still debated (14–15). The red giant phase ends when the mass–loss has almost exhausted the gas of the star. The effective temperature of the star then increases until its radiation becomes sufficient to ionize the CSE into a planetary nebula, before the final evolution of the latter into a white dwarf. The transition objects between the AGB and planetary nebulae, "pre–planetary nebulae" (PPN), are especially interesting, both for the understanding of stellar evolution and for their particular structure and physical and chemical conditions.

Precise estimates in the solar neighbourhood (see e.g. 16) have shown that AGB CSEs are the main contributors to return of matter from stars to the interstellar medium. Therefore, they play a major role in chemical galactic evolution, being the main source of enrichment of the interstellar gas in ^{12}C, ^{14}N, ^{13}C and the s–isotopes. As already mentioned, they are also contributing mostly to the injection of new grains, O– and C–rich, into the interstellar medium, with uncertain, but open, consequences on interstellar chemistry and solid particles of the early solar system.

The understanding of the chemical processes in these highly structured and rapidly evolving objects, relies firstly on a good estimate of the physical parameters in their different layers (Section 2), and on the accumulation of observational information on molecular abundances (Section 3). An overview of the chemical processes is presented in Section 4, together with a discussion of their time rates which are compared to the characteristic evolution time of the different shells. Sections 5 and 6 deal with the chemistry of internal layers : chemical thermodynamical equilibrium gives the abundance in the photosphere and in the dense layers which prolongate it. However, non–equilibrium processes prevail in most of the complex and inhomogeneous regions which are crossed by large scale motions and by shocks induced by pulsations, and where grains and the circumstellar wind form. Then the chemistry can be relatively inactive in the smoother and freely expanding outer layers as long as they are shielded from interstellar UV radiation. Section 7 discusses how the latter finally destroys all the circumstellar molecules by photodissociation and more marginally by photoionization. Observations and theoretical modeling have shown that before this destruction is complete, there exists an area of very active photochemistry able to achieve the synthesis of complex molecules through ionic and radical reactions. This is the most active and successful part of the modeling of circumstellar chemistry which is reported in Section 8.

Previous reviews on circumstellar chemistry will be found in references (1–4), while references (5–13) review the general properties of CSEs and of their evolution.

2. PHYSICAL PARAMETERS

2.1 Overview

A general view of the structure of the physical parameters of CSEs is sketched in Figure 1. A complex region lies above the photosphere, where pulsations periodically raise masses of gas which are believed to partly return to the photosphere, and to partly be ejected to give mass-loss ; the levitated gas cools, allowing dust to form. The dust is pushed outwards by radiation pressure carrying away the gas by friction. Then, the ejected gas and dust move away practically in free expansion with a velocity in the range 10–20 km/s. The temperature and density progressively decrease in these intermediate layers formed of stable molecules and dust. The conditions approach those of interstellar molecular clouds, and finally the molecules are photodissociated by the interstellar UV radiation. Before the gas is completely reduced to atomic form, photodissociation generates unstable radicals and ions which can further react to produce complex unstable molecules.

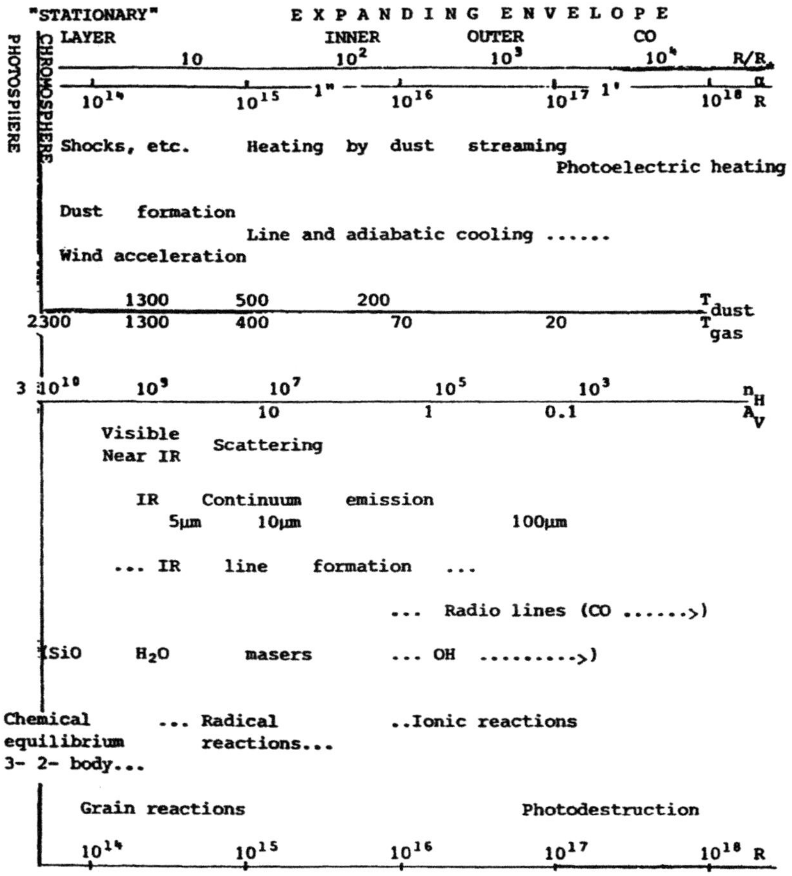

Figure 1. Structure of a massive envelope (IRC+10216). Assumed parameters : d = 200 pc, $R_* = 1000\ R_\odot = 7\ 10^{13}$ cm, $\dot{M} = 4\ 10^{-5}\ M_\odot$/yr. In the outer envelope : $n_H = 5\ 10^{37}\ r^{-2}\ cm^{-3}$, $A_V = 2.5\ 10^{-22}\ N_H$. The range of maser emission is indicated for similar O–rich envelopes, it is obviously irrelevant for C–rich shells such as IRC+10216. From ref. 2.

2.2 Internal layers

As already mentioned, the structure and the conditions are very complex and uncertain in the layers above the photosphere until the wind is fully accelerated. Gas motions in these regions are related to pulsations and wind acceleration, probably with shocks. Grain formation which takes place in the range 1500–500 K, will be discussed in Section 6. The temperature law depends on different heating and cooling processes (see e.g. 17), and remains quite uncertain. It should be noted that there is some evidence that this inner envelope could often be asymmetric (e.g. IRC+10216), which could be related to the presence of companions as identified in o Ceti (Mira).

2.3 Density

When the mass–loss is isotropic, the density n at radius r is simply related to the local value $\dot{M}(r)$ of the mass–loss rate :

$$\dot{M} = n\ \mu\ 4\pi\ r^2\ V_e \tag{1}$$

where μ is the mean molecular weight and V_e the expansion velocity. Thus, if the mass–loss rate is constant, the density obeys a simple r^{-2} law in the outer envelope where V_e has reached its constant asymptotic value.
It is obvious that this simple r^{-2} density law is not valid in the inner envelope and in the region of dust acceleration, there being many cases known where the mass–loss is clearly asymmetric and/or has varied with time.

2.4 Temperature

The temperature is more difficult to estimate in CSEs, even in the outer envelope with well defined velocity and density laws. There are few and limited observational diagnostics of the temperature.
In the more outer layers, important for millimeter observations, one has generally to rely on theoretical modeling of the kinetic temperature (18, 19). In the regions shielded from interstellar UV radiation, the main heating process has been identified as friction due to grain streaming through the gas under the action of radiation pressure. Comparable amounts of cooling are produced by the adiabatic expansion and by the CO rotational lines.
For a CSE with $\dot{M} \sim 10^{-5} M_\odot/yr$ such as IRC+10216, the drift velocity of the grains through the gas, v_d, is $\sim$ 1 km/s, and one has the typical temperature law

$$T \ \sim\ 100\ K(r/10^{16} cm)^{-0.7} \tag{2}$$

However, such a law is valid only for mass–loss rates in the range of $10^{-5} M_\odot$/yr. It has been shown (20) that for $\dot{M} \sim 10^{-7} M_\odot$/yr with $v_d \sim$ 10 km/s, T_K can be an order of magnitude larger than given by Eq. (2). On the other hand, T_K should be much lower for $\dot{M} \sim 10^{-4} M_\odot$/yr ($v_d \sim$ 0.3 km/s), such as in extremely cold OH/IR stars.

2.5 Ultra–Violet Radiation

Photochemistry due to interstellar UV radiation is probably the major chemical process in CSEs (Sections 7, 8). On average the outer layers of CSEs are thus submitted to the "standard" UV galactic field emitted by remote O and B stars. Local enhancement for CSEs in the vicinity of O and B stars does not normally exceed a factor $\sim$ 2–5, with possible spatial asymmetries ; the only such case proved is NML Cyg (21).
The effect of this UV radiation on inner layers of CSEs, crucially depends on dust shielding. However, the UV properties of circumstellar dust are still poorly known, and one generally assume that they are similar to those of interstellar dust, with comparable dust to gas ratios. H_2O provides an additional continuum UV opacity in O-rich CSEs (22, 23). Absorption in UV lines of H_2 and CO blocks the UV radiation in limited spectral ranges, and particularly insures a very efficient self-shielding of these molecules (24).
Photodissociation by chromospheric UV radiation is important in stars such as the supergiant α Ori (25). The effect of internal UV is obviously prominent in planetary nebulae, and possibly in some pre–planetary nebulae.

3. MOLECULAR ABUNDANCES

3.1 Difficulty of Observational Derivation

3.1.1 Radio Lines

Most of the identified circumstellar molecules have been detected by their millimeter rotational lines. For optically thin lines, the abundance of the emitter is basically proportional to the intensity of the line. However, the average of the emission on the telescope beam can encompass regions of quite different abundances and provides only a mean abundance. In addition the actual formula for the abundance

$$f_i = \frac{n_i}{n_{H_2}} = \frac{W\ V_e\ \Phi_B^2\ exp(E_u/kT_R)\ Z\ (T_R)}{\nu\ S_{ul}\ \mu_0^2\ r_i\ \dot{M}/D^2} \tag{3}$$

shows the different requirements to derive abundances and the corresponding uncertainties (V_e is the expansion velocity, Φ_B the angular diameter of the telescope beam, E_u the energy of the upper level of the line, Z the partition function, ν the line frequency and D the distance of the source) :
The integrated area W is affected by noise and the difficult calibration of the radiotelescope.
The excitation temperature T_R is approximately determined from line ratios or from theoretical modeling. It varies with r and one has to use some average value.
The permanent dipole momentum μ_o and to a less extent the line strength S_{ul} may be unknown, and thus only guessed, for exotic molecules.
The mass–loss rate $\dot{M}$ supposed to be constant, is always uncertain. However, the uncertainty is smaller for the ratio $\dot{M}/D^2$.

- The extension r_i of the emission is generally determined by photodissociation. In most cases it has to be somewhat guessed.
- The above formula is only valid for optically thin lines. The use of optically thick lines implies an always uncertain modeling of radiative transfer.

3.1.2 Infrared Spectroscopy

This is basically an absorption spectroscopy on the line of sight in front of the region of continuum dust emission. The abundance is then proportional to the equivalent width of the line. However, the line must be optically thin, and the regions of continuum emission and of line absorption have to be well separated.
The main additional difficulties are then : the identification of the lines for exotic species; the signal–to–noise ratio for low abundances and/or weak infrared sources; blends with circumstellar or telluric lines; etc...

3.2 Overview of Molecular Abundances

A general view of the molecules presently detected in CSEs is presented in Table 1. Their number is about twice smaller than the larger set of molecules detected in interstellar clouds. Generally speaking, CSEs are much less massive than prominent millimeter sources in interstellar molecular clouds, and the detectability of a given species requires as a rule that its abundance be several orders of magnitude larger in CSEs than its detection limit in interstellar clouds. This conclusion is examplified in O–rich CSEs (Section 3.5) where less than 20% of the known interstellar molecules have been detected. However, the presence of a single source, IRC+10216 (CW Leo), tempers this conclusion quite a lot ; the overabundance of carbon makes the chemistry much richer, and this massive CSE ($\dot{M} \sim 10^{-5} M_\odot$/yr), is exceptionally near (100–150 pc). Thus, IRC+10216 is one of the most important sources of millimeter radioastronomy, where many new cosmic molecules have been discovered, in particular long chains (cyanopolyynes and radicals) and cyclic molecules (SiC_2 and C_3H_2).

3.3 Molecular Abundances in IRC+10216 (CW Leo)

The molecules detected in IRC+10216 are reported in Table 2 with an order of magnitude of their abundances. Because of different problems discussed in Section 3.1, the accuracies of the displayed abundances are certainly not better to a factor 3, and possibly 10 in some cases.
Compared with the interstellar medium, there are very few oxygen bearing molecules (CO, SiO, possibly HCO^+) because C/O > 1 and most of the oxygen is locked in CO. The abundances of most of the other molecules are larger than in the interstellar medium (by about one order of magnitude). There are no deuterium compounds since deuterium has been destroyed in the star.

Table 1. Molecules Detected at Radio Wavelengths in Circumstellar Envelopes (adapted from refs. 26, 27)

Molecule	Approximate Number of Sources C-rich	O-rich
CO	150	150
CN	15	
CS	15	3
SiO ("Thermal")	4	30
SiS	7	1
SiC_2	7	
SO		7
SO_2		15
H_2S	1	15
OCS		1
HCO^+	3	2
HCN	80	13
CH_3CN	3	
HNC	5	2
NH_3	2	1
HC_3N	6	
HC_5N	4	
HC_7N	2	
HC_9N	1	
$HC_{11}N$	1	
C_2H	7	
C_3H	2	
C_3H_2	3	
C_4H	4	
C_5H	1	
C_6H	1	
C_3N	5	
C_2S	1	
C_3S	1	
$HSiC_2$ (HSC_2) (?)	1	
NaCℓ	1	
AℓCℓ	1	
KCℓ	1	
AℓF	1	
SiC	2	
C_4Si	1	
CP	1	
OH (maser)		1500
H_2O (maser)	4	200
SiO (maser)		140

A few small stable molecules are most abundant and contain most the main elements (CO, C_2H_2, CS, SiS, SiO, etc). This is not surprising since a similar situation is found in the photosphere where thermodynamical chemical equilibrium is achieved, and it is believed that these stable molecules found in the outer shells have survived from the photosphere in the ejected envelope (see Sections 5 and 6).

Some simple unstable radicals (CN, C_2H, SiC, etc) are very probably direct products of photodissociation.

There are a lot of complex carbon molecules, as expected in a C–rich environment. However, at least in the molecules detected, this complexity implies only acetylenic chains in most cases, and a few examples of small rings. Indeed, a large proportion of these unstable and exotic species were discovered in this source, even before laboratory detection, in particular the cyanopolyynes ($HC_{2n}CN$) and the radicals C_nH (with $n \geq 3$).

Table 2. Observed abundances in IRC+10216 updated from table 1 of ref. (1) and from (28) (see also 29–35), and in OH231.8+4.2 (36). All abundances are uncertain by factors > 3. $x(CO)/H_2 \sim 6\ 10^{-4}$. Molecules denoted with a * have only been detected in the infrared.

$x(i)/H_2$	IRC+10216						OH231.8+4.2
	$C_2H_2^*$						SO_2
10^{-5}	HCN						
							SO
					CN		NH_3, OCS
10^{-6}	NH_3				C_3^*		H_2S
					C_2H		HCN, CS
		SiS			C_4H		
			CS	HC_3N			
	$C_2H_4^*$	C_2Si			C_3N		HNC
10^{-7}	CH_4^*	SiO		HC_5N	C_5^*		SiO, HCO^+
	HNC			HC_7N	C_6H	$C\ell A\ell$	
					C_3H_2		
			C_2S		C_3H		
10^{-8}	CP		C_3S	$HC_{11}N$	C_5H	$FA\ell$	
		SiH_4^*,SiC					
	CH_3CN						
			H_2S				
10^{-9}						$C\ell Na$	
	(HCO^+)	C_4Si				$C\ell K$	

3.4 Trends in Other Carbon–Rich Envelopes

CO is detected in about 150 C–rich CSEs (see e.g. 37 and references therein), and HCN in about half of them. Other C–bearing molecules such as HC_3N, CS, CN, C_2H, C_2Si, $H^{13}CN$, HNC, C_4H, etc.., as well as SiS, are easily detected with large radiotelescopes in prominent sources such as CIT6 (RW LMi), CRL2688, CRL618, CRL3068. The abundances exhibit the same trends as in IRC+10216 (Table 2), with however some significant variations, such as abundant HC_3N in CRL618 (38), still to be explained. Molecules such as HCN, CN and CS have been detected even in CSEs with $\dot{M} \sim 10^{-7} M_{\odot}/\mathrm{yr}$ such as Y CVn (as well as their ^{13}C isotopic varieties in this peculiar ^{13}C-rich source) (39). The ratio of the line intensities CN/HCN displays large variations with $\dot{M}$ (40).
It is not surprising that the molecular abundances are rather different in the planetary nebulae with molecular shells such as NGC 7027 or IRAS 21282+5050 : molecules easily photodissociated such as HCN are relatively less abundant, others such as HCO^+ or C_3H_2 can be strongly enhanced.

3.5 Abundances in Oxygen–Rich CSEs

The chemistry of O–rich CSEs is much less rich, and in addition there is no massive O–rich CSE as close as IRC+10216. Accordingly, few molecules have been detected except for the major species : CO for carbon, H_2O and OH, together with CO for oxygen, SO, SO_2 and H_2S for sulfur, SiO for silicon, NH_3 for nitrogen (N_2 being undetectable). However, a few C–bearing molecules, besides CO, are relatively common, in particular HCN. Other unexpected species such as CS, SiS, HCO^+ prove the existence of non equilibrium processes (shocks, photochemistry, etc). The PPN OH231.8+4.2 is particularly rich in various molecules which could result from an active shock chemistry.

3.6 Spatial Variations of the Abundances

The analysis of the spatial distribution of the molecules in CSEs is crucial for identifying the chemical processes which form or destroy them. However, the angular resolution available in millimeter radioastronomy (12" at 230 GHz for a 30 m telescope, a few arcseconds in practice with millimeter interferometers) limits such studies to relatively nearby sources, preferentially with large mass–loss rates, such as IRC+10216. It should be noted that the shape of radio lines by itself can give a direct indication of the angular extension of their emission : in particular optically thin lines display a double peak structure characteristic of expanding envelopes when their diameter is larger than the telescope beam.
The most complete studies, and the most relevant ones for the understanding of circumstellar chemistry, have been performed by Bieging and Rieu with the Hat Creek interferometer on IRC+10216 (41). They have found in particular that the emission of C_2H, HNC, C_3N and HC_3N comes mainly from a hollow shell (with however a non zero central intensity for HC_3N). This is a direct proof that these molecules are mainly formed in the outer layers of the envelope. This was clearly expected and partially

proved by single disk studies for C_2H created by photodissociation of C_2H_2 (as well as CN formed from HCN). The synthesis of HNC and HC_3N in the outer shells was less obvious. Its proof is important to give credit to the models of formation of complex species by photosynthesis discussed in Section 8.
There are also indications for external formation of C_4H, C_3H and SiC ; but on the other hand SiS, SiO, CP and metal halides appear to have a reduced spatial extension. For all molecules the extension of the emission is limited by photodissociation. Thus, its measurement is a check of photodissociation models (e.g. CO, HCN, SiO, OH, etc, see Section 7).

4. TIME SCALES AND OVERVIEW OF CHEMICAL PROCESSES

One of the main characteristics of circumstellar chemistry is the very short time constant for evolution which can be described by

$$\tau_{dyn} = r/V_e \tag{4}$$

For a typical value of the expansion velocity $V_e = 15$ km/s, $\tau_{dyn} = 10^3$ yr at $r = 5\ 10^{16}$ cm.
A destruction reaction will be important only if $\tau_{reac} \lesssim \tau_{dyn}$. This defines a radius r_f outside of which the considered destruction reaction can be considered as "frozen out". The value of r_f gives also an idea of the region where the reaction is important for the formation of other molecules. The explicited value of r_f is

$$r_f = 1.3\ 10^{16}\ cm\ \ M_{H-5}\ \ k_{-10}\ \ x_{i-5}\ /\ V_{e15}^2 \tag{5}$$

where $\dot{M}_{-5}$, k_{-10}, x_{i-5} and V_{e15} are the hydrogen mass–loss rate, the reaction rate, the abundance of the reactant i (N_i/N_H) and the expansion velocity, in units of $10^{-5} M_\odot$/yr, 10^{-10} cm^3 s^{-1}, 10^{-5} and 15 km/s respectively. The variations of τ_{reac} with r are sketched in Figure 2, where the values of r_f can be visualized. Typical values of k_{-10} are $\sim$ 10 for ionic reactions, $\sim$ 0.1–1 for radical reactions without activation energy ; but they are very small at low temperature when an activation energy is present, so that r_f is then quite small.
Also shown in Figure 2 are a typical photodissociation rate (see Section 7), and a rough value of the collision rate with the grains (Section 6). It is immediately seen that the reactions with the grains are important only in the inner layers not reached by photodissociation in such conditions.
It is also seen that there is a possible overlap between the regions of photodissociation and of active reactions of ions or radicals to the condition that the product $k_{-10}\ x_{i-5} \gtrsim 1$, i.e. reactions of ions with species more abundant than 10^{-6}, or of radicals (and atoms) with species of abundance $\sim 10^{-5}$–10^{-4} (for $\dot{M} \gtrsim 10^{-5}\ M_\odot$/yr) (see Section 8).

5. CHEMISTRY IN THE INNER LAYERS

5.1 Thermodynamical chemical equilibrium

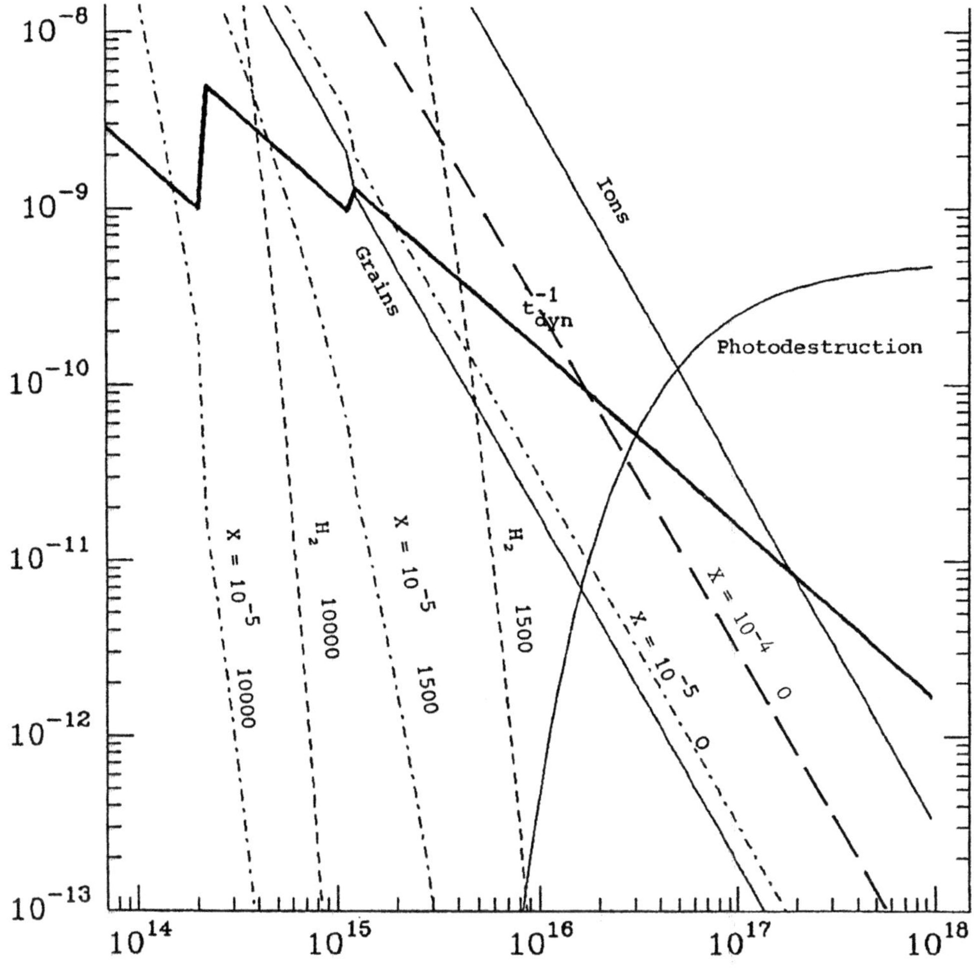

Figure 2. Time constants and reaction rates (sec^{-1}) in the envelope of IRC+10216 with the model of Figure 1. The assumed typical rates are the following : photodissociation 5 10^{-10} exp(-7 $10^{16}/r$) ; grains 6 $10^{-17}n_H$; ions $10^{-9}Xn_H$ with $X = 10^{-5}$; reactions with H_2 5 $10^{-12}n_H$ exp($-T/T_B$) with $T_B = 1500$ and 10000 K ; reactions with other abundant species (e.g. C_2H_2) $10^{-11}Xn_H$ exp($-T/T_B$) with $X = 10^{-5}$ and $T_B = 0$, 1500 and 10000 K, and with $X = 10^{-4}$ and $T_B = 0$. From ref. 2.

Gas phase chemical equilibria in O-rich(left panel) and C-rich(right panel) molecular envelope with chemical compositions of H/C/N/O/Si/P/S=12.0/8.55/ 7.93/8.77/ 7.55/ 5.43/ 7.21 and 12.0/ 9.07/ 9.25/ 8.77/7.55/ 5.43/ 7.21(logarithmic scale), respectively. Gas pressure is kept to be log P_g =-3.0 throughout.

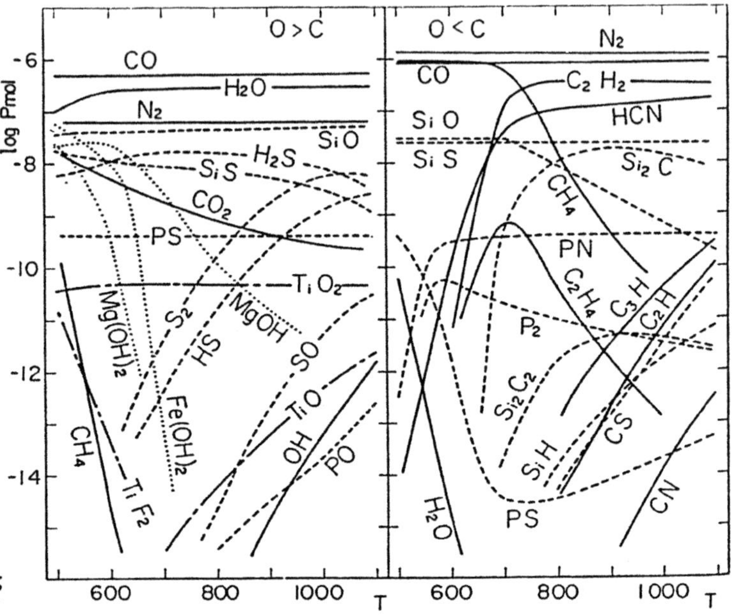

Figure 3. Example of abundances at thermal equilibrium (from Tsuji, ref. 46, p.414)

Local thermodynamical chemical equilibrium is achieved when chemical reactions between all species are faster than the evolution time constants of the medium. This implies large temperatures to overcome the activation energies, and large densities because 3-body collisions are necessary to allow fast association reactions (mainly H_2 formation). Such conditions are certainly met in the photosphere where LTE abundance determination is routinely included in stellar atmosphere modeling. They should hold in colder layers above the photosphere as long as the H–H_2 thermodynamical equilibrium is achieved, e.g. $T \geq 1500$ K and $n \geq 10^{11}$ cm^{-3} (see Glassgold and Huggins (42) for a discussion of the atomic H abundance).
The calculation of equilibrium molecular abundances is straightforward since it only requires the knowledge of the element abundances and of the enthalpy of formation ΔH of the molecular species, as well as an estimate of their partition functions. However, because of the exponential dependence, even a small error in ΔH yields a large error on the abundances at the low temperatures of CSEs (a factor 2 for 0.1 eV at 1500 K). The ΔHs of nearly all unstable species (radicals, cyanopolyynes, etc...) are not known with such an accuracy.
Classical references for photospheres are the works of Tsuji (43). The main difficulty of their extension to inner layers of the envelope above the photosphere is the knowledge of the physical parameters T and n. Such model calculations were performed e.g. in refs. (44–46). Their main result (see e.g. Figure 3) is that most unstable species observed in CSEs cannot be formed there at equilibrium : most radicals (including those abundant at higher temperature in photospheres, such as CN), HNC, NH_3, cyanopolyynes

(except may be HC_3N), etc. However, the use of different thermodynamical data for cyanopolyynes leads Shmeld et al. (47) to the opposite conclusion that they could come from thermal equilibrium in these inner regions.

5.2 Non–LTE Processes

As already stated, thermal chemical equilibrium stops in the very inner layers of CSEs, probably when recombination reactions are no longer effective (1, 42, 45). However, chemistry is still very active in these warm and dense regions. The main chemical processes, not completely independent, are there radical reactions with hydrogen, grain formation and shock chemistry. Grain processes are discussed in Section 6. Radical or atom reactions with closed shell molecules have only moderate activation energies (typically 5000–10000 K). Accordingly, they are important with abundant species for temperatures down to 1000 K, or even lower. The most abundant "radical" is certainly by far atomic hydrogen. The abundance of all radicals and subordinated species in the inner layers of CSEs is thus dominated by reactions with hydrogen,

$$A\bullet + H_2 \rightleftharpoons AH + H\bullet \tag{6}$$

e.g. Goldreich and Scoville (18)

$$\begin{aligned} O + H_2 &\rightleftharpoons OH + H \\ OH + H_2 &\rightleftharpoons H_2O + H \end{aligned} \tag{7}$$

The abundance of H is extremely uncertain. If one takes only the freezing–out of 3–body recombination, it is quite large ($H/H_2 \sim 10^{-2}$–10^{-3}) (42). However, it is very likely that grains constitute an efficient sink for H, while H_2 could be dissociated by shocks or even by chromospheric UV radiation. Anyway, if H/H_2 exceeds 10^{-4}–10^{-5}, the ratios A/AH should be at equilibrium with H/H_2 in most cases :

$$\frac{n(A)}{n(AH)} = \alpha(T)\ exp(-\Delta E/kT)\ \frac{n(H)}{n(H_2)} \tag{8}$$

where ΔE is the energy of the reaction (6) and α(T) is the ratio of the partition functions not extremely different from unity and slowly varying with T. As ΔE is typically ~1 eV ~ 10000 K for most reactions, the effect of the exponential factor is quite dominent at T ~ 1000 K. Accordingly, all radicals whose reaction with H_2 is exothermal (i.e. those to which the H binding energy is larger than 4.5 eV) (OH, O, CN, C_2H, etc...) are completely destroyed in the inner layers of CSEs, even if they are abundant in the photosphere. On the other hand, hydrids such as H_2S, SH, NH_3, etc.., whose reaction with H is exothermal (H binding energy $<$ 4.5 eV), are destroyed into radicals or atoms if the H abundance is not very small ($\geq 10^{-4}$–10^{-5}) e.g.

$$H_2S + H \longrightarrow SH + H_2 \xrightarrow{H} S + 2H_2 \tag{9}$$

Accordingly, the detection of large abundances of H_2S or NH_3 in relatively outer layers of CSEs could mean that the H abundance is quite small.
The uncertainties on the physical parameters of the inner layers, as well as on grain formation, make the modeling of their chemistry very difficult beyond these qualitative conclusions. The existence of shocks associated to supersonic motions, such as those generated by pulsations (see e.g. 48) or high velocity winds, opens further possibilities in particular the dissociation of H_2 and especially of CO which could explain the observations of HCN and CS in O–rich CSEs (Section 8.3.c). However, detailed information on such shocks is still lacking, as well as the modeling of their chemical consequences.

6. GRAIN PROCESSES

6.1 Infrared Observations and Grain Properties

As in the interstellar medium, grains contain about half of the mass of heavy elements in CSEs. They play a fundamental role in circumstellar chemistry, since they directly form from the gas, taking away most molecules except H_2, CO, and generally H_2O. Grains are responsible for the bulk of the infrared emission of CSEs which contains a rich information about the composition and the location of the grains. Ground and space observations are quite complementary in the progression of our knowledge of the circumstellar dust.
Early observations in the atmospheric windows proved the importance of circumstellar dust (see e.g. the IRC survey at 2.2 μm (49)); gave insights on its composition through its spectral features, such as silicates at 10 μm (50), SiC at 11.4 μm (51) or ice at 3.1 μm (52); allowed the first modelings of the dust amount and of its distribution.
The results of the IRAS mission since 1983 have brought a new dimension to our knowledge of CSEs. They are the main class of sources of the IRAS Point Source Catalog (53) which gives their far IR flux in at least one of the four IRAS broad bands at 12, 25, 60 or 100 μm for 80000 of them. This completely new knowledge of their far IR emission has prompted multiple studies of CSEs and modelings of their IR emission. The IRAS far infrared colours contain useful information about the mass–loss rate and the dust composition. They allow for instance to roughly distinguish between O– and C–rich CSEs, and to infer the λ dependence of the dust emissivity law in the far infrared, proving that circumstellar silicates are generally rather amorphous than crystalline. Relatively anomalous IR colours have allowed the identification of CSEs with special structures : hollow shells of pre–planetary nebulae, multishell, etc.
Another extremely important source of information on CSEs and their dust composition is the IRAS Low Resolution Spectra (LRS). Its published atlas (54) provides 7.7–22.6 μm spectra of 5425 sources, $\sim$ 80% of which are CSEs. The complete LRS data base contains useful information on much more sources. The 10 classes and 72 subclasses of the LRS atlas allowed a first classification of CSEs, which will certainly be refined in the future. Prominent absorption or emission features are attributed to : i) silicates (9–12 and 15–21 μm), whose strong absorption in extreme CSEs was recently modeled (55) in order to infer dust properties and the mass–loss rate. ii) Broad SiC emission at 11.4 μm in C–rich CSEs. iii) Emission attributed to large aromatic molecules (PAH) at 11.3 μm

(and 8.6 μm) in sources where UV radiation is present (essentially young planetary nebulae as concerns evolved objects).
Other important recent developments related to dust composition and hence to circumstellar chemistry are : i) IR absorption (56) and emission (57) of ice in massive and cold CSEs. ii) The presence of polycyclic aromatic compounds (PAH) in young planetary nebulae, mainly C–rich (see e.g. 58) and in some pre–planetary nebulae. Up to now PAHs have been detected only through their near IR fluorescence (3.3, 6.2, 7.7, 8.6, 11.3 μm, etc) following UV, or possibly visible, illumination. This explains why their IR bands are strong only in planetary and warm pre–planetary (Red Rectangle, HR4049) nebulae, but quite weak in CRL2688 (59), the archetype of pre–planetary nebulae. Thus, it is still unknown whether PAHs preexist with large abundances in regular C–rich CSEs or are mostly formed in the transition stages to planetary nebulae. Some PAHs are also present (59) in the few C–rich pre–planetary nebulae displaying a strong emission band at 21 μm (60), but the origin of this 21 μm emission is still unclear.

6.2 Grain Formation. Overview

Grain formation is probably the major problem in circumstellar chemistry. It mainly determines the properties of circumstellar dust itself and of its infrared emission, with possible inferences on the properties of interstellar dust. It is also crucial for the chemistry of the circumstellar gas since it completely determines the abundances of all the refractory elements in the gas of the outer shells : metals, silicone, carbon (outside of CO), and even possibly sulfur and partially oxygen.
It is not surprising that many efforts have been devoted to this problem and in particular to its theoretical modeling. However, because of the extreme complexity of the real problem, these theoretical models are still much too simple to really account it, and direct observations are probably more important for our knowledge of dust formation. The following section reviews the different steps and problems arising in the modeling of the formation of circumstellar grains : the thermodynamical equilibrium theory of solid condensation does not fully apply, but it is a useful guide for the hierarchy of condensation of the different products. The nucleation of the first solid particles is a real problem and is certainly delayed with respect to thermal equilibrium. However, the homogeneous nucleation theory cannot apply to the extremely inhomogeneous composition of the circumstellar gas. Every aspect of the generalization to this inhomogeneous situation should be fully explored : e.g. the effect of the overwhelming abundance of hydrogen, the possibility that solid particles of more refractory species such as SiC are used as germs for the condensation of more volatile compounds such as solid carbon, etc. Additional difficulties are the complexity and the uncertainty of the physical parameters, and of the time evolution of the involved regions, as well as the related problems such as the formation of the stellar wind. Similar questions are still present in the final stages of dust formation with grain–grain collisions which probably determine their size distribution through coagulation or shattering, and with the final interactions between the gas and the grains : possible accretion of volatile compounds such as H_2O in the cold layers, possibility of catalytic reactions on the grains, of return of matter from the grains to the gas, etc.

The information on dust formation from infrared observations much depends on their angular resolution. When the source is unresolved as in the case of almost all present observations, the information is only indirect and relies much on modeling : spatial distribution inferred from the temperature distribution involved by the global infrared spectrum, composition from spectral features, size distribution from polarization properties, etc. Much more direct information is expected from the advent of high angular resolution in infrared observations, which should really show where dust really forms and how its properties evolve in the envelope with the distance to the star, as demonstrated by the first results of infrared interferometry such as those of ref. (61) in IRC+10216.

6.3 Dust Formation. From Thermodynamics to Nonhomogeneous Nucleation Theory

Most of early studies of star dust formation are based on thermodynamical equilibrium of a slowly cooling gas, which progressively yields the condensation of solids beginning with the most refractory compounds allowed by the gas composition. Such condensation sequences have been published for compositions, temperatures and pressures relevant for dust formation in O– and C–rich CSEs (62–67). In the carbon rich case, the first solid to appear could be SiC, followed by solid carbon whose equilibrium condensation from C_2H_2 is always located about 1800 K. For the O–rich gas, the most stable solids are $A\ell_2O_3$ and other refractory oxides or the corresponding silicates, such as $Ca_2\ A\ell_2\ SiO_7$, which condensate at equilibrium about 1500 K or slightly above, and then the more abundant magnesium silicates and metallic iron (1400–1100 K).
However, there are many indications and reasons for non equilibrium condensation of dust in CSEs : the analysis of the infrared emission shows that the maximum temperature of dust is much smaller (by $\sim$ 300–500 K) than the equilibrium condensation temperature, both for carbon dust in C–rich CSEs and for the bulk of silicates in O–rich ones. The remaining of important quantities of carbon compounds (C_2H_2, HCN, etc) in the outer gas of C–rich CSEs, and even of SiO in O–rich ones, proves that the condensation is far from being complete as it would be foreseen by the thermodynamical equilibrium. On theoretical grounds, it is well known that the need to form solid germs prior to the condensation introduces large delays in the condensation of initially pure gas. The corresponding theory is well established for pure compounds ("homogeneous nucleation", see e.g. 63, 68, 69). The lowering of the condensation temperature is easily understood by the requirement to form clusters of a critical size (typically 10–20 monomers) which are less stable than the bulk solid because "surface" effects reduce the binding energy per atom. The determination of the actual condensation temperature is then relatively simple and precise for a pure compound when the free energy of its clusters with various numbers of atoms is accurately known.
However, the circumstellar gas is a very complicated mixture of the different elements entering various kinds of molecules. The intermediate "clusters" that one has to consider in the corresponding "nonhomogeneous nucleation theory" are no longer polymers of the hypothetical monomers of the solid, which generally do not even exist in the gas, but complex large molecules of composition different from that of the solid (see e.g. 67). For instance for carbon condensation the overwhelming abundance of hydrogen makes that both the basic molecules of the gas (C_2H_2, etc) and the "clusters" (PAHs

or acetylenic chains) (and even the solid itself) are hydrogenated compounds. For the formation of O–rich dust, the situation is even more complicated, since there is nothing ressembling a silicate molecule in the gas, and it is not easy to foresee what are the important intermediate "clusters".

Several discussions of the formation of carbon dust with PAHs as intermediate steps have recently been presented (67, 70–72). However, the problem is more complex than a simple extension of the classical homogeneous nucleation theory with a discussion of the stability of more complex clusters. It is clear that one has also to consider kinetic aspects, by introducing the whole network of chemical reactions relevant for all the implied molecules. Therefore, not only the thermodynamical data of the large intermediate molecules ("clusters") are required, but also the various reaction rates. A particular difficulty concerns the reactions within the solid phase. Their slowness, at least inside the grains, can prevent the transformation of compounds previously condensed to more stable forms at low temperature.

To give a full idea of the formidable difficulty of the whole problem, one should add that :

A composite onion shell structure of the grains is likely, with successive condensations of layers of more and more volatile compounds. The first formed grains can serve as nuclei for the condensation of subsequent materials. For instance, it has been proposed that SiC grains are used for the condensation of carbon (44, 73).

The temperature and the density are quite uncertain in the region of dust formation with pulsations, shocks and inhomogeneities (Section 2.2). Moreover, dust, when formed, plays a fundamental role in the acceleration of CSE winds and in the gas heating.

Gas abundances are non–LTE. In particular overabundant atomic hydrogen interacts with "clusters", and its abundance is probably determined by its destruction on the grains (and possibly on the "clusters").

Because of the interaction with radiation, the internal temperatures of the grains and of the "clusters" are different from the kinetic temperature of the gas, and they vary with their sizes. Besides, the temperatures of small grains and "clusters" strongly fluctuate when they absorb a photon or accrete species with large binding energies.

6.4 Grain Growth, Size Distribution and Return to the Gas

As long as the density is large enough so that $\tau_{grain} < \tau_{dyn}$, grains can continue to accrete gas species outside of the region of their initial formation. The particular case of ice condensation on silicate grains was discussed by Jura and Morris (74). It occurs below ~ 110 K, and only with large mass–loss rates ($\sim 10^{-4} M_\odot$/yr) in order to ensure enough collisions with the grains in such outer layers.

It is likely that circumstellar grains present a distribution of sizes as in the interstellar medium. Since the grain ejection velocities depend on their sizes, collisions between grains are relatively frequent in the inner layers. Following the relative grain velocities, one can expect a size growth by coagulation or partial volatilization and shattering, which could generate a size distribution similar to the interstellar medium (75) and also changes in the internal structure of the grain.

Besides, grain–grain collisions (possibly generating for instance HC_7N in CRL2688 (76)), matter can be returned from grains to the gas if grains are reheated, or in shocks (possibly generating SO_2 in OH231.8+4.2 (97)), or by the formation of volatile compounds in surface reactions (e.g. CH_4 in reaction of H with graphite (77)).

6.5 Tentative Summary of the Chemical Composition at "Mid–Envelope"

In massive CSEs, one can well distinguish between the inner envelope where grains are formed and accelerated and the outer envelope where photochemistry is active (Figures 1 and 2). A clear view of the chemical composition at their separation, "mid–envelope", is important to provide the initial conditions for the external photochemistry.
One typically finds there :

- Non condensable species frozen out from the photosphere : H_2, He, CO, N_2, as well as most H_2O except for very large mass–loss rates.
- Species left out from grain condensation : C_2H_2 and other hydrocarbons such as CH_4, HCN, CS, SiS, SiO, possibly a limited amount of HC_3N, (H_2S ? or S ?), etc.
- Stable molecules formed by other processes such as shocks or grain reactions : NH_3 ? ; HCN and other C–bearing molecules in O–rich CSEs (Section 8.3.c).

7. PHOTODISSOCIATION BY INTERSTELLAR UV RADIATION

7.1. General case

As already stated (Section 2.5), it is believed that for most molecules except CO and H_2, photodissociation follows continuum absorption of a UV photon. The dissociation rates in the unshielded standard UV interstellar field are in the range 10^{-9}–10^{-11} s^{-1} (78). However, many rates are uncertain, especially for radical and exotic molecules, and even for SiO; the UV intensity can be enhanced with respect to its standard value, and the dust shielding is somewhat uncertain. There is some evidence that the far UV opacity of CSE dust is not very different from the interstellar one. Although one can work out a more elaborate modeling of the radiative transfer, one often uses a simple model for the UV intensity

$$I(r) = I(\infty)\ exp(-d/r) \tag{10}$$

where the shielding parameter d slightly depends on the molecule with a typical value (see e.g. 23)

$$d = 2.5\ 10^{16}\ (\dot{M}/10^{-5} M_\odot/yr)\ cm \tag{11}$$

Subsequent photodissociations generate some kinds of sequences (1, 79), e.g.

$$\begin{array}{l} C_2H_2 \rightarrow C_2H \rightarrow C_2 \rightarrow C \rightarrow C^+ \\ HCN \rightarrow CN \rightarrow C \rightarrow C^+ \\ CO \rightarrow C \rightarrow C^+ \end{array} \tag{12}$$

The case of $H_2O \rightarrow OH \rightarrow O$ is of particular interest since it determines the OH abundance and, hence, the OH maser emission rather universal in O–rich CSEs. It has been shown in its modeling (22, 80) that the UV shielding through continuum absorption of H_2O is not negligible with respect to dust shielding.

7.2. Photodissociation of CO

CO is the most easily observable molecule in CSEs (Table 1). The CO photodissociation is also important because it is the major source of C^+, especially in the regions of O–rich CSEs, where photochemistry is active.

It is well known that the interstellar and circumstellar photodissociation of CO operates through line absorption, with a large unshielded rate (see e.g. 81) and a strong line shielding. The circumstellar photodissociation has been modeled by Mamon, Glassgold and Huggins (24). They found that for $\dot{M} \gtrsim 10^{-6} M_\odot/\mathrm{yr}$, the photodissociation radius was roughly proportional to $\dot{M}^{1/2}$ (Figure 4).

The self-shielding is obviously smaller for ^{13}CO. However, the resultant fractionation is probably negligible (24) because of the reaction

$$^{13}C^+ + {}^{12}CO \longrightarrow {}^{13}CO + {}^{12}C^+ \qquad (13)$$

On the other hand, the abundance of $C\,^{18}O$ and $C\,^{17}O$ can be reduced with respect to that of $C\,^{16}O$.

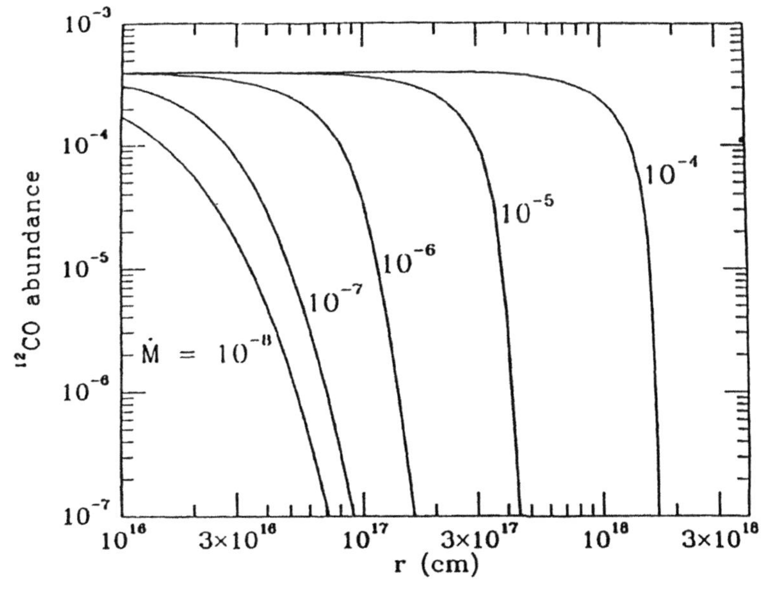

Figure 4. The fractional abundance of CO calculated as a function of radius for various mass–loss rates (Mamon et al. ref. 23).

8. SYNTHESIS BY PHOTOCHEMISTRY

8.1 Ions in CSEs

Ions are known to play a central role in interstellar chemistry, and the radio lines of HCO^+ for instance are quite strong in interstellar sources. However, until recently, molecular ions had remained undetected in CSEs. Nevertheless, theoretical modeling has shown that ions are present in CSEs, at a level not much lower than in the interstellar medium, and they can play an important role in circumstellar chemistry as well (1, 23, 82-84).

In regular (AGB) CSEs, the main sources of ionization are cosmic rays and photoionization by the interstellar UV radiation of C_2H_2 (leading to $C_2H_2^+$ or $C_2H_3^+$) and of C (see Eq. 13). Subsequent ionic reactions produce a lot of secondary ions (see e.g. Figure 5).

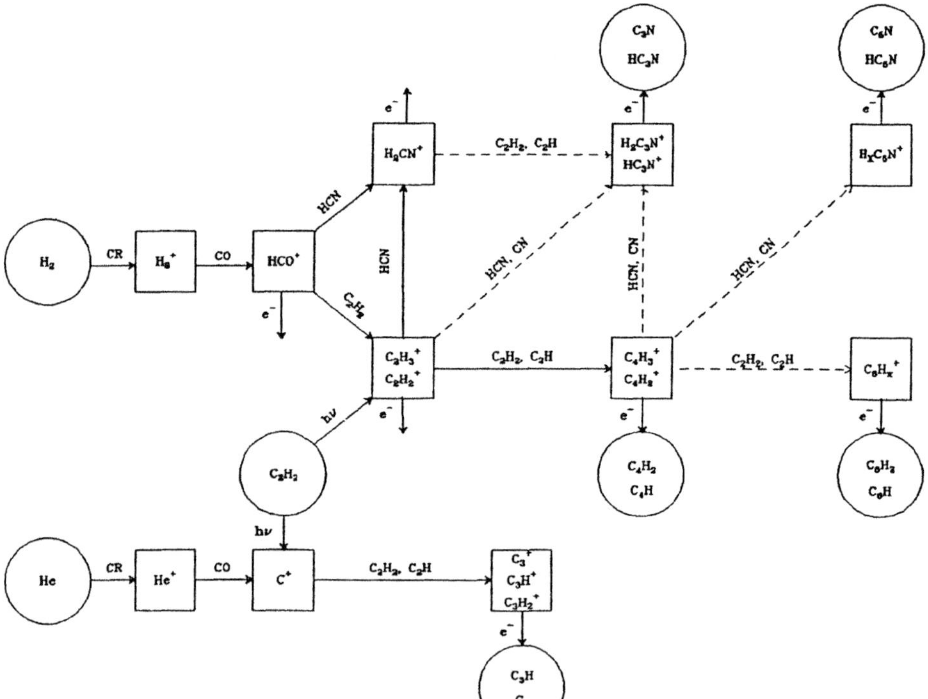

Figure 5. Schematic diagram of ion-molecule chemistry in a C-rich CSE. Neutral species in circles, ions in squares: Solid lines indicate reactions pathways based on measured reaction rates, dash lines are possible pathways which require reactions that have not yet been measured. The main primary ions are H_3^+ (from cosmic rays), $C_2H_2^+$ (photoionization of C_2H_2), and C^+ (mainly the end product of the C_2H_2 photochemical sequence). From ref. 82.

The most abundant ions are probably $C_2H_2^+$ in C–rich CSEs, and H_3O^+ and HCO^+ in O–rich CSEs, with abundances $\sim 10^{-7}$. The main route for HCO^+ in O–rich CSEs is

$$CO \xrightarrow{h\nu} C \xrightarrow{h\nu} C^+ \xrightarrow{H_2O,OH} H_3O^+ \xrightarrow{CO} HCO^+ \tag{14}$$

Smaller amounts of HCO^+ are formed in C–rich CSEs from cosmic rays as in the interstellar gas :

$$H_2 \xrightarrow{CR} H_2^+ \xrightarrow{H_2} H_3^+ \xrightarrow{CO} HCO^+ \tag{15}$$

The tentative detection of HCO^+ in IRC + 10216 (85-86) is in agreement with the order of magnitude of the predicted abundance, $\lesssim 10^{-9}$.
Much stronger ionizations are encountered when there is an internal source of ionization, such as chromospheric UV radiation in the supergiant α Ori (35) or the central stars of planetary nebulae. In both cases the resultant ions are mainly atomic. However, strong HCO^+ radio lines have been detected in planetary nebulae with prominent molecular shells (NGC 7027, CRL 618, IRAS 21282+5050 (86, 38, 87), as well as in a few other CSEs (CIT 6 and VY CMa (86)), without a clear identification of the formation process of HCO^+.

8.2 Photosynthesis through Ionic Reactions

The fact that the relatively large amounts of ions produced by photoionization could induce the synthesis of complex molecules by a chain of ionic reactions similar to that of the interstellar chemistry was first proposed by Glassgold and Huggins (1). Detailed modelings were performed in C–rich CSEs (82–84, 88), and in O–rich ones (23, 89). The main products which could be formed in that way are in particular (see Figure 5) :

i) C_3H and C_3 (83), from the reaction

$$C^+ + C_2H_2 \rightarrow C_3H^+ + H \tag{16}$$

ii) C_4H and larger acetylenic hydrocarbons (82–84, 88), from

$$\begin{aligned} C_2H_2^+ + C_2H_2 \quad &\rightarrow \quad C_4H_3^+ + H \\ &\rightarrow \quad C_4H_2^+ + H_2 \end{aligned} \tag{17}$$

iii) HC_3N (82–84, 88), from

$$C_2H_2^+ + HCN \longrightarrow H_2C_3N^+ \xrightarrow{e^-} H_3CN \tag{18}$$

However, the amount of HC_3N producted in such a way is about a factor 3 smaller than observed, and a radical reaction with CN could be more efficient (Section 8.3).

iv) HNC (83), from

$$C_2H_2^+ + HCN \longrightarrow H_2CN^+ + C_2H \xrightarrow{e^-} HNC,\ HCN \tag{19}$$

in very good agreement with the abundance of HNC observed in IRC+10216.

v) SiC_2 from $Si^+ + C_2H_2$ or C_2H (82).

vi) C_2S and C_3S from $CS + C_2H_2^+$ (88).

8.3 Photosynthesis through Radical Reactions

8.3.a C–rich CSEs

Although radical reactions have smaller rates than ionic ones, they can be more important in some cases because of the relatively large abundances of radicals which are directly created by photodissociation and are not destroyed by dissociative recombination. Various radical reactions have been recently considered by Howe and Millar (88) in C–rich CSEs. The most important possibilities are in particular

i) HC_3N, from

$$CN + C_2H_2 \rightarrow HC_3N + H \tag{20}$$

The very large reaction rate ($2\ 10^{-10}$ cm^3 s^{-1}) seems to ensure that this is the dominent formation route for HC_3N, in good agreement with the observations. Similar reactions of CN with C_xH_2 could perhaps form HC_5N, HC_7N, etc.

ii) C_4H and C_6H, from

$$C_{2n}H_2 + C_2H \longrightarrow C_{2n+2}H_2 + H \xrightarrow{h\nu} C_{2n+2}H \tag{21}$$

which could be at least as efficient as the ionic reactions (17).

iii) SiC_4 from $SiC_2 + C_2H_2 \rightarrow SiC_4 + H_2$?

8.3.b S–bearing molecules in O–rich CSEs

Radical reactions of atomic S with OH were proposed a long time ago (90, 91) to produce large quantities of SO and SO_2 in O–rich CSEs :

$$\begin{aligned} S + OH &\rightarrow SO + H \\ SO + OH &\rightarrow SO_2 + H \end{aligned} \tag{22}$$

Detailed modeling (91, 89) predicts that a large fraction of sulfur in the gas phase should be in the form of SO and SO_2 in the outer layers of OH/IR stars. The relatively small intensity of the radio lines of these molecules (92, 93) shows that an important, and probably major, fraction of sulfur is locked up in the grains. The shape of the observed lines of SO_2 (93) seems in agreement with such a formation in the outer layers of the CSEs, although one will have to wait for a detailed mapping to confirm it.
The case of H_2S, detected in the same stars (93), is less clear. There are some indications from the line profile and from the high excitation temperature, that the emission could come from the inner layers. However, it is difficult to imagine a detailed chemical scenario explaining all the features of the emission of H_2S and of SO_2, and taking into

account the orders of magnitude expected for the possible chemical processes (grain reactions, equilibrium with H/H_2, photodestruction, etc).

8.3.c Carbon–bearing molecules in O–rich CSEs

Thermal equilibrium predicts negligible amounts of all C–bearing molecules in O–rich CSEs outside of CO. However, relatively large abundances of HCN and CS are observed (94–96 and Section 3.5).
The first proposed explanation for HCN, i.e. photosynthesis from reactions between the photodissociation products of the primary molecules CO and N_2, was proved to fail by several orders of magnitude. Most of the other considered processes also proved to be not efficient enough, e.g. reaction of NH_3 (itself to be explained) with C^+ (89, 95).
A possible explanation would be the reaction of N with CH_x radicals producted by the photodissociation of CH_4 (89). However, it requires very large amounts of CH_4, which are not expected at all by thermal equilibrium calculations in the vicinity of the photosphere. Accordingly, it seems that some non–equilibrium process is required to extract carbon from CO in the inner layers (shocks ?) and to incorporate it there, either directly into HCN or into some intermediate product such as CH_4.

9. CONCLUSION

The main conclusion emerging from the observational evidences and the theoretical discussions is that photochemistry could be responsible for nearly all the exotic molecules observed in CSEs. The recent quantitative developments of circumstellar photochemistry are rather impressive, probably because the conditions and the time evolution are much better defined in CSEs than in interstellar clouds. With the improvements of the observations (especially as concerns angular resolution), of the data on reaction rates and of the modeling, one can hope to work out satisfactory models for most molecules in the various types of sources.
The situation is much more difficult for the chemistry of the inner layers. Even a qualitative understanding still requires a lot of efforts. The first goals are obviously a better knowledge of the physical parameters, of the observational abundances there and of the dust formation. The expected developments in infrared angular resolution and spectroscopy will clearly be crucial.

Acknowledgements

I wish to thank Mrs C. Douillet, V. Demailly and M.C. Pantalacci for their editorial work.

REFERENCES

1. A.E. Glassgold and P.J. Huggins (1985): M–Type stars, ed. by H.R. Johnson and F. Querci (NSF-CNRS 1988) p.291
2. A. Omont: In Mass Loss in Red Giants, ed. by M. Morris and B. Zuckerman (Reidel, Dordrecht 1985) p.269
3. A. Omont: In Astrochemistry, IAU Symp. N. 120, ed. by M.S. Vardya and S.P. Tarafdar (Reidel, Dordrecht 1987) p.357
4. T.J. Millar: In Rate Coefficients in Astrochemistry, ed. by T.J. Millar and D.A. Williams (Kluwer Academic Pub. 1988) p.287
5. Physical Processes in Red Giants, ed. by I. Iben and A. Renzini (Reidel, Dordrecht 1981)
6. Mass Loss in Red Giants, ed. by M. Morris and B. Zuckerman (Reidel, Dordrecht 1985)
7. B. Zuckerman: Ann. Rev. Astron. Astrophys. 18, 263 (1980)
8. Evolution of Peculiar Red Giants, ed. by H.R. Johnson and B. Zuckerman Cambridge University Press (1989)
9. M–Type stars, ed. by H.R. Johnson and F. Querci (NSF-CNRS 1988)
10. From Miras to Planetary Nebulae: Which Path for Stellar Evolution ?, ed by M.O. Menessier and A. Omont (Editions Frontieres, Gif-sur-Yvette 1990)
11. Late Stages of Stellar Evolution, ed. by S. Kwok and S.R. Pottasch (Reidel, Dordrecht 1987)
12. I. Iben and A. Renzini: Ann. Rev. Astron. Astrophys. 21, 271 (1983)
13. I. Iben: Quat. J. Roy. Astr. Soc. 26, 1 (1986)
14. T. de Jong: Astron. Astrophys. Lett. 223, L23 (1989)
    In From Miras to Planetary Nebulae: Which Path for Stellar Evolution ?, ref. 10 p.289
15. B. Zuckerman and R.J. Madalena: Astron. Astrophys. Lett. 223, L20 (1989)
16. S.G. Kleinmann, M. Jura: Astrophys. J. 341, 359 (1989)
17. A.G.G.M. Tielens: Astrophys. J. 271, 702 (1983)
18. P. Goldreich, N.Z. Scoville: Astrophys. J. 205, 144 (1976)
19. J. Kwan, R.A. Linke: 254, 587 (1982)
20. M. Jura, C. Kahane, A. Omont: Astron. Astrophys. 201, 80 (1988)
21. M. Morris, M. Jura: Astrophys. J. 264, 546 (1983)
22. P.J. Huggins and A.E. Glassgold: Astron. J. 87, 1828 (1982)
23. G.A. Mamon, A.E. Glassgold, A. Omont: Astrophys. J. 323, 306 (1987)
24. G.A. Mamon, A.E. Glassgold, P.J. Huggins: Astrophys. J. 328, 797 (1988)
25. A.E. Glassgold: In Astrochemistry, IAU Symp. N. 120, ed. by M.S. Vardya and S.P. Tarafdar (Reidel, Dordrecht 1987) p.379
26. H. Olofsson: In (Sub)millimeter Astronomy, ed. by P.A. Schaver and K. Kjar (ESO Publications, Garching 1985)
27. H. Olofsson: Space Sci. Rev. 47, 145 (1988)
28. M. Guelin and J. Cernicharo: in The Physics and Chemistry of Interstellar Molecular Clouds, Zermatt, ed. by G. Winnewisser and J.T. Armstrong. Lecture Notes in Physics (Springer–Verlag, Heidelberg, 1989)

29. J. Cernicharo, M. Guelin, H. Hein, C. Kahane: Astron. Astrophys. 181, L9 (1987)
30. J. Cernicharo, M. Guelin, K.M. Menten, C.M. Walmsley: Astron. Astrophys. 181, L1 (1987)
31. M. Guelin, J. Cernicharo: Astron. Astrophys. 183, L10 (1987)
32. M. Guelin, J. Cernicharo, G. Paubert, B.E. Turner: Astron. Astrophys. 230, L9 (1990)
33. M. Ohishi et al.: Astrophys. J. Lett. 345, L83 (1990)
34. K.H. Hinkle, J.J. Keady, P.F. Bernath: Science 241, 1319 (1988)
35. P.F. Bernath, K.H. Hinkle, J.J. Keady: Science 244, 562 (1989)
36. M. Morris, S. Guilloteau, R. Lucas, A. Omont : Astrophys. J. 321, 88 (1987)
37. C. Loup, T. Forveille, J.F. Paul and A. Omont: Astron. Astrophys. Supp. in preparation
38. V. Bujarrabal, J. Gomez-Gonzalez, R. Bachiller, J. Martin-Pintado: Astron. Astrophys. 204, 242 (1988)
39. C. Kahane, M. Jura, A. Omont: in preparation
40. C. Loup, T. de Jong, A. Omont, T. Forveille, M.A.T. Groenewegen: in preparation
41. J.H. Bieging, Nguyen-Q-Rieu: Astrophys. J. Lett. 329, L107 (1988)
In From Miras to Planetary Nebulae: Which Path for Stellar Evolution ?, ref 10 p.164
42. A.E. Glassgold, P.J. Huggins: Monthly Notices Roy. Astron. Soc. 203, 517 (1983)
43. T. Tsuji: Ann. Tokyo Astron. Obs. 2nd Ser. 9, 1 (1963)
Astron. Astrophys. 23, 411 (1973)
44. E.M. McCabe, R.C. Smith, R.E.S. Clegg: Nature 281, 263 (1979)
45. S. Lafont, R. Lucas and A. Omont: Astron. Astrophys. 106, 201
46. T. Tsuji: In Astrochemistry, IAU Symp. N. 120, ed. by M.S. Vardya and S.P. Tarafdar (Reidel, Dordrecht 1987) p.409
47. I.K. Shmeld, V.S. Strelnitskii, L.V. Gurvich: Sov. Astron. Lett. 11, 254 (1985)
48. P.R. Wood: In From Miras to Planetary Nebulae: Which Path for Stellar Evolution ?, ref. 10 p.67
49. G. Neugebauer, R.B. Leighton: NASA SP 3047 (1969)
50. N.J. Woolf and E.P. Ney: Astrophys. J. (Letters) 155, L181 (1969)
51. R. Treffers, M. Cohen: Astrophys. J. 188, 545 (1974)
52. B.T. Soifer, S.P. Willner, R.W. Capps, R.J. Rudy: Astrophys. J. 250, 631 (1981)
53. IRAS, Point Source Catalogue: US Government Publication Office (1985)
54. IRAS, LRS Catalogue, eds F.M. Olnon and E. Raymond: Astron. Astrophys. Supp. 65, 607 (1986)
55. W. Schutte, A.G.G.M. Tielens: Astrophys. J. 343, 369 (1989)
56. R.G. Smith, K. Sellgren, A.T. Tokunaga: Astrophys. J. 334, 209 (1988)
57. A. Omont, H.M. Moseley, T. Forveille, W.J. Glaccum, P.H. Harvey, L. Likkel, R.F. Loewenstein and C.M. Lisse: Astrophys. J. Lett. 355, L27 (1990)
58. M. Cohen et al.: Astrophys. J. 302, 737 (1986)
59. M. Cohen: private communication
60. S. Kwok, K.M. Volk and B.J. Hrivnak : Astroph. J. Lett. 345, L51 (1989)
61. W.C. Danchi, M. Bester, C.G. Degiacomi, P.R. McCullough, C.H. Townes: Astrophys. J. Lett. 359, L59 (1990)
62. J.S. Lewis: this volume

63. E.E. Salpeter: Ann. Rev. Astron. Astrophys. 15, 267 (1977)
64. L. Grossman, J.W. Larimer: Rev. Geophys. Space Phys. 12, 71 (1974)
65. S.S. Barshay, J.S. Lewis: Ann. Rev. Astron. Astrophys. 14, 81 (1976)
66. A.G.G.M. Tielens: in From Miras to Planetary Nebulae: Which Path for Stellar Evolution ?, ref. 10, p. 186
67. A.G.G.M. Tielens: in Carbon in the Galaxy: Studies from Earth and Space, ed. J. Tarter, (NASA-CP) in press
68. E.E. Salpeter: Astrophys. J. 193, 579 (1974)
69. B.T. Draine: Astr. Space Sci 65, 313 (1979)
70. R. Keller: in Polycyclic Aromatic Hydrocarbons and Astrophysics, eds A. Leger, L.B. d'Hendecourt, N. Boccara (Reidel, Dordrecht 1987), p.63
71. H.P. Gail and E. Sedlmayr: in Physical Processes in Interstellar Clouds, eds G.E. Morfill and M. Scholer (Reidel, Dordrecht 1987), p.275
72. M. Frenklach, E.D. Feigelson: Astrophys. J. 341, 372 (1989)
73. M. Frenklach, C.S. Carmer, E.D. Feigelson: Nature 339, 196 (1989)
74. M. Jura, M. Morris: Astroph. J. 292, 487 (1985)
75. P. Bierman and M. Harwit: Astrophys. J. Lett. 241, L105 (1980)
76. M. Jura and H. Kroto: Astrophys. J. 351, 222 (1990)
77. M.S. Barlow, J. Silk: Astrophys. J. 215, 800 (1977)
78. E. F. van Dishoeck: In Rate Coefficients in Astrochemistry, ed. by T.J. Millar and D.A. Williams (Kluwer Academic Pub. 1988) p.49
79. P.J. Huggins. A.E. Glassgold: Astrophys. J. 252, 201 (1982)
80. N. Netzer, G.R. Knapp: Astrophys. J. 323, 734 (1987)
81. C. Letzelter, M. Eidelsberg, F. Rostas, J. Breton, B. Thieblemont: Chem. Phys. 114, 273 (1987)
82. A.E. Glassgold, R. Lucas, A. Omont: Astron. Astrophys. 157, 35 (1986)
83. A.E. Glassgold, G.A. Mamon, A. Omont, R. Lucas, : Astron. Astrophys. 180, 183 (1987)
84. L.A.M. Nejad, T.J. Millar: Astron. Astrophys. 183, 279 (1987)
85. R. Lucas, M. Guelin: Kona Symposium, Submillimeter and Millimeter Astronomy (1989)
86. S. Guilloteau, A. Omont, R. Lucas: in preparation
87. L. Likkel, T. Forveille, A. Omont, M. Morris: Astron. Astrophys. 198, L1 (1988)
88. D.A. Howe and T.J. Millar: Monthly Notices Roy Astr. Soc. 244, 444 (1990)
89. L.A.M. Nejad, T.J. Millar: Monthly Notices Roy Astr. Soc. 230, 79 (1988)
90. J.M. Scalo, D.B. Slavsky: Astrophys. J. 239, L73 (1980)
91. D.B. Slavsky, J.M. Scalo: preprint (1986)
92. R. Lucas, A. Omont, S. Guilloteau, Nguyen-Q-Rieu: Astron. Astrophys. 154, L1 (1986)
93. A. Omont, M. Morris, R. Lucas, S. Guilloteau: in preparation
94. S. Deguchi, M.J. Claussen, P.F. Goldsmith: Astrophys. J. 303, 810 (1986)
95. E. Nercessian, S. Guilloteau, A. Omont, J.J. Benayoun: Astron. Astrophys. 210, 225 (1989)
96. M. Linqvist, L-A Nyman, H. Olofsson, A. Winnberg: Astron. Astrophys. 205, L15 (1988)
97. J.M. Jackson, Nguyen-Q-Rieu: Astrophys. J. Lett. 335, L83 (1988)

FIGURE CREDITS

Figure 1. Mass Loss in Red Giants, ed. by M. Morris and B. Zuckerman (Reidel, Dordrecht, 1981), p.269.

Figure 2. Astrochemistry, IAU Symp. N. 120, ed. by M.S. Vardya and S.P. Tarafdar (Reidel, Dordrecht 1987) p.362.

Figure 3. Astrochemistry, IAU Symp. N. 120, ed. by M.S. Vardya and S.P. Tarafdar (Reidel, Dordrecht 1987) p.414.

Figure 4. G.A. Mamon, A.E. Glassgold, P.J. Huggins: Astrophys. J. 328, 797 (1988).

Figure 5. A.E. Glassgold, R. Lucas, A. Omont: Astron. Astrophys. 157, 35 (1986) p.38.

The Chemistry of Supernova 1987A

A. DALGARNO AND S. LEPP
Harvard University
Center for Astrophysics
60 Garden Street
Cambridge, MA 02138 USA

1 Introduction

Approximately 170,000 years ago the core of a blue supergiant star in the Large Magellanic Cloud collapsed and the star exploded, becoming a supernova. The light from the supernova reached the earth on Febuary 23, 1987 and the supernova was named SN1987A. It is the brightest supernova since the invention of the telescope providing a unique opportunity to bring to bear the full power of modern astronomical equipment to study the expanding nebula. SN1987A is only the second supernova for which the progenitor star has been observed. It is the first supernova to be observed in all wavelength bands: radio, infrared, optical, ultraviolet, X-ray and γ-ray, and it is the first supernova from which neutrinos have been detected.

Truly a once in a lifetime opportunity, SN1987A has already taught us much about supernovae. The neutrinos provided the first direct evidence of the formation of a neutron star. The early emergence of X-rays and γ rays has changed our views of the evolution of the ejecta. Another surprise was the detection of molecules in the ejecta. The molecular species carbon monoxide, probably silicon monoxide and possibly the molecular ion CO^+ have been identified in the ejecta of SN1987A. The existence of molecules in such an apparently hostile environment is an interesting problem of chemistry and it means that chemistry can be used to investigate the conditions in the ejecta.

In this lecture we will review the life of a massive star, its internal structure and what happens when it becomes a supernova. Next we discuss some of the observations of SN1987A and what can be inferred about the ejecta environment. We discuss the formation and destruction of CO molecules in the ejecta environment and what may be learned about the supernova from the observations of CO. Finally, we discuss the additional research needed to understand the supernova and we consider the possible future of SN1987A.

2 Birth of a Supernova

A supernova is created in the death of a massive star. Massive stars are formed from collapsing clouds. As the cloud collapses the density and temperature rise until the pressure balances the gravitational attraction. With the increased temperature the cloud radiates away energy and allows the collapse to continue. The center of the cloud continues to increase in temperature until the temperature becomes so high nuclear burning of hydrogen

J. M. Greenberg and V. Pirronello (eds.), Chemistry in Space, 199–209.

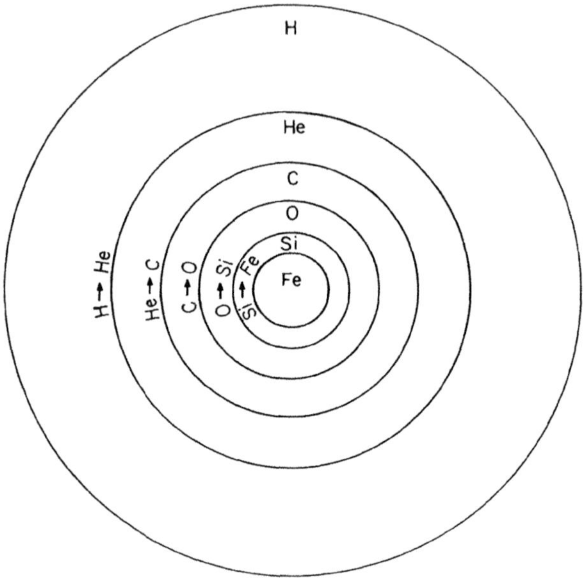

Figure 1: A simplified diagram of the structure of a massive star immediately preceding core collapse. The nuclear burning occurring on each shell is also shown.

is initiated in the core of the the cloud and a star is born. At this time an equilibrium exists as the star burns hydrogen into helium in its core generating energy to balance the radiation emitted. A massive star spends most of its life in this hydrogen burning phase.

When all the hydrogen in the core is converted to helium the core contracts and the temperature rises. Hydrogen burning continues on the surface of the helium core. As more mass is added on to the He core the temperature and density increase. Eventually the temperature and density are high enough to allow nuclear burning of helium. Helium is converted to carbon releasing energy and a new equilibrium is formed.

When the helium in the core is completely converted to carbon the core again contracts and heats up. The He burning continues on a shell around the carbon core and the hydrogen burning continues on a shell around the He layer. This process of burning followed by core collapse continues with successive cores of H, He, C, O, Si and finally Fe. Just before the star undergoes core collapse the structure of the star consists of a core surrounded by many shells like the skins of an onion. At the center is a core of Fe surrounded by a shell of Si surrounded by successive shells of O, C, He and H. On the interface between each shell nuclear burning continues to convert the lighter element to the heavier. The structure is illustrated in Fig. 1.

Iron is the most tightly bound of the elements and so energy cannot be released by nuclear burning of iron. Instead the iron core collapses until the electrons fill all available states. The electrons, spin one-half particles, are fermions and only one electron may occupy each state. This electron degeneracy pressure supports the core until the mass of the core exceeds the Chandrasekhar limit, which is about 1.4 $M_\odot$. At this mass the electron degeneracy pressure can no longer resist the gravitational forces and the core collapses. As the core collapses electrons combine with protons to form neutrons and neutrinos. The neutrinos escape the core and carry energy away, which cools the core and allows the collapse to continue. Additional cooling comes when Fe nuclei are broken up into alpha particles and neutrons. The collapse continues until the core reaches nuclear matter densities, when the collapse stops abruptly. The core rebounds and deposits about one percent of the energy of the collapse into a shock wave which propagates out toward the surface of the star. Most of the energy is released in the form of neutrinos as the newly formed neutron star cools. The core is so dense that the neutrinos do not escape directly but are thermalized and escape over a few seconds. The first evidence of SN1987A to reach the earth was this burst of thermal neutrinos; the neutrino burst carried 10^{53} ergs of energy.

The energy released by SN1987A was enormous. In the largest nuclear weapon exploded about 19 kg (approximately 8.5 lbs) of mass was converted into energy; the supernova converted 20,000 times the mass of the earth. If just one percent of the energy released could be captured and converted into electricity it would be enough to keep the world supplied with electricity at our present rate of consumption for approximately 10^{27} years, that is 10^{17} times the age of the universe. (Of course, if we insist on increasing our consumption by 10% each year then its only enough energy to last about 600 years.)

As the shock propagates into the shells surrounding the now collapsed core it compresses and heats the material to temperatures of 10^6 K. Some of the energy of the shock is lost to explosive nuclear burning, the rest is split between heating and outward motion. Over several days much of the thermal energy is converted into kinetic energy of the expanding shell. The lower density outer parts of the progenitor star are accelerated to higher velocities than the higher density inner region. The ejecta eventually settle into an homologous free expansion with material moving at a constant velocity. The radius of material moving at a velocity V_0 is given by $R(t) = V_0 t$ where R is the radius and t the time since the explosion. The velocity ranges from 500 to 10000 kms^{-1}.

Initially the ejecta are layered as was the progenitor star. The inner shell of the ejecta consists of material which has been explosively nucleosynthesised up to the iron group surrounded by successive shells consisting of mostly Si, O, C, He and H. Much of the iron group shell consists of ^{56}Ni which is unstable and beta decays to ^{56}Co with a half life of 6 days. The decay of ^{56}Ni deposits approximately 10^{49} ergs s^{-1} into the innermost shell, the dynamical effects of which we discuss below. The ^{56}Co decays to ^{56}Fe with a half life of 77 days and provides a significant source of heat and ionization to the ejecta over several half lives.

3 Observations

The optical light curve for SN1987A is shown in Fig. 2 (McCray 1989). There is a bright flash as the shock reaches the surface of the star. Initially the ejecta is optically thick and energy deposited by the shock in the inner regions slowly propagates outwards. As the supernova expands the surface area increases allowing energy to escape more easily leading to the slow rise to a secondary peak in the light curve. The secondary peak is also driven by the energy from radioactive decay of ^{56}Ni. The size of this second peak is related to the

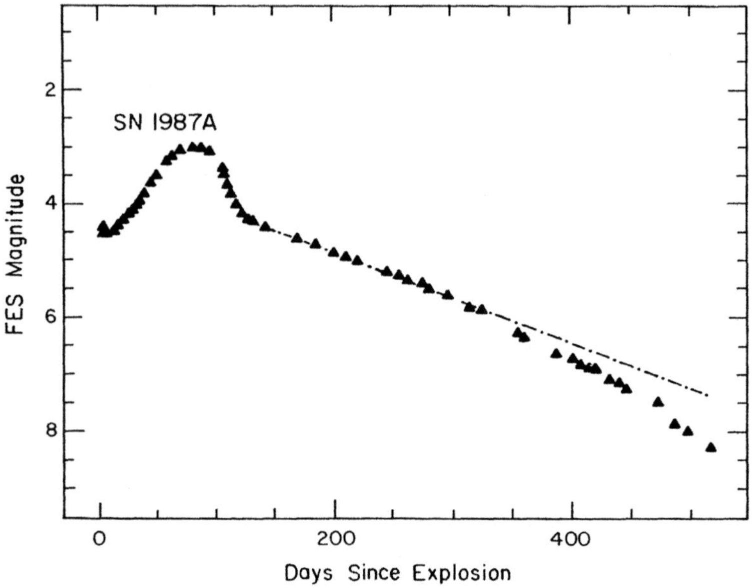

Figure 2: The optical light curve of SN1987A. The x axis is the luminosity of the supernova in magnitudes from the FES camera on the *International Ultraviolet Explorer* satellite (from McCray 1989).

total energy deposited and may be used to determine that a mass of about 0.07 $M_\odot$ ^{56}Ni was ejected by the supernova. As the ejecta expands the material becomes less optically thick and after about 100 days the ejecta has reached a steady state with the time for radiation to escape from the inner shells short compared to the time for the radius to change. For a time the light curve follows the 77 day half-life characteristic of the energy deposited from the decay of ^{56}Co.

The decay of ^{56}Co produces γ rays and an energetic positron. At times less than 300 days the γ-rays produced in the center cannot escape the shell and they Compton scatter off free and bound electrons producing X-rays and fast electrons. The X-rays are rapidly absorbed by photoionizing inner shell electrons of iron producing additional fast electrons The fast electrons are slowed by collisions with thermal electrons, by ionization of atoms and by excitation of electronic states in atoms.

One of the great surprises of SN1987A was the emergence of X-rays in July, 1987 (Dotani et al. 1987, Sunyaev et al. 1987). The X-rays were of the spectrum expected (McCray, Shull, and Sutherland 1987) but were observed nearly 4 months earlier than had been predicted. In the standard model the source of the X-rays, the radioactive decay of ^{56}Co, occurs at the innermost shell of the ejecta and for the X-rays to escape they must travel through all the material of the ejecta. The early emergence of the X-rays indicates that some of the ^{56}Co producing the X-rays has been mixed outward nearer the outer portions of the ejecta. A mechanism for this mixing is a Rayleigh-Taylor instability. The energy

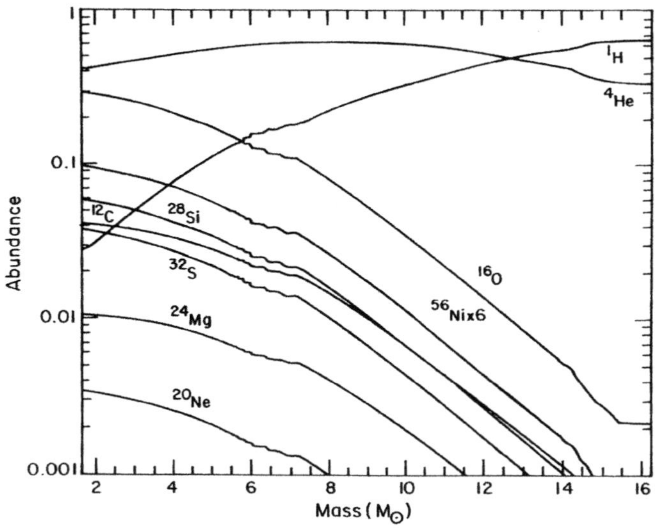

Figure 3: The fractional abundance by mass versus the mass interior (from Pinto and Woosely 1988).

input by the decay of ^{56}Ni heats the inner portions of the ejecta which are then accelerated into the lighter outer regions (Arnett, Fryxell and Müller 1989, Ebisuzaki, Shigeyama and Nomoto 1989).

Fig. 3 shows the fractional abundance as a function of the mass interior for a mixed model by Pinto and Woosley (1988). The mass interior is equivalent to radius from the center. This model has been artificially mixed to reproduce the optical light curve and the early emergence of γ-rays and X-rays. In it the stratified structure of the progenitor star is no longer evident except in the concentration of heavy elements toward the center. The mixing is probably neither as complete nor as uniform as suggested by this picture in that the Rayleigh-Taylor instabilities cause the heavy material to send fingers into the lighter material rather than achieve complete mixing. The effect of mixing on molecule formation is discussed below.

4 Molecular Spectra

Initially the envelope was opaque and its spectrum featureless. As the supernova expanded, the envelope became transparent, the continuum emission gave way to emission lines and the explosive products of SN1987A could be seen. Fine-structure lines of Ni, Ni^+, and Co^+ appeared as did vibrational bands of CO (Oliva, Moorwood and Danziger 1987, Rank et al. 1988, Aitken et al 1988, Spyromilio et al. 1988, Meikle et al. 1989). An excess emission between 8 and 9 μ has been tentatively attributed to SiO (Rank et al. 1988, Aitken et al 1988) and a feature at 3.26μ to CO^+ (Petuchowski et al. 1989).

Table 1: CO temperature and mass at various times[1]

Epoch (days)	Temperature (K)	Mass ($M_\odot$)
192	3000	1.7×10^{-5}
255	1800	4.7×10^{-5}
284	1600	1.2×10^{-4}
349	1200	8.0×10^{-5}

[1] from Spyromilio et al. (1988)

The 5-3, 4-2, 3-1, and 2-0 bands of CO could be distinquished. By assuming that the vibrational levels of CO are populated according to thermal equilibrium,

$$\frac{n(v)}{n(0)} = exp(-(E_v - E_0)/kT)\,,$$

where E_v is the energy of the vibrational level v, an excitation temperature T can be derived by fitting the emission profile. Given the excitation temperature, the total mass of CO can be derived from measured emission intensities. Spyromilio et al. (1988) obtained the temperatures and masses shown in Table 1. The fits are less satisfactory for days 284 and 389 possibly due to the presence of other emission lines and possible due to departures from thermal equilibrium as the electron density in the ejecta diminishes, the principal population mechanism being

$$\mathrm{e} + \mathrm{CO}(v) \rightarrow \mathrm{e} + \mathrm{CO}(v')\,.$$

The derived CO velocities are about 1200 - 2000 km s^{-1} showing that molecular formation and excitation is taking place in the core of the ejecta.

5 Ionization Distribution

Before considering the chemistry we need a picture of the ionization distribution in the ejecta. The ejecta are subjected to γ-rays and fast positrons from the decay of ^{56}Co. The γ-rays are degraded into X-rays which are preferentially absorbed by the heavy elements. Auger processes produce multiply-charged ions which capture electrons in radiative recombination,

$$\mathrm{X}^{\mathrm{m}+} + \mathrm{e} \rightarrow \mathrm{X}^{(\mathrm{m}-1)+} + \mathrm{h}\nu\,,$$

and in dielectronic recombination

$$\mathrm{X}^{\mathrm{m}+} + \mathrm{e} \rightarrow \mathrm{X}^{(\mathrm{m}-1)+*} \rightarrow \mathrm{X}^{(\mathrm{m}-1)+} + \mathrm{h}\nu\,.$$

When the most rapidly recombining ion becomes neutral, charge transfer

$$\mathrm{X}^{\mathrm{m}+} + \mathrm{Y} \rightarrow \mathrm{X}^{(\mathrm{m}-1)+} + \mathrm{Y}^{+}$$

quickly removes the multiply-charged ions. The absence of emission from Fe^{2+} (Meikle et al. 1989) may be attributed to the rapid removal of Fe^{2+} by reactions such as

$$Fe^{2+} + O \rightarrow Fe^{+} + O^{+} .$$

The detailed structure of the ejecta after the multiply-charged ions have been reduced to singly-charged species depends on the elemental abundances and the rate coefficients for the charge transfer of singly-charged ions. We expect that processes involving complex systems such as

$$O^{+} + S \rightarrow O + S^{+}$$

will usually be very fast with rate coefficients approaching in excess of $10^{-9} cm^3 s^{-1}$. Qualitatively ionization will flow from elements of high ionization potentials like O and C to elements of low inization potentials like Si, Fe, Co and Ni. Because O is more abundant than carbon and the other metals, charge transfer processes such as

$$O^{+} + C \rightarrow O + C^{+}$$

will leave oxygen partly ionized until the fractional ionization falls below the fractional abundance of carbon. At later times carbon and oxygen are more nearly neutral because they can charge transfer to the less abundant sulfur and silicon.

The helium ions He^{+} may not undergo rapid charge transfer. The presence of He^{+} ions high in the terrestrial ionosphere shows that

$$He^{+} + O \rightarrow He + O^{+}$$

cannot be very fast. Inspection of the calculated potential energy curves (Augustin et al. 1973) suggests that it may be quite slow with a rate coefficient unlikely to exceed $10^{-12} cm^3 s^{-1}$. The rate coefficients for the charge transfer of He^{+} with C, S, and Si are unknown.

6 Chemistry

The physical environment in which the molecules are formed and destroyed is very different from that in interstellar clouds. Dense interstellar clouds have densities of the order of $10^3 cm^{-3}$ and they are cold with temperatures of 10 – 20 K and they are weakly ionized with a fractional ionization of about 10^{-7}. The gas is hydrogen-rich and it contains grains on whose surfaces molecular hydrogen can form. The gas phase chemistry is driven by ionization produced by energetic cosmic rays. The supernova ejecta have densities at 100 days of about $10^{11} cm^{-3}$ and they are hot with temperatures of several thousand degrees and the fractional ionization is probably between 0.01 and 0.1. The gas is hydrogen poor and metal rich and grains did not develop before 400 days.

Early in the life of SN1987A, the temperature is too high for molecules to survive. Carbon monoxide will be destroyed by electron impact

$$e + CO \rightarrow e + C + O$$

and by atom impact

$$\mathrm{He + CO \rightarrow He + C + O}\,.$$

When the heating by γ-rays is no longer sufficient to overcome the cooling by adiabatic expansion and by radiative processes the ejecta cools. As the temperature falls below 5000 K, collision-induced dissociation becomes inefficient, the dissociation energy of CO being 11.09 eV.

When the CO was first detected, the density was of the order $10^{11}\,\mathrm{cm}^{-3}$, a density that is too low for three-body recombination such as

$$\mathrm{C + O + He \rightarrow CO + He}\,.$$

There are no grains so formation must be initiated by a radiative process.

The obvious candidate (Lepp, Dalgarno and McCray 1988, Petuchowski et al. 1989) is the radiative association of C^+ and O,

$$\mathrm{C^+ + O \rightarrow CO^+ + h\nu}\,.$$

Then CO may be formed by

$$\mathrm{CO^+ + O \rightarrow CO + O^+}\,.$$

However it turns out that the radiative association of C^+ and O has a small rate coefficient of order $10^{-18}\mathrm{cm}^3\mathrm{s}^{-1}$ (Dalgarno, Du and You 1989). A more attractive possibility is the radiative association of C and O,

$$\mathrm{C + O \rightarrow CO + h\nu}$$

which has a rate coefficient of order $10^{-17}\mathrm{cm}^3\mathrm{s}^{-1}$ (Dalgarno, Du and You 1989). Another possibility is a sequence initiated by radiative attachment

$$\mathrm{e + O \rightarrow O^- + h\nu}$$

which has a rate coefficient of order $10^{-15}\mathrm{cm}^3\mathrm{s}^{-1}$, two orders of magnitude faster then the radiative association of C and O. It can be followed by

$$\mathrm{O^- + C \rightarrow CO + e}$$

or

$$\mathrm{O^- + O \rightarrow O_2 + e}$$
$$\mathrm{O_2 + C \rightarrow CO + O}\,.$$

However the effectiveness of the negative ion sequence is mitigated by photodetachment

$$\mathrm{O^-} + h\nu \rightarrow \mathrm{O + e}$$

and by mutual neutralization

$$\mathrm{O^- + He^+ \rightarrow O + He}\,.$$

Depending on the degree of mixing of hydrogen back into the core region, reactions such as

$$\begin{aligned} e + H &\rightarrow H^- + h\nu \\ H^- + H &\rightarrow H_2 + e \end{aligned}$$

and

$$\begin{aligned} He^+ + H &\rightarrow HeH^+ + h\nu \\ HeH^+ + H &\rightarrow H_2^+ + He \\ H_2^+ + H &\rightarrow H_2 + H^+ . \end{aligned}$$

the initial steps of which are radiative, can form H_2. At 2000K, the rate coefficients of the initiating reactions, radiative attachment to form H^- and radiative association to form HeH^+, are comparable in magnitude. The effectiveness of the H^- sequence is diminished by photodetachment and mutual neutralization. Neither photodissociation nor dissociative recombination of HeH^+ is significant and if most of the electrons are produced by the ionization of helium, the sequence initiated by the formation of HeH^+ is the larger source of H_2 molecules. Then at the high temperatures of the supernova, OH is formed by

$$O + H_2 \rightarrow OH + H$$

and CO by

$$C + OH \rightarrow CO + H$$

in a chemistry like that of a dissociative shock (Neufeld and Dalgarno 1989).

The carbon monoxide molecule can be destroyed by photoabsorption,

$$\begin{aligned} CO + h\nu &\rightarrow C + O \\ CO + h\nu &\rightarrow CO^+ + e , \end{aligned}$$

by fast electron impact

$$\begin{aligned} e + CO &\rightarrow e + C + O \\ &\rightarrow e + CO^+ + e , \end{aligned}$$

and by the chemical reaction

$$He^+ + CO \rightarrow He + C^+ + O .$$

This last reaction is the major destruction mechanism and the density of He^+ is a crucial parameter in the calculation of the abundance of CO. It is uncertain because we do not know the rate coefficients for the charge transfer of He^+ with the metals. We do know that

$$He^+ + He \rightarrow He_2^+ + h\nu$$

is too slow despite the high abundance of He.

Detailed calculations (Lepp, Dalgarno and McCray 1990) show that the observed abundance for CO can be reproduced by the chemistry provided the element distribution in the ejecta is neither the unmixed or the fully-mixed model but lies in between, much as would be expected as a consequence of the Rayleigh-Taylor instabilities. The observed abundances could also be reproduced in the mixed model if any of the rate coefficients for the charge transfer of He^+ with C, S and Si were large. The observed abundance could be substantially exceeded in the unmixed model.

The chemistry yields significant quantities of other molecules CS, SiO, H_2, O_2, CH, SiS and OH, but it fails to produce enough CO^+ which is rapidly removed by dissociative recombination

$$CO^+ + e \rightarrow C + O$$

and by charge transfer

$$CO^+ + O \rightarrow CO + O^+ .$$

Depending on the detailed assumptions the chemical model underestimates CO^+ by factors from 10^3 to 10^5 at 192 days. The ratio of the observed emissions increases with time. Although a decreasing ratio of CO to CO^+ could be explained by the greater efficiency of electrons in exciting vibrational levels in the molecular ion, we suspect the emission of 2.26μ is not from CO^+.

The intensities diminish with time because the temperature and also the density are decreasing. The future for molecular spectroscopy in the infrared is bleak unless an energy source, such as a pulsar, reheats the gas. A flattening of the optical light curve may be expected as the heating from the decay of ^{56}Co falls is replaced from the heating of the more slowly decaying ^{57}Co and ^{44}Ti (Kumagai et al. 1989). The supernova chemistry may also lead to more complex molecules and the formation of grains. As the temperature decreased below the condensation temperature grains may have been formed (Dwek 1988). Continuum emission in the infrared has been interpreted as evidence that dust formation has taken place (Danziger et al. 1989) or that existing dust has been reheated (Roche et al. 1989).

Acknowledgments. This work was supported by the National Science Foundation, Division of Astronomical Sciences, under Grant AST-86-17675.

7 References

Aitken, D.A, Smith, C.H., James, S.D., Roche, P.F., Hyland A.R. and McGregor, P.J. 1988 *M. N. R. A. S.* **235**, 19P.

Arnett, D., Fryxell, B. and Müller, E. 1989 *Ap. J. (Letters)* **341**, L63.

Augustin, S.D., Miller, W.H., Pearson, P.K. and Schaefer, H.F. 1973 *J. Chem. Phys.* **58**,2845.

Dalgarno, A., Du, M.L. and You, J.H. 1989 *Ap. J.* in press.

Danziger, I. J., Gouiffes, C., Bouchet, P. and Lucy, L.B. 1989 *I.A.U. Circ.*, No. 4746.

Dotani, T., Hayashida, K., et al. 1987 *Nature* **330**, 230.

Dwek, E. 1988 *Ap. J.* **328**, 699.

Ebisuzaki, T., Shigeyama, T. and Nomoto, K. 1989 *Ap. J. (Letters)* **344**, L65.

Kumagai, S., Toshikazu, S., Nomoto, K., Itoh, M., Nishimura, J and Tsuruta, S. 1989 *Ap. J.* **345**, 412.

Lepp, S., Dalgarno, A. and McCray, R. 1988 *Bull. A.A.S.* **20**, 671.

Lepp, S., Dalgarno, A. and McCray, R. 1990 to be published.
McCray, R. 1989 in *Molecular Astrophysics*, Ed. T. Hartquist (Cambridge University Press).
McCray, R., Shull, J. M. and Sutherland, P.G. 1987 *Ap. J. (Letters)* **317**, L73.
Meikle, W.P.S., Allen, D.A., Spyromilio, J. and Varani, G.-F. 1989 *M. N. R. A. S.* **238**,193.
Neufeld, D. and Dalgarno, A. 1989 *Ap. J.* **340**, 869.
Oliva, E., Moorwood, A.F.M. and Danziger, I.J. 1987 *The Messenger* **50**, 18.
Petuchowski, S.J., Dwek, E., Allen, J.E. and Nuth, J.A. 1989 *Ap. J.* **342**, 406.
Pinto, P.A. and Woosley, S.E. 1988 *Nature* **333**, 534.
Rank, D.M., Pinto, P.A., Woosely, S.E., Bregman, J.D., Witteborn, F.C., Axelrod, T.S. and Cohen, M. 1988 *Nature* **331**, 505.
Roche, P.F., Aitken, D.K., Smith, C.H. and James, S.D. 1989 *Nature* **337**, 533.
Spyromilio, J., Meikle, W.P.S. Learner R.C.M. and Allen, D.A. 1988 *Nature* **334**, 327.
Sunyaev, R., Kaniovsky, A., et al. 1987 *Nature* **330**, 227.

SMALL GRAINS AND LARGE AROMATIC MOLECULES

A. LÉGER, L. D'HENDECOURT, L. VERSTRAETE, P. EHRENFREUND
Groupe de Physique des Solides, Université Paris VII
Tour 23 - 2, place Jussieu
75251 Paris Cédex 05
FRANCE

ABSTRACT : Recently, the interpretation of the so-called "unidentified" infrared emission bands has drastically improved. These bands are attributed to the fundamental vibrations of large molecules known in organic chemistry as Polycyclic Aromatic Hydrocarbons (PAH's). In this paper, we review the main arguments which led to this identification and focus our attention to some recent IR spectroscopic studies which allow us to interpret further the main characteristics of IR astronomical spectra. We show that a mixture of compact and partially dehydrogenated PAH's can be the carrier of the main emission features.

1. Introduction

IR spectral features of the interstellar matter are:

(i) absorption bands
- 9.7 μm (silicates)
- 3.4 μm (saturated hydrocarbons)
- 3.0 - 3.1 μm (H_2O ice)
- 4.67, 6.0, 6.8 μm... (ices)

and

(ii) emission bands
- 9.7 μm (silicates, seen only very close to stars)
- 3.3, 6.2, 7.7, 8.6, 11.3 μm (unidentified IR emission features).

Fig. 1 illustrates this, showing two spectra typical of absorption and emission by the interstellar medium. The stricking fact is the almost complete absence of overlap between the two sets of features. At first glance, it is not clear that we are looking at the same material !

An interesting case is the comparison between the 3.4 μm band attributed to aliphatic C-H bonds seen in absorption and the 3.3 μm one assigned to aromatic C-H bonds seen only in emission (Fig. 2). Thus, there must be a *fundamental difference between the physical states* of these two closely related chemicals to account for this difference.

J. M. Greenberg and V. Pirronello (eds.), Chemistry in Space, 211–225.

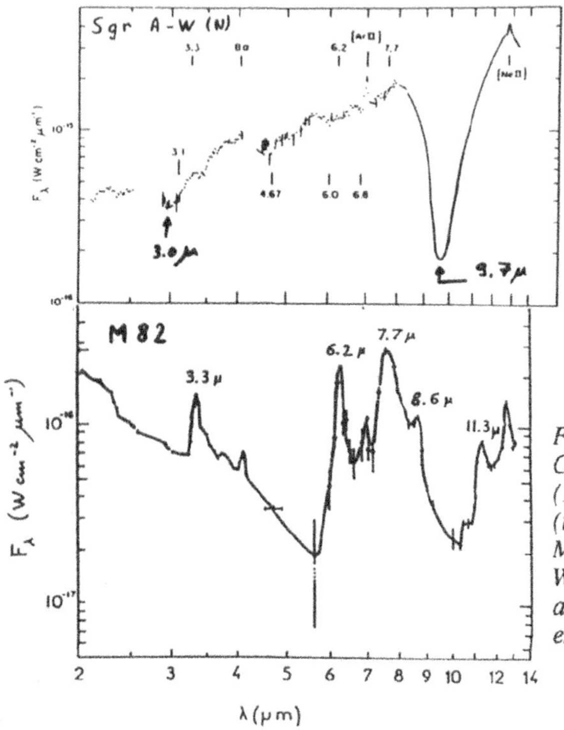

Fig. 1 : Spectrum of the Galactic Center adapted from Aitken (1981), typical of absorption (upper part) and of the Galaxy M82 central part, adapted from Willner et al. (1977) and Gillett et al. (1975), typical of emission(lower part).

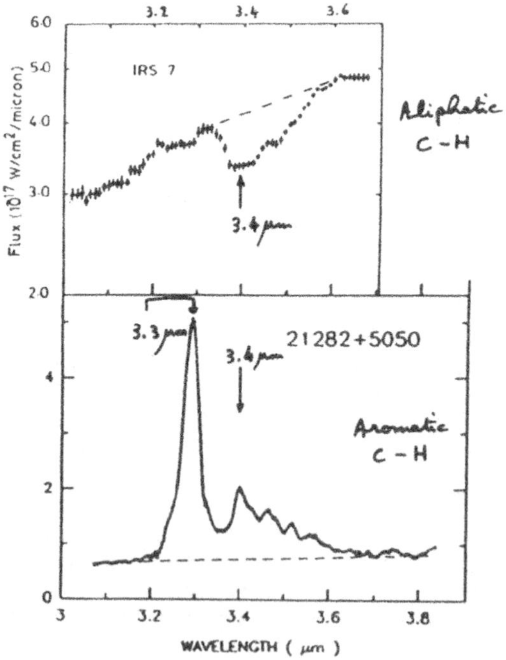

Fig. 2. Hydrocarbon feature in absorption towards the Galactic Center (Butchard et al., 1986) (left part) and hydrocarbon features in emission (de Muizon et al., 1986).

Furthermore, the short wavelength component (3-30 μm) of this IR emission of the interstellar (IS) medium cannot be explained by standard models of IS dust because classical grains are much too cold [Pajot et al. 1986, Boulanger et al. 1985].

Our understanding of the problem was greatly improved when Andriesse (1978) and then Sellgren (1984) showed that thermal emission by very small grains (~ 50 atoms) during the transient heating following the absorption of individual UV photons can explain color temperatures of ~ 10^3 K in spatial regions where the grain equilibrium temperature could not be above 60 K.

Discussing the nature of these grains and their necessary stability against sublimation, Léger and Puget (1984, thereafter LP) showed that large planar molecules of Polycyclic Aromatic Hydrocarbons (PAHs) were good candidates. Then, using the benefit of having a precise model, they tried to calculate explicitly the IR emission from such species. At that time, the laboratory IR data on large (~ 50 atom) PAHs were scarce and LP used the largest molecule whose IR absorption was known : coronene ($C_{24}H_{12}$). The similarity between the spectrum they obtained and the observed IR features strongly suggested that a mixture of PAHs could be the carrier of the emission. The whole set of the observed emission bands (3.3 - 6.2 - 7.7 - 8.6 - 11.3 μm) could be associated with the IR active vibrational modes of these molecules. In addition, they were able to explain the *intensity* of the features which had always been a difficulty in the earlier attempts to interpret the bands (Allamandola, 1984). In fact, as early as 1981, Duley and Williams had pointed out that two of the bands occur at wavelengths typical of aromatic C-H bonds but they could not account simply for the whole spectrum and the emission mechanism problem was not addressed.

In 1985, Allamandola et al. repeated the conclusion of LP by considering other PAHs spectra and pointed out that these molecules are expected to be ionized in strongly irradiated regions (e.g. reflection nebula).

The hypothesis that PAHs molecules are an hitherto undetected component of the IS matter responsible for its near IR emission has many implications. First, it means that they process 10-30% of the total energy of a galaxy (Pajot et al., 1986, Puget 1987, Ghosh and Drapatz, 1987). Their abundance, as inferred from the intensity of the observed bands, brings them to the level of the most abundant free organic molecules detected in the universe (LP, 1984). They have been proposed as candidates for the carriers of the Diffuse Interstellar Bands in the visual (van der Zwet and Allamandola, 1985 ; Léger and d'Hendecourt, 1985 ; Crawford et al., 1985), the red luminescence observed in several reflection nebulae (d'Hendecourt et al., 1986) and the heating of the interstellar H_I gas by photoelectronic effect (d'Hendecourt and Léger, 1987).

It is therefore important to settle the identification of PAHs in the observed IR spectra on grounds as safe as possible. LP have emphasized that they expected the interstellar PAHs to be a large family of species whereas their spectroscopical comparison was done with a single molecule (coronene). The aim of the present paper is to extend this comparison to other relevant PAHs and to describe, with precision, the emission mechanism.

Few examples of small and large PAHs are reported in Fig. 3.

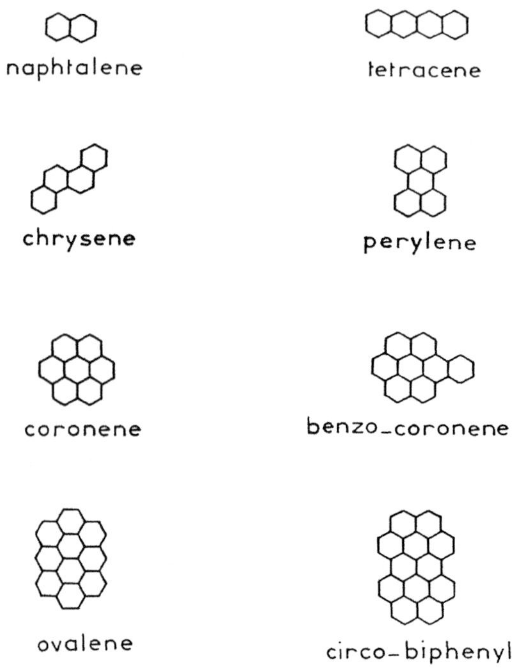

Figure 3 : Some PAH molecules. Conventionally, H atoms are not represented, they are at the periphery and saturate the free carbon bonds.

2. IR emission mechanism from isolated molecules

2.1. IR EMISSION OF A LARGE MOLECULE AT TEMPERATURE T

The emission spectral intensity of a physical system, at temperature T, is :

$$I_\lambda = B_\lambda(T)\, \varepsilon_\lambda(T) \qquad (1)$$

where $B_\lambda(T)$ is the black body emission at wavelength λ and ε the emissivity of the system.

The second law of thermodynamics implies that the molecule emissivity is equal to its absorption coefficient at the same temperature, $A_\lambda(T)$ (Kirchhoff law, see Reif, 1965). For an assembly of molecules, the latter quantity is easier to measure than their emission. In an *absorption experiment,* A_λ is connected to the incident flux ϕ_o and the transmitted one ϕ_λ by :

$$A_\lambda = \frac{\phi_o - \phi_\lambda}{\phi_o} = 1 - \exp(-N\sigma_\lambda), \qquad (2)$$

where N (mol cm^{-2}) is the molecular column density and σ_λ (cm^2 mol^{-1}) the cross section per molecule. In the optically thin limit ($N\sigma_\lambda << 1$) the same molecule assembly emits, at temperature T, an intensity:

$$I_\lambda = B_\lambda(T)\, N\, \sigma_\lambda(T) \qquad (3)$$

The spectral power emitted, per molecule, is obtained by integrating over all the directions :

$$P_\lambda = 4\pi\, B_\lambda(T)\, \sigma_\lambda(T) \qquad (4)$$

If the absorption cross sections of PAH ions were measured in the laboratory at temperature T, one could calculate exactly their emission at that temperature through (4). In practice, most of the measurements available have been made for neutral molecules and at room temperature. So, a first approximation will be the use of $\sigma_\lambda^{neutral}$ (300 K) instead of $\sigma_\lambda^{ion}(T)$. We shall discuss the consequences of this approximation at the end of the paper.

2.2. IR EMISSION DURING A THERMAL SPIKE

When a molecule absorbs a UV photon, its internal energy U is increased by $h\nu_{UV}$ and when it emits a IR or a visual photon, U decreases by the corresponding amount. At each time, the vibrational temperature T of the system can be defined and is connected to the internal energy U (when the system is in its electronic ground state) by :

$$U(T) = \int_0^T C(T')\, dT' , \qquad (5)$$

where C(T) is the specific heat of the molecule. During the IR emission, the ion cools as the energy is emitted, according to :

$$P(T)\, dt = - C(T)\, dT, \qquad (6)$$

where $P(T) = \int P_\lambda(T)\, d\lambda$. These expressions apply as an average, but for a given molecule the quantization of the emission must be taken into account.

The IR emission of a large interstellar molecule is then the sum of spectral emissions given by expression (4) at temperatures T, during the thermal spikes represented in Fig. 4.

We have now the ingredients to calculate the IR emission of an assembly of PAH ions irradiated by an interstellar UV flux as performed by Puget et al. (1985) or Désert, et al. (1986). Given the UV flux and the absorption cross sections of the different species, the rate of energy absorption by the molecules can be calculated. Each molecule with energy U, temperature T (eq. 5), emits visual and IR fluorescence, according to eq. (4). As a result, its internal energy and temperature decrease according to eq. (6). Up to now, the calculations have been done in the absence of visible fluorescence. We shall do so in this paper, but it is clear that the rôle of the fluorescence channel of energy decay should be considered in the future, including the Poincaré fluorescence (Léger et al., 1988a).

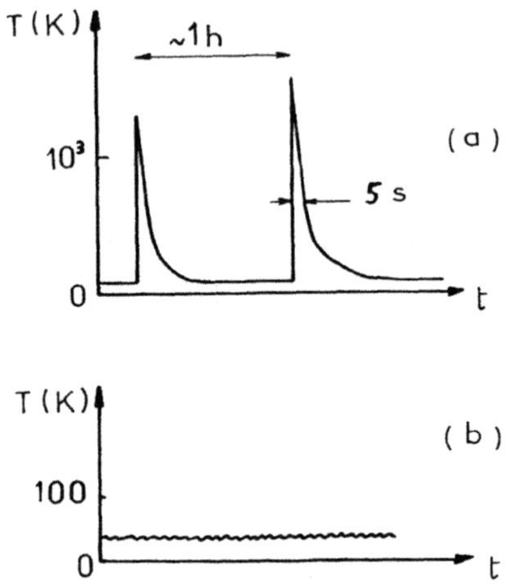

Figure 4. : (a) Thermal evolution of very small grains or large molecules with N ~ 50 atoms and (b) of classical grains (a ≳ 100 Å, N ≳ 10^5 atoms) in radiation field ~ 10^3 times the diffuse ISRF.

2.3. IR COOLING TIME

In the IS medium, the main cooling processes for a hot PAH are IR radiation and visual fluorescence. Let us evaluate the cooling time if IR emission is acting alone.

The emitted power P and the specific heat C rules the cooling according to (4) and (6). A cooling time t_{IR} can be defined as :

$$t_{IR} = \frac{C(T)\,T}{P(T)}$$

Integrating (4) over the different wavelengths gives :

$$P = 4\pi \sum_i B\,(\lambda_i, T)\, A_i^c\, N_c\,, \qquad (7)$$

where B(λ,T) is the Planck function, $A_i^c = \sigma_i^c\, \Delta\lambda_i$ the integrated cross section per carbon of the ith band and N_c the number of carbon atoms in the molecule. Using laboratory measured values of A_i and C(T) (Léger, d'Hendecourt and Défourneau, 1989) the IR cooling time t_{IR} derives. More precisely, one can calculate the time for half cooling during the emission process. For a molecule with hydrogen coverage factor x_H of 30% which is typical of a reflexion nebula, it is found :

$T = 1500$ K → $t_{1/2} = 5.1$ s

$T = 1000$ K → $t_{1/2} = 9.6$ s

It is independent of the molecule size because both the emitted power and the specific heat are proportional to the number of atoms.

2.4. MEAN TEMPERATURE OF EMITTING PAHS

The emission of photons and the cooling are ruled by equations (4) and (6).

For two bands, of a same atomic group (e.g. C-H), that occur at different wavelengths, equation (4) relates the absorptions $A_i = \sigma_i\, \Delta\lambda_i$, to the emission, $P_i = P_{\lambda_i}\, \Delta\lambda_i$, through the Planck function. Considering the νC-H (3.3 μm) and γC-H (11.3 μm) modes, one can define a *mean* temperature, T_c, *if both bands were emitted by the same single size molecules,* as :

$$\frac{P_{11.3}}{P_{3.3}} = \frac{B_{11.3}\,(T_c)}{B_{3.3}\,(T_c)}\,\frac{A_{11.3}}{A_{3.3}}\ , \qquad (8)$$

where P_i is the total power emitted in band i. Note that P_i is observed, A_i is measured in the laboratory for different species so that T_c can be determined.

2.5. MEAN SIZE OF EMITTING MOLECULES

For each molecule size and absorbed UV photon energy, a peak temperature (T_p) is reached and the IR emission follows. The characteristics of the actual emission can be calculated by summing over the different distributions and spectra (Puget et al., 1985 ; Désert et al., 1986).

It would be most useful to invert the problem and deduce the nature and size distribution of molecules from observations. Unfortunately, presently we have not enough constraints to do so, and *the use of mean quantities is useful as a first step* to get orders of magnitudes of the parameters, although it can be misleading.

For instance, a mean molecule size can be deduced from the color temperature T_c of section 2.4 as follows.

Using a model emission, one can show that for a reasonable range of values, T_c is close to the temperature at which the molecule, cooling from T_p, has lost half of its energy :

$$U(T_c) \approx 0.5\ U(T_p) \qquad (9)$$

Using the relation (5), one finds for $T_c \sim 600$ K :

$$T_c \approx 0.7\ T_p\ ^{(*)} \qquad (10)$$

So, the measurement of a color temperature implies a mean value for T_p and then $U(T_p)/(3N\text{-}6)$.

On the other hand, the peak temperature is related to the absorbed photon energy $h\nu_{UV}$ by : $U(T_p) \approx h\nu_{UV}$, as the initial energy content of the molecule is mostly negligible for species emitting in the $\lambda < 20\ \mu m$ range (Fig. 4). An estimate of the mean absorbed energy gives a value for $\langle U(T_p)\rangle$.

Therefore, there is a single number N of atoms in the molecule that is compatible with these two constraints.

Considering the situation in a *reflection nebulae*, the illuminating star has a temperature $T_* \sim 20000$ K and a known spectrum. The UV absorption of a PAH mixture has been recently measured (Léger et al., 1988c), yielding a mean energy of absorbed photon in this radiation field of $\langle h\nu_{abs}\rangle = 9$ eV ($\lambda = 1400$ Å). With $T_c = 600$ K, a mean number of atoms is found :

$$\langle N\rangle \approx 90 \qquad (11)$$

They are really big molecules when compared to individually identified interstellar molecules.

(*) More precisely, the integration during the cooling process of a PAH gives T_c/T_p equal to 0.74 - 0.71 - 0.66 for T_c = 500 - 600 - 700 K respectively.

Again, it must be emphasized that this number is derived under the restrictive approximation that a single size of molecules is responsible for both emissions at 3.3 and 11.3 µm. Actually, *a size distribution is expected.*

3. Laboratory IR spectra of PAHs. Comparison with astronomical data

It is important to measure, in the laboratory, IR spectra of PAHs with sizes as comparable as possible to that determined above (N = 90 atoms). This could indicate whether the identification proposed by LP using coronene data stands with a more extended data base.

We have performed such measurements. Most are with PAHs in a solid state (molecular crystals) and one in an isolated state (rare gas matrix isolation). We first discuss the former.

3.1. IR ABSORPTION MEASUREMENTS

PAH molecules larger than coronene (N=36) are very difficult to synthesize and we are grateful to Prof. Clar for having provided us with most of them. The corresponding crystalline solid grains are ground and dispersed in CsI (mid IR) or polyethylene (far IR) pellets in a classical way. IR spectra are taken with a Bomen FTIR spectrometer using Hg-Cd-Te and bolometer detectors respectively.

Typical mid-IR spectra are reported in Fig. 5 and are used to calculate the emission spectra.

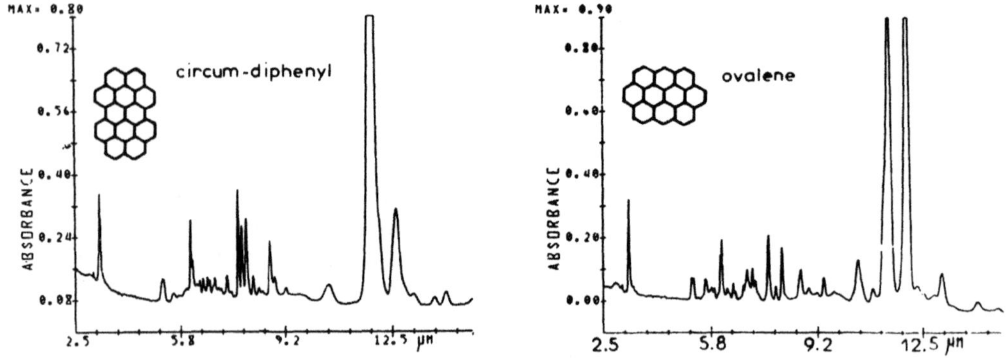

Figure 5 : Typical IR absorption spectra measured for large PAH molecules in the solid state phase (crystals dispersed in CsI pellets). Resolution is 1 cm^{-1}, T = 300 K and absorbance or optical depth is $N\sigma_\lambda$.

3.2. CALCULATED EMISSION FOR COMPACT PAHS

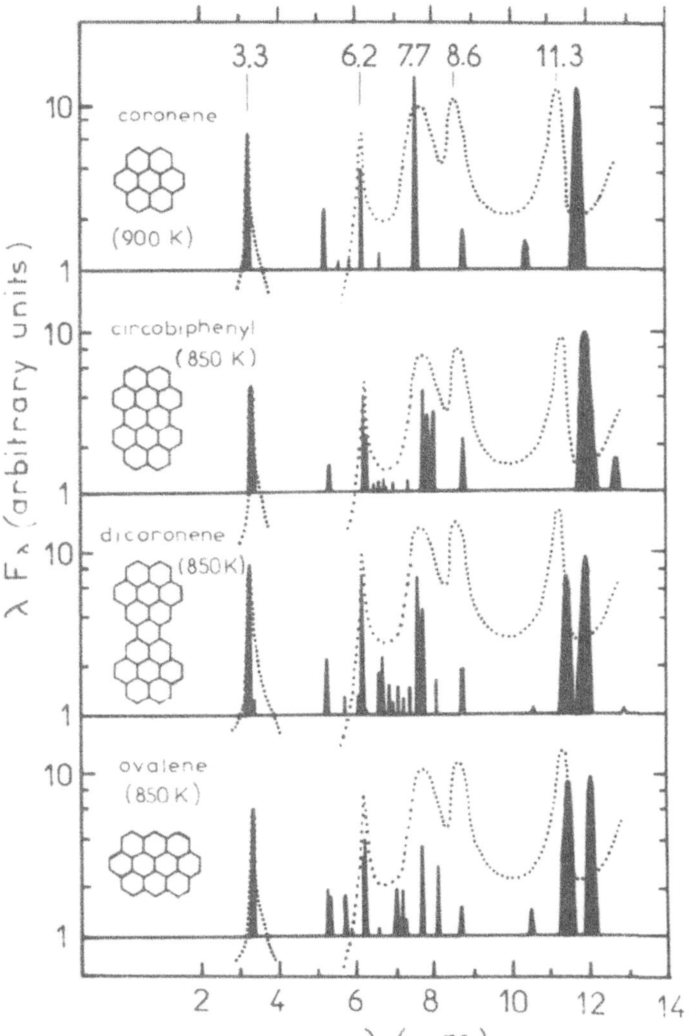

Figure 6 : Emission spectra of several compact PAHs calculated from their absorption spectra measured in the laboratory (in solid phase) using relation (4). The spectrum expected for the reflection nebula NGC 2023 is reported for comparison (dotted line). It is deduced from observations by Sellgren et al. (1985) but for the 6.2μm band which was not resolved and whose line profile is supposed to be similar to that of M82 (Willner et al., 1977). The only free parameter in the calculation is the emission temperature which is adjusted to reproduce the observed ratio of the CH bands at 3.3 and 11.3 μm (adopted T is in brackets). Note the accurate agreement of bands at 3.3 and 6.2 μm between the observation and compact PAH spectra. For the rest of the spectra comparison, see text.

The calculated emissions of several large compact PAHs are reported in Fig. 6 and are compared with an astronomical observed spectrum. As mentioned above, the fit has only one adjustable parameter : the emission temperature. *The similarity between the spectra strongly suggests that the "Unidenfied" IR Emission Bands can be explained by a mixture of compact PAHs.* Coronene is therefore not the only "miracle" molecule whose spectrum mimics the observations.

The 3.3 μm (3030 cm^{-1}) band corresponds to an aromatic C-H stretch. The 6.2 μm (1610 cm^{-1}) absorption is "highly characteristic" of C=C stretch in aromatic rings (Bellamy, 1966, p. 69) ; its presence in most astronomical spectra and the absence of 6.7 μm (1500 cm^{-1}) absorption indicates that most rings are in a compact arrangement.

The agreement for the 3.3 and 6.2 μm features is quite good with any of the PAH spectra.

A band is observed in all the calculated spectra at 5.3 μm. This region of the spectrum has not yet been well observed for technical reasons. *We strongly recommend in the future to look for this feature in astronomical objects.*

In the 6.5-8.5 μm range, the main bands of the compact PAHs appear within the width of the broad 7.7 μm feature observed in nebulae. This region is important as it permits to discriminate between different aromatic compounds. It seems possible to explain the observed broad band at 7.7 μm by a mixture of compact PAHs.

The 8.85 μm (1130 cm^{-1}) band is an in-plane aromatic C-H bend whereas the 11.9 μm (840 cm^{-1}) is an out-of-plane aromatic C-H bend. The exact position of this last mode depends upon the number of adjacent H which are on a ring : solo (11.0-11.6 μm, 910-860 cm^{-1}), duo (11.6-12.5 μm, 860-800 cm^{-1}) or trio (12.4-13.3 μm, 810-750 cm^{-1}) (Bellamy, 1966). The observed spectra have a peak at 11.3 μm (885 cm^{-1}) and a broad structure up to 13.5 μm (Cohen et al., 1985 ; de Muizon et al., 1986). This indicates the presence of different hydrogen sites with a predominance of solo.

In conclusion, from the point of view of the analytic IR spectroscopy, the observation of the whole set of the 3.3, 6.2 and 11.3 μm bands *is highly characteristic of aromatics.* This is one of the key arguments in favor of the PAH model.

4. Abundance : The most abundant known organic molecules in the gas phase

The evaluation of the abundance of PAHs in the Interstellar Medium directly results from the interpretation of its Near IR and Far IR emissions.

Grains and molecules absorb the incoming stellar flux according to their respective abundance and specific cross section. As they re-emit in separate domains of wavelength, their emission fluxes F_i can be measured separately. The total absorption cross section per H atom of the IS matter $\sigma_{tot,H}$ results from the measured extinction curve and an adopted albedo. The PAH cross section per carbon atom σ_{PAH} has been measured for different mixtures of PAH molecules (Léger et al., 1988c) and although they are not the IS mixture these measurements give us our present best way to estimate the cross section of IS species. If U_v is the local radiation field density, $N_{C,PAH}$ the number of carbon atoms in PAH molecules, F_{PAH} and F_{tot} the near IR flux attributed to PAHs and total flux, one reads :

$$\frac{F_{PAH}}{F_{tot}} = \frac{\int c\, U_\nu\, \sigma_{PAH}(\nu)\, N_{C,PAH}\, d\nu}{\int c\, U_\nu\, \sigma_{tot,H}(\nu)\, N_H\, d\nu}$$

The denominator for $N_H = 1$, also called $L_{IR,H}$, has been estimated to 5.7 10^{-31} W/H atom in the diffuse H_I interstellar medium by Puget (1988). Mathis et al. (1984) give a slightly lower value : 4.2 10^{-31} W/H atom. The observed value (total IR emission) measured by Boulanger and Pérault (1987) is 6.1 10^{-31} W/H atom. The good agreement between the predicted value and the observed one confirms that the quantities involved (σ_{tot}, U_ν) are rather well known.We attribute the diffuse IS emission up to the longest limit (15 µm) of the IRAS 12 µm band to PAHs because we have spectroscopical evidences for it . The emission in the 25 and 60 µm bands is also attributed to thermal fluctuations but we have no spectroscopical information on the chemical nature of the particules that we propose to call very small grains (VSG). By abundance of PAH we refer only to the first species. Using astronomical spectra in reflection nebulae and other galactic objects (Sellgren et al., 1985 and Cohen et al., 1985) and the IRAS 12 µm band shape, we estimate the emission of PAHs with respect to the flux in this band as : $F_{PAH} = 1.1\nu\, F_\nu$ (12 µm). For PAHs irradiated by the ISRF (U_ν), we get the numerator for one carbon : 2.24 10^{-27} W/C atom. If we attribute 20% of the average Cirrus emission to PAHs corresponding to 1.14 10^{-31} W/H atom. A carbon abundance in these molecules results : $N_{C,PAH}/N_H$ =5.1 10^{-5} or 14% of the cosmic carbon. Abundances in other regions are reported in table 1.

We need a large but plausible amount of the cosmic carbon to account for the observed bands. This means that the PAH model is able to explain *quantitatively* their intensity.

	$\frac{F_{PAH}}{F_{total\ IR}}$	$\frac{M^C_{PAH}}{M^C_{cosmic}}$	$\frac{S_{PAH}}{S_{grains}}$
Reflection Nebula	10%	7%	3.1
Diffuse Interstellar Medium	20%	14%	6.2
R. Cor. Austr*	50%	35%	15.0

Table 1 : Ratios of Near IR to Far IR emissions of different objects and deduced fraction of the cosmic carbon involved in PAHs. The relative geometrical surface (S) of these molecules to that of large grains in classical models (Draine and Lee, 1984) emphasizes their possible rôle in the Interstellar Medium.
* A specific molecular cloud, see Leene (1986).

Inspection of Table 1 indicates that :

(1) Involving several percent of the cosmic carbon, *PAHs are the most abundant free organic molecules known at that date in the gas phase.* The abundance of few other

detected interstellar molecules is reported in Table 2. PAHs are in third position, over three orders of magnitude above molecules which are easily observed in radioastronomy (H_2CO, HCN, ...). It is remarkable that such abundant species have remained undetected for a long time. This is due to their *lack of simple signature in the radiowavelengths* (J and K ~ 100). It is interesting to point that if this range had remained the only tool to detect IS molecules, we would have missed the most abundant free organic molecules !

(2) PAH geometrical surface is comparable or larger than that of classical grains. This points to their possible rôle in interstellar chemistry (Omont, 1986 ; Lepp and Dalgarno, 1988).

Molecule	n_{at}/n_H
H_2	1
CO	$2\ 10^{-4}$
PAHs	$5\ 10^{-5}$
H_2CO	$5\ 10^{-8}$
NH_3	$5\ 10^{-8}$
HCN,	$2\ 10^{-8}$
HC_9N	10^{-9}

Table 2 : Typical abundance of different interstellar molecules *expressed as a number of atoms included in the molecule* versus the total number of hydrogen nuclei in regions where they are observed (data partially from Duley and Williams, 1984).

5. Conclusion

The proposed identification of PAHs in astronomical IR spectra as presented by LP (1984) was not obtained by looking for the best spectrum in a IR Atlas. The presence of large PAHs was inferred from discussing the possible composition of interstellar very small particules on stability arguments and the spectroscopic agreement was obtained afterwards, as a check.

The possibility of explaining at once (i) the position of the whole set of observed bands by a single family of molecules and (ii) the mechanism of their emission are strong arguments in favour of this explanation.

Although a recent attempt has been made to explain the mid IR emission (20 µm : Duley and Williams, 1988), we do not think that their alternative HAC large grain model can explain the high yield of the UV → near IR (3 - 15 µm) conversion. It would require that the energy of an incident photon can remain localized in a molecule bound to the rest of the grain during the time of a few seconds required for the IR emission (section 2.3). This would be in conflict with the failures of all attempts made in physical chemistry to localize energy in molecules for times longer than 10^{-11} s (e.g. Rettschnick, 1980, Hutchinson et al., 1983 ; Wild et al., 1985, ...).

However, the PAH model does not give a one to one spectroscopic identification with a given molecule but an identification with a family of molecules which have characteristic bands. Some points are still not quite clear such as the precise interpretation of the 11.3 μm band.

The content of this molecular family has to be made more precise in the future. Advances in that direction have been presented here : *compact PAHs give a better spectroscopic agreement than non-compact ones.*

In this explanation, *the fact that the molecules are free flyers is essential* in order to give the high conversion yield from UV to Near IR. It is needed to explain that 30% of the interstellar radiation is emitted by this process.

This identification of an important component of the IS Medium, which was ignored before, has numerous implications on the physics of the medium (heating of the H_I gas, extinction curve, red luminescence...) and on its chemistry. It also gives new ideas for explaining the mystery of the Diffuse Interstellar Bands in the visual. For a review of these implications, see Puget and Léger (1989).

References

- Aiken D.K.,1981 : in *Infrared Astronomy,* IAU Symposium n°96, eds Wynn Williams C.G. and Cruikshank D.P., Reidel, p. 217

-Allamandola L.J., 1984 : in *"Galactic and Extragalactic IR Spectroscopy",*, Eds Kessler M.F. and Phillips J.P., Reidel
- Allamandola L.J., Greenberg J.M. and Norman C.A., 1979 : Astron. & Astrophys. **77,** 66
- Allamandola L.J., Tielens A.G. and Barker J.R., 1985 : Astrophys. J. Lett. **290,** L25
- Andriesse C.D., 1978 : Astron. & Astroph. **66,** 169
- Avonis P., Gebart W. and El Sayed M.A., 1977 : Chem. Rev. **77,** 793
- Barker J.R., Allamandola L.J. and Tielens A.G.G.M., 1987 : Asroph. J. Let. **315,** L61
- Bellamy L.J., 1966 : *"IR Spectra of Complex Molecules",* Wiley
- Benson S.W., 1965 : J. of Chem. Educat. **42,** 503
- BernardJ.P., d'Hendecourt L. and Léger A., 1988 : Astron. & Astrophys. (submitted)
- Birks J.B., 1970 : *"Photophysics of Aromatic Molecules",* Wiley-Interscience
- Blanco A., Borghesi, Fonti S., Orofino V., Bussoletti and Colangeli L., 1987: communication at the Conference : *"Dust in the Universe",* Manchester (G.B.)
- Bondibey V.E., 1984 : An. Rev. Phys. Chem. **35,** 591
- Boulanger F., Baud B., van Albada G.D., 1985 : Astron. & Astrophys. Let. **144,** L9
- Boulanger F. and Pérault M., 1987 : Ap. J. **330,** 964
- Butchart L., McFadzeau A.D., Whittet D.C.R., Geballe T.R. and Greenberg J.M., 1986 : Astron. & Astrophys. Let. **154,** L5
- Clar E., 1964 : *"Polycyclic Hydrocarbons",* Academic Press
- Cohen M., Allamandola L.J., Tielens A.G., Bregman J., Simpson J.P., Witteborn F.C., Wooden D. and Rank D., 1986 : Astroph. J. **302,** 737
- Cohen M., Tielens A.G. and Allamandola L.J., 1985 : Astroph. J. Let. **299,** L93
- Crawford M.K., Tielens A.G.G.M., Allamandola L.J., 1985 : Astrophys. J. Lett. **293,** L45
- Cyvin B.N., Brunvoll J. and Cyvin S.J., 1984 : Spectrosc. Let. **17** (9), 559
- Cyvin S.J., 1982 : J. of Mol. Struct. **79,** 423

- d'Hendecourt L. and Léger A., 1987 : Astron. & Astrophys. Lett. **180**, L9
- d'Hendecourt L.B., Léger A., Olofsson G. and Schmidt W., 1986 : Astron. & Astrophys. **170**, 91
- de Muizon M., Geballe T.R., d'Hendecourt L.B. and Baas F., 1986 : Astrophys. J. Let. **306**, L105
- Désert X., Boulanger F. and Shore S., 1986 : Astron. & Astroph. **160**, 295
- Donn B., 1968 : Astroph. J. Let. **152**, L129
- Duley W.W. and Williams D.A., 1981 : M.N.R.A.S. **196**, 269
- Duley W.W. and Williams D.A., 1984 : *"Interstellar Chemistry"*, Acad. Press
- Duley W.W. and Williams D.A., 1988 : MNRAS **231**, 969
- Dweck E., Sellgren K., Soifer B.T. and Werner M.W., 1980 : Astroph. J. **238**, 140
- Ghosh S.K. and Drapatz S., 1987 : in *"Polycyclic Aromatic Hydrocarbons and Astrophysics"*, Ed. Léger, d'Hendecourt and Boccara, Reidel
- Giard M., Pajot F., Lamarre J.M., Serra G., Andouze J., Caux E., Combes M., Gispert R, Léger A. and Rouan D., 1988 : Astron. & Astroph. Let. **201**, L1
- Hutchinson J.S., Reinhardt W.P. and Hynes J.T., 1983 : J. Chem. Phys. **79**, 4247
- Kroto H.W., Heath J.R., O'Brien S.C., Curl R.F., Smalley R.E., 1985 : Nature **318**, 162
- Krumhansl J. and Brooks H., 1953 : J. Chem. Phys. **21**, 1663
- Leene A., 1986 : Astr. Astrophys. **154**, 295
- Lepp S. and Dalgarno A., 1988 : Astroph. J. **324**, 553
- Léger A. and d'Hendecourt L., 1985 : Astron. & Astrophys. **146**, 81
- Léger A. and Puget J.L., 1984 : Astron. & Astrophys. Lett. **137**, L5 (LP)
- Léger A., Boissel P. and d'Hendecourt L., 1988a : Phys. Rev. Lett. **60**, 921
- Léger A., Boissel P., Désert X. and d'Hendecourt L.B., 1988b : Astron. & Astrophys. (accepted)
- Léger A., Verstraete L., Dutuit O., d'Hendecourt L.B., Schmidt W., 1988c : in *Interstellar Dust*, IAU Symposium 135, eds. L.J. Allamandola and A.G.G.M. Tielens (in press)
- Léger A., d'Hendecourt L., Défourneau D., 1989 : Astron. & Astrophys. **216**, 148
- Mathis J.S., Mezger P.G. and Panagia N., 1984 : Astron. Astroph. **128**, 212
- Omont A., 1986 : Astron. & Astrophys. **164**, 159
- Oref I. and Rabinovitch B.S., 1979 : Accounts of Chem. Res. **12**, 167
- Pajot F., Boissé P., Gispert R., Lamarre J.M., Puget J.L. and Serra G., 1986 : Astron. & Astrophys. **157**, 393
- Puget J.L., Léger A. and Boulanger F., 1985 : Astron. & Astrophys. Let. **142**, L19
- Puget J.L., 1987 : in *"Polycyclic Aromatic Hydrocarbons and Astrophysics"*, Ed. Léger, d'Hendecourt and Boccara, Reidel
- Puget J.L., 1988 : in *Comet to Cosmology*, ed. Lawrence, Springer-Verlag, p. 113
- Puget J.L., Léger A., 1989 : Annual Review of Astronomy and Astrophysics, eds Burbidge et al, Annual Rev. Inc., USA
- Reif F., 1965 : *"Fundamentals of Statistical and Thermal Physics"*, McGraw Hill
- Rettschnick R.P.H., 1980 : in *Radiationless transitions,* Acad. Press, p. 185
- Roussel J.C., 1984 : private communication
- Sellgren K., 1984 : Astrophys. J. **277**, 623
- Sellgren K., Allamandola L.J., Bregman J.D., Werner M.W. and Wooden D.H., 1985 : Astroph. J. **299**, 416

- Smalley R.E., 1983 : An. Rev. Phys. Chem. **34**, 129
- Tielens A.G.G.M., Allamandola L.J., Barker J.R. and Cohen M., 1987 : in *Polycyclic Aromatic Hydrocarbons and Astrophysics*, eds. Léger, d'Hendecourt and Boccara, Reidel, p. 273
- van der Zwet G.P. and Allamandola L.J., 1985 : Astron. & Astrophys. **146**, 77
- Wild W., Seilmeier A., Gottfried N.H. and Kaiser W., 1985 : Chem. Phys. Let. **119**, 259
- Willner S.P., 1984 : in *"Galactic and Extragalactic IR Spectroscopy"*,, Eds Kessler M.F. and Phillips J.P., Reidel
- Willner S.P., Soifer B.T., Russell R.W., Joyce R.R. and Gillett F.C., 1977 : Astrophys. J. Let. **217**, L121

PHYSICAL, CHEMICAL AND OPTICAL INTERACTIONS WITH INTERSTELLAR DUST

J. MAYO GREENBERG
Laboratory Astrophysics
Huygens Laboratory
University of Leiden
2300 RA Leiden, The Netherlands

ABSTRACT. Broad band and spectral properties of interstellar dust are used to define the major mass component of the solid particles. Physical and chemical interactions with the gas, with ultraviolet photons and with cosmic ray particles are discussed. Both active and passive roles of dust are displayed theoretically and observationally and comparisons made with laboratory data on photoprocessed and heated molecular ice mixtures. Optical properties of the organic refractory dust mantle component are used to calculate the extinction, scattering and polarization by a cylindrical silicate core-organic refractory mantle grain model for comparison with observations.

1. Introduction: Observational and other constraints on interstellar dust

The interstellar dust is sufficiently complex not to be uniquely defined, as a molecule can be, simply by its spectroscopic properties. It is necessary to adopt a global view to achieve any degree of confidence in a model.

First of all there is a set of observational constraints on the dust itself. Second there are consistency requirements based on including the physical and chemical interactions with the other components of the interstellar medium.

The major observational properties of the dust are:

1) Wavelength dependence of extinction
2) Wavelength dependence of linear and circular polarization
3) Correlation between total extinction and gas (hydrogen)
4) Variations of the above depending on different types of environments
5) Broadband scattering and absorption: scattering assymetry factor

   $g = < \cos \theta >$, albedo $\alpha = \dfrac{\text{scattering cross section}}{\text{total (extinction) cross section}}$.

6) Infrared absorption spectra
7) Infrared emission spectra

The major interstellar medium connections are:

J. M. Greenberg and V. Pirronello (eds.), Chemistry in Space, 227–261.

1) Elemental abundances
2) Ultraviolet radiation
3) Cosmic (particle) rays
4) Magnetic fields
5) Molecular abundances

The major physical events and processes are:
1) Dust production by stars
2) Growth (accretion) in the ISM
3) Destruction in the ISM: supernova, star formation
4) Heating
5) Energetic processing (non destructive) by ultraviolet and cosmic rays
6) Grain alignment
7) Cloud lifetimes

In the following sections a number of the above will be discussed aimed particularly at explaining the role of dust in cloud chemistry. A grain model will be derived which satisfies the major observed properties of particles which account for the dominant mass component of the dust population.

2. The size of interstellar dust.

It is well established that there are at least three basic populations of interstellar dust. There are several different theories of what each of these constituents must be to explain the typical interstellar extinction and polarization characteristics and variations thereof in various regions of space (Greenberg, 1978; Mathis, Rumpl and Nordsieck, 1977; Duley and Williams, 1988). However there is general agreement that the major bulk of the particles is in the tenth micron size range, the remainder being in the hundredth micron or smaller size range down to molecular dimensions of tens of nanometers. What is not generally agreed upon are the sources and the precise chemical and physical compositions of the components. I shall come back to a more detailed discussion of this in later sections. For the moment let us restrict ourselves to some qualitative remarks about the tenth micron component, which is responsible for the visual part of the extinction curve (See Fig. 1.1).

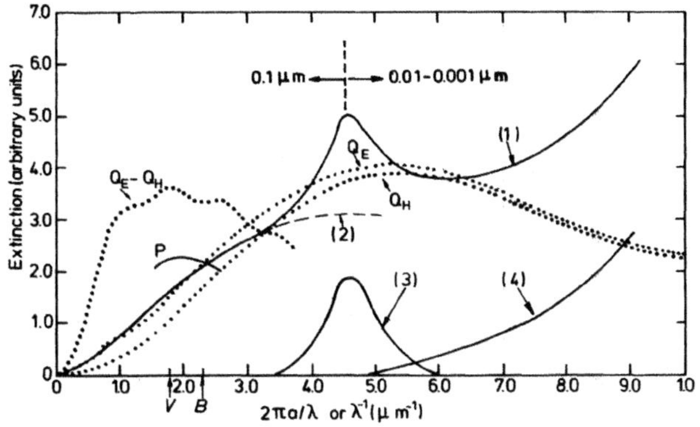

Fig. 1.1. Comparison of theoretical extinction curves for aligned circular cylinders with index of refraction $m = 1.3\text{-}0.05i$ with observed average interstellar extinction (curve 1) and polarization (p). Curves 2), 3) and 4) are separated contributions to the total extinction by classical-size ($a \simeq 0.15$ μm) and small ($a \simeq 0.01$ μm) particles. Curve 3) is the « 2200 Å» hump. Curve 4) is called the far-ultraviolet (FUV) contribution. Q_E and Q_H are the extinction efficiencies (per unit length) for cylinders lined up along the perpendicular to the electric vector of incident linear polarized radiation. Theoretical extinction is $(Q_E + Q_H)/2$ and polarization is $(Q_E - Q_H)/2$.

The wavelength dependence of the extinction A(λ) determines the size distribution which we characterize by a mean radius $\bar{a}$,

$$A_{vis}(\lambda) \rightarrow \bar{a} \tag{2.1}$$

The total visual extinction A(V) (V ≃ 500 nm) along a line of sight L is given by (for spheres)

$$\frac{A(V)}{L} = \frac{N_d \pi \bar{a}^2}{L} Q_V(\bar{a})\,(1.086) \tag{2.2}$$

where N_d is number of particles per unit area, $Q_V(\bar{a})$ is the mean extinction efficiency (the blocking area of the grain ÷ the geometrical area) in the visual and the factor 1.086 is a conversion factor from optical depth to extinction in magnitudes. For the sizes and wavelengths in question the extinction efficiency is somewhat greater than unity: $Q_V(\bar{a}) \simeq 1.4$.

Combining equations 2.1 and 2.2 we deduce the total volume of dust per unit space volume

$$\frac{N_d 4/3\, \pi a^3}{L} = (1.086)\,\frac{4\,\bar{a}}{3Q_V}\,\frac{A(V)}{L} \simeq \frac{\bar{a}\,A(V)}{L} \tag{2.3}$$

Thus for an estimate of the dust mass density we have

$$\rho_d \simeq \bar{a} \frac{A(V)}{L} \rho \qquad (2.4)$$

where ρ is the density of the dust material, so that a cloud of 1 pc radius and 2 magnitudes of total extinction consisting of 0.1 μm particles with density 2 g cm^{-3} has a dust mass density $\rho_d \simeq 10^{-23}$ g cm^{-3}. For a general average of 1 mag/kpc the dust density is $\rho_d \simeq 10^{-26}$ g cm^{-3} which is of the order of 1% of the total gas density. More precisely, it is observed that the column density of hydrogen and the extinction are related by (Spitzer 1978)

$$\frac{N_H}{A(V)} = 1.9 \times 10^{21} \text{ cm}^{-2} \qquad (2.5)$$

which gives on the average

$$\rho_d/\rho_H = 2.8 \frac{\rho}{1 \text{ g cm}^{-3}} \frac{\bar{a}}{10^{-5} \text{ cm}} 10^{-3} \qquad (2.6)$$

A more detailed calculation of this mass ratio is not generally required; i.e., using complete dust models, the results are somewhat model dependent but not critically so.

3. Cosmic abundance constraints

In Table 1 are shown several author's deductions of the relative abundances of the most abundant elements in space which may be consituents of solids (with the exception of He) in interstellar space. We note that while there are differences from author to author depending on how the estimate is arrived at, there are no major discrepancies. The values given in Table 1 are presumed to have prevailed when the solar system was formed 4.6×10^9 years ago and there are certainly differences from that time to be expected in the current epoch but these are mostly limited to isotopic abundances. The mean cosmic abundance in our galaxy has changed little in the past 4.6×10^9 years (see Matteucci, this volume). We shall henceforth refer to these solar system results as cosmic abundances when referring to the values in Table 1 with application to the current interstellar medium.

Table 1. Relative cosmic (solar system) abundances of the most common elements

	(1)	(2)	(3)	(4)
H	1	1	1	1
He	0.069	0.068	0.081	0.079
C	3.71(-4)	4.17(-4)	4.45(-4)	4.90(-4)
N	1.18(-4)	0.87(-4)	0.91(-4)	0.98(-4)
O	6.76(-4)	6.92(-4)	7.40(-4)	8.13(-4)
Mg	0.32(-4)	0.399(-4)	0.396(-4)	0.380(-4)
Si	0.32(-4)	0.376(-4)	0.368(-4)	0.355(-4)
S	0.16(-4)	0.188(-4)	0.189(-4)	0.162(-4)
Fe	0.25(-4)	0.338(-4)	0.331(-4)	0.467(-4)

(1) A.G.W. Cameron (1973) Sp. Science Reviews 15, 121-146.

(2) A.G.W. Cameron (1982) in "Elements and Nuclidic Abundances in the Solar System", ed. C. Barnes, R.N. Clayton and D.N. Schramm (Cambridge, Cambridge Univ. Press), 23.

(3) E. Anders and Mitsuru Ebihara (1982), Geochimica & Cosmochimica Acta 46, 2363-2380.

(4) N. Grevesse (1984), Physica Scripta T8, 49-58.

Equation 2.6 in combination with the cosmic abundances may very simply be used to exclude certain *conceivable* grain materials if assumed to be solid homogeneous spheres. Mean particle sizes depend on the optical properties of the material assumed and will be stated within an accurancy of about 20% which will be sufficient for the following demonstration. In selecting mean particle sizes for the visual extinction by dielectrics we have that $\bar{a} \simeq \pi^{-1}$ (m-1) μm where m is the real part of the index of refraction (Greenberg 1978). For non-dielectrics (metals) $\bar{a} \simeq (0.5/4\pi)$ μm (Greenberg 1968).

In Table 2 we have used the cosmic abundances derived by Cameron (1982) and we see, for example, that silicates are inadequate by a factor 0.17 to provide the full extinction while molecules containing oxygen and carbon are more than adequate. In fact, if they were not, there would be no gas atoms to produce the wide variety of molecules observed by radio astronomers.

Table 2. Cosmic abundance limitations on grain materials. Availability factor <1 or >1 implies inadequate or adequate abundance respectively as a single dust component.

Material	Size $\bar{a}$ (μm)	Density ρ(g cm^{-3})	$(\rho_d/\rho_H) \times 10^3$ Required Eq. 2.6	$(\rho_d/\rho_H) \times 10^3$ Available by C.A.	Factor of availability
Iron	0.05	7.8	11	1.9	0.17
Quartz[(a)]	0.1	2.6	7.3	2.2	0.30
Olivine[(b)]	0.1	3.3	9.3	2.9	0.31
Water Ice	0.15	1	4.2	12.4	2.95
Graphite	0.05	2.15	3.0	5.0	1.7

(a) SiO_2: limited by Si
(b) $(Mg,Fe)_2\ SiO_4$: limited by Mg,Fe

4. Infrared spectrum

The first positive identification of an important solid interstellar dust component was that showing the silicate absorption at about 9.7 μm. The earliest attempts to observe the 3.07 μm (see later that this wavelength is not invariant) of solid H_2O were negative (Danielson et al. 1965, Knacke et al. 1964). But it is clear from cosmic abundance arguments that silicates alone can not account for the observed amount of extinction. Various suggestions were made to get out of this dilemma. One of these was that although H_2O may form as a mantle on the dust, it could be photodissociated by ultraviolet radiation and even if oxygen remains as a dominant element in the molecular composition the amount of H_2O would be reduced enough to make it unobservable (Greenberg, 1971a). Incidentally this was one of the first recognitions that ultraviolet in space could play a role in the chemistry of interstellar dust. However, we now have more than 50 examples of the "3 μm" ice band of solid H_2O in interstellar dust ranging from protostellar sources (see Gillett and Forrest, 1973), to molecular clouds (see Whittet et al., 1983) and OH IR sources. How can we reconcile the existence and non-existence of ice with a dust model? This will be taken up in later sections. For the moment let us see how much may be revealed of the chemical composition of interstellar dust by some examples of infrared spectra.

In Fig. 4.1 is shown a representative description of where various molecular groups absorb in the region of the infrared which is known as the "fingerprint region". Unfortunately the region between about 5 and 8 μm is totally absorbed by the earth's atmosphere so that only from space or very high flying aircraft may such spectra be obtained. Fig. 4.2 shows a series of spectra of various objects. Note that in the BN source, the H_2O ice band is substantial while in the galactic center source SgA the ice band is absent and only a relatively weak 3.4 μm feature is seen.

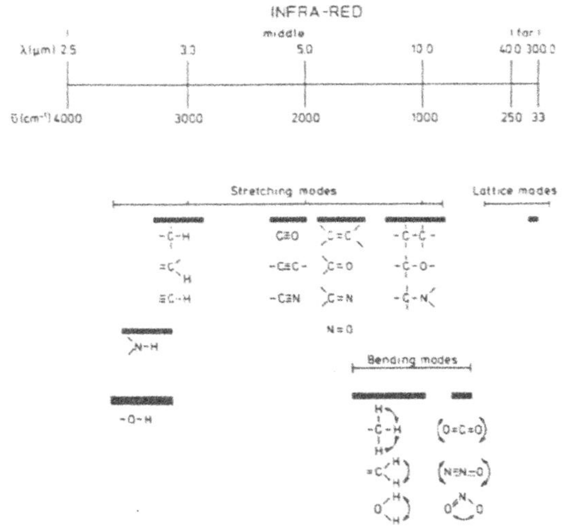

Fig. 4.1. Characteristic wavelengths or wavenumbers for various molecular groups in solids.

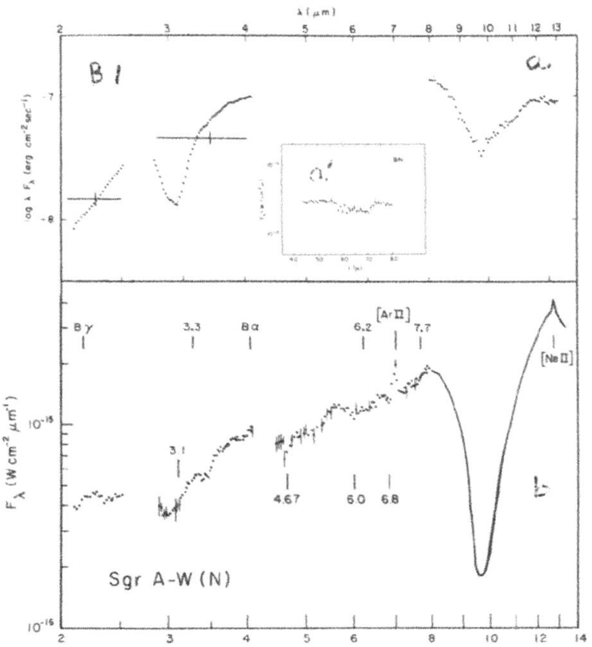

Fig. 4.2. a) The BN source in Orion; b) The galactic center.

Somehow, this 3.4 μm feature must be evidence for a sufficiently thick mantle to account for the total extinction since the silicate alone has been shown to be inadequate. As shown later the 3.4 μm feature is due to complex organic molecular mantles and has been observed not only towards the galactic center (Wickramasinghe and Allen, 1980; Butchart et al. 1986) but also in more general regions (Tapia et al. 1989) and must, as was predicted (Greenberg et al., 1977; Greenberg, 1982a), be characteristic of grains throughout the galaxy.

Fig. 4.3 the deeply imbedded IR source W33A reveals some interesting and imortant spectra feature and observation from the Kuyper high flying aircraft observations in the 6-8 μm region will also be seen to reveal some important aspect of dust photochemical evolution (see section 7) where further comparison with laboratory spectra will be made.

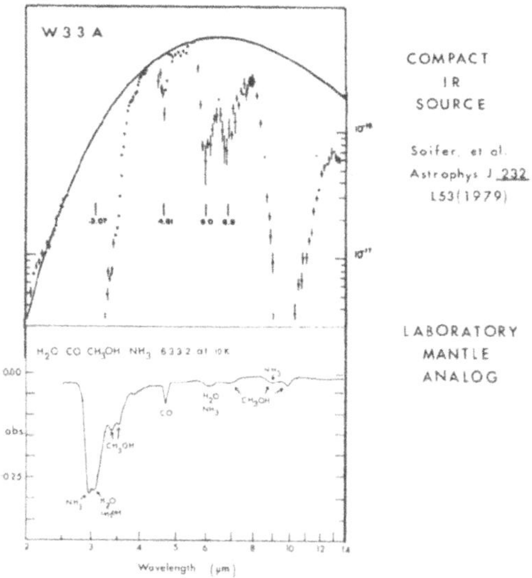

Fig. 4.3. Comparison of the infrared spectrum of a deeply imbedded source with that of a non-irradiated mixture of ices. The silicate absorption is not simulated in the laboratory spectrum.

5. Dust mantle accretion and destruction

The observational evidence that interstellar dust has mantles of ices in molecular clouds and no ices in diffuse clouds clearly indicates the formation and destruction of these mantles.

If we exclude heating of grains in star forming regions, the general mean tenth micron grain temperature in molecular clouds as well as in diffuse clouds is $T_d \leq 15$ K (Greenberg 1971b). Thus every common atom or molecule (excepting H and He) will stick and remain on the grains. The rate of grain growth in a medium of hydrogen density n_H is given by

$$\frac{da}{dt} = \frac{f_A\, n_H\, \bar{m}_A \bar{v}_A}{4\rho} \text{ cm s}^{-1} \tag{5.1}$$

where n_H is the density of H in all form ($n_H = n_H + 2\, n_{H2}$, f_A is the fraction of condensable atom/molecular species in the gas (relative to n_H), $\bar{m}_A$ and $\bar{v}_A$ are the mean mass and velocity of these species, and ρ is the mantle density. Thus, if unrestricted, the mean mantle thickness growth in a cloud lifetime of 5×10^7 yrs. is (in the case of no gas depletion), for $f_A = 10^{-3}$ (see table 1), $n_H = 10^3$ cm^{-3}, $\bar{m}_A = 14.5\, m_H$ (no molecules), $v_A = 10^4$ cm s^{-1} ($T_{gas} \simeq 50$ K), $\rho = 1$ g cm^{-3}

$$\Delta a = 2.7\ \mu m \text{ in } 5 \times 10^7 \text{ yrs.} \tag{5.2}$$

Since, in fact, the maximum particle size is limited by cosmic abundance to be only about $\bar{a} \simeq \sqrt{2.95} \times 0.15 = 0.22\ \mu m$ (see table 2 for H_2O) this implies, first of all, a complete depletion of the condensable gas many times over. Secondly it would imply dust whose extinction would be independent of wavelength throughout the visible and whose total value in the visual would be more than 100 times what is observed; i.e. a 1 pc diffuse cloud ($n_H = 10^3$ cm^{-3}) would have a ridiculously large total extinction of the order of 200 magnitudes!. This kind of possibility was already noted by Oort and Van de Hulst (1946) (without the cosmic abundance constraint) and was the reason they proposed a mechanism based on cloud-cloud collisions to limit the growth of grains. It turns out that if there is a collision dominated destruction mechanism- thus proportional to the grain area - one may derive a very useful size distribution law for the grain mantles. Balancing the rates at which grains grow to a certain size and are depleted from this size one may write (Greenberg 1964)

$$\dot{a} \frac{dn(a)}{da} + P_D(a)a^2 = 0 \tag{5.3}$$

where $P_D(a)a^2$ is the destruction probability per unit time for particles of size a. The solution of this is

$$n(a) = n_o \exp\left[-(P_D/3\dot{a})a^3\right] \tag{5.4}$$

Using reasonable destruction rates *outside* of molecular clouds leads to a mean size representative of average extinction curves (Greenberg 1979). This result is ecouraging but a fully detailed description of this size destribution processing remains to be given and *furthermore* it is based on grain destruction which takes place *outside* of molecular clouds (see Draine and Salpeter 1979a, 1979b). It is the basis for the cyclic evolution theory of dust mantles *in and out* of molecular clouds (Greenberg 1982a, Greenberg 1986) but it doesn't answer the question of why the grains *in* molecular clouds do not grow too large. Because of the rate of destruction and erosion of dust in diffuse clouds silicate cores without mantles would be destroyed faster than they are created in the outflow of cool stars so that whatever mantles are created in molecular clouds must have a degree of permanence to serve as protective "ablation shields" when the grains are exposed to diffuse cloud destruction processes (Greenberg, 1986). Both the limitation of mantle growth and the creation of protective grain mantle material will be discussed in section 7.

6. Ultraviolet and cosmic ray processing

The two major energetic sources for chemical processing of interstellar dust are ultraviolet photons ($E_{h\nu} \gtrsim 6eV$) and low energy cosmic rays ($E \leq 2$ MeV). The photon lower energy limit is a rough definition of that required to break or ionize molecules in the solid dust (lower energies are also possible). The upper limit on the cosmic ray energy is defined by the fact that the grains are so small that high energy cosmic ray particles - with the exception of Fe ions - deposit relatively little energy during the collision. This will be seen in more detail in this section. First let us consider the ultraviolet photoprocessing and then compare it with the processing of dust by cosmic rays.

a) Ultraviolet photoprocessing

The ultraviolet photon flux in diffuse clouds is

$$\Phi_{h\nu} = 10^8 \text{ cm}^{-2} \text{ s}^{-1} \qquad h\nu > 6 \text{ eV} \tag{6.1}$$

For a 0.1 μm interstellar grain the ultraviolet absorption efficiency is $Q_{abs} \simeq 1$ so that almost every impacting photon is absorbed. The total energy deposited in the grain is thus

$$\frac{d\, E_{h\nu}}{dt} = Q_{abs}\, \Phi_{h\nu}\, \overline{E}_{UV}\, \pi a^2 \tag{6.2}$$

where $\overline{E}_{UV}$ is the mean energy in the ultraviolet photon spectrum. To a fair approximation the UV photon energy distribution is given by $n(E) \approx E^{-3}$ (Greenberg 1978) for $6 < E < 13.6$ eV so that $\overline{E}_{UV} = 8.3$ eV. Combining Eqs. 6.1 and 6.2 and assuming a 0.1 μm grain one gets

$$\frac{d\, E_{h\nu}}{dt} = 0.26 \text{ eV s}^{-1} \tag{6.3}$$

Assuming a mean molecular bond pair per unit volume of 8×10^{-24} cm^{-3}, a 0.1 μm grain contains a number of bonds

$$N_b = \frac{4/3\ \pi\ a^3}{8 \times 10^{-24}} = 5.2 \times 10^8 \tag{6.4}$$

Using a mean bond energy $E_b \geq 6$ eV (most bond energies are < 8 eV), the time required for every bond to be exposed to a bond breaking photon in diffuse clouds (D.C.) is

$$\tau_{UV}^{D.C.} = \frac{6 \times 5.2 \times 10^8}{0.26}\ s = 400\ yrs \tag{6.5}$$

In dense clouds photoprocessing time increases with increasing attenuation of the ultraviolet due primarily to the dust absorption. If the size distribution of the dust remains constant throughout the cloud *and* if the mean albedo and angular scattering distribution of the dust is known this attenuation may be calculated for some simple cloud geometries. However, the ultraviolet within the cloud is given not only by the radiation penetrating from outside but also by internal sources (Norman and Sil, 1980; Prasad and Tarafdar, 1983; Greenberg, 1971). Furthermore, the external radiation may itself be supplemented by local hot stars. Individual cases should be individually considered. However, a few general remarks and results are relatively easily indicated which show the signifiance of ultraviolet photoprocessing of dust within dense clouds.

Consider, first, what happens if the particle size distribution in the dense cloud is the same as in a diffuse cloud. The large (0.1 μm) particles (L) have an albedo of about $\alpha_L = 0.5$ in the U.V. The small particles (S) have an albedo $\alpha_S \simeq 0$ in the U.V. Separating the *extinction* (given here as optical depth τ) contributions of the large and small particles and multiplying each by $(1-\alpha)$ gives the net *absorption* optical depth.

$$\tau_{abs} = \tau^L{}_{ext}\ (1-\alpha_L) + \tau^S{}_{ext}\ (1-\alpha_S) \tag{6.6}$$

Since the small particles are taken to be pure absorbers, the *scattering* distribution in the UV is dominated by the large particles. It is well known that the larger the particle the more it scatters toward the forward direction; i.e., it has a high value of its scattering asymmetry factor $g = \langle\cos\ \theta\rangle$. As g approaches unity we may neglect multiple scattering effects to first order so that the attenuation is given directly by the absorption. The ultraviolet radiation field is attenuated approximately as $\exp(-\tau_{abs})$ (See Chlewicki and Greenberg, 1984).

For the mean extinction curve the saturation level of the large particle extinction in the UV ($\lambda \leq 250$ nm) is $\tau^L{}_{ext}$ (UV) $\simeq$ 1.5 $\tau^L{}_{ext}$ (V). There is a wide variety of possible *extra* contributions to the UV extinction by the small particles $\tau^S{}_{ext}$ but a *rough* mean at 220 nm is about $\tau^S{}_{ext} \simeq \tau(V)$ with a subsequent drop to a minimum at 170 nm of substantially less than $\tau(V)$ and a rise in the far UV to as much as $\tau^S_{ext} \approx 2\ \tau(V)$. How much attenuation the small particles contribute in the deep interiors of molecular clouds is open to considerable speculation. If, as is likely, the small particles tend to accrete onto the large particles their effect is reduced as one penetrates deeper and deeper into the cloud. Thus the mean ultraviolet attenuation within a cloud is probably less than that given by

$$\tau_{abs}\ (UV) \simeq (0.5)\ (1.5\ \tau_{ext}\ (V)) + \tau_{ext}\ (V)$$

$$= 1.75\ \tau_{ext}\ (V) \qquad (6.7)$$

where $\tau_{ext}\ (V) = (1.086)^{-1}\ A(V)$

The reduction of the ultraviolet by a factor of 10^5 corresponds to a cloud visual extinction at the center of 7 magnitudes, or a total extinction of 14 magnitudes - a dark cloud indeed.

The additional ultraviolet in a cloud arrises from either young stellar wind shocks (Norman and Silk 1980) or from penetrating cosmic radiation (Prasad and Tarafdar 1983, Greenberg 1971b). These two contributions have each been estimated as about $2\text{-}3 \times 10^3\ cm^{-2}\ s^{-1}$.

With the exception of the densest cloud cores one expects that the ultraviolet photoprocessing time scale is seldom greater than the cloud lifetime. For example, at the center of a 1 pc thick cloud of density $n_H = 5 \times 10^3$ the photoprocessing time scale is 5×10^5 yrs and, because of the general ubiquitous internal sources of UV, the photoprocessing time scale even in a dense molecular cloud is probably $< 5 \times 10^6$ years. A caveat is that *local* conditions may deviate from those assumed here. However, these ultraviolet photoprocessing time scales are very useful as guidelines.

b) Cosmic ray processing

The major contribution to the cosmic ray energetic processing of interstellar dust grains is by particles whose energy is of the order of 1 MeV. Unfortunately our current knowledge of the spectra of cosmic rays does not extend much below energies of hundreds of MeV. Only indirectly may we obtain some estimates of the cosmic ray particle flux at low ($\approx$ MeV) energies. One way to deduce this is from the ionization rate coefficient ζ in the interstellar medium. Upper and lower bounds on this were defined by Spitzer (1968) with $6.8 \times 10^{-18}\ s^{-1} < \zeta \leq 1.17 \times 10^{-15}\ s^{-1}$. More recently Spitzer (1978) argued in favor of $\zeta \simeq 1 \times 10^{-17}\ s^{-1}$ while Lepp and Dalgarno (1989) suggest a value $\zeta < 5 \times 10^{-17}\ s^{-1}$ with this limit dictated by gas phase molecular schemes. For want of anything more definite this latter value will be used to derive an upper limit on the flux of low energy cosmic ray photons which are the primary cause of this ionization. We shall consider the particles to be either entirely 1 MeV or 2 MeV protons. What follows is based primarily on Bethe (1933) as used by Spitzer (1968).

Given the ionization rate coefficient ζ, the proton flux is given by

$$\Phi_p = \zeta/\sigma_{IP} \tag{6.8}$$

Where σ_{IP} = ionization cross section for low energy protons in collisions with hydrogen. This cross section is given by the Bethe formula (Bethe, 1933) valid for E > 0.3 Mev

$$\sigma_{IP} = \frac{1.23 \times 10^{-20}}{\beta^2} \{\log[\beta^2/(1-\beta^2)] + 6.2 - 0.43\, \beta^2\} \tag{6.9}$$

where $\beta = v/c$. For protons in the low MeV range an additional ionization produced by secondary electrons leads to an increase in the *effective* cross section by a factor 1.67 (Dalgarno + Griffing, 1958). Incorporating this factor in Eq. 6.9 and using Eq. 6.8 with $\zeta < 5 \times 10^{-17}\ s^{-1}$ leads to proton fluxes

$$\Phi^1 < 2.4\ cm^{-2}\ s^{-1} \qquad (E_p = 1\ Mev)$$

or (6.10)

$$\Phi^2 < 3.4\ cm^{-2}\ s^{-1} \qquad (E_p = 2\ Mev)$$

We now derive the respective ionization energy input for such particles passing through interstellar grains. For a solid of density ρ and thickness 2a the ionization loss by a low energy proton is

$$\Delta E_p = 2a \frac{4\pi\, n_e\, e^4}{mc^2\, \beta^2} \ln \left[\frac{2mc^2\beta^2}{\bar{E}(1-\beta^2)} - \beta^2\right] \tag{6.11}$$

where n_e = volume density of electrons $\approx \rho/2m_H$ (assuming $\approx 1/2\ \bar{M}$ electrons for each mean molecule $\bar{M}$), m = electron mass, $\bar{E}$ = mean ionization energy $\approx$ 30 ev. For a sphere of radius a the mean thickness is $\approx 1/2\ 2a$. A density $\approx 2\ g\ cm^{-3}$ corresponds to the organic refractory mantle components (see sections 7,8) while $\rho \approx 1\ g\ cm^{-3}$ corresponds to the outer icy mantle. Since it is only the mantles which are under consideration here and since *combined* they constitute about half the total grain thickness there is no significant error if we take the product $\rho a = 0.1\ g\ cm^{-2}$. Finally we obtain from Eq. 6.11 a mean ionization energy loss in a grain which, for $\beta \ll 1$, reduces to

$$\Delta\bar{E}_p = 1.43 \times 10^3\ E^{-1}\ [4.29 + \ln E]\ eV \quad (E\ in\ MeV) \tag{6.12}$$

For 1 MeV and 2 MeV protons this gives 6.13 KeV and 3.56 KeV respectively. Note how much the ionization *decreases* with increasing proton energy. The *relative* effectiveness of 1 MeV and 2 MeV protons expressed in terms of the product $\Phi_p\ \Delta\bar{E}_p$ is 14.7 KeV $cm^{-2}\ s^{-1}$ or 12.1 KeV $cm^{-2}\ s^{-1}$, which differ only by about 20%.

Rather than directly computing the cosmic ray chemical processing time we shall compare it with the ultraviolet photoprocessing time as a ratio of the rates of energy deposition given by

$$R_\tau = \frac{\tau_{CR}}{\tau_{h\upsilon}} = \frac{\Phi_{UV}\ \bar{E}_{UV}\ \pi\ a^2}{\Phi_p\ \Delta\bar{E}_p\ \ \pi\ a^2} \qquad (6.13)$$

In diffuse clouds, using an average of 2 and 1 MeV protons,

$$R_\tau = \frac{10^8 \times 8.3}{13.4 \times 10^3} = 0.6 \times 10^5 \qquad (6.14)$$

In molecular clouds the cosmic ray energy absorbed by grains (even assuming no reduction in the low energy cosmic ray flux at 1-2 MeV) becomes comparable to the ultraviolet only if the ultraviolet is reduced by a factor 1.7×10^{-5}. This could occur in a cloud whose visual optical depth at the center is about 7 magnitudes and in which there is *no other source* of ultraviolet. However, the cosmic rays themselves are a source of ultraviolet equal to about Φ_{UV} (C.R.) $> 2 \times 10^3$ cm^{-2} s^{-1} which is *twice* that given by a reduction of 1.7×10^{-5} so that for cosmic rays to contribute significantly to grain chemistry there must be some local excess sources of MeV particles. The solar neighborhood is a good example of this as also are probably regions close to flare stars. However in the most general regions of space, including the cores of molecular clouds, it appears that the ultraviolet radiation is the dominant source of chemical and physical processing of dust. Note that the ionization rate assumed by Prasad and Tarafdar (1983) is lower than the one assumed here by a factor of about 1.5 so that both the cosmic ray flux deduced here, as well as the ultraviolet flux, are about 1.5 times higher than they would deduce.

7. Mantle evolution: photochemistry, heating

The conditions for photochemistry of dust - that grain mantles accrete and that ultraviolet photons are sufficiently abundant to break the molecules in the mantles in times shorter than molecular cloud lifetimes - is supported by both theoretical and observational arguments.

The study of photochemical evolution of dust mantles may be made under laboratory conditions simulating the basic physical conditions of interstellar space. In the Leiden laboratory the mantle is simulated by thin layers of molecules deposited at low temperature - generally $T \simeq 12$ K - on a metal or other surface in a vacuum chamber maintained at pressures $\leq 3 \times 10^{-8}$ tor and irradiated by vacuum ultraviolet lamps with high enough fluxes to produce substantial chemical modifications on time scales of the order of minutes to hours. The details of the equipment and its operation have been described elsewhere (Hagen et al. 1979). Other laboratories are now engaged in similar research (see e.g., Allamandola and Sandford, 1988).

We will here show a few illustrative examples of how laboratory results can be used to provide data for direct comparison with astronomical observations. The first example shows the affects of heating alone. The second shows the effect of photochemistry plus heating.

a) Heating

An example of two effects of heating on simple mixtures of ices is shown in Fig. 7.1 where a laborotory H_2O + CO mixture is followed from deposition at 12 K through a sequence of heating steps. As expected, the CO fraction decreases with increasing temperature because of its much higher volatility than H_2O. Accompanying the decrease in CO one observes a change in the shape of the 3.07 absorption indicating approach to crystallinity with increasing temperatures.

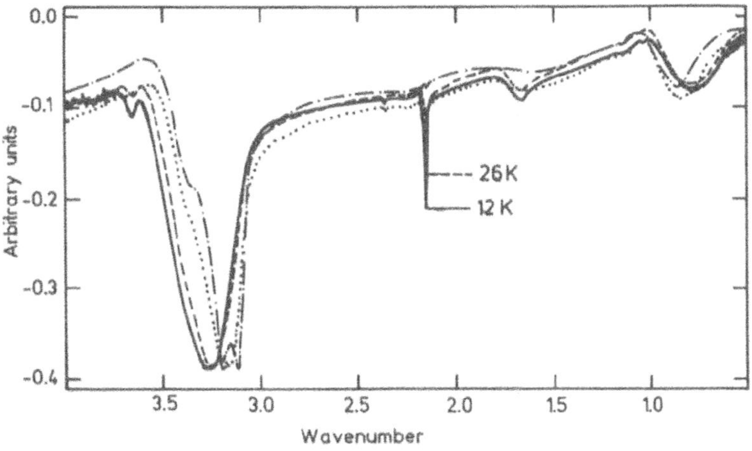

Fig. 7.1. Infrared spectra of an H_2O:CO mixture deposited at 12 K and subjected to warmup. Solid: T = 12 K, dashed: T = 26 K, dotted: T = 120 K, dash-dot: T = 126 K. At 12 K the ice is very amorphous and at T = 126 K the shape of the absorption is close to that of crystalline ice.

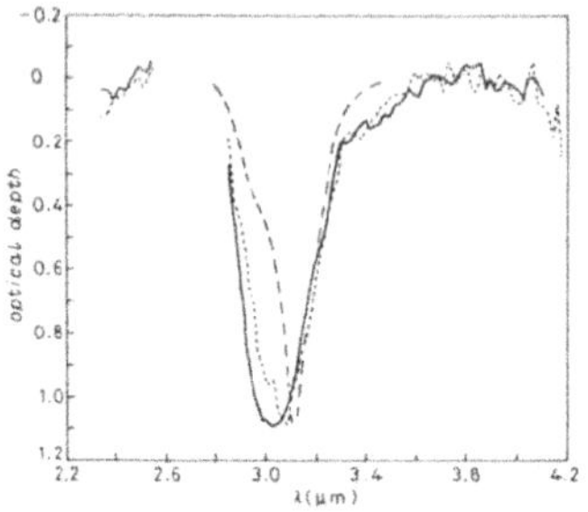

Fig. 7.2. Comparison of the observed spectra of HL Tau (dots), Elias 16 (solid) and the absorption of crystalline ice (dashes) all normalized to the same optical depth.

A comparison of the observed ice bands in Elias 16 and in HL Tau shown in Fig. 7.2 shows that the latter exhibits a history of heating. Concomitantly, the CO in HL Tau is unmeasurable while that in Elias 16 indicates a $CO:H_2O$ ratio of ≈ 0.15 (Whittet et al. 1985, Van de Bult et al. 1985). Figs. 7.3 and 7.4 show how well the observed shape of the ice bands can be matched by appropriate calculations of absorption by non spherical (required because of polarization) particles with mantles of ices.

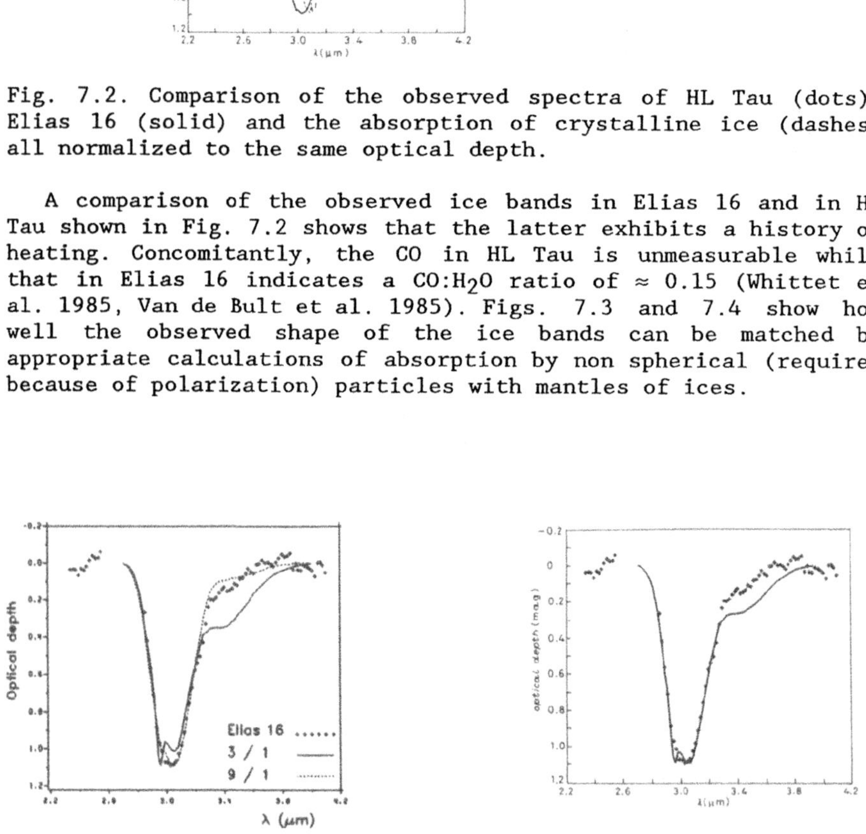

(a) (b)

Fig. 7.3a) Comparison of *absorption spectrum* of $H_2O:NH_3 = 3:1$ and 9:1 12 K (unannealed) mixture with Elias 16. b) Comparison of *absorption by cylinders* having unannealed mantles of $H_2O:NH_3$ = 3:1 (---) with Elias 16 (+++). Cylinder compound core a_{cc} = silicate core plus organic refractory mantle (a_{cc} = 0.12 μm). Mantle thickness Δa_m = 0.012 μm. Note how cylinder (elongated particles) reduces "effective" absorption at 2.95 μm due to NH_3.

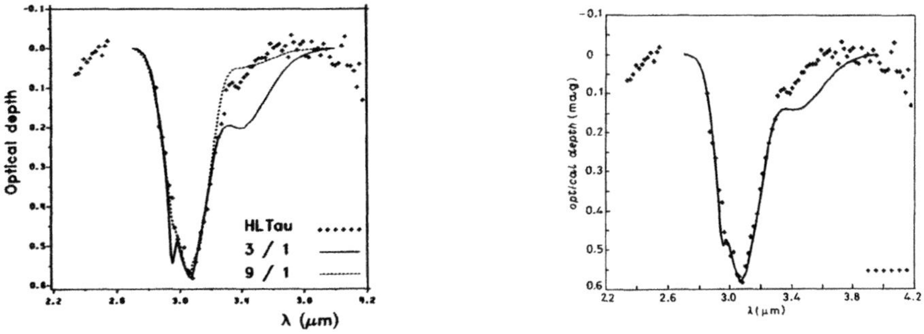

(a) (b)

Fig. 7.4a). Comparison of *absorption spectrum* of $H_2O:NH_3$ = 3:1 and 9:1 50 K (annealed) mixture with HL Tau. b) Comparison of *absorption by cylinders* having annealed mantles of $H_2O:NH_3$ = 3:1 (---) with HL Tau (+++). Cylinder compound core a_{cc} ≡ silicate core plus organic refractory mantle (a_{cc} = 0.12 μm). Mantle thickness Δa_m = 0.012 μm. Note how cylinder (elongated particle) reduces "effective" absorption at 2.95 μm due to NH_3.

In the case of HL Tau and Elias 16 we are dealing with more or less uniform types of grains in each environment, the former being in an accretion disk close to the star, the latter being an average over a single molecular cloud. However there exist many examples of objects towards which the line of sight contains grains with a variety of mantle thicknesses and previous heating histories. In Fig. 7.5 are shown a few such examples. For example the BN absorption shows features characteristic of both strongly annealed ice - even crystalline - and very cold amorphous ice mixtures and, as had been noted earlier (Greenberg 1982) the dust close in must be denuded of ice with all the volatile mantles evaporated off. None of these absorption bands can be characterized by a *single* form of ice whether pure or a mixture. They all have features corresponding to a variety of formation and evolution histories.

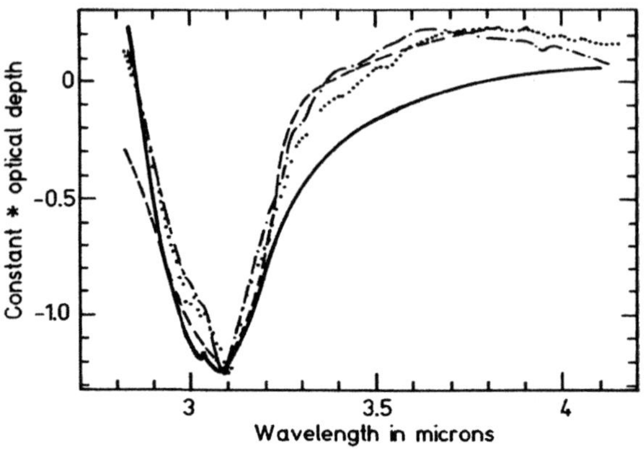

Fig. 7.5. "Ice" absorption spectra of several young stellar objects normalized to the same optical depth: BN (dash dot); GL2136 (dots); GL 989 (solid); and a star undergoing mass loss, OH 0739-14 (dash).

b) Photochemistry and heating

An example of the effects of ultraviolet photolysis and subsequent heating of an ice mixture is shown in Fig. 7.6. Among other things we note how photolysis reduces the CH_4, increases the CO_2 and modifies the Co band by adding a short wavelength wing at about 2170 cm^{-1}. Following warmup, the CO is strongly reduced while the wing structure remains and has since been identified.

One astronomical source which is an excellent example of just such solid state photochemistry is the protostellar object W33A. The spectrum shown in Fig. 4.3 has several major points of interest: (1) both the 3.1 μm ice band and the 9.7 μm silicate band are exceedingly strong indicating a very strong total extinction, (2) an absorption at $\approx$ 4.61 μm around the wavelength characteristic of CO, (3) an absorption at about 6.8 μm which will be discussed later.

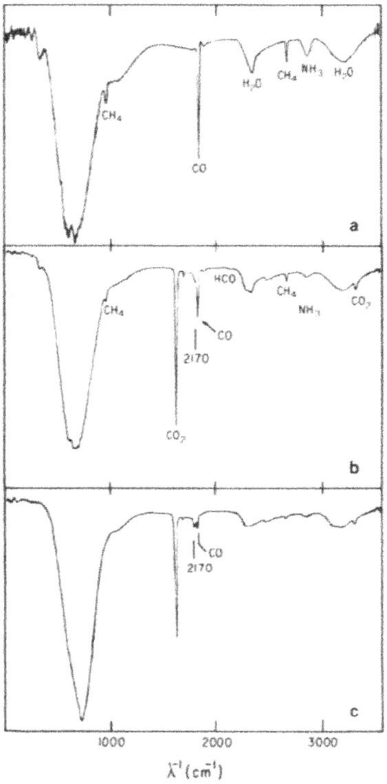

Fig. 7.6. Infrared spectrum of the mixture $H_2O:CO:CH_4:NH_3$ = 6:2:1:1: *a)* after deposition, no ultraviolet irradiation, temperature 12 K; *b)* after 300 h ultraviolet irradiation, temperature 12 K; *c)* warm-up after irradiation to 95 K.

The "4.6 μm" band was shown by a higher resolution observation (Lacy et al. 1984) to be a superposition of two absorption features, one peaking at 4.67 μm (2140 cm^{-1}) and assigned to solid CO, the other peaking at 4.62 μm (2165 cm^{-1}). Laboratory photochemical experiments carried out at the Leiden Laboratory on various mixtures led to the tentative identification of the 2165 cm^{-1} absorption as probably due to a neutral molecule containing CN labeled "XCN". Its final identification makes a interesting story. First of all it appeared that a CO/NH_3 mixture was *much* more efficient in producing the band than any mixture such as CH_4/NH_3 not containing oxygen. Secondly it occurs after a very short photolysis time indicating that only a very few steps are required to generate the molecule. These two facts immediately eliminated such candidates as, for example, methyl isocyanide CH_3NC and strongly suggested the presence of oxygen in the molecule. The key to the positive identification was in *laboratory* use of isotopically labelled carbon (^{12}C, ^{13}C) and nitrogen (^{14}N, ^{15}N) and measuring the shifts in the peak absorption associated with these species. This led ultimately to the *ion* OCN^- and was the first evidence that non ionizing ultraviolet radiation may lead to ions in grain mantles. Figure 7.7 shows the comparison between the laboratory and the observed spectrum. We note that the sample had to be heated as would happen in the dust in the protostellar cloud (near enough to the star) and the CO has all but disappeared compared with its original abundance while the OCN^- remains. A subsequent analysis of the 6.86 μm absorption in W33A has led to the suggestion that it is produced by NH_4^+, the counterion to OCN^-.

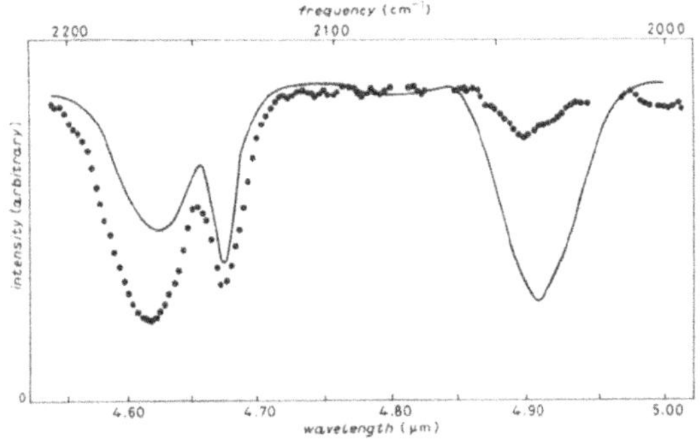

Fig. 7.7. Laboratory spectrum (——) of a frozen mixture of simple molecules, including H_2S, CO and NH_3, which has been subjected to UV irradiation at 10 K and then warmed to 110 K. The spectrum of W33A (•••) is shown for comparision with its baseline flattened. The peaks out at the left, are, respectively, OCN^- and CO. The peak at 4.9 μm is OCS.

The 6.86 μm absorption in not so unambiguously identified with NH_4^+ as the OCN^- is with the 4.62 μm absorption so that other identifications have still be to considered. As shown in the lower spectrum of Fig. 7.8 the feature is rather like that of methanol (Tielens et al. 1984). However, the methanol absorption in the wing of the 3.1 μm ice band indicates a factor of at least 5 times less methanol than needed to account for the 6.8 μm absorption. From a positive part of view however the evidence for at least $\approx$ 5% methanol in the W33A spectrum appears to be well supported by the observations (Baas et al., 1988).

c) Organic refractory residues

As we noted earlier the absence of an H_2O ice band can not, for abundance constraint reasons, be evidence for a lack of mantles on the silicates. When a well photolyzed ice mixture is warmed up in the laboratory there always remains a residue which is certainly

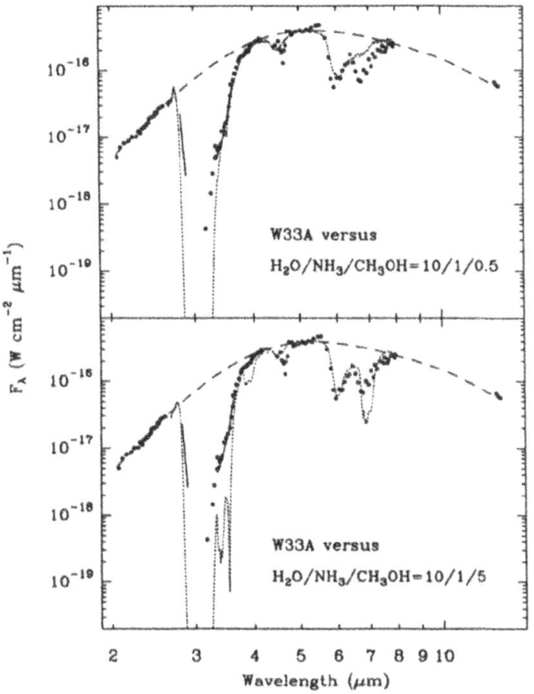

Fig. 7.8. Comparison of the 2-15 μm spectrum of W33A with astrophysical ices containing 5% and 50% methanol respectively (dashed lines). The data for W33A were taken from Soifer et al. (1979). The solid line represents the observations (Baas et al. 1988). The continuum was defined by a polynomial fit.

refractory; i.e., non-volatile, by interstellar standards in the sense that it remains indefintely at ≈ 300 K which is much higher than any but the hottest dust near young hot stars. Such residues are characterized by infrared absorption in the "3.4 μm" region characteristic of C-H stretches in CH_2 and CH_3 groups sequentially appearing (CH_3, CH_2, CH_3, CH_2) with increasing wavelength. Although it was long ago conjectured that organics constituted the major bulk of interstellar dust, the first observation of this was in the galactic center. A comparison of this 3.4 μm interstellar absorption with various laboratory residue spectra (Fig. 7.9) shows significant differences. The reason for this is most likely that in order to see the feature in the interstellar dust it has to be observed along a line of sight spanning many different histories, or stages of dust evolution, and is an average over many residues with different amounts of radiation and heating. The strength of the 3.4 μm absorption in space has been estimated at about 400 cm^2 gm^{-1} which is quite low compared with, say H_2O ice.

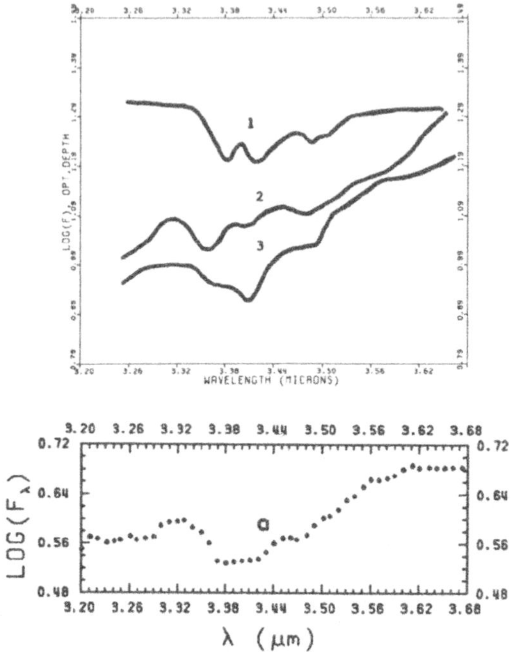

Fig. 7.9. Top- "3.4 μm" laboratory residue spectra for three different initial mixtures: 1) CH_4, 2) H_2O:CO:CH_4:NH_3 = 2:2:1:2 3) H_2O:CO:NH_3 = 5:5:1. Bottom - the 3.4 μm absorption in the galactic center (IRS 7). Note how the characteristic CH_3 (≈ 3.4 μm, ≈ 3.5 μm) and CH_2 (≈ 3.45 μm, ≈ 3.52 μm) absorptions shift and change in strength from one laboratory residue to the other.

d) Grain explosion and chemical evolution from dust and gas interaction

The limiting of grain mantle accretion in molecular clouds was proposed some time ago (Greenberg and Yencha 1972) to occur as a result of photolysis in low temperature ices. It was noted that radicals could be stored in sufficient abundance to provide an explosion mixture. All that was needed was a triggering mechanism. The original triggering mechanism suggested was caused by sudden temperature increases resulting from grain-grain collisions. Since then another possible trigger is thought to the impulsive heating due to penetration of grains by cosmic ray iron nuclei (Leger, Jura and Omont, 1985). In either case a sudden jump in temperature by only about 10 K appears to be adequate to lead to a runaway chain reaction as demonstrated in the laboratory (d'Hendecourt et al. 1982). The maximum mantle thickness limited by the time between triggering (events which is of the order of 10^6 yrs at n_H = 10^4 cm^{-3}) is ≈ 0.03 μm or corresponds to a mean mantle thickness of ≈ 0.015 μm; i.e. 15% of the mean core size (See next section). This means that molecules produced in or accreted on the grains are sporadically released back into the gas phase. Such a process must be invoked to explain why the HCO_2^+ abundance is observed to be at least 100 times more than predicted by the pure gas phase reaction of Herbst + Leung (1986) and is even 10000 times higher than predicted from the scheme of De Jong, Dalgarno and Boland (1980).

The results shown in Fig. 7.10 (Breukers et al 1990) are deduced from the gas phase reaction network of the De Jong et al. (limited to H, C, O, N, atoms and metals) in combination with solid grain chemistry similar to that of d'Hendecourt et al. (1985). The solid grain chemistry consists of: 1) Accretion of gas phase species, 2) Surface reactions (primarily with H), 3) Photoproduction of CO_2 (not considered by d'Hendecourt et al. 1985), 4) Desorption of volatile mantles - including CO_2-by grain-grain triggered explosions. Initial conditions include :1) depleted gas phase abundances determined by grain composition (Greenberg, 1982) and approximately equal to those of ζ Oph.; 2) cloud density $n_0 = 2 \times 10^4$ cm^{-3}, 3) $\Phi_{h\nu} = \Phi_{h\nu}{}^{D.C.}\ e^{-2A(V)}$ (A(V) = 4), 4) $T_{gas} = T_{dust} = 10$ K.

We see immediately in Fig. 7.10, that both CO_2 and HCO_2^+ are orders of magnitude higher than produced by pure gas phase reactions. Also, to be noted is that the solid state CO_2 is about as abundant as CO, which was observed by d'Hendecourt and Muizon (1989). It thus appears that the basic combination needed to produce both the CO_2 in the solid *and* the abundant HCO_2^+ in the gas is the photoproduction of CO_2 in the grain *and* the explosion release of CO_2 back into the gas. As had already been demonstrated, the H_2O abundance in the dust is high compared with CO whereas the H_2O abundance in the gas is (whether calculated or observed) low compared with CO (not shown here).

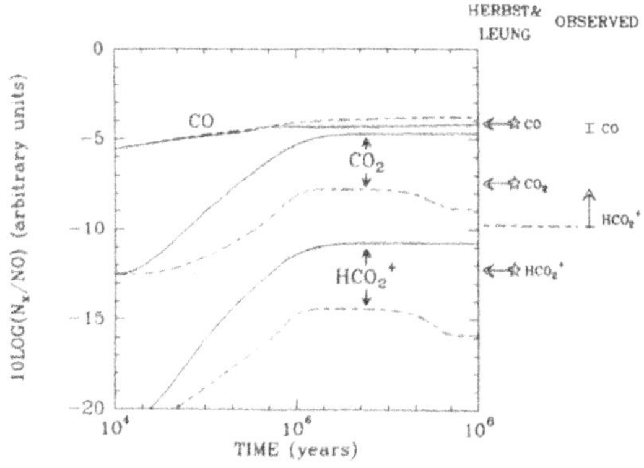

Fig. 7.10. Chemical evolution of CO, CO_2 and HCO_2^+ from pure gas phase model (dashed line) and combined dust gas model (continuous line) as fractional abundance vs. time. At the right of the figure the predictions of Herbst & Leung and the observed abundances are indicated.

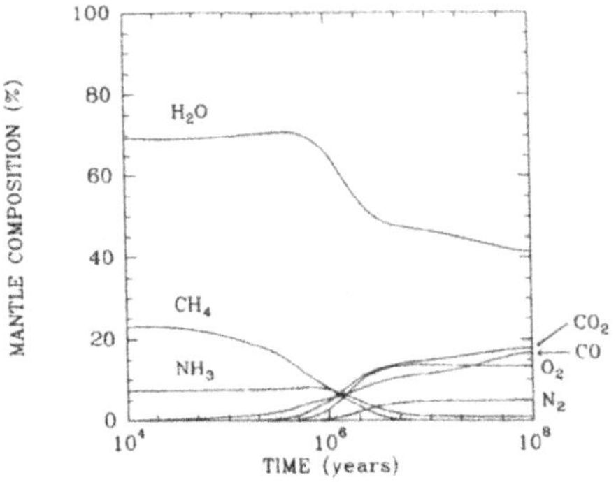

Fig. 7.11. Evolution of the most important mantle species, given in percentages for the combined dust/gas model.

8. Grain model

Observations of interstellar circular polarization are regarded as one of the essential constraints on theoretical models of interstellar grains. Although the measurements are difficult, and the data are limited to only a handful of stars, all of them consistently show that the sign of circular polarization changes at a wavelength λ_c, which is always close to the peak wavelength of linear polarization, λ_{max}. This effect was initially discovered in several lines of sight with λ_{max} close to the general average value of 5500 Å (Kemp and Wolstencroft 1972; Martin, Illing and Angel 1972) and was later demonstrated for lines of sight in which λ_{max} varied from 4000 Å to 8000 Å (Martin and Angel 1977). Mie theory calculations (Martin 1972, 1974) apparently indicated that the observed condition $\lambda_c \approx \lambda_{max}$ could be satisfied only if the particles responsible for the interstellar polarization were almost entirely transparent; i.e., the imaginary part m'' of their index of refraction ($m = m' - im''$) was low at visible wavelengths. Martin's criterion, which effectively imposed an upper limit on the value of $m'' \simeq 0.1$ in the visible, appeared to have immediate implications for theoretical models of grains. A complementary restriction on the properties of particles at visual wavelengths can be derived from the observations of the visual albedo, which yield an average value of ≈ 0.6 (Savage and Mathis 1979), and thus imply that at least one of the populations contributing to visual extinction must be significantly absorptive. Using Martin's calculations Mathis (1983) derived (for $m' = 1.5$) an upper limit of $m'' = 0.03$ (!) and with this result ruled out the three-population model of Greenberg (Mathis 1986) because of the unacceptably high visual albedo (at least 0.9 with Mathis's upper limit on m'').

It is readily demonstrated that this interpretation of Martin's Mie theory calculations was incorrect and that the relationship between the observed λ_{max} and λ_c in fact provides no fundamental limits on the absorption of the constituent materials of interstellar grains. Application of Kramers-Krong relations eliminates the low *m*" "circular polarization criterion". This approach to interstellar circular polarization was first suggested by Kemp (1972) and Kemp and Wolstencroft (1972, 1974) and subsequently given a very clear and elegant formulation by Shapiro (1975), who also showed that a strongly absorbing conductor (magnetite with room temperature optical constants) could reproduce a coincidence of λ_c with λ_{max}. It is surprising that, with very few exceptions, this argument has received little attention in subsequent publications on interstellar polarization one exception being Martin (1975b).

a) Optical constants of astrophysical solids and Kramers-Kronig relations

The optical constants of most materials likely to be the constituents of interstellar grains cannot be derived directly from experimental measurements, and a "synthesis" of various laboratory results is usually required.

The cores of large grains in the three-population model are assumed to consist of silicates (Hong and Greenberg 1980). Since the mantle dominates the optical properties of the particles, any refractory dielectric could be assumed for the cores with no significant effect on Mie theory results. However, some form of amorphous silicate material is suggested by the observations of the 9.7 μm SiO feature. We have adopted the optical constants of silicates in the UV and in the visual from experimental measurements for olivine (Huffman and Stapp 1973), with a small fixed amount of absorption (ε" = 0.12) added in the visual to account for the amorphous structure of interstellar grains. In the near infrared, we have used the data obtained for amorphous silicates by Day (1979). The resulting set of optical constants is very similar to the properties of "cosmic silicate" derived by Draine and Lee (1984).

The derivation of the optical constants for grain mantles is more complex, because there are as yet no direct laboratory measurements of the index of refraction similar to the data used for silicates. The mantles in the diffuse medium are assumed to consist of the refractory component of complex molecular mixtures initially accreted in molecular clouds and subsequently photoprocessed (Greenberg 1982a). The formation of such refractory residues in irradiated molecular ices has been shown experimentally; however, the doses of radiation applied in the laboratory have so far been limited to many orders of magnitude less than those received by grains in the diffuse medium. Materials studied in the laboratory are therefore representative of grain mantles which exist shortly after the particles emerge from molecular clouds (first - generation organic refractory), but their optical properties must differ from those of more highly photoprocessed mantle material in the diffuse medium. Existing laboratory measurements of refractory residues must consequently be taken only as general indications of the changes induced in organic refractory grain mantles by UV irradiation over times of 10^8 yr.

We have assumed that the elemental compositon of the mantles in the diffuse medium is likely to be close to that given by the ratios C:O:N:H = 2:1:1:2. The value of 1:1 adopted for the C:H ratio, which is particularly important in determining the optical properties of the mantle, is consistent with the number derived from the observed strength of the 3.4 μm feature (Greenberg 1982a), although this estimate depends on what fraction of carbon in the organic refractory is assumed to form aliphatic bonds. The high value assumed for the C:O ratio is based on the chemical analysis of laboratory samples of organic residues (Agarwal et al. 1985), which indicate an increase in the carbon-to-oxygen ratio from 1:4 in the initially accreted volatile material to the more than 1:1 in first - generation organic refractory. The wavelength - dependence of electronic absorption is obviously determined by the specific molecular composition of the material; however, a survey of the spectra of complex organic molecules indicates that the strongest features are likely to occur in the far - UV; i.e., for photon energies greater than 7 - 8 eV (Calvert and Pitts 1966; Silverstein and Bassler 1967). The relative oscillator strength in the near - UV is determined by the fraction of carbon $\underline{sp}^2$ bonds, which will give rise to π-electron absorption below 6 eV. In the samples presented in this paper, we have assumed that the π-electron absorption in the mantles is relatively weak. The near - infrared properties (1 μm < λ < 20 μm) have been derived from laboratory spectra of first - generation organic refractory (Schutte, private communication); no attempt was made to derive the optical constants at longer wavelengths. We allowed the imaginary part of the dielectric constant ($\varepsilon = \varepsilon' - i\varepsilon''$; $\varepsilon = m^2$ so that $\varepsilon' = m'^2 - m''^2$ and $\varepsilon'' = 2m'm''$) at visible wavelengths to be varied in order to analyze the effects of various assumptions about the visual absorptivity of the material on the extinction and polarization. The calculations were simplified by assuming that ε'' remained constant down to energies of $\approx$ 1 eV; a more realistic model of a dielectric should include a decline of the value of ε'' toward longer wavelengths.

General physical relationships, which must be satisfied in any optical medium, can be used to derive a consistent set of optical constants from qualitative predictions about the characteristics of astrophysical solids, such as those outlined above for the organic refractory mantles. The most important of these general relations is contained in the Kramers-Kronig formulae, which for the dielectric constant take the form

$$\varepsilon'(\omega) - 1 = \frac{2}{\pi} \int_0^\infty \frac{\omega' \, \varepsilon''(\omega')}{\omega'^2 - \omega^2} d\omega' \tag{8.1}$$

$$\varepsilon''(\omega) = \frac{2}{\pi} \int_0^\infty \frac{\omega \, [\varepsilon'(\omega') - 1]}{\omega'^2 - \omega^2} d\omega' . \tag{8.2}$$

The Kramers-Kronig formulae applied to any frequency-dependence function which describes the response of a physical system to an oscillating signal ($\varepsilon - 1 \propto \chi$, the dielectric susceptibility) are equivalent to the presentation of causality in the system's time-dependent response (Titchmarsh 1937). The discussion of interstellar circular polarization provides an example of the importance of this equivalence. Although the index of refraction does not represent directly the electromagnetic response of the medium ($m = \sqrt{\epsilon}$), it satisfies the mathematical conditions necessary for the derivation of the Kramers-Kronig formulae, and Eqs. 8.1 and 8.2 can be rewritten with m substituted for ε.

The Kramers-Kronig formulae lead to several further relations, some of which are of practicual use in deriving the optical constants of materials. For $\omega = 0$, Eq. 8.1 yields

$$\varepsilon'(0) - 1 = \frac{2}{\pi} \int_0^\infty \frac{\varepsilon''(\omega')}{\omega'} d\omega' . \qquad (8.3)$$

Since the static dielectric constant is often known for solids of astrophysical interest Eq. (8.3) can serve as a useful constraint on the oscillator strength assumed for a specific material throughout the wavelength range of valence electron absorption.

At high frequencies, the valence electrons behave collectively as a free-electron plasma. The plasma frequency characterizing that behavior can be derived from an asymptotic form of the Kramers-Kronig relations. If $\varepsilon''(\omega)$ converges sufficiently rapidly [$\varepsilon''(\omega) \to 0$ faster than $(\omega^2 \ln\omega)^{-1}$ for $\omega \to \infty$], one obtains an important sum rule (Bassani and Altarelli 1983):

$$\frac{4\pi e^2}{m_e} n_e n_a = \frac{2}{\pi} \int_0^\infty \omega \, \varepsilon''(\omega) \, d\omega, \qquad (8.4)$$

where n_e is the effective number of valence electrons per atom, n_a is the volume density of atoms, e and m_e represent the electric charge and the mass of the electron. If the integration in Eq. 8.4 is carried out only up to some finite frequency, ω_0, the sum rule yields the number of electrons that are optically active for $\omega \le \omega_0$. This makes it a powerful tool in assigning the absorptions measured in experiments to specific electronic transitions or, as in our study, in linking the optical constants of a material to its assumed chemical composition.

Eqs. 8.1-8.4 have been used in calculating the optical constants of the mantle material. The contribution of electronic transitions in the UV to the polarizability of the material was calculated as a sum of 40 Lorentz oscillators. The imaginary part of the dielectric constant was derived from the Clausius-Mossotti equation. After setting the value of ε'' in the visible, the real part of the dielectric constant is derived from Eq. 8.1. The total oscillator strength in all transitions (both given directly by the Lorentzians and implicitly represented by the added absorption in the visible) is constrained by the sum rule in eq. 8.4. Since the distribution of high-energy transitions is uncertain, the integration in Eq. 8.4 was carried out only up to 30 eV. In applying the sum rule, we have taken into account that in many materials, transitions above 30 eV may contribute as much as 20%-30% of the value of the integral at $\omega_0 = \infty$ so that for $\omega_0 = 30$ eV, the sum rule may be expected to recover approximately 75% of the total number of valence electrons. In using Eq. 8.4, we have assumed a specific gravity of 1.5 g cm^{-3} for the mantle material, which corresponds to $n_a = 7.5 \times 10^{22}$ and the total number of valence electrons per cm^3 of 2.6×10^{23}. One of the most important implications of the Kramers-Kronig relations applied to the organic refractory material is the relatively high value of m', which at visual wavelengths has a value of ≈ 1.7 for moderately absorbing materials. Although this number is significantly higher than the values expected for saturated organic molecules, which are in the range 1.4 - 1.5 (*CRC Handbook of Chemistry and Physics* 1975), it is in good agreement with the index of refraction measured for amorphous CH with a similar number of valence electrons per unit volume (Fink et al. 1984).

b) Extinction and polarization cross sections

We have derived the extinction and polarization cross section from Mie theory solutions for infinite cylinders given by Lind and Greenberg (1966) for homogeneous particles and by Shah (1970) for the core-mantle case. These algorithms require significantly less computer time than Mie theory solutions for finite-size nonspherical grains and can therefore be used to study in detail the influence of the size distributions as well as the alignment mechanism on polarization. Comparison with results obtained for prolate spheroids shows that cross sections calculated for infinite cylinders provide a good approximation to the wavelength - dependence of extinction and polarization even for axial ratios as low as 2, although the polarization-to-extinction ratio can be overestimated (Rogers and Martin 1979).

We denote the maximum and minimum extinction cross section for linearly polarized light passing through a cloud of aligned particles by Q_1 and Q_2. These factors can be obtained by averaging the efficiencies of derived directly from Mie theory (calculated as a function of the angle between the direction of propagation of the incident wave and the symmetry axis of the cylinder) over all orientations of the spinning particles. Geometrical projection factors for arbitrarily oriented particles have been derived by Greenberg (1968) and Hong and Greenberg (1980) and have been used in our calculations without substantial changes. The extinction and polarization for an initially unpolarized beam passing through a path length $\underline{L}$ in the interstellar medium are given by

$$A_\lambda = 1.086 \text{ L } n_d \, \sigma_d \, \frac{Q_1 + Q_2}{2} , \tag{8.5}$$

$$\Delta m_p = 1.086 \text{ L } n_d \, \sigma_d \, (Q_1 - Q_2), \tag{8.6}$$

where n_d is the column density of the grains, σ_d is the geometric cross section of the particles (by definition, $\sigma_d = 2al$ for a cylinder of radius a and length l, Δm_p is the difference in magnitudes between the values of extinction observed for the two orientations of the electric vector. Eq. 8.6 is valid only for those lines of sight which have a constant orientation of the magnetic field and otherwise provides an upper limit on the observed polarization.

If the difference in optical depths for the two perpendicular orientations of the electric vector is small (this condition is satisfied for most of observed lines of sight), the percentage polarization, p, is given by

$$p = L \, n_d \, \sigma_d \, \frac{Q_1 - Q_2}{2} . \tag{8.7}$$

The ratio of maximum percentage polarization is extinction, p_{max}/A_V, which can be compared directly with observations, can then be calculated from

$$\frac{p_{max}}{} = \frac{1}{1.086} \, \frac{(Q_1 - Q_2)_{max}}{(Q_1 + Q_2)_V} . \tag{8.8}$$

We note that the polarization on average has its maximum close to λ_V, which is about where the extinction exhibits a change in curvature (Greenberg 1978).

Circular polarization results from linear birefrigence of the interstellar medium; i.e., the dependence of the phase lag substained by the transmitted wave on the orientation of the electric vector. The phase lag of the transmitted wave is given by the imaginary part of the forward scattering amplitude (van de Hulst 1957) and can be formally included in the extinction cross section defined as a complex quantity through the optical theorem (van de Hulst 1957):

$$C_{ext} = \frac{4\pi}{k^2} \, S(0), \tag{8.9}$$

where $S(0)$ is the complex forward scattering amplitude and $k = 2\pi/\lambda$. In this approach, $\text{Re}[Q_1 - Q_2]$ represents the linear dichroism of the interstellar medium, and $\text{Im}[Q_1 - Q_2]$ its linear briefringence.

The ratio of Stokes parameters, V/I, which is normally used as a measure of the observed circular polarization, is given by the expression (Martin 1972; Serkowski 1962)

$$\frac{V}{I} = G\, n_d^2\, \sigma_d^2\, \frac{\mathrm{Re}[Q_1 - Q_2]}{2}\, \frac{\mathrm{Im}[Q_1 - Q_2]}{2} \quad , \tag{8.10}$$

where the factor G accounts for the path length as well as geometric effects of the variation of the direction of alignment along the line of sight.

Eq. 8.5 - 8.10 can easily be generalized for a size distribution of particles. The distribution adopted in our calculations is of the form

$$n(a) = n_d\, f_0 \exp\left[-5\left(\frac{a - a_c}{a_i}\right)^3 \right], \tag{8.11}$$

where n_d is the total number density of grains, a is the particle radius, and $n(a)da$ is the number of particles per unit volume with radii in the range $[a, a+da]$. The parameter a_i describes the "cutoff" of the distribution, a_c is the size of the core, which is assumed to be constant for a given distribution. The normalizing factor, f_0, and the average particle size, $\langle a\rangle$, are given by the formulae (Greenberg 1968)

$$f_0 = \frac{3 \times 5^{1/3}}{\Gamma(\frac{1}{3})\, a_i} = \frac{1}{0.5222\, a_i} \quad , \tag{8.12}$$

$$\langle a\rangle = a_c + \frac{\Gamma(\frac{2}{3})}{\Gamma(\frac{1}{3})\, 5^{1/3}}\, a_i = a_c + 0.2956\, a_i \; . \tag{8.13}$$

We have used the same distribution for homogeneous grains, with a_c treated as an arbitrary parameter that could be adjusted to provide the best match to the observed wavelength dependence of extinction and linear polarization.

We have considered the case of (1) perfect spinning (PSA) for illustration. Partial Davis-Greenstein alignment (DG) due to paramagnetic relaxation with collisional spin-up gives similar results.

c) Interstellar circular polarization

Examples of the wavelength - dependence of linear and circular polarization derived from Mie theory cross sections for infinite cylinders have been collected in Fig. 8.1 which shows the results for core-mantle grains with moderately absorbing mantles [the visual absorptivity given by $m''(V) \approx 0.15$]. The size distribution which best matches the observed wavelength dependence of linear polarization $(\lambda_{max})^{-1} \approx 1.75$ μm^{-1} is given by $a_c = 0.05$ μm and $a_i = 0.05$ μm and has an average particle radius of 0.095 μm. The difference between results for partial and complete spinning alignment is rather small. The larger width of the peak in the DG curve reflects the dependence of the degree of alignment on particle size.

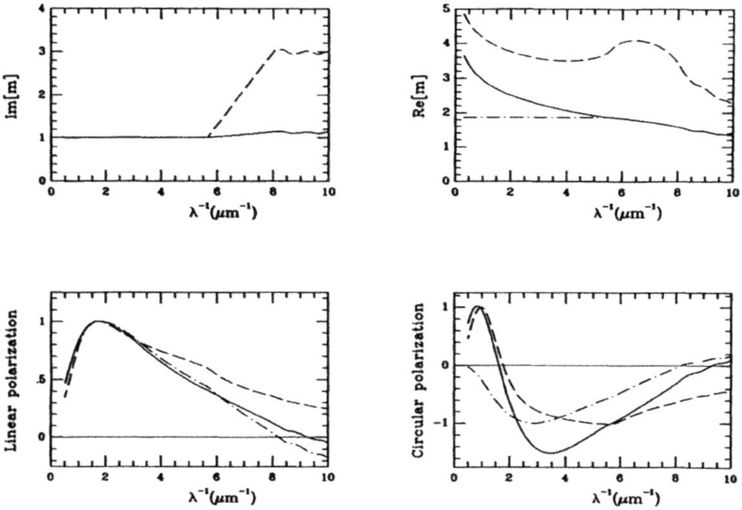

Fig.8.1 Kramers - Kronig interpretation of circular polarization. Different lines represent three sets of optical constants ($m = m' - im''$) shown in upper left (m'') and upper right (m') panels. The dash-dot curve deliberately violates the Kramers-Kronig relations by having a constant value of m' from the near ultraviolet to the infrared. The calculations of polarization are for homogeneous cylinders with the size distribution adjusted to match the observed average value of the peak wavelength of linear polarization.

d) Implications for grain models

The only remaining restriction on the visual optical properties grain materials is provided by the observed albedo. Model materials whose results have been collected show that the $m''(V) = 0.15$ marks an approximate lower limit on the absorptivity of grain mantles allowed by the observations (the albedo at λ_V is 0.7 for the core-mantle grains alone and 0.6 after averaging over all three populations). The albedo obtained for the three-population model using core-mantle spheres (in order to avoid artificially high albedo produced by two-dimensional infinite particles; Greenberg, van de Bult and Allamandola 1983) is not significantly different in the visual from the prediction of the MRN model.

One of the consequences of adopting the optical constants of grain mantles described here is a reduction in the total of volume of core-mantle grains relative to what is required for mantles with a lower index of refraction. The difference is due to an increase in the volume efficiency of extinction, α, which is defined as the ratio of the extinction cross section to the volume of the particle, $\alpha = C_{ext}/V$, and is essentially independent of particle shape (Greenberg and Hong 1974). The value of α at 5500 Å is 1.3 x 10^5 cm^{-1} for an organic refractory mantle and only 6.8 x 10^4 cm^{-1} for particles with an index of refraction of 1.45 which appears to be typical of a material consisting of saturated organic molecules. Since the specific gravity of the two types of materials is similar, and carbon makes up an almost identical fraction of the mass in each of them (the difference in the optical properties is due mainly to the change in the C:H ratio), the predicted depletion of carbon in core-mantle grains drops almost in inverse proportion to the value of α. For m' = 1.45, the mantles must contain ≈ 45% of cosmic carbon (using cosmic abundances tabulated by Greenberg 1978), whereas for the organic refractory the carbon depletion is only 26%. Since as much as 25% of carbon is required to match the observed strength of the 2200 Å hump - if it is due to the π-electron plasmon in graphite spheres - the total depletion of carbon required by the three-population model is reduced from an uncomfortably high value of ≈ 70% to only ≈ 50%, which agrees much better with the low depletion of carbon suggested by the observations (see Whittet 1987 and references therein). It is perhaps worth noting that in spite of the low required depletion of carbon, all the polarization and most of the extinction are produced by the major carbon - bearing core mantle grains in our model.

e) Dust-gas depletions.

An important consequence of the dust is that it depletes the gas phase species in different proportions than their cosmic abundance ratio thus leaving not only smaller absolute numbers in the gas but also different relative numbers.

First of all we note the well known fact that the rocky species (Si, Mg, Fe) are almost 100% depleted in the dust (except in very energetic events). For the rest let us restrict ourselves to the oxygen-carbon depletion.

Independent of the grain model used, the absolute amount of carbon bound up in solids is greater than the amount of oxygen. Using the grain model here suggested, the dust in diffuse clouds contains between 10 and 20% of the available oxygen and 40 to 60% of the carbon. Using the initial abundances of Cameron (1982) this leaves an average diffuse cloud gas ratio $[C/O]^{DC}_{gas}$ = 0.36 as compared with $[C/O]_{CA}$ = (6.9/4.2) = 0.6. Taking the Elias 16 dust as representative of that in molecular clouds the amount of oxygen and carbon bound in solid H_2O and CO leaves average gas phase oxygen and carbon abundances of $[O]^{MC}_{gas}$ = 5.1 x 10^{-4} and $[C]^{MC}_{gas}$ = 2.0 x 10^{-4} respectively. These values give $[C/O]^{MC}_{gas}$ = 0.39 which is slightly higher than the diffuse cloud ratio but still significantly lower than the cosmic abundance ratio.

Acknowledgement

This work was partially supported by NASA grant # NGR 33-018-148.

REFERENCES

Aannestad, P. A., Greenberg, J. M. (1983), Ap. J. 272, 551.

Agarwal, V. K., Schutte, W., Greenberg, J. M., Ferris, J. P., Briggs, R., Connor, S., van de Bult, C. P. E. M., Baas, F. (1985), Origins of Life 16, 21.

Allamandola, L. J., Sandford, S. A. (1988), in M.E. Bailey and D.A. Williams (eds), Dust in the Universe, Cambridge Univ. Press. 229.

Allen, D. A., Wickramasinghe, D. T. (1981), Nature, 294, 239.

Anders, E., Mitsuru Ebihara (1982), Geochimica, Cosmochimica Acta 46, 2363.

Baas, F., Grim, R. J. A., Geballe, T. R., Schutte, W., Greenberg, J. M. (1988), in M.E. Bailey, D.A. Williams (eds), Dust in the Universe, Cambridge Univ. Press, 5.

Bassani, F., Altarelli, M. (1983), in E. E. Koch (ed.) Handbook of Synchroton Radiation Amsterdam 1.

Bethe, H. (1933), Handbuch der Physik, Springer, Berlin, 24, Part I, 518.

Breukers, R.J.L.H., d'Hendecourt, L. B., Greenberg, J. M. (1990), Proceedings Pacifichem Symposium "Chemistry and Spectroscopy of Interstellar Molecules".

van de Bult, C. E. P. M., Greenberg, J. M., Whittet, D. C. B.(1985), M.N.R.A.S. 214, 289.

Butchart, I., McFadzean, A. D., Whittet, D. C. B., Geballe, T. R., Greenberg, J. M. (1986), Astron. Astrophys. 154, 15.

Calvert, J., Pitts, J. N. (1966), Photochemistry, J. Wiley, New York.

Cameron, A. G. W. (1973), Sp. Science Review 15, 121.

Cameron, A. G. W. (1982), in C. Barnes, R. N. Clayton, D.N. Schramm (eds), Elements and Nuclidic Abundances in the Solar System, Cambridge Univ. Press, 23.

Chlewicki, G., Greenberg, J. M. (1984), M.N.R.A.S. 210, 791.

Chlewicki, G. (1985), Ph.D. Thesis, University of Leiden.

Dalgarno, A. Griffing, G. W. (1958), Proc. Roy. Soc. (London), Ser. A, 248, 415.

Danielson, E. R., Woolf, J. N., Gaustad, J. E. (1965), Ap.J. 141, 116.

Davis, L., Greenstein, J. (1951), Ap. J. 114, 206.

Day, K. L. (1979), Ap. J. 234, 158.

Draine, B. T., Lee, H. M. (1984), Ap. J. 285, 89.

Duley, W. W. (1988), in M. E. Bailey, N.A. Williams (eds), Dust in the Universe, Cambridge Univ. Press, 209.

Geballe, T. R., Baas, F., Greenberg, J. M., Schutte, W. (1985), Astron. Astrophys. 146, L6.

Gillet, F. C., Forrest, W. J. (1973), Ap.J. 179, L483.

Greenberg, J. M. (1966), in Spectral Classification and Multicolor Photometry, Willmer Brothers 291.

Greenberg, J. M. (1968), in B. Middlehurst, L.H. Aller (eds), Stars and Stellar Systems, 7, Chicago Univ. Press., 221.

Greenberg, J. M. (1969), Physica 41, 67.

Greenberg, J. M. (1971a), in M. A. Gordon, L. E. Snyder (eds), Molecules in the Galactic Environment, John Wiley and Sons, 94.
Greenberg, J. M. (1971b), Astron. Astroph. 12, 240.
Greenberg, J. M., Yencha, A. J. (1972), in J. M. Greenberg and H. C. van Hulst, (eds), Interstellar Dust and Related Topics, Proceedings of IAU Symposium no. 52, D. Reidel, Dordrecht 3.
Greenberg, J. M. (1974), Astrophys. Sp. Sc. 39, 9.
Greenberg, J. M., Hong, S. S. (1974), in G.B. Fields, A.G.W. Cameron (eds), The Dusty Universe, (Neale Watson) p. 131.
Greenberg, J. M. (1978), in J. A. M. McDonnell (ed), Cosmic Dust Wiley Ch. 4, 187.
Greenberg, J. M. (1979), in B.E. Westerlund (ed), Stars and Star Systems, Reidel, Dordrecht, 173.
Greenberg, J. M. (1982), in L. L. Wilkening (ed), Comets, University of Arizona Press 131.
Greenberg, J. M. (1982), in J. E. Beckman, J. P. Philips (eds), Submillimetre Wave Astronomy, Cambridge Univ. Press 261.
Greenberg, J. M., van de Bult, C. E. P. M., Allamandola, L. J. (1983), J. Phys. Chem. 87, 4243.
Greenberg, J. M. (1986), in F. Israel (ed), IRAS Symposium "Light on Dark Matter", D.Reidel, Dordrecht 177.
Grevesse, N. (1984), Physica Scripta T8, 49.
Hagen, W., Allamandola, L. J., Greenberg, J. M. (1979), Astr. Sp. Sc. 65, 215.
d'Hendecourt, L. B., Allamondola, L. J., Greenberg, J. M. (1985), Astron. Astrophys. 152, 130.
d'Hendecourt, L. B., Jourdain de Muizon, M. (1989), Astron. Ap. 223, L5.
Herbst, E., Leung, C. M. (1986), M.N.R.A.S. 222, 689.
Hong, S. S., Greenberg, J. M., (1980),Astr. Apo. 88, 194.
Huffman, D. R., Stapp, J. L. (1973), in J. M. Greenberg, H. C. van Hulst (eds), Interstellar Dust and Related Topics, IAU Symp. 52, (Dordrecht Reidel), p. 297.
Van de Hulst, H. C. (1957), Light Scattering by Small Particles, Wiley, New York.
Jones, R. V., Spitzer, L. (1967), Ap.J., 147, 943.
de Jong, T., Dalgarno, A., Boland, W. (1959), Astron. Ap. 91, 68.
Kemp, J. C. (1972), in J. M. Greenberg, H. C. van Hulst (eds), Interstellar Dust and Related Topics, IAU Symp. 52, (Dordrecht Reidel) p. 181.
Kemp, J.C., Wolstencroft, R. D. (1972), Ap.J. (Letters), 176,L115.
Kemp, J. C., Wolstencroft, R. D. (1974), in T. Gehrels (ed), Planets, Stars and Nebulae Studied with Photopolarimetry, (Tucson University of Arizona Press), p. 938.
Knacke, R. F., Gaustad, J. E., Gillett, F. C., Stein, W. K. (1969), Ap. J. 155, L189.
Lacy, J. H., Baas, F., Allamandola, L. J., Persson, S. E., McGregor, P. J., Lonsdale, C. J., Geballe, T. R., van de Bult, C. E. P. M. (1984), Ap. J. 276, 533.
Landau, L. D., Lifshitz, E. M. (1960), Electrodynamics of Continuous Media (New York, Pergamon).
Léger, A., Jura, M., Omont, A. (1985) Astron. Astrophys. 144, 147.
Lind, C. A., Greenberg, J. M. (1966), J. Appl. Phys. 37, 3195.
Martin, P. G. (1975), Ap. J., 202, 389.

Martin, P. G., Angel, J. R. P. (1977), Ap. J., 207, 126.
Martin, P. G., Illing, R., Angel, J. R. P. (1972), M.N.R.A.S., 159,191.
Mathis, J. S., Rumpl, W., Nordsieck, K. H. (1977), Ap. J. 217, 425 (MRN).
Mathis, J. S., (1983), Ap. J. 267, 119.
Mathis, J. S. (1986), Ap. J. 308, 281.
Norman, C., Silk, J. (1980), Ap. J. 238, 158.
Prasad, S. S., Tarafdar, S. P. (1983), Ap. J. 267, 603.
Rogers, C., Martin, P. G. (1979), Ap. J. 228, 450.
Savage, B. D., Mathis, J. S. (1979) Ann. Rev. Astron. Astrophys. 17, 73.
Serkowski, K. (1962), Adv. Astr. Ap., 1, 289.
Shah, G. A. (1970), M.N.R.A.S. 148, 93.
Shapiro, P. S. (1975), Ap. J. 201, 151.
Silverstein, R. M., Bassler, G. C. (1967), Spectrometric Identification of Organic Compounds, Wiley, New York.
Soifer, B. T., Puetter, R. C., Russell, R. W., Willner, S. P., Harvey, P. M., Gillett, F. C. (1979), Ap.J. (Letters) 232, L53.
Spitzer, L. (1968), Diffuse Matter in Space, Wiley, New York.
Spitzer, L. (1978), Physical Processes in the Interstellar Medium, Wiley, New York.
Tapia, M., Persi, P., Roth, M., Ferrari-Toniolo, M. (1984), Astron. Ap. 225, 488.
Tielens, A. G. G. N., Allamandola, L. J., Bregman, T., Goebel, T., d'Hendecourt, L. B., Witteborn, E. C. (1984), Ap.J. 287, 697.
Titchmarsh, E. C. (1937), Introduction to the Theory of Fourier Integrals, Oxford Univ. Press, New York.
Whittet, D. C. B., Bode, M. F., Longmore, A. J. Baines, D. W. T., Evans, A. (1983), Nature, 303, 218.
Whittet, D. C. B. (1984), M.N.R.A.S. 210, 479.
Whittet, D. C. B., Longmore, A. J., Mcfadzean, A. D. (1985), M.N.R.A.S. 216, p 45.

PHYSICAL AND CHEMICAL EFFECTS INDUCED BY FAST IONS IN ICES OF ASTROPHYSICAL INTEREST

Valerio Pirronello
Dipartimento di Fisica
Universita' della Calabria
87036 Rende
Italy

ABSTRACT. It is given an introductory presentation of the main effects induced by the irradiation with energetic (keV-MeV) ions in ices and ice mixtures of astrophysical interest through the description of a selection of experimental results, obtained in the laboratory in a sort of "simulation" of some conditions encountered in space, and it is also given a brief presentation of a selected number of applications of such effects to some astrophysical scenarios.

1. Introduction

In these lectures I will try to let you have a glance of a relatively new area of research in fundamental physics which has important applications in astrophysics and which could be identified as that of "the interaction of energetic particles with molecular ices". My plan is to describe some of the most relevant, in my opinion, experimental results (whenever possible in chronological order) and to show some of the applications to space scenarios. These lectures then do not want and could not be an exhaustive review of the argument (each of its sections would require a review or a book) but an introduction to this fascinating area and its investigation that was started in 1978 by Walter L. Brown, Louis J. Lanzerotti, John M. Poate and W.M. Augustyniak of the Bell Laboratories of Murray Hill in New Jersey.

When they published the first paper on the erosion of water ice bombarded by MeV ions (Brown et al., 1978), sputtering of solids by impinging fast particles was already a studied phenomenon occurring in metals and semiconductors. That first paper in 1978 by Brown and co workers opened by the way a new field because ices are (as it is well known) completely different from conductors and semiconductors: they are molecular solids with binding energies among molecules much smaller than binding energies among atoms inside each molecule, and they are electrical insulators, which means that they are also bad thermal conductors because any contribution to thermal conductivity from electrons is suppressed as they are tightly bound to molecules (typical band gaps are of the order of 10 eV). These two facts make a great difference with respect to other materials as it is evident from

J. M. Greenberg and V. Pirronello (eds.), Chemistry in Space, 263–303.

the measured yields that for ices are some orders of magnitude higher than those obtained for metals and semiconductors. The ices we will speak about are those of astrophysical interest made by the most abundant elements (hydrogen, carbon, oxygen and nitrogen) like H_2O, CO, CO_2, CH_4, NH_3 and so on. We will not discuss here then about ices formed by condensed atomic rare gases, even if they should deserve some attention because their simplicity can give valuable hints for some of the possible mechanisms of the involved processes. For the interested reader it can be useful to consult for instance O'Shaughnessy et al. (1986) and references therein.

Fast particles penetrating into a solid loose part of their energy by elastic and inelastic collisions with atoms and molecules of the target material.

Some of the effects induced by the energy loss of impinging energetic particles in solids that are relevant in astrophysics are:

a) sputtering (i.e. the ejection of atoms and molecules by the solid during irradiation);
b) changes in the chemical composition of the target material;
c) changes in the structure of the solid from amorphous to crystalline and/or from crystalline to heavily disordered;
d) changes in the optical properties of the target material.

The schematic subdivision shown above about the principal effects of the irradiation of solids is given only for the sake of simplicity, because the mentioned effects, of course, do not occur independently from each other, they take place simultaneously and with deep interconnections (it will be clearer later), for instance: sputtering has a molecular character and some of the ejected species during bombardment are the product of chemical reactions induced by the projectiles (see e.g. Brown et al., 1982a; Pirronello et al., 1982a; Pirronello et al., 1988b,c) also changes of the optical properties of the irradiated material come from changes in its structure like in the case of crystalline regrowth around the ion track (see e.g. Brown et al., 1978; Pirronello et al., 1981) or in its chemical composition (see e.g. Moore et al., 1983).

The astrophysical interest for such interactions is due to the fact that in several environments of our Universe particles bombard solid materials and in particular ices often at low temperatures and almost always in low density conditions. Just for the sake of simpleness I will mention that these interactions occur, for instance, when cosmic rays hit either the nuclei of comets in the Oort cloud or dust grains in diffuse interstellar clouds and icy mantles on grains in dense ones, when particles from stellar wind irradiate circumstellar grains, when ions trapped in the magnetosphere of the giant planets bombard the surface of satellites embedded in them or the icy grains of rings and so on.

2. The Experiments

2.1. THE EQUIPMENT

Typical experimental equipment of the type used for instance at the AT&T Bell Laboratories to measure the effects of the ion penetration in ices is shown in Fig. 1. It is substantially constituted by a cylindrical ultrahigh vacuum (UHV) chamber in the center of which it is situated a target holder. From one of the apertures in the middle plane the beam coming from a Van de Graaff accelerator or from an ion gun is allowed to enter. The target can be cooled and maintained at temperatures as low as about 7 K by means of a continuous flow of liquid helium on its reverse side, allowing in this way to perform experiments that, at least from this point of view, can closely simulate the extreme temperature conditions in dense interstellar clouds. A copper net biased to a high negative voltage (≅ -1.3 kV) is often used to prevent secondary electrons to escape from the target (a fact that makes difficult the correct evaluation of the dose of the bombarding ions). Cryo pumps are used to maintain clean high vacuum conditions (to better than 1.2×10^{-9} torr) in the chamber together with differential pumping by turbo pumps in the beam line to prevent gas diffusion from the ion source of the Van de Graaff accelerator. Inside the scattering chamber a copper can, surrounding the target, may be cooled to approximately the liquid nitrogen temperature to provide additional pumping of the residual gas and to trap (at least some) species released from the target that would be detected only after interacting with the chamber walls.

Analyses of the effects induced by the ion bombardment are performed "in situ" by means of several techniques. Species that are released from the target in the gas phase are detected using mass spectrometry (in Figure 1 the probe of a quadrupole mass spectrometer is shown flanged to the UHV chamber. At the entrance of the probe there is an electron impact ionizer because it is well known that only a very small minority of sputtered atoms and molecules are ionized while the overwhelming majority is neutral). When a Van de Graaff accelerator is available an annular solid state detector (not shown in Figure 1) is placed in the vacuum chamber at the beam entrance in order to measure the flux of scattered particles of the beam from the target material to perform the so called Rutherford Backscattering analysis (RBS). This is a technique that, allowing accurate measurements of the thickness of the bombarded layer and of its atomic composition, has been widely used to measure total erosion yields. There has often been added to this equipment a spectrophotometer to measure either in the infrared, in the visible or in the ultraviolet changes in the optical properties of the irradiated material or in its chemical composition (see e.g. Moore and Donn, 1983; Benit et al., 1988).

The sample preparation of the layer to be bombarded consists simply in a deposition of gas admitted from a manifold (possibly through a movable doser having amicrochannel plate at its end), onto the substrate surface cooled by the liquid helium flow. The thickness

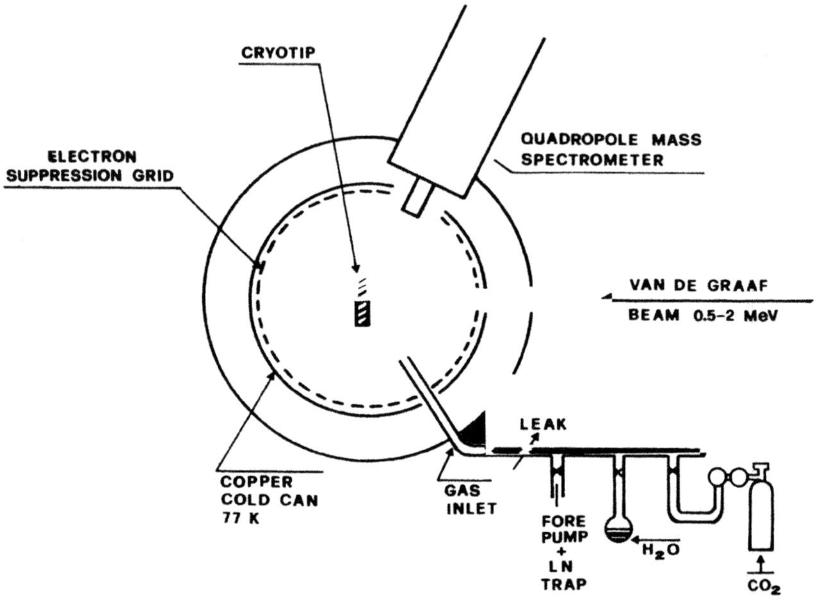

Figure 1. Typical experimental set up to study the effects of the interaction of energetic ions with ices

of the film can be measured with high accuracy by means of RBS, by the interference fringes of light reflected from the ice and the cold substrate surfaces or in an indirect way just by the pressure changes in the fixed volume of the manifold. The process of deposition, by the way, has to be done following some important rules (see e.g. Rice, 1975) if it is intended to build an amorphous film, the type of ice encountered in dense interstellar clouds (Leger et al., 1979; Hagen et al.,1981) and when accreting intimate mixtures of frozen gases, if it is required to irradiate homogeneous mixtures without segregation of species.

Even if layers of all thicknesses can be and have been bombarded by ions I think that, whenever possible, it should be preferred to irradiate films that are thin when compared with the range of the incident particle in the material they are made. This choice assures that all of the measured quantities will be relative to a precise energy of the projectile (that in this case looses in the layer only a very small fraction of its initial kinetic energy). On the contrary, in the case of non negligible thicknesses, the measured yields are often averages over orders of magnitude changes in the deposited energy.

2.2. SELECTED EROSION MEASUREMENTS

A relevant number of experiments on sputtering of ices have been performed in the last decade. Here I will only recall some of them that I consider of particular relevance for the development of the subject and for applications to astrophysical environments.

All the results I will speak about here have been obtained using beam currents in the range in which the measured yields change linearly with the flux of impinging particles. This choice assures that only single particle effects are produced, a mandatory requirement for any possible application to astrophysical scenarios where the interval of time between two successive arrivals of ions may be of days, weeks, months or years.

Table 1 shows the erosion yield "Y" (defined as the average number of ejected molecules per each impinging ion) of homogeneous layers of astrophysical ices irradiated with 1.5 MeV helium ions at low temperatures about 10 K, together with the molecular binding energy. The experimental technique employed to get them is that of RBS, a technique that is sensible to the atomic composition of the analyzed layer, but not to the existence of chemical bonds among atoms.

TABLE 1. Erosion yields of ices irradiated by 1.5 MeV helium ions at 10 K

Ice	Binding energy (eV)	Y
H_2O	.52	8
NH_3	.31	140
CH_4	.18	120
CO_2	.27	120
SO_2	.38	30

As it is evident huge values are obtained. They are orders of magnitude higher than those relative to conductors and semiconductors and are not at all well described by the so called collision cascade model (see below).

In order to gain a better insight into the processes responsible for such high erosion yields, several measurements were performed varying the experimental conditions (for the same combination target-projectile, in fact, the total amount of eroded material may depend strongly on several parameters such as the energy of the impinging ions, the temperature, the thickness of the ice and so on) and other measurements were performed monitoring not only the total yield but also some of the characteristics of the ejected particles, such as its energy and angular distribution.

2.2.1. Yield Dependence on the Substrate Temperature

The first important parameter (that is also fundamental for any astrophysical application) is certainly the temperature of the film. In Figure 2 (taken from Brown et al., 1982) it is shown for three

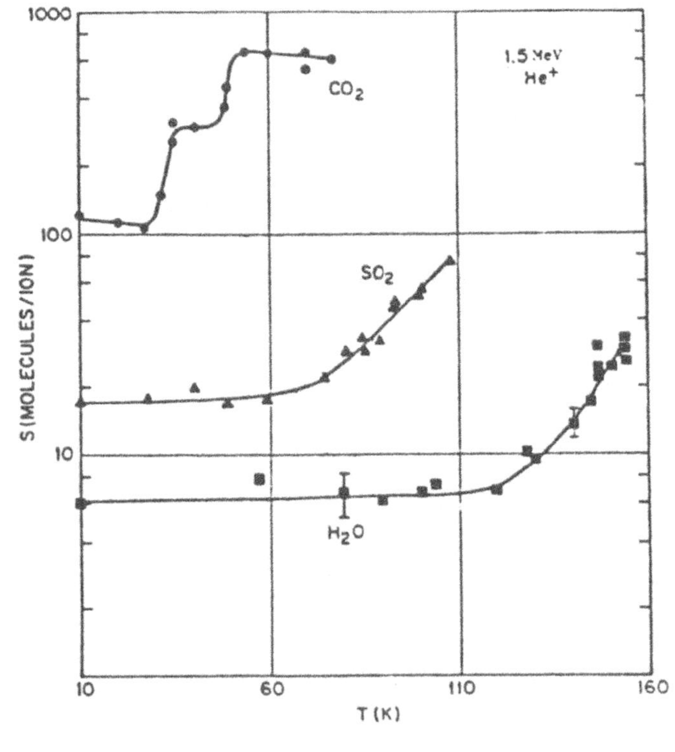

Figure 2. Temperature dependence of the total erosion yields for different frozen gas

different ices hit with a 1.5 MeV He^+ beam the total erosion yield as function of the substrate temperature "T". The case of H_2O and SO_2 is typical of almost all ices. At low "T" values there is a temperature independent region and after a threshold value a strongly non linear dependence on T, that is not due to the normal sublimation and is suggestive of a growing importance of a process of enhanced diffusion of species by radiation, species not necessarily equal only to those initially deposited in the layer. The case of CO_2 seemed, already at the first sight, more intriguing. The highly reproducible step-like structure observed in the total erosion yield, in fact, suggested that perhaps more than one chemical specie other than carbon dioxide were produced and released at well determined temperatures. These species are probably O_2 and CO and it is noteworthy that such first evidences of chemistry going on during bombardment of molecular ices came from RBS measurement, a technique not sensible to chemical structures. We will see later , for instance already for water ice and by means of a more appropriate technique, that erosion occurs with the production and migration of species not initially in the layer, but with the difference that new species are released not at selected temperatures, as in the case of carbon dioxide.

2.2.2. Yield Dependence on the Energy of Projectiles

The other very important parameter to be varied when bombarding any solid is the energy of impinging ions. As a consequence of its variation, in fact, the energy deposited by each single projectile in the irradiated layer changes. This is a quantity parameterized by the so called stopping power "dE/dx" (the energy lost by ions per unit path length inside the solid), a well known tabulated function (see e.g. Ziegler, 1980), that is given by the linear superposition of two independent contributions, the "nuclear" one $(dE/dx)_{nucl}$ and the "electronic" one $(dE/dx)_{el}$ i.e.

$$(dE/dx)_{tot} = (dE/dx)_{nucl} + (dE/dx)_{el}$$

The first of the two contributions is due to elastic collisions among the impinging particle and the "core" (nucleus plus inner shell electrons) of the target atoms; the second one is due to ionizations, excitations and charge exchange reactions induced by the impinging particle interacting with the electronic cloud of the target atoms. The relative importance of these processes is given for a H^+ ion penetrating in H_2O in the vapor phase in Fig. 3 (from Miller and Green, 1973). Roughly, one can say that the electronic contribution becomes dominant over the nuclear contribution when the speed of the projectile is equal or higher than the average orbital speed of the external electrons of atoms.

Effects induced by fast particles bombarding solid materials depend on their stopping power. Opposite to the case of metals and semiconductors, for which the total sputtering yield is proportional to the "nuclear stopping power", irradiation of water ice at MeV energies and in the temperature independent region shows that the erosion process is substantially connected with the "electronic stopping" but

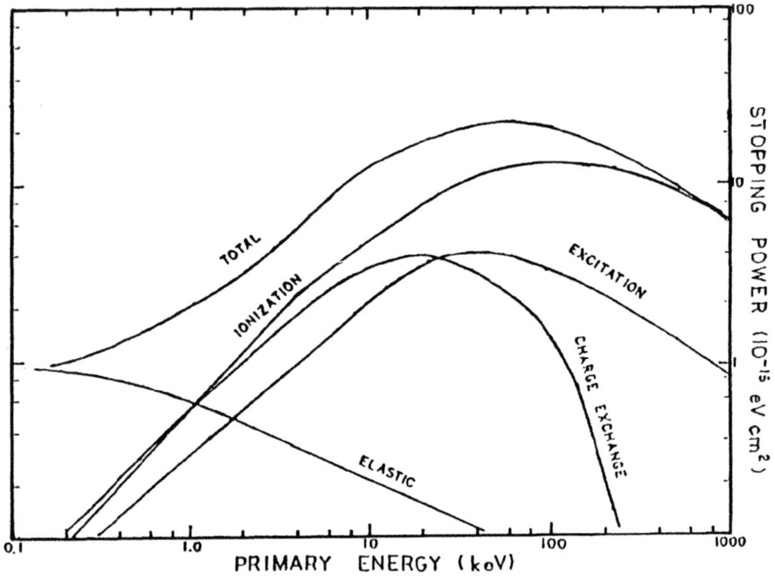

Figure 3. Contribution of the various processes to the energy loss of protons in water vapor as a function of their kinetic energy.

in a non linear way (Brown et al., 1982a) as it can be seen in Fig. 4. Such a relevant non linearity is a fundamental hint for understanding which is the mechanism to be invoked to explain both on a qualitative and quantitative ground the erosion process of molecular icy solids of astrophysical interest.

2.2.3. Yield Dependence on the Projectile Charge State

The energy loss of ions in matter depends also on their charge state. After a path length of some tens of Angstroms the particle impinging on the solid reaches its "equilibrium charge state" (i.e. the situation in which beam particles, before loosing a relevant part of their initial energy, have got a constant distribution among all possible charge state). It is then conceivable that very superficial effects (as it is sometime sputtering) will depend considerably on the ion charge state, while on the contrary "bulk" effects (here the term bulk means only regions in the solid where the

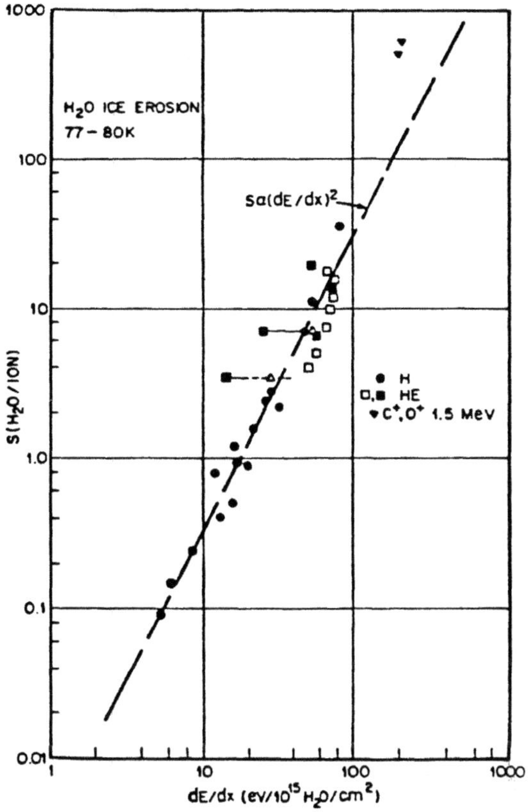

Figure 4. Erosion yield of water ice as a function of the stopping power of impinging ions

equilibrium charge state has been reached) will be independent on the initial charge of the projectile.

To investigate the erosion yield dependence on the charge state the experimental set up has been modified (Evatt, 1980) by adding a thin (120 nm thick) carbon film upstream the vacuum chamber to act as an electron stripper. For instance in the case of a 1.5 MeV Helium beam, when the charge state of He ions after passing through the foil was " +2 ", the measured erosion yield of water ice resulted about .3 times higher than what is obtained with the singly ionized beam coming directly from the Van de Graaff accelerator, a too high yield to be accounted for only on the base of the energy loss suffered by the He ions (≅ 50 keV) in the carbon film that can produce only a too small increase in the stopping power of helium in the ice (remember that at these energies we are beyond the maximum in the stopping cross section, so a decrease in the kinetic energy of the ion produces an increase in the stopping).

2.2.4. Yield Dependence on the Layer Thickness

No thickness dependence in the total yield (the one measured with RBS) has been observed in water ice bombarded in the temperature independent region with MeV light ions till few hundred Angstrom thick layers, while it certainly exists, as we will see later, for species formed under bombardment.

2.2.5. Energy Distribution of Ejected Molecules

Another type of experiments that can give valuable hints to discover the prominent mechanism of erosion has been performed measuring the time of flight of ejected molecules to obtain their energy distribution. The detector used is still the mass spectrometer and the bombarding beam has been suitably pulsed in order to allow accurate timing between ejection from the solid and detection from the quadrupole of the monitored species.

The energy spectra of the release of D_2O molecules from D_2O ice bombarded with 1.5 MeV He^+, 1.5 Mev Ar^+ and 50 keV Ar^+ obtained by Brown et al. (1984) are shown in Figure 5. The reason of using different ions having different kinetic energies is to change considerably the relative importance of electronic and nuclear stopping, the ratio between the two contributions is in fact about "10^3" for the helium ions, about "10" for the argon ions at MeV energies and a bit more than ".33" for the argon ions at 50 keV. In the figure it is also shown an attempt (dashed line) to fit the data with predictions from the "collision cascade" model (see below) obtained by Thompson (1965). This fit describes well the energy spectrum of ejected D_2O molecules when the projectiles are 50 keV Ar^+, as it was expected even if a very low value of the binding energy of molecules (U = .055 eV) had to be used as fitting parameter. The surprising fact is that it also describes the high energy tail of the spectra obtained when bombarding the heavy water ice target with ions that have a predominant component of electronic loss of energy in the solid.

The interpretation of the low energy portion of the spectrum of sputtered molecules by 1.5 MeV He^+ is not easy: it is certainly not maxwellian.

Similar type of results have been obtained independently by Haring et al. (1983, 1984a,c) for H_2O, NH_3 and CO, using only ions in the keV range but looking also at the kinetic energy distribution of synthesized species (see the section on chemical effects induced by the energy loss of impinging ions in ices) such as N_2, H_2, O_2, and CO_2.

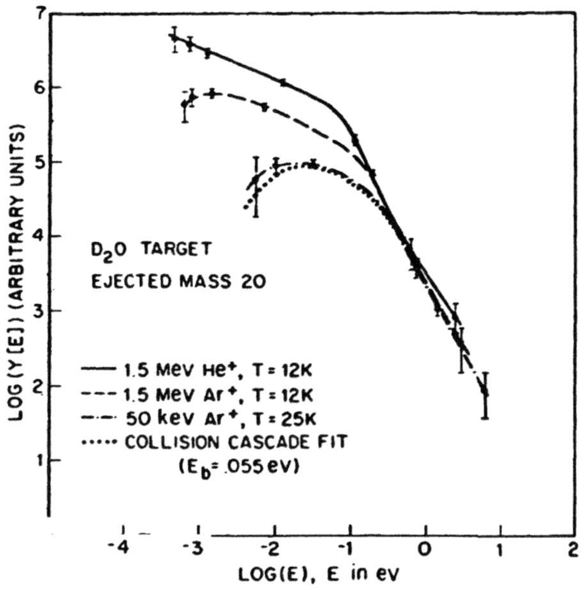

Figure 5. Energy spectra of water ice molecules sputtered by light and heavy ions

They, however, had clear evidence in the case of H_2 ejected by NH_3 bombarded by 6 keV He^+ that sputtering had a sort of evaporation behaviour, in fact the low energy portion of the distribution could be fitted by a maxwellian distribution at 16 K (Haring et al. 1984c).

2.2.6. Angular Distribution of Ejected Molecules

The data we have been referring before have been obtained at a perpendicular incidence of the beam on the surface of the ice layer. When the incidence angle "ϑ" of the helium beam bombarding at 10 K a frozen layer of carbon monoxide has been changed a peculiar dependence from it has been found for the erosion yield by Brown and co workers (1984)

$$Y \cong K \cos\vartheta^{-1.6}$$

where K is a constant.

Such a dependence is not coming only and simply from the increased sublimation due to the growth of the area produced by the intersection between the cylindrical hot region, that is hypothetically formed (see below) around the ion track, and the interface between the solid and the vacuum because such an area increases (and hence the yield should scale) as $(\cos\vartheta)^{-1}$, but also from the diffusion of species that escape laterally from the hot cylinder whose surface becomes closer and closer to that of the ice when the incidence direction of the beam becomes grazing.

3. Chemical Effects

3.1. THE FIRST EXPERIMENTS

As it has been already told it became soon apparent for instance when measuring by RBS the erosion yield of frozen CO_2 (see Figure 2) that some chemical effects induced by the energy loss of impinging ion were occurring in the bombarded sample. Such first hint became experimental evidence when the probe of a quadrupole mass spectrometer was flanged into the scattering chamber and water ice was irradiated with helium ions having kinetic energies in the MeV range at different temperatures of the substrate where the ice had been accreted (Brown et al., 1982a). The quadrupole in fact detected upon bombardment not only H_2O molecules coming from the target but also the newly formed (because not originally deposited) $H_2$and O_2; practically no other species were detected or better H and OH were just the expected cracking fraction of water molecules crossing the electron impact ionizer put in front of the quadrupole so their presence was an artifact of the detection system and they were not coming from the target. Roughly in the same period qualitative results on the same matter were obtained at the nominal temperature of 77 K using ions in the keV range by Ciavola et al. (1982).

Data taken by Brown et al. (1982a) are shown in Figure 6 as a function of the substrate temperature. D_2O instead of H_2O was used in order to improve the signal to noise ratio, a standard trick in mass spectroscopy. The behaviour of the three monitored species is remarkably different with the temperature. While the signal from D_2O shows a temperature independent region in its yield and only at values of T higher than a critical one a steep increase (very similar to the behaviour of the total erosion yield measured in backscattering and shown in Figure 2), the other two released species, i.e. molecular hydrogen (mass 4) and molecular oxygen (mass 32) rise monotonously with the temperature even if with different slopes. This different behavior is probably due to the fact that while water molecules are ejected only from a tiny superficial layer the other two are produced in the whole film and escape along the hot cylinders around the ion track. In this way the temperature dependence in the production yield (number of molecules produced and released per each passing particle) can be

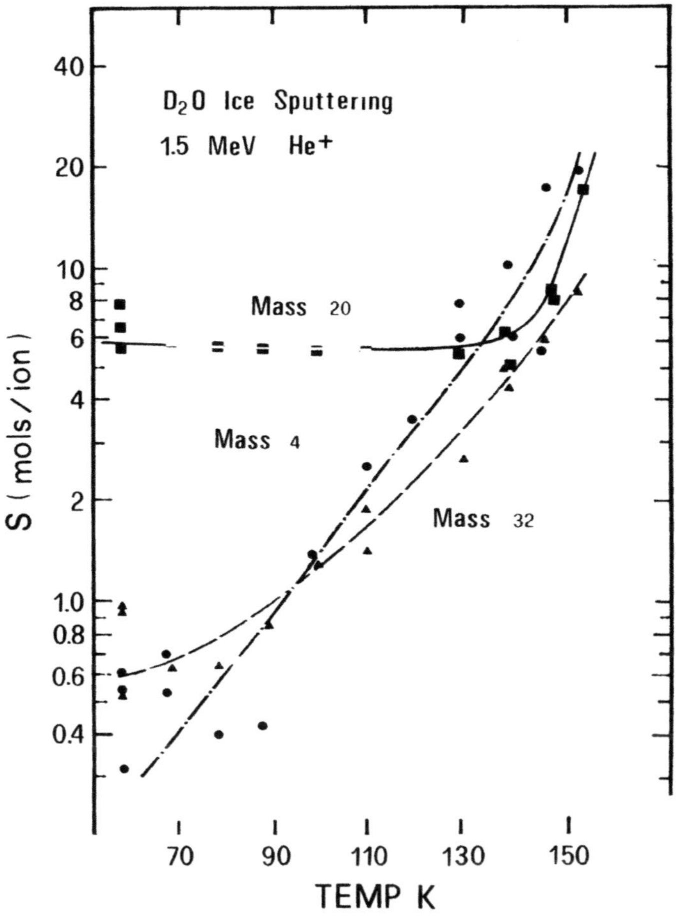

Figure 6. Temperature dependence of the release of D_2O, D_2 and O_2 from D_2O ice irradiated with 1.5 MeV He^+

regarded as due to both a temperature dependence in the formation of molecules and to a temperature dependence in their diffusion in the energized region of the film.

A question came up at once when we were performing those experiments: "Do H_2 and O_2 form inside the icy film, migrate toward the surface, escape and are then detected by the quadrupole or rather radicals are formed in the solid, escape and then recombine when in the gas phase or even (worse for astrophysical importance of the processes investigated) are the signals at mass 4 and 32 just produced by the fragmentation of D_2O molecules in the electron impact ionizer put at the entrance of the probe?". Well the answer to this multiple question is that we can be sure that newly formed species originate in the ice, in fact, species ejected in the gas phase do not interact anymore before being detected because at the density conditions existing in the scattering chamber during such experiments (to better than 1.2×10^{-9} torr) the mean free path of molecules is practically infinite and we can be equivalently sure that they do not form in the ionizer because such an effect is not observed when the water ice film is allowed to sublime from the substrate without having suffered any irradiation, in this case in fact only water molecules and their cracking fractions are detected., but no H_2 or O_2 molecules.

This first success triggered the interest toward the study of other chemical effects. It was natural then to move on and to accrete on the cold substrate (in the jargon of low temperature physicists "the cold finger") mixtures of different ices, prepared in the gas phase to obtain an "intimate" mixing.

After the pioneering of Berger (1961) and Oro (1963), who in the early sixties irradiated layers of water, ammonia and methane and qualitatively observed (using an analysis *not performed in situ*) the formation of new molecules, Pirronello, Brown, Lanzerotti, Marcantonio and Simmons in 1982 irradiated at 9 K a frozen film (1.8×10^{18} mol cm^{-2} thick) made of a mixture of water and carbon dioxide with a total dose of about 10^{15} He^+ cm^{-2} at 1.5 MeV of kinetic energy obtaining a relevant production of formaldehyde. On the average the production yield was

$$Y \cong 3.7$$

The huge amount of synthesized formaldehyde had not been released during bombardment like molecular hydrogen and molecular oxygen did. In fact, not being mobile enough it remained trapped in the frost layer until the end of irradiation and the detection happened only when the temperature of the cold finger was raised. Formaldehyde did not escape at the temperature at which it would have occurred if it were alone but, on the contrary, was released at the onset of the sublimation of water at about 165 K. When the experiment was repeated with a dose of projectiles four times higher the average production yield of H_2CO only doubled showing the beginning of a saturation effect.

That experiment gave some noteworthy information:

a) this type of processes can produce "complex" (it will be more and more true later) organic molecules in relevant amounts;

b) synthesized molecules can be divided roughly in two classes according to their mobility, those which can escape from the layer during bombardment (like H_2 and O_2) and those that remain trapped (for instance H_2CO);

c) the saturation effect observed gives evidence of the fact that at the increase of the concentration of the synthesized molecules the probability of forming more of them diminishes with the fluence (probably because of the deep processing of the residual layer, an interesting part of the whole investigation) and also that the chance of destroying already formed species increases;

d) the use of isotopically labelled compounds (^{16}O and ^{18}O) in the deposited water and carbon dioxide allowed to monitor the origin of oxygen atoms in the synthesized formaldehyde, a fact that can shed some light in the mechanisms responsible for the formation of new molecules in frozen gas mixtures under ion irradiation (see Roessler, 1986).

3.2. SOME OTHER RESULTS ON MOLECULAR SYNTHESIS

Very interesting work has been performed by several groups on the synthesis of molecules in pure ices and ice mixtures bombarded by fast ions.

About one year after the paper on the formation of formaldehyde in H_2O/CO_2 by Pirronello et al. (1982a) had appeared Moore, Donn, Khanna and A'Hearn (1983) published a very interesting paper in which using IR spectroscopy identified the signatures of C_2H_6, CO_2, CO, NO_2, NO and CH_4 (weak) in various mixtures containing H_2O, NH_3, CH_4, N_2, C_3H_8, CO and CO_2 irradiated with 1 MeV protons at 20 K, while the major effect seen by Moore (1984) was the synthesis of SO_3 when irradiating SO_2.

Synthesis of CO_2, O_2, C_2, C_2O and C_2O_2 in pure CO, of N_2 and N_2H_4 in pure NH_3, of H_2, O_2, H_3O, O_2H, H_2O_2 in pure H_2O, of HCO, CO_2 O_2 and C_2 in H_2O/CO mixtures and of N_2, HCN (?) and N_2H in NH_3/CO mixtures when irradiated with argon, helium or molecular hydrogen ions at 20 K have been reported by Haring et al. (1983, 1984a,b,c) using mass spectroscopy.

Ejection of H_2O, O_2, H_2 and H from water ice irradiated by protons and neon ions in the important energy range between .5 and 6 keV has been investigated by Bar-Nun and co workers (1985).

Chemical effects in ices (pure H_2O and pure CH_4, but also binary mixtures of $H_2O + CO_2$, $H_2O + NH_3$ and $CO_2 + NH_3$) have been studied with keV ions by Benit et al. (1986, 1987, 1988), using an apparatus and techniques that before had allowed them to show formation of carbon dioxide and carbon monoxide in silica implanted with C ions (Rocard and Bibring, 1982).

Recently Pirronello et al. (1988b,c) have studied several mixtures constituted by H_2O and CD_4, irradiating them with MeV helium ions under different conditions and finding, as expected, very rich

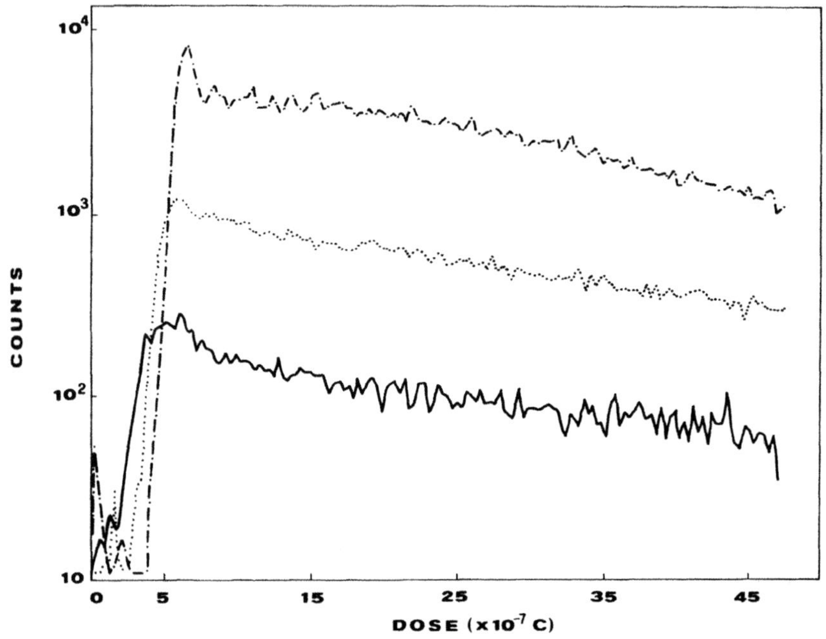

Figure 7. Dose and thickness dependence of the release of HD from an even mixture of H_2O + CD_4 irradiated with MeV ions

chemical effects. In particular it was changed the relative abundance of components, the thickness of the layer, the irradiation temperature and the dose of impinging ions. In Figure 7 it is reported the dose dependence of the "HD" (formed by a H atom from water and a D atom from methane) production and release for films of three different initial thicknesses: 1.9×10^{17} mol cm^{-2} (full line), 5.7×10^{17} mol cm^{-2} (dotted line), 1.9×10^{18} mol cm^{-2} (dot-dashed line). As it can be seen the ejection in the gas phase is not prompt, it occurs after a delay probably due to the formation of a necessary build up of the concentration of molecular hydrogen before release or (a hypothesis I am proposing now) it could be due to the formation of suitable paths (may be dendritic crystalline regrowth, that were suggested to occur around the ion track from changes of the surfaces topography measured in RBS already by Brown and co workers. in 1978) in which I think that the mobility of molecular hydrogen is increased with with respect to that in amorphous ice. From the fact that the HD release (that incidentally is similar to that of H_2 and D_2) changes dramatically with fluence (it is a logarithmic plot) it can be inferred the fast change in composition and structure of the deposited mixture. Figure 7 also shows also a clear dependence of the amount of released HD on the layer thickness.

If molecular hydrogen is an example (perhaps the best one) of mobile species, in Table 2 (from Pirronello et al., 1988c) it is shown, in terms of counts from the quadrupole, the average yield of what we

TABLE 2. Formaldehyde production (counts) for three different values of the ion flux, the layer thickness and the substrate temperature

Changing parameter	1	2	3
J	3530	1550	1020
t	320	1550	5560
T	1550	1370	1580

could call "heavy formaldehyde" D_2CO (a specie that is not mobile enough to escape and then remains in the ice film) produced in the same type of mixtures changing the dose of He^+ ions "J" ($1.1x10^{15}$ cm^{-2}, $2.0x10^{15}$ cm^{-2}, $4.1x10^{15}$ cm^{-2}), the temperature of the substrate "T" (8 K, 25 K, 32 K) and the thickness of the icy film "t" ($1.9x10^{17}$ cm^{-2}, $5.7x10^{17}$ cm^{-2}, $1.9x10^{18}$ cm^{-2}).

Inspection of this Table allows some deductions. The dose dependence (first line) shows that the concentration of heavy formaldehyde diminishes with the increase of the fluence of projectiles which is suggestive of the fact that already at the lowest fluence we used in this experiment we had overcome the maximum in the production of this specie, differently with respect to the case of formaldehyde production in the H_2O/CO_2 mixture (Pirronello et al., 1982a), but in agreement with results of Lanzerotti et al. (1985, 1987a) showing how in pure methane already after about a dose of 10^{15} cm^{-2} helium ions the concentration of hydrogen is strongly diminished (see below) a fact that makes more difficult producing formaldehyde, but does not of course stop its destruction. From the second row of the same Table it is clear that the amount of formaldehyde synthesized scales linearly with the initial thickness of the icy film, showing that a uniform production of this specie through the entire sample is occurring. The result shown in the third row says that, at least in the investigated range, the production of H_2CO does not depend on the temperature of irradiation. For the sake of completeness, it must be added that in these experiments in which mixtures of water and methane have been bombarded by swift ions also other more complex organic species have been detected, like for instance methanol and perhaps acetaldehyde.

IR spectroscopy, employed in some of the mentioned papers, together with mass spectrometry are probably the most important techniques to detect synthesized species in irradiated samples. They are quite complementary to each other in the sense that one (mass analysis) gives only information on molecules released in the gas phase (and one has to use the "escamotage" to let sublime the whole film warming up the substrate to detect molecules that remain trapped in the icy layer after formation), while the other (IR spectroscopy) analyzes only what is in the solid and cannot give any information on sputtered species because you never have a enough high optical depth from species in the gas at the density (UHV) you want to operate to avoid significant condensation of gas onto the cold finger with contamination of the sample you are studying. There are advantages and disadvantages in each of them. Mass spectrometry, for instance, detects different molecules having the same mass as equal (a problem that can be often circumvented by depositing isotopically labelled species) and, furthermore, it does not say anything on the important question about when species that remain trapped in the layer are formed, whether upon bombardment or during the forced warm up of the sample when frozen radicals can recombine. IR analysis answers to the last question positively, in the sense that shows that species are formed already during irradiation, but it has the general drawback that IR analyses need high doses of projectiles ($>10^{14}$ cm^{-2}) to synthesize enough molecules to be over the threshold of detectability (a reasonable signal to noise ratio), a fact that can impede to monitor species that exist only transiently in the initial phases of irradiation.

Another technique used to analyze the irradiated sample is gas or liquid chromatography, a *non in situ* type of analysis that can deeply change the target situation because the irradiated sample is first melt, evaporated or dissolved (Berger, 1961; Oro, 1963).

3.3 THE RESIDUE

One of the interesting physical and chemical effects of the irradiation by fast (keV-MeV) ions of some carbon bearing ices like CH_4, CO and mixtures containing them is that there is a preferential release in the gas phase of oxygen or hydrogen with the transformation (in the case of ices rich in both hydrogen and carbon it is sometimes and perhaps quite improperly called polymerization) of the originally deposited film of frozen gas into a nonvolatile carbon-rich layer that can be stable at room temperature or even at higher ones, the so called "residue" (Cheng and Lanzerotti, 1978; Moore and Donn, 1982; Moore et al., 1983; Calcagno et al., 1983; Foti et al., 1984; Lanzerotti et al., 1985; Lanzerotti et al. 1987a). The process really consists in the progressive destruction of C-H (Tombrello, 1987) or C-O bonds, in the loss of the more volatile component and in the formation of the more stable C-C bonds with a final result, if the irradiation continues for very long, of a complete carbonization of the sample.

These residues have been analyzed either optically, or by the simple Rutherford backscattering or by means of the nuclear reaction ^{3}He (d,p) ^{4}He, that allows accurate measurements of the deuterium loss from the material in consideration. A dependence of the effect on the initial thickness that is important for astrophysical applications has been found for instance for irradiated CD_4 (Lanzerotti et al., 1987a). For the thinnest films bombarded (only few hundred angstroms) a complete erosion of the films occurs when MeV He^+ ions are used, with no residue formation. Thicker films show loss of both hydrogen and carbon but with a ratio between the hydrogen to carbon content that decreases at increasing doses of projectiles. The rate at which this ratio changes is quite high (with an effective cross section of about $\cong 5\times10^{-16}$ cm^2) at the beginning of the irradiation (till a dose of about 10^{15} ions cm^{-2}), then it diminishes and the effective cross section becomes $\sigma \cong 10^{-17}$ cm^2.

4. The Mechanisms Responsible for Physical and Chemical Effects

4.1. EROSION MECHANISMS

The full understanding of the processes, that are responsible for the experimental results we have shown, has not yet been reached. In this section we will present some of the current ideas that try to give an interpretation of the collected data.

Before considering erosion of ices by MeV light ions, let's spend just few words about sputtering of materials irradiated by light and heavy ions in the keV range. For their technological importance mainly conductors and semiconductors had been already studied in great detail when the first paper on water ice appeared (Brown et al., 1978) and the so called yield "Y" (for them defined as the average number of atoms instead of molecules ejected per each impinging ion) had been measured and as a function of several parameters.

From the large amount of data collected (see e.g. Behrisch, 1981, 1983 and references therein), to describe quantitatively the ejection of atoms from solids irradiated in the keV range it has been built the "collision cascade model". In it sputtering is interpreted as being due to the direct transfer of momentum via elastic collisions among the impinging ion and the nuclei plus core electrons of the target atoms; these atoms (called recoils), once displaced, collide with their neighbors giving rise to the cascade of collisions to which the name of the model refers. In the region where such collisions occur, while some of the atoms are permanently displaced producing the so called defects, some have a finite probability to gain a component of momentum toward the outside, then if they are close to the surface of the solid and have received enough energy can overcome the surface barrier and escape, see the schematic in Figure 8.

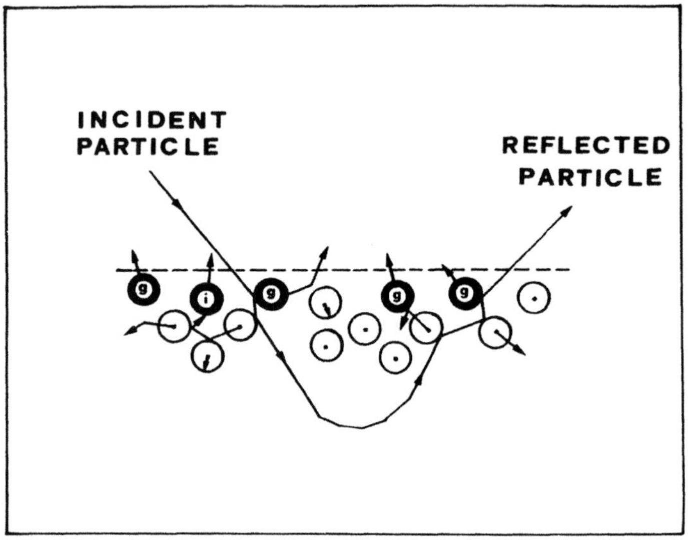

Figure 8. Schematic of the sputtering process in the collisional regime

A theoretical description of the mentioned processes, given by means of the elegant transport theory by Sigmund (1969) and revisited successively by others (see e.g. Falcone, 1989), estimates (in the so called linear regime) the sputtering yield from metals and semiconductors bombarded by keV ions as

$$Y \cong K\ (dE/dx)_n (n\sigma U)^{-1}$$

where:

$(dE/dx)_n$ is the so called "nuclear stopping power", i.e. the energy lost per unit path length by the ion in the material through elastic collisions, a tabulated quantity;

n is the number density of the material;

σ is the momentum transfer cross section;

U is the surface binding energy;

K is a correction factor.

The other important result, that we should like to recall here (we already mentioned it in the discussion of Figure 5), is that one obtained by Thompson (1965), i.e. the energy distribution of sputtered atoms in the collisional regime is given by

$$\Phi(E) \cong kE/(E+U)^3$$

where k is a constant.

When the energy density released in the solid overcomes a certain threshold value the collisional regime is no more linear, collisions redistribute completely the kinetic energy among recoils, thermalizing the deposited amount of energy and giving rise to the so called "thermal spike" (Vineyard, 1976).

Opposite to the collisional regime, in which it is easy to understand what energizes target atoms, displaces them and in some cases allows them to be sputtered (i.e. the direct transfer of momentum among partners in elastic collisions), in molecular insulators, irradiated by MeV light ions, even if it is clear from experiments that at those speeds of projectiles the primary loss of energy occurs via electronic processes, it is still not well known how such an energy, originally deposited into the electronic system, is transferred to the nuclei in such an efficient way to displace them and produce the type of phenomena we are considering.

Let's try to understand first the experimental data given in Fig. 4 on water ice erosion in the temperature independent region that show a quadratic dependence from the electronic stopping power (Brown et al. 1982a).

We can imagine that, around each primary event of ionization and/or excitation occurring in molecules of the target placed along the track of the penetrating ion, a region (we can call it the Spherical Region of Influence or "SRI"), develops and in which atoms, molecules, radicals of the target are energized (see Figure 9). For each primary event an energy release "W" takes place that (in analogy with the gas phase case, without paying too much attention to some conceptual difficulties into defining ionization in the condensed phase and considering also secondary electrons) is roughly given by:

$$W \cong \langle E_i \rangle + \nu \langle E_e \rangle + \langle \varepsilon \rangle$$

where

$\langle E_i \rangle$ = average energy required per ionization

$\langle E_e \rangle$ = average energy of excitation

$\langle \varepsilon \rangle$ = kinetic energy of subexcitation electrons

ν = average number of excitations per ionization

On the average, the amount "W" (20 or 30 eV) does not depend dramatically on the type and the kinetic energy of incident ions but, on the contrary, it is almost constant with respect to them (Fano, 1946); the average distance among primary events is then

$$\lambda = W/(dE/dx)$$

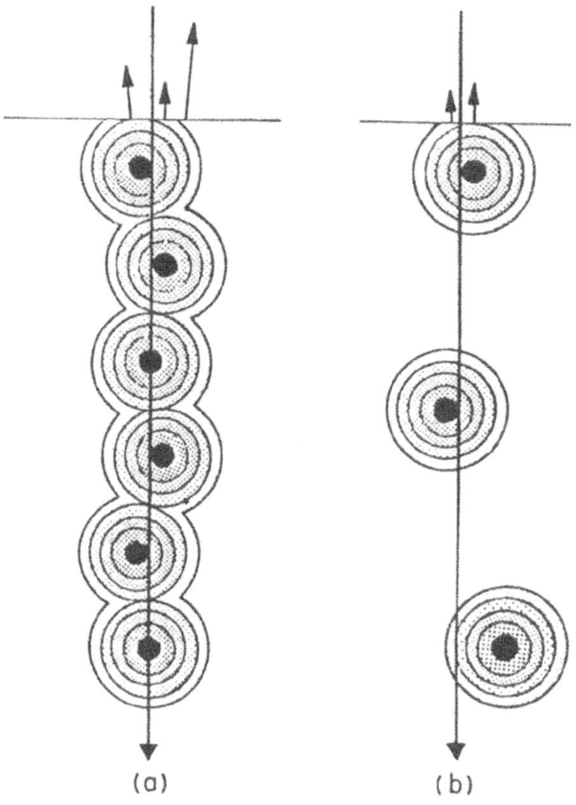

Figure 9. Schematic view of the spherical regions around primary ionization events in which target molecules are energized

Following this set up of the problem, we can then interpret the laboratory results of Figure 4 as relative to a case in which the value of "λ" is smaller or much smaller than the radius of the SRI generated by the impinging ions in that target material at those energies (see Figure 9, section b). If such an idea is correct, then when "λ" is larger or much larger than the "size" of the SRI (see Figure 9, section a) the total erosion yield should scale linearly with the electronic stopping power and not quadratically. This is, as a matter of fact,what happens for N_2 irradiated with hydrogen and helium ions of different energies (Brown et al., 1986), where the transition from the linear to the quadratic behavior occurs at a value of the electronic stopping between 10 and 40×10^{15} eV cm^{-2}. A case not easy to reconcile with the simple picture just given is that of irradiated water ice (Brown et al., 1982a) and that of frozen CO (Johnson and Brown, 1983), where the non linearity extends to the lowest values of the stopping investigated.

A further important hint to understand the mechanism of erosion is provided by the data on the energy of molecules ejected during bombardment shown in Figure 5. As it can be seen the high energy tail of the spectrum is quite well fitted with the Thompson's predictions (Thompson, 1965) of the energy distribution of particles sputtered via collision cascades. By the way, the overwhelming majority of molecules are released at much lower kinetic energies and with a distribution that is not maxwellian but might be regarded, "I think", as a superposition of several maxwellian distributions, each characterized by a suitable value of "temperature". This fact, together with the consideration that at the end most of the energy lost by passing ions has to be transformed into heat, suggests that in certain type of materials (and ices are just this type because are good thermal insulators) a thermal spike can be generated by the energy released by a projectile. The generation of a thermal spike is possible if the energy deposited remains confined locally for long enough (10^{-11} seconds) to be thermalized. If it occurs then a steep thermal transient will be produced with initial temperatures that, in the core region (around the place where the ionization or excitation event occurred), can reach thousand degrees. Molecules close to the surface, or if they are mobile enough migrating from inside, can then acquire enough energy to overcome the surface barrier and evaporate. In this case the yield (see e.g. Johnson and Evatt, 1980) can be given by

$$Y = 2\pi \int_0^{t_{max}} \int_0^{r_{max}} \Phi(r,t)\ r\ dr\ dt$$

where

t = time

r = distance measured on the interface solid-vacuum from the center of the transiently heated region

t_{max} = time duration of the spike (before temperature drops below the value that can give evaporation)

r_{max} = maximum size of the cross section of the hot cylinder in which during t_{max} evaporation occurs

$\Phi(r,t) = N\ p/(2\pi MRT)^{1/2}$ is the flux of evaporating molecules

N = Avogadro number
p = equilibrium vapor pressure
M = mass of evaporating species
T = transient value of temperature

From the integral expression of the yield given above one can easily understand the origin of the square dependence on the stopping power in the cylindrical geometry. The total yield is, in fact, proportional to the size of the cross section between the hot cylinder and the interface solid-vacuum (that reasonably is proportional to the deposited energy) and to the time duration of the spike (that is also proportional to the size of the cross section and hence again to "dE/dx").

We have probably made some steps toward the understanding of the process, but still the main problem remains: how in each of the expanding SRI created by single events (mainly ionization or excitation) energy is transferred from the electronic system, where it has been deposited, to motion of the nuclei belonging to the target material? What mechanism energizes them till the point that some can overcome the surface barrier to be sputtered away?

One of the possible mechanisms, that could at least in principle energize nuclei and that is known to be responsible for the formation of nuclear tracks for instance in mica, has been considered initially by Brown et al. (1982a): the so called "Coulomb explosion". In it external electrons of target atoms and molecules are thought to be displaced at right angles with respect to the trajectory of the incoming fast ion, in such a way to form a negative cylindrical sheet surrounding the positive distribution of the atom cores along the axis of the cylinder. In such a configuration the electric field would be radial everywhere but in the vicinity of the ice surface, where a net axial component of the field toward the outside would accelerate target ions to escape. It can be shown that the cylindrical geometry, also in this case, gives the right square dependence of the yield on the stopping power A serious difficulty for this mechanism to work is represented by the high mobility of electrons that not remaining confined for long enough in the negative sheet would come back to neutralize the positive ions.

A different model was proposed by Seiberling et al. (1982) and is known as the "thermalized Coulomb explosion". In this model the Coulomb explosion is used only to set in motion the nuclei of atoms and the molecules, then collisions thermalize the energy distribution. This model, however, suffers of the same difficulties of the standard Coulomb explosion and furthermore gives a steeper dependence of the yield from the stopping power steeper than quadratic, also for light ions, not in agreement with measurements on astrophysical ices.

A promising mechanism calls into play chemistry and more exactly calls into play "dissociative recombination", a well known process that occurs, for instance, in the gas phase but that can also occur in solids if there are not present too many scavengers to catch free electrons. It takes place with a relatively high rate constant (values of the dissociative recombination rate constants relative to some species are given in Table 3) when a positive molecular ion recombines with an electron (in the solid the electron could be that

TABLE 3. Dissociative recombination rate constants at 100K

Molecular ion	Rate constant ($cm^3\ s^{-1}$)
H_3O^+	1.1×10^{-6}
H_2O^+	1.3×10^{-6}
OH^+	1.4×10^{-7}
CH_4^+	1.3×10^{-6}
CH_3^+	1.2×10^{-6}
CH_2^+	8.7×10^{-7}
$C\ H^+$	3.3×10^{-7}
$N\ H^+$	1.9×10^{-7}

initially stripped by the ionization process, or better that produced by an electron-hole creation, induced by the penetrating fast ion), forming initially an intermediate excited state of the molecule that then dissociates into two neutral products. They then carry away the energy excess setting in motion by collisions their neighbors.

What was considered a problem for the Coulomb explosion mechanism, i.e. the fast neutralization of the positive "core" of the cylinder, is an advantage of the chemical mechanism just mentioned because it implies that a non negligible fraction of the electron-hole pairs formed by the incident ion quickly recombine, transferring kinetic energy to atoms and molecules.

4.2. MECHANISMS FOR CHEMICAL EFFECTS

Most of the chemical effects that have been reported in the previous section may be understood in terms of reactions involving neutral species and molecular ions produced directly by impinging particles or by secondary electrons in the energized region around the primary events that cause the energy loss of incident ions.

In the case of small "dE/dx", when "SRI" are well separated from one another (see e.g. Figure 9, section a), the ejection of molecules upon bombardment happens when SRI are generated close enough to the surface, a fact that occurs with a probability that is roughly given by the ratio between the average size of SRI "$< r >$" and the mean free path between primary ionization events "λ_i", i.e.

$$p\ (\text{ejection}) \cong < r >/\ \lambda_i$$

Molecules synthesized in SRI too deeply inside the solid do not escape upon bombardment but remain trapped till when they are allowed to migrate, reach the surface and are released in the gas phase. When it happens it is seen that the total amount of synthesized molecules scales with the thickness of ice (see for example the results relative to D_2CO shown in Table 2) because they were formed throughout the whole icy film. In the laboratory it can be easily obtained raising the temperature of the target, in interstellar conditions it can occur for instance either through the chemical explosion of the mantle (Greenberg, 1973; d'Hendecourt et al. 1982) or through the interaction between icy mantles of grains and the heavy component of cosmic rays (Leger et al., 1985).

If, on the contrary, the value of the stopping power is not small and SRI merge into each other (see Figure 9, section b), mobile species can escape from all depths inside the film as in the case of D_2 and O_2 (see Figure 6) from D_2O bombarded by MeV helium ions (Brown et al., 1982a).

Contributions to the formation of new species in solids may come also from more exotic processes, those that are usually catalogued in the domain of the so called "hot atom chemistry" (see for an extensive review the paper from Roessler, 1986) and that are suggested to occur also in the synthesis of formaldehyde (Pirronello et al., 1982a)

The main differences with gas phase reactions in interstellar conditions are that certainly not only two body collisions occur and that also endothermic reactions (for instance between neutral molecules) can take place in the energized regions (or one can say in the high temperature region around the ion track, if the "thermal spike model" applies).

5. Astrophysical Applications

The type of processes described in the previous sections, i.e. sputtering and molecule formation in icy solids, occur continuously in the Universe and applications of the laboratory results to several situations of astrophysical interest have been performed. The scenarios we will speak here are: the solar system, the circumstellar and the interstellar medium.

5.1. SOLAR SYSTEM

The astrophysical scenario that has received more attention is of course the Solar System. Relevant effects of the interaction of energetic ions with solids inside it take place , for instance, when solar wind particles impinge on the surface of satellites and planets without atmosphere and on interplanetary dust, but the effects we are most interested in (because of the presence of ices) are those that occur when the plasma trapped in the magnetosphere of the giant planets bombard the icy surface of satellites or the grains of the rings, that are embedded inside them.

5.1.1. Satellites

When plasma particles impinge on the surface of satellites (if they can overcome the screening effect produced by the intrinsic magnetic field the object may have) they induce sputtering, chemical modifications (see e.g. the preferential ejection of more volatile species of Haff et al., 1979) physical changes (like the evidence of dendritic crystalline regrowth given by Brown et al. already in 1978; faster destruction of smaller icy grains, Clark et al., 1983) and albedo alteration (Nash and Fanale, 1977; Eviatar et al., 1985), that can give observable differences between the trailing hemisphere of satellites (this is, in fact, preferentially bombarded by ions trapped in the magnetosphere of the giant planet, when satellites are locked in a gravitational spin-orbit coupling, because the magnetic field co rotating with the planet has a higher angular velocity than the satellite) and the leading one (Clark et al., 1984).

Ejected atoms and molecules may follow three different "fates", according to the different conditions presents in the satellites (see Figure 10, taken from Brown et al., 1982b):

a) if no atmosphere is present or only a very thin one

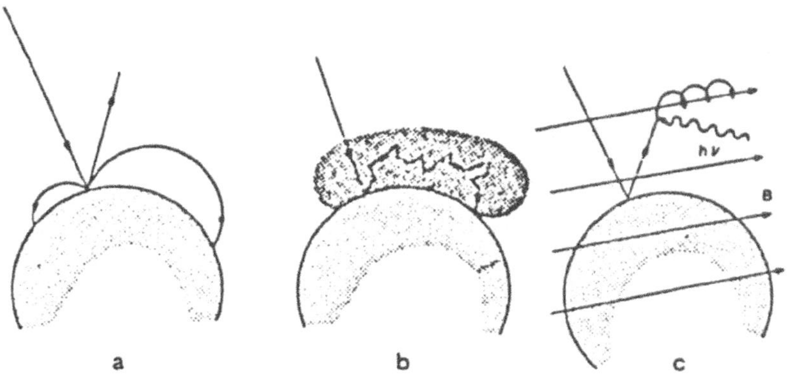

Figure 10. Possible fates of sputtered molecules from satellites

molecular species sputtered from the satellite surface follow ballistic trajectories (see Figure 10, section a) that, according to the speed and the angle of ejection, transport them toward other regions of the surface or shoot them into space. In the first case sputtered particles during the time of their flight form a very thin atmosphere around the object and when they hit its surface, at the end of their trajectories, can induce again sputtering. In the second case, when their speed is higher than the escape value, either remain orbiting around the satellite or, if they are able to enter in the region of space where the dominant gravitational field is that of the giant planet, may become part of one of the observed tori;

b) if a thick atmosphere is present sputtered species do not move on ballistic trajectories because their mean free path is much smaller than the thickness of the atmosphere itself. Almost no molecules can be directly ejected away from the gravitational field of the satellite and they contribute into enriching its atmosphere (Figure 10, section b);

c) in a thin atmosphere sputtered molecules once ionized for instance by ultraviolet photons are swept away by the passing magnetic field of the giant planet (Figure 10, section c).

Ice sputtering has been considered in several satellites (using available ion flux, either estimated or measured "in situ" by spacecrafts), like in the saturnian and jovian satellites (see e.g. Lanzerotti et al., 1978; Cheng and Lanzerotti, 1978; Haff et al., 1979; Johnson et al., 1981; Cheng et al., 1982a; Johnson et al., 1982; Watson, 1982; Sieveka and Johnson, 1982; Lanzerotti et al.,1983; Johnson et al., 1983a; Lanzerotti et al., 1984; Cheng et al., 1986). A great attention has been devoted to Io, also from the point of view of the interaction between ions and icy surfaces for its peculiarly interesting environment created by the presence in it of continuously active volcanoes, that make it be a unique case in the solar system (see e.g. Matson et al., 1974; Haff at al., 1981; Lanzerotti et al., 1982; Melcher et al., 1982; Lanzerotti and Brown, 1983; Matson and Nash, 1983; Cheng, 1984; Johnson et al., 1984; Sieveka and Johnson, 1984; Sieveka and Johnson, 1985; Chrisey et al., 1987; Strazzulla et al., 1987; Chrisey et al., 1988). More recently attention has been paid also to the effects induced by energetic ions on the surfaces of the satellites of Uranus by Lanzerotti et al. (1987b).

5.1.2. Rings

The presence around planets of ring systems, whose icy grains have sizes ranging from the micron scale to tens of centimeters or more, and the fact that they are embedded in the magnetosphere suggests that the interaction processes, we are dealing with, take place also in that environment.

Erosion induced in ring grains by ion impacts certainly contributes to the enrichment of the magnetospheric plasma, a process that is made easier by the absence of gravity (apart from that of the planet, that by the way does not interfere). It was in 1982, in fact, that Cheng, Lanzerotti and Pirronello (Cheng et al., 1982a), looking for a significant source of hydrogen to try to explain the presence in the vicinity of Saturn's rings of the H ring atmosphere (discovered by rocket-based measurements (Weiser et. al., 1977) and later confirmed by Pioneer 11 (Judge et al., 1980)), tried to see if sputtering of water ice by charged particles trapped in the planet magnetosphere could account for it. In order to estimate the loss rate of low energy ions in Saturn's inner magnetosphere, they used the ion phase space densities "f" of Krimigis et al. (1981) and, assuming steady inward radial diffusion (a consistent interpretation of the measured negative gradient of "f" when the ion source is in the outer magnetosphere) at fixed adiabatic invariants "μ" and "J", they used the transport equation given by Schulz and Lanzerotti (1974) in terms of the particle drift shell parameter "L"

$$L^2 (D_{LL} L^{-2})' F' + D_{LL} F'' + D_{LL} (F')^2 - \tau^{-1} = 0$$

where

F = ln f
f = phase space density of particles
L = McIlwain particle drift shell parameter
D_{LL} = radial diffusion coefficient
τ = lifetime against particle loss
' = derivative with respect to L

They got for the average lifetime against loss of particles

$$2 \times 10^6 \text{ s} < \tau_{ave} < 3 \times 10^8 \text{ s}$$

according to the chosen radial coefficient dependence on L. Taking then an omnidirectional flux of protons of about 10^7 cm^{-2}s^{-1}, an average velocity of 5×10^8 cm s^{-1} Cheng et al. (1982a, b) evaluated between $L \cong 4.5\ R_S$ and $L \cong 8\ R_S$ (with R_S the radius of Saturn) a total loss rate of particles per second "η" of

$$3 \times 10^{22} \text{ s}^{-1} < \eta < 4 \times 10^{24} \text{ s}^{-1}$$

Using the assumption that the proton loss in that region was dominated by the sweeping effects of Dione, Thetys and the E-ring (consisting primarily of micron sized icy particles) and that such a loss was due to impacts on ice, using the experimental data on water ice irradiation by Brown et al. (1982a), they obtained a source of hydrogen that was about three orders of magnitude lower than required and then clearly inadequate to explain the presence of the ring atmosphere of hydrogen.

To counterbalance this negative result, Cheng et al. (1982a) estimated that sputtering of ice could be important for producing the heavy ion plasma torus, found by Frank et al. (1980) and for which they evaluated that an ion source $\cong 10^{23}$-10^{24} s^{-1} was required within a distance between 4 to 8 R_s. Each water molecule released by sputtering in the magnetosphere is expected to be dissociated and/or ionized to yield, eventually, an oxygen ion. Cheng et al. (1982a) obtained an oxygen source between $3x10^{22}$ and $4x10^{24}$ s^{-1}, i.e. of the required magnitude, and let them suggest that the heavy ion plasma torus could have been mainly formed by O^+ rather then by N^+, as it was equivalently consistent with voyager data (Bridge et al., 1981).

The most important result of that paper was, by the way, the evaluation of the lifetime of E-ring icy particles against erosion induced by irradiation of ions trapped in Saturn's magnetosphere. The erosion rate of ice surfaces in that environment was estimated to be of about 10^{-4}µm yr^{-1}, with a resulting lifetime of E-ring grains of

$$T < 10^4 \text{ years}$$

Such a short lifetime either says that we are observing the ring in a very special and improbable moment of its life or that it is required continuous a replenishment of fresh material from some sources. The envisaged source could be the bombardment by meteoroids of the Enceladus, as it is also suggested by the increased brightness of the E ring in the vicinity of the orbit of the satellite (Terrile and Tokunaga, 1980). The same result was addressed later by Haff et al. (1983) and Hill (1984).

5.1.3. Comets

A large amount of observational evidences has confirmed in the last forty years the Whipple (1950) hypothesis that comets, or better cometary nuclei, are made of a matrix of solid materials mixed with ices. Such nuclei have been, during their long life (they probably formed when the protosolar nebula was generating the solar system, roughly 4.6 x 10^9 years ago), exposed to the irradiation energetic particles (mainly cosmic rays in the Oort cloud and solar wind particles in the "vicinity" of the sun. It was recognized already several years ago (see e.g. Shul'man, 1972; Donn, 1976) that such a long bombardment can alter significantly the pristine chemical composition of cometary nuclei, at least in their very external layers, a fact of considerable interest because it is connected with the possibility of gaining, through observations performed now, information on the environment in which cometary nuclei originally formed.

On the basis of experimental results (e.g. Brown et al., 1982a; Pirronello et al., 1982a) on erosion and chemical modifications induced by fast ions in ices and ice mixtures some applications to cometary physics have been carried out. It has been for instance investigated the production of molecular hydrogen (Pirronello et al.,

1982b), and the formation of formaldehyde in cometary nuclei exposed to the irradiation of low energy cosmic rays till saturation (Pirronello et al. 1984). General discussion on physical and chemical effects have also been given (see e.g. Strazzulla et al., 1982; Johnson et al., 1983c, Ryan and Draganic, 1986). The problem of the formation of a non-volatile comet crust due to decomposition of the matrix and preferential loss of the most volatile component (that should be in competition with the crust formation process occurring for the preferential sublimation of volatiles at any perihelion passage of the nucleus) has been correctly addressed by Donn (1976) and Johnson et al. (1987).

Very recently Pirronello and Lanzafame (1989), on the basis of a previous idea (Pirronello, 1985), proposed a method (the only one up to now to our knowledge) to get experimental information on the spectrum of the low energy ($E < 1$ GeV) cosmic rays. Such a spectrum is almost completely unknown, a part from some theoretical estimates obtained by Morfill et al. (1976) on the basis of the transport theory. Measurements of the cosmic ray flux, performed in the vicinity of the earth i.e. well inside the heliosphere (the region whose electrodynamical properties are controlled by the presence of the magnetic field, carried in the interplanetary medium by the solar wind), suffer from a strong screening effect. This screening is particularly effective the lightest and less energetic ions, the overwhelming majority. For this reason these measurements are not representative of what pervades the Galaxy and impinge on interstellar clouds. The method proposed relies on a simple idea. Comets are objects that stay in the so-called Oort cloud (i.e. outside of the heliosphere) for most of their long life and there they are irradiated without any screening by cosmic rays of all energies. Such a bombardment leaves a permanent recording, as chemical modifications of the external layers of the nuclei, modifications that depend on the suffered irradiation. When the comet is then occasionally injected into the inner solar system, its nucleus starts to release in the gas phase by sublimation molecules of the frozen matrix. It releases then also species synthesized by the irradiation of energetic ions and that remained trapped in the layer without escaping during irradiation at low temperatures (as in the laboratory simulation of Pirronello et al., 1982a). Several observations of the abundances either of synthesized molecules or of their photodissociation products, performed when the comet (a new or a very long period one) is at various distances from the Sun during its inbound trajectory, should then give, at least in principle, quantitative information on the energy spectrum of cosmic rays that hit the nucleus. Assuming still the Morfill et al. (1976) spectrum, Pirronello and Lanzafame (1989) deduced (see Figure 11, full line) the concentration of formaldehyde (relative to the case of a uniform distribution of H_2CO inside the nucleus) versus the heliocentric distance in the coma of an hypothetical comet, entering the inner solar system after irradiation in the Oort cloud. In the same Figure 11 (dashed line) it is shown the effect the so called "anomalous component" of cosmic rays would have if it had an interstellar origin, a problem that could in this way receive an answer.

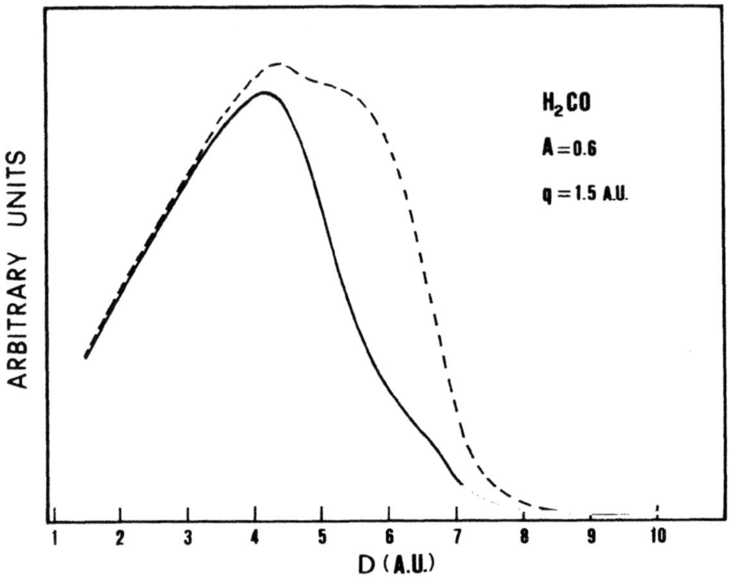

Figure 11. Concentration of formaldehyde in the coma of an hypothetical comet entering in the inner solar system after a long irradiation outside the heliosphere of unscreened cosmic rays

5.2. CIRCUMSTELLAR ENVIRONMENTS

The study of the interaction of icy grains or icy mantles on grains in circumstellar environments (other then the solar one) is particularly difficult, due to the lack of reliable estimates of the energetic (keV-MeV) component of the winds.

An almost unique possibility is offered by stars undergoing a flaring activity because, conceivably, the mechanism of stellar flares is not too dissimilar from the solar ones, even if these are much less energetic. Hence for these stars it is, in principle, possible to deduce the required wind particle flux by observations of their lightcurves; from them one can evaluate the energetics of flares and by analogy with the well studied solar phenomena obtain the sought flux.

Among stars undergoing a flaring activity a special position is occupied by T Tauri stars. These are young pre-main sequence stars, surrounded by a circumstellar nebula of gas and dust (of roughly 1000 AU in radius), whose properties have been already described in detail already several years ago (Herbig, 1962). In 1981, Worden and co-workers deduced observationally (in the way we shortly described above) the flux of fast ($E > 10$ MeV) protons to try to explain the observed early solar system abundance anomalies. Strazzulla et al.

(1983) then considered the effect this flux would have had when impinging on circumstellar ice grains, discovered by Cohen (1975) only around the very young star of the same type HL Tau, a star still embedded in its prestellar cocoon. The fact that the 3.1 μm signature of ice was observed only in the youngest star of the group suggested that it was residual material left over from the contraction of the nebula from which the star formed (Cohen, 1975). Some mechanism should then cause the destruction of ice, if it is not observed in more evolved T Tauri stars. The problem still remains even if it is now known that HL Tau nebula is disk-like and seen edge on (Cohen, 1983). Strazzulla et al. (1983) investigated the possibility that the required destruction process was erosion of ice grains induced by the energetic component of the intense T Tauri wind. They calculated the lifetime of ice grains as a function of their distance from the central star. The interesting result was that sputtering was, as expected, dominant over sublimation in the outer portion of the nebula and that in a time comparable with the total duration of the T Tauri stage (≅ a few million years) all of the ice present in the nebula could be eroded. The amount of molecular hydrogen synthesized was also evaluated. A subsequent application to the solar nebula during the T Tauri phase,

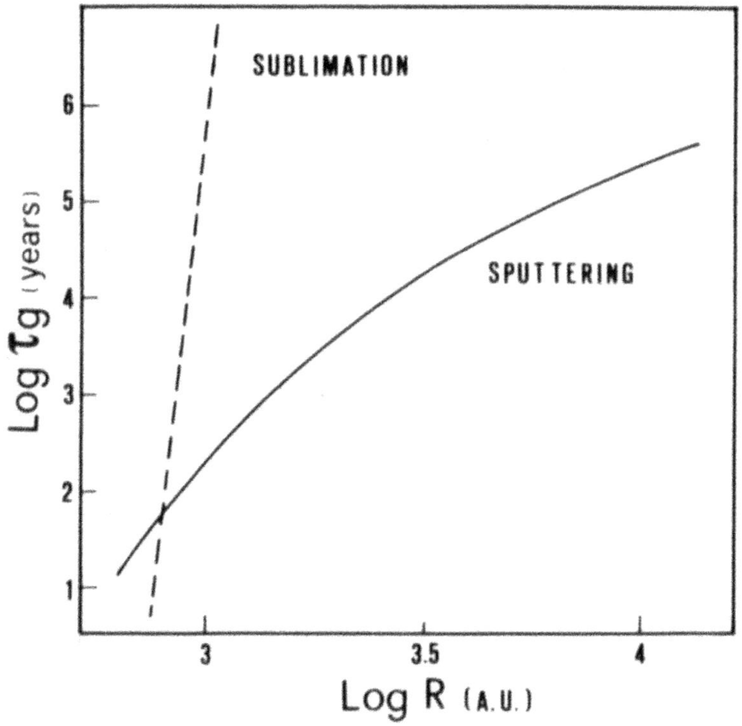

Figure 12. Stability of ice grains during the FU Ori stage

crossed by the sun was also made (Strazzulla, 1985). More recently Pirronello (1987) studied the stability of ice in a T Tauri nebula during an FU Orionis flare up, an eruptive stage probably lasting no more than fifty or hundred years, that (according to the nice statistical analysis performed by Herbig in 1977) can be recursively crossed by a star during its T Tauri phase with a periodicity between 3×10^3 and 10^4 years. The lifetime against sublimation and sputtering of ice grains, of the average size (≅ .3 μm in radius) observed by Cohen (1975), during the FU Ori stage is shown in Figure 12. It can be seen that in less than 100 years (the assumed time duration of the FU Ori stage) during such a phase all of the ice present is destroyed till the edges of the nebula.

A possible conclusion that can be reached, if Herbig statistical analysis is correct, is then that ice is still observed around HL Tau not only because it is seen edge on, but also because such a star is so young that has not yet undergone its first FU Ori type flare up. This result seems to be supported by observation of Di Martino et al. (1989).

5.3. INTERSTELLAR MEDIUM

The interaction of energetic particles with interstellar grains has been taken into account in the past in several occasions (see e.g. de Jong and Kamijo, 1973; Draine and Salpeter, 1979a, b), mainly focusing attention to the refractory component of them.

In the last few years some of the most important problems related with the presence in dense interstellar clouds of "dirty ice" mantles on grains and their irradiation by cosmic rays have been considered.

In 1985 Leger and co workers (leger et al., 1985) tackled the important problem of the restoration in the gas phase of molecules from grain mantles. Some processes are, in fact, required to counterbalance accretion that, with a very short timescale compared to the lifetime of clouds, would eliminate all species from the gas, in contrast with observations. They investigated the evaporation of the icy mantle induced by the energy loss of impinging heavy cosmic rays, either from the whole grain, if small, or from the hot spot formed at the interface ice vacuum by the cylindrical thermal spike for large grains. The other process (proposed by Greenberg in 1973 and proved experimentally by d'Hendecourt et al. in 1982), they took into account, was the chemical explosion of the dirty ice mantles, previously photolyzed to saturation with ultraviolet photons and hence rich of radicals. Such radicals can remain "frozen" at interstellar temperatures until they are set in motion and recombine by a triggering event. This event could possibly be also the local heating due to the energy loss of an heavy cosmic ray. The results of their calculation showed that these processes, induced by the heavy component of cosmic rays, could help considerably to restore species from the solid to the gas phase especially under some circumstances, like for instance in the case of frozen mantles rich in CO.

The other area in which the interaction of cosmic rays with

ices can be relevant in interstellar clouds is that of the chemical modification induced in grain mantles. Most of the experimental results, mentioned in the previous sections, could be used to account also for chemical reaction taking place in the "bulk" of the mantle, together with those occurring in the gas phase and on the surface of grains. However, probably the most important problem in astrochemistry is that relative to the formation of molecular hydrogen in interstellar clouds for at least three reasons:

1) H_2 is the most abundant molecule in the Universe;
2) H_2 is the molecule that, once ionized by UV photons in diffuse clouds or by cosmic rays in dense clouds, triggers the gas phase ion-molecule reaction chains that form most of the other species observed in space;
3) H_2 and some of its daughter molecules are important for the cooling of interstellar clouds, for their collapse and fragmentation, hence for the formation of stars and for the structure and the dynamics of galaxies.

The problem of the production of molecular hydrogen would not exist if it could form by radiative association of two hydrogen atoms in the gas phase. This process, however, does not produce any appreciable amount of H_2 because just formed molecules should release about 4.5 eV to become stable. Emission of a photon is a slow process in this case because radiative relaxation involves forbidden vibration rotation transitions. The presence of a third body is required to absorb the energy excess. Several authors and in particular Hollenbach and Salpeter (1971) have evaluated the role of catalysts played by interstellar grains covered with crystalline mantles. In their model H atoms can be adsorbed on the surface of such mantles and, if they are mobile enough even at 10 K (the temperature of grains in dense clouds), they can scan by tunneling all adsorption sites until two of them encounter and form a H_2 molecule. This molecule, after releasing to the grain the 4.5 eV, is restored to the gas phase by the subsequent local thermal transient. At the beginning of the eighties new observations and theoretical studies questioned the validity of a critical choice. Leger et al. (1979) and Hagen et al. (1981), comparing astronomical and laboratory infrared spectra of water ice, showed that interstellar dust mantles had to be amorphous in structure. Smoluchowski (1981, 1983) showed with a rigorous quantum mechanical calculation that mobility of adsorbed H atoms on an amorphous surface is orders of magnitude lower than on a crystalline one. Consequently also the formation of molecular hydrogen was lower than that estimated previously by several orders of magnitude. In this case other mechanisms may become competitive. We (Pirronello and Averna, 1988) calculated, on the basis of our laboratory results and by means of a Monte Carlo simulation, the production rate "R" (cm^{-3} s^{-1}) of H_2 induced in grain mantles by the bombardment of low energy cosmic rays, at several depths inside a small dense cloud of the type modeled by Boland and de Jong (1984). Results are shown in Figure 13, where the production rate of molecular hydrogen (dashed lines refer to our simulation, the full lines refer to our calculations we performed for comparison following the approach of Smoluchowski) is reported versus the temperature of the icy surface and versus the depth measured in magnitudes. The conclusion of the paper was that at the temperatures appropriate for the core of the dense clouds the synthesis of H_2 due to cosmic rays interacting with the icy mantles was dominant over the association of H adatoms if the approach of Smoluchowski to the mobility on amorphous surfaces is correct.

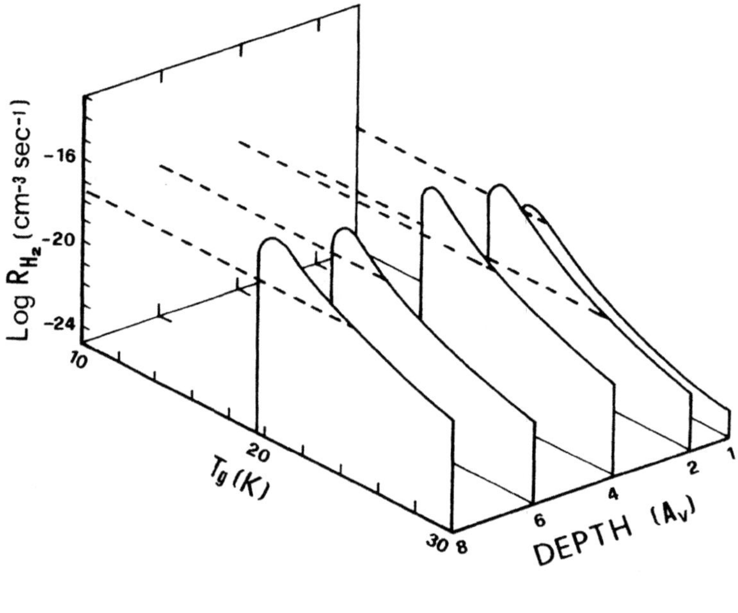

Figure 13. Production rates of H_2 inside a cold dense cloud by cosmic rays impinging on grain mantles and by association of H adatoms

In the described paper we used the Mathis model of grains (Mathis et al., 1977); new results (Averna et al., 1989) obtained considering the Greenberg model of grains (see e.g. Greenberg, 1978) that have an intermediate organic refractory mantle, still bearing hydrogen, seem to give still more relevance to the mechanism involving fast particles.

6. Conclusions

As I told in the introduction the aim I had was not to give an extensive review, but to let you have through these lectures a glance to a new and fascinating field in order to introduce it to those of you who did not know anything about, to help to clarify ideas to those who already knew something and may be to start a new interest in some of you.

I hope I have been at least partially successful.

7. References

Averna, D., Pirronello, V., Brown, W.L., Lanzerotti, L.J.: 1989, (in preparation)

Bar-Nun, A., Herman, A.G., Rappaport, M.L.: 1985, Surface Sci. **150**, 143

Behrisch, R. (ed): 1981, "Sputtering by Particle Bombardment I", Springer-Verlag, Berlin, Fed. Rep. of Germany

Behrisch, R. (ed): 1983, "Sputtering by Particle Bombardment II", Springer-Verlag, Berlin, Fed. Rep. of Germany

Benit, J., Bibring, J-P., Della Negra, S., Le Beyec, Y., Mendenhall, M., Rocard, F., Standing, K.: 1987, Nucl. Instr. Meth. **B19/20**, 838

Benit, J., Bibring, J-P., Della Negra, S., Le Beyec, Y., Rocard, F.: 1986, Radiation Effects **99**, 105

Benit, J., Bibring, J-P., Rocard, F.: 1988, Nucl. Instr. Meth. **B32**, 349

Berger, R.: 1961, Proc. Nat. Acad. Sci. USA **47**, 1434

Boland, W., de Jong, T.: 1984, Astron. Astrophys. **134**, 87

Bridge, H., Belcher, J., Lazarus, A., Olbert, S., Sullivan, J., Bagenal, F., Gazis, P., Hartle, R., Olgivie, K., Scudder, J., Sittler, E., Eviatar, A., Siscoe, G., Goertz, C., Vasyliunas, V.: 1981, Science **212**, 217

Brown, W.L., Augustyniak, W.M., Marcantonio, K.J., Simmons, E., Boring, J.W., Johnson, R.E., Reimann, C.T.: 1984, Nucl. Instr. Methods **B1**, 307

Brown, W.L., Augustyniak, W.M., Simmons, E., Marcantonio, K.J., Lanzerotti, L.J., Johnson, R.E., Reimann, C.T., Boring, J.W., Foti, G., Pirronello, V.: 1982a, Nucl. Instr. Methods **198**, 1

Brown, W.L., Lanzerotti, L.J., Johnson, R.E.: 1982b, Science **218**, 525

Brown, W.L., Johnson, R.E.: 1986, Nucl. Instr. Methods **B13**, 295

Brown, W.L., Lanzerotti, L.J., Poate, J.M., Augustyniak, W.M.: 1978, Phys. Rev. Lett. **40**, 1027

Calcagno, L., Foti, G., Strazzulla, G.: 1983, Lettere al Nuovo Cimento **37**, 303

Cheng, A.F.: 1984, J. Geophys. Res. **89**, 3939

Cheng, A.F., Haff, P.K., Johnson, R.E., Lanzerotti: 1986, *Interactions of Magnetospheres with Icy Satellite Surfaces*, in Satellites, Burns J.A. and Matthews M.S. eds., p. 403, Arizona Univ. Press, Tucson, U.S.A.

Cheng, A.F., Lanzerotti, L.J.: 1978, J. Geophys. Res. **83**, 2597

Cheng, A.F., Lanzerotti, L.J., Pirronello, V.: 1982a, J. Geophys. Res. **87**, 4567

Cheng, A.F., Lanzerotti, L.J., Pirronello, V.: 1982b, *On the Lifetime of E-ring Grains and their Nature*, in Sun and Planetary Systems, Fricke and Teleki eds., p. 251, Reidel Publ. Co., Dordrecht, The Netherlands

Chrisey, D.B., Johnson, R.E., Boring, J.W., Phipps, J.A.: 1988, Icarus **75**, 233

Chrisey, D.B., Johnson, R.E., Phipps, J.A., McGrath, M.A., Boring, J.W.: 1987, Icarus **70**, 111

Ciavola, G., Foti, G., Torrisi, L., Pirronello, V., Strazzulla, G.: 1982, Radiation Effects **65**, 167

Clark, R.N., Brown, R.H., Owensby, P.D., Fanale, F.D.: 1984, Icarus **59**, 265

Clark, R.N., Fanale, F.P., Zent, A.P.: 1983, Icarus **56**, 233

Cohen, M.: 1975, M.N.R.A.S. **173**, 279

Cohen, M.: 1983, Astrophys. J. Lett. **270**, L69

de Jong, T., Kamijo, F.: 1973, Astron, Astrophys. **25**, 363

d'Hendecourt, L.B., Allamandola, L.J., Baas, F., Greenberg, J.M.: 1982, Astron. Astrophys. **109**, L12

Di Martino, M., Mantegazza, L., Pirronello, V.: 1989, *On the Energetic Component of FU Ori Wind*, in Dusty Objects in the Universe, Bussoletti, E., Vittone, A., eds., Reidel Publ. Co., Dordrecht, The Netherlands

Donn, B.: 1976, *The Nucleus*, in The Study of Comets, Donn, B., Mumma, M., Jackson, W., A'Hearn, M., Harrington, R. eds., p. 611, NASA SP-393, Washington, U.S.A.

Draine, B.T., Salpeter, E.E.: 1979a, Astrophys. J. **231**, 77

Draine, B.T., Salpeter, E.E.: 1979b, Astrophys. J. **231**, 438

Evatt, R.: 1980, Master's Thesis, University of Virginia, Charlottesville

Eviatar, A., Bar-Nun, A. Morris, P.:, 1985, Icarus **61**, 185

Falcone, G.: 1989, La Rivista del Nuovo Cimento (in press)

Fano, U.: 1946, Phys. Rev. **70**, 44

Foti, G., Calcagno, L., Sheng, K., Strazzulla, G.: 1984, Nature **310**, 126

Frank, L.A., Burek, B.G., Ackerson, K.L., Wolfe, J.H., Mihalov, J.D.: 1980, J. Geophys. Res. **85**, 5695

Greenberg, J.M.: 1973, in "Molecules in the Galactic Environment", Gordon, M.A., Snyder, L.E., eds., p.93, John Wiley and sons

Greenberg, J.M.: 1978, *Interstellar Dust*, in Cosmic Dust, McDonnell, J.A.M. ed., p. 187, John Wiley and sons

Hagen, W., Tielens, A.G.G.M., Greenberg, J.M.: 1981, Chem. Phys. **56**, 367

Haff, P.K., Eviatar, A., Siscoe, G.: 1983, Icarus **56**, 426

Haff, P.K., Watson, C.C., Tombrello, T.A.: 1979, *Ion erosion of the Galilean Satellites of Jupiter*, in Proc. 10th Lunar Plan. Sci. Conf. p. 1685, Lunar Plan. Inst., Houston, U.S.A.

Haff, P.K. Watson, C.C., Yung, Y.L.: 1981, J. Geophys. Res. **86**, 6933

Haring, R.A., Haring, A., Klein, F.S., Kummel, A.C., De Vries, A.E.:

1983, Nucl. Instr. Meth. **211**, 529
Haring, R.A., Kolfschoten, A.W., De Vries, A.E.: 1984a, Nucl. Instr. Meth. **B2**, 544
Haring, R.A., Pedrys, R., Oostra, D.J., Haring, A., De Vries, A.E.:, 1984b, Nucl. Instr. Meth. **B5**, 476
Haring, R.A., Pedrys, R., Oostra, D.J., Haring, A., De Vries, A.E.: 1984c, Nucl. Instr. Meth. **B5**, 483
Herbig, C.H.: 1962, Adv. Astr. Ap. **1**, 47
Herbig, C.H.: 1977, Astrophys. J. **217**, 693
Hill, T.W.: 1984, Adv. Space Res. **4**, 149
Hollenbach, D.J., Salpeter, E.E.: 1971, Astrophys. J. **163**, 155
Johnson, R.E., Boring, J.W., Reimann, C.T., Barton, L.A., Sieveka, E.M., Garrett, J.W., Farmer, K.R., Brown, W.L., Lanzerotti, L.J.: 1983a, Geophys. Res. Lett. **10**, 892
Johnson, R.E., Brown, W.L.: 1983b, Nucl. Instr. Meth. **209/210**, 469
Johnson, R.E., Lanzerotti, L.J., Brown, W.L., Augustyniak, W.M., Mussil, C.: 1983c, Astron. Astrophys. **123**, 243
Johnson, R.E., Cooper, J.F., Lanzerotti, L.J., Strazzulla, G.: 1987, Astron. Astrophys. **187**, 889
Johnson, R.E., Evatt, R.: 1980, Radiation Effects **52**, 187
Johnson, R.E., Garrett, J.W., Boring, J.W., Barton, L.A., Brown, W.L.: 1984, J. Geophys. Res. Suppl. **89**, B711
Johnson, R.E., Lanzerotti, L.J., Brown, W.L.: 1982, Nucl. Instr. Meth. **198**, 147
Johnson, R.E., Lanzerotti, L.J., Brown, W.L., Armstrong, T.P.: 1981, Science **212**, 1027
Judge, D., Wu, F., Carlson, R.: 1980, Science **207**, 431
Krimigis, S.M., Armstrong, T.P., Axford, W.I., Bostrom, C.O., Gloeckler, G., Keath, E.P., Lanzerotti, L.J., Carbary, J.F., Hamilton, D.C., Roelof, E.C.: 1981, Science **212**, 225
Lanzerotti, L.J., Brown, W.L.: 1983a, J. Geophys. Res. **88**, 989
Lanzerotti, L.J., Maclennan, C.G., Brown, W.L., Johnson, R.E., Barton, L.A., Reimann, C.T., Garrett, J.W., Boring, J.W.: 1983b, J. Geophys. Res. **88**, 8765
Lanzerotti, L.J., Brown, W.L., Augustyniak, W.M., Johnson, R.E., Armstrong, T.P.: 1982, Astrophys. J. **259**, 920
Lanzerotti, L.J., Brown, W.L., Johnson, R.E.: 1985, *Laboratory Studies of Ion Irradiation of Water, Sulfur Dioxide and Methane Ices*, in Ices in the Solar System, Klinger, J., Benest, D., Dollfus, A., Smoluchowski, R., eds., p. 317, Reidel Publ. Co. Dordrecht, The Netherlands
Lanzerotti, L.J., Brown, W.L., Marcantonio, K.J.: 1987a, Astrophys. J. **313**, 910 910
Lanzerotti, L.J., Brown, W.L., Maclennan, C.G., Cheng, A.F.,Krimigis, S.M., Johnson, R.E.: 1987b, J. Geophys. Res. **92**, 949
Lanzerotti, L.J., Brown, W.L., Marcantonio, K.J., Johnson, R.E.: 1984, Nature **312**, 139
Lanzerotti, L.J., Brown, W.L., Poate, J.M., Augustyniak, W.M.: 1978, Geophys. Res. Lett. **5**, 155
Leger, A., Jura, M., Omont, A.: 1985, Astron. Astrophys. **144**, 147
Leger, A., Klein, J., de Cheveigne, S., Guinet, C., Defourneau, D.,

Belin, M.: 1979, Astron. Astrophys. **79**, 256
Mathis, J.S., Rumpl, W., Nordisieck, K.H.: 1977, Astrophys. J. **217**, 425
Matson, D.L., Johnson, T.V., Fanale, F.P.: 1974, Astrophys. J. **192**, L43
Matson, D.L., Nash, D.B.: 1983, J. Geophys. Res. **88**, 4771
Melcher, C.L., Le Poire, D.J., Cooper, B.H., Tombrello, T.A.: 1982, Geophys. Res. Lett. **9**, 1151
Miller, J.H., Green, A.E.S.: 1973, Radiation Res. **54**, 343
Moore, M.H.: 1984, Icarus **59**, 114
Moore, M.H., Donn, B.: 1982, Astrophys. J. **257**, L47
Moore, M.H., Donn, B., Khanna, R., A'Hearn, M.F.: 1983, Icarus **54**, 388
Morfill, G.E., Voelk, H.J., Lee, M.A.: 1976, J. Geophys. Res. **81**, 5841
Nash, D.B., Fanale, F.P.: 1977, Icarus **31**, 40
Oro, J.: 1963, Nature **197**, 971
O'Shaughnessy, D.J., Boring, J.W., Phipps, J.A., Johnson, R.E.: 1986, Nucl. Instr. Meth. in Phys. Res. **B13**, 304
Pirronello, V.: 1985, *Molecule Formation in Cometary Environment*, in Ices in the Solar System, Klinger, J., Benest, D., Dollfus, A., Smoluchowski, R., eds., p. 261, Reidel Publ. Co., Dordrecht, The Netherlands
Pirronello, V.: 1987, Nucl. Instr. Meth. **B19/20**, 959
Pirronello, V., Averna, D.: 1988a, Astron. Astrophys. **196**, 201
Pirronello, V., Brown, W.L., Lanzerotti, L.J., Lanzafame, G., Averna, D.: 1988b, *Formaldehyde Formation in $H_2O/^{13}CD_4$ Frozen Mixtures by Swift Ions*, in Dust in the Universe, Bailey, M.E., Williams, D.A., eds., p. 281, Cambridge Univ. Press, Cambridge, United Kingdom
Pirronello, V., Brown, W.L., Lanzerotti, L.J., Maclennan, C.G.: 1988c, *HD and CO Release during MeV Ion Irradiation of H_2O/CD_4 Frozen Mixtures at 9 K*, in Experiments on Cosmic Dust Analogues, Bussoletti, E. et al. eds., p. 287, Kluwer Acad. Publ., Dordrecht, The Netherlands
Pirronello, V., Brown, W.L., Lanzerotti, L.J., Marcantonio, K.J., Simmons, E.: 1982a, Astrophys. J. **262**, 636
Pirronello, V., Strazzulla, G., Foti, G.: 1982b, Astron. Astrophys. **118**, 341
Pirronello, V., Lanzafame, G.: 1989, Astrophys. J. **342**, 527
Pirronello, V., Strazzulla, G., Foti, G., Brown, W.L., Lanzerotti, L.J.: 1984, Astron. Astrophys. **134**, 204
Pirronello, V., Strazzulla, G., Foti, G., Rimini, E.: 1981, Astron. Astrophys. **96**, 267
Rice, S.A.J.: 1975, Top. Curr. Chem. **60**, 109
Rocard, F., Bibring, J-P.: 1982, Phys. Rev. Lett. **48**, 1763
Roessler, K.: 1986, Radiation Effects **99**, 21
Ryan, H.P., Draganic, I.G.: 1986, Astr. Space Sci. **125**, 49
Schulz, M., Lanzerotti, L.J.: 1974, "Particle Diffusion in the Radiation Belts", Springer-Verlag, New York
Shul'man, L.M.: 1972, in The Motion, Evolution of Orbits and Origin of Comets, Chebatarev, G.A., Kazimirchak Polonskaya, E.I. and Marsden, B.G. eds., p. 265, Reidel Publ. Co., Dordrecht, The Netherlands
Sieveka, E.M., Johnson, R.E.: 1982, Icarus **51**, 528

Sieveka, E.M., Johnson, R.E.: 1984, Astrophys. J. **287**, 418
Sieveka, E.M., Johnson, R.E.: 1985, J. Geophys.Res. **90**, 5327
Sigmund, P.: 1969, Phys. Rev. **184**, 383
Smoluchowski, R.: 1981, Astr. Space Sci. **75**, 353
Smoluchowski, R.: 1983, J. Phys. Chem. **87**, 4229
Strazzulla, G.: 1985, Icarus **61**,48
Strazzulla, G., Pirronello, V., Foti, G.: 1982, Astron. Astrophys. **123**, 93
Strazzulla, G., Pirronello, V., Foti, G.: 1983, Astrophys. J. **271**, 255
Strazzulla, G., Torrisi, L. Coffa, S., Foti, G.: 1987, Icarus **70**, 379
Terrile, R., Tokunaga, A.: 1980, Bull. Am. Astron. Soc. **12**, 701
Thompson, M.W.: 1965, Phil. Mag. **18**, 377
Tombrello, T.A.: 1987, Nucl. Instr. Meth. **B24/25**, 517
Vineyard, G.H.: 1976, Radiation Effects **29**, 245
Watson, C.C.: 1982, *The Sputter Generation of Planetary Coronae: Galilean Satellites of Jupiter,* in Proc. 12th Lunar Plan. Sci. Conf., p. 1569, Lunar Plan. Inst., Houston, U.S.A.
Weiser, H., Vitz, R., Moos, H.: 1977, Science **197**, 755
Whipple, F.L.: 1950, Astrophys. J. **111**, 375
Worden, S.P., Schneeberger, T.J., Khun, J.R., Africano, J.L.: 1981, Astrophys. J. **244**, 520
Ziegler, J.F.: 1980, "Stopping Cross Sections for Energetic Ions in All Elements", Pergamon Press, New York

Chemical Analyses with the Infrared Space Observatory (ISO)

Martin Harwit
Laboratory for Astrophysics
National Air and Space Museum
Smithsonian Institution
Washington, D.C. 20560, USA

ABSTRACT. The Infrared Space Observatory (ISO) is a European Space Agency (ESA) mission to be launched in the early 1990's. It will be able to carry out astronomical observations at wavelengths ranging from 3 µm to 200 µm. A complement of instruments permitting imaging photometry, spectrophotrometry, spectrophotometric polarimetry as well as spectrometry both at high and at low resolving power, should provide an observer a wide variety of options for the study of chemical species and chemical processes. Proposals to carry out observations with this instrument will be solicited from the astronomical community in ESA member states, and from U.S.astronomers through the National Aeronautics and Space Administration (NASA). No decision has yet been made on wider solicitation.

1. Introduction

The purpose of the Infrared Space Observatory (ISO) will be to observe astronomical sources at wavelengths ranging from 3 to 200 µm in the infrared. Planned for operating in the mid-1990's, ISO is expected to have a useful lifetime of 18 months limited primarily by the 2300 liters of superfluid helium which the spacecraft will carry aloft. While a portion of the observing time is guaranteed to the teams of scientists who have been involved in the plans for and construction of the scientific instruments, approximately 2/3 of the observing program will be determined by proposals from the general community.

J. M. Greenberg and V. Pirronello (eds.), Chemistry in Space, 305–319.

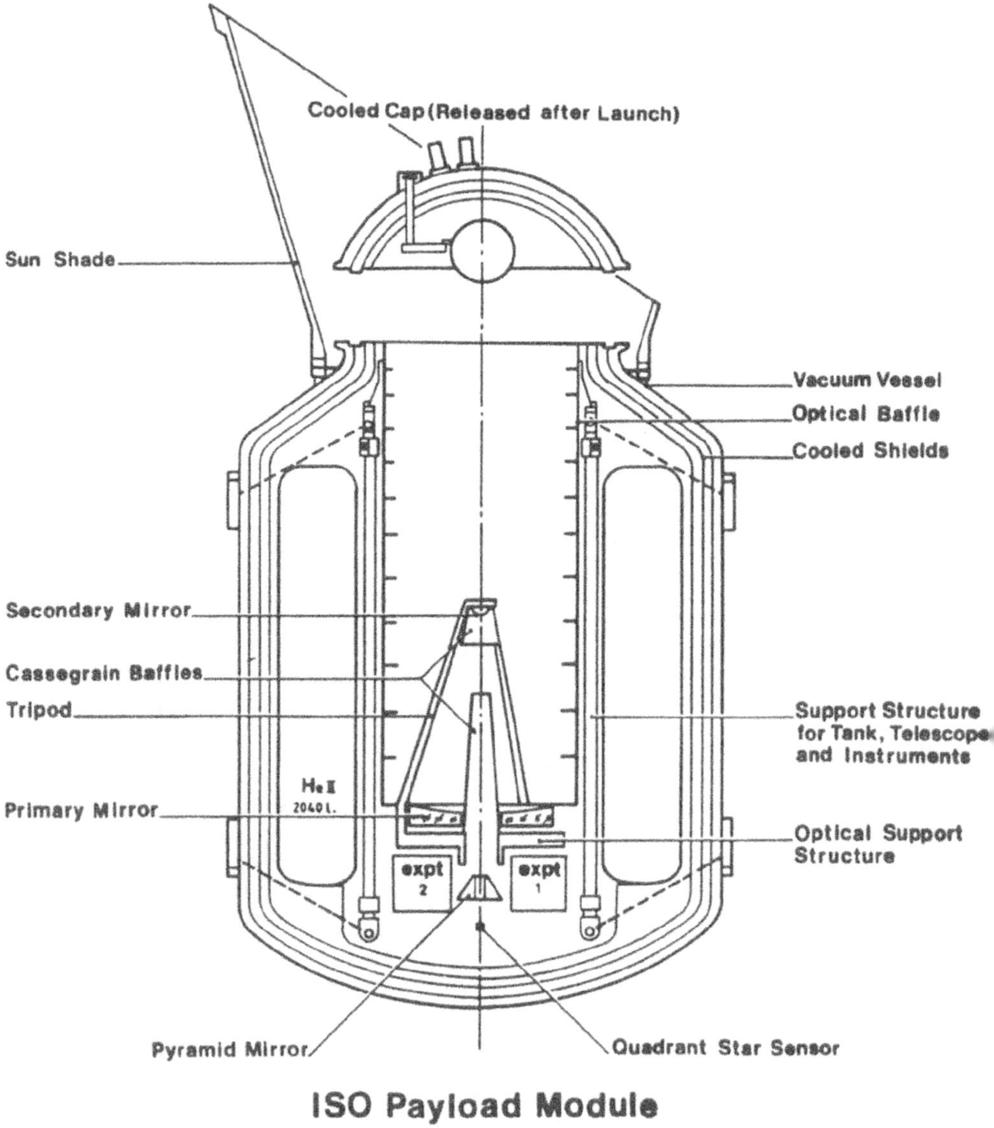

Figure 1. ISO Payload Module

ISO is a natural successor to the Infrared Astronomical Satellite (IRAS). It is designed to have a longer lifetime, cover a wider spectral range of the infrared at greater sensitivity, and provide both higher angular and spectral resolution, as well as a first opportunity to carry out polarimetric observations from space. Its 60 cm diameter primary is expected to provide a diffraction-limited image at 5 µm wavelengths. It will operate in tandem with a Cassegrain-configured secondary mirror, which will transmit the radiation to a pyramidally shaped mirror, each face of which will reflect radiation onto one of the four instruments on board. This Ritchey-Chretien system will have a focal ratio f/15 and will provide each instrument with a 3 arc minute unvignetted field of view on the sky. Figure 1 shows this optical configuration and schematically indicates the locations of two of these four instruments. The spacecraft housing this telescope will have a mass of 2400 kg, a height of 5.3 and a width of 2.3 m. Carried aloft by an Ariane 44 P rocket launched from Kourou in French Guiana, it will enter a highly eccentric 24-hour orbit with apogee 70,000 km and perigee 1000 km at low inclination to the Earth's equatorial plane. The entire spacecraft will be pointed with a jitter of less than 2.7 arc seconds, an absolute pointing accuracy of ~ 12 arc seconds and an after the fact pointing accuracy determinable with a precision of 7 arc seconds, by means of star trackers, sun sensors, gyroscopes and reaction wheels.

The purpose of ISO is to study all possible types of astronomical sources, including planetary system objects, stars, interstellar gas and dust, and strong extragalactic sources. The broad range of analytical instruments comprising this payload will permit a wide variety of chemical investigations as part of the overall scientific program.

2. Scientific Instruments

The complex of four scientific instruments comprises a camera (CAM) a photometric system (PHT), a short wavelength spectrometer (SWS) and a long wavelength spectrometer (LWS). Normally only one of these will be operational at a time, though the camera at times will be operated in parallel with other instruments.

a) Camera: CAM comprises two detector arrays each of which has 1024 elements arrayed in a 32 x 32 matrix. The pixel field of view can be varied in steps to provide 1.5, 3, 6 and 12 arc second fields per element. The 2-5 μm range is serviced by an InSb array, while longer wavelength coverage between 5 and 17μm is provided by Si:Ga elements. Before impinging on one or the other of these arrays, which can be only used one at a time, the radiation can be made to pass through a variety of narrow or wide-band filters as well as through linear polarization analyzers. While overlapping a wavelength range that partially is observable with ground-based instrumentation, CAM will exhibit higher sensitivity than large ground-based telescopes, especially for low-surface-brightness extended objects, and also higher sensitivity, by approximately two orders of magnitude, than IRAS attained.

b) Photometer: The photometric system (PHT) is the most complex instrumentation package aboard. It comprises a set of photometric detectors, covering the range from 3 to 200μm, like the camera operating with a wide variety of broad and narrow filters; an Ebert-Fastie spectrometer with spectral resolving power R = 100 over the 3-16 μm range; a far infrared camera operating between 30 and 200 μm; and a mapping array covering wavelengths between 18 and 28 μm. This instrument also will have the ability to analyse for linear polarization.

Sensitivity in the photometric mode will be a couple of orders of magnitude higher than for IRAS, and the wavelength range covered will extend both to shorter and longer wavelengths than covered by IRAS. At the short wavelength extreme, the sensitivity also will exceed that of ground-based instruments of bigger aperture. Detectors of six different types will be utilized to cover the wide range of wavelengths.

c) Short Wavelength Spectrometer: SWS will cover the wavelength range from 2.5 to 45 μm with detectors of four different types. Two different modes of operation will be available, a grating mode with spectral resolving power $R=10^3$ and a grating/Fabry-Perot mode providing a spectral resolving power $R = 2 \times 10^4$. The spatial field of view covered will be 14" x 20" and 20" x 30", respectively at wavelengths below and above 28 μm. Tentative estimates suggest that, in one hour's observing time, grating measurements will permit observation of 0.1 to 1 Jy regions, respectively at 4 and 40μm. Similarly, an hour's observation in the high resolution mode should permit detection of sources at roughly 30 Jy continuum levels.

d) <u>Long Wavelength Spectrometer</u>: LWS will cover the range 45 to 180 μm with a field view of 1.65 arc minutes and a resolving power that will be R = 200 in the grating mode and R = 10^4 in the grating/Fabry-Perot mode. Three different types of detectors will be utilized to provide coverage of the entire spectral range. In a period of an hour, lines as faint as 10^{-20} W cm^{-2} or continua as weak as 10 Jy are currently projected to be detectable in the high resolution mode. At low resolution, the corresponding figures are expected to be 3 x 10^{-22} W cm^{-2} and 0.3 Jy.

3. Comparison to Observations from Other Existing Platforms

Spectral observations to determine chemical composition of astronomical sources, as revealed through their infrared spectra, are not new. Ground-based observations at wavelengths where the atmosphere is transparent have been available for many years, and airborne spectral measurements were initiated in the mid-1970s. However, several species such as H_2O vapor and O_2, are all but unobservable through the atmosphere, even at altitudes of 12km reached by airplanes, and many other species whose emission lines or bands fall into atmospheric O_2, H_2O or CO_2 bands, also become unobservable. A primary task of ISO will be to make those species available to direct spectroscopic observation. On IRAS, low resolution spectra could be obtained (cf IRAS Science Team, 1986) at wavelengths between 7.5 and 22 μm, with lines as weak as ~3 x 10^{-16} W cm^{-2}, accessible at a resolving power which in practice appeared to be order R~20. Much of this spectral range will be covered by the Ebert-Fastie spectrometer in the photometric instrument,at higher spectral resolution and sensitivity; and, at even higher spectral resolution, the SWS will cover a wider spectral range though with a smaller field of view.

ISO will, therefore, have many capabilities not available to astrochemistry heretofore; but they will have to be used judiciously to use the time in orbit to best purposes, leaving to ground-based and airborne instruments those tasks that do not require an expensive orbiting instrument.

4. Chemical Analyses

Infrared observations can provide information on the chemical makeup of interstellar or interplanetary grains and gases, of the atmospheres of stars and planets, and of the surfaces of smaller solar system bodies, if the spectra obtained can be uniquely associated with gaseous or solid matter and quantitatively interpreted. Normally, that interpretation requires the availability of atomic, molecular or minerological data that can be only obtained in the laboratory or through detailed theoretical modelling and computation.

Spectra very roughly can be divided into continuum and discrete -- line or band -- spectra. Among the continuum spectra we recognize:

i) sources of synchrotron radiation, such as pulsars or possibly nuclei of active galaxies;
ii) sources of bremsstrahlung or free-free emission, such as X-ray stars or high-temperature, dust-free main-sequence stellar atmospheres;
iii) free-bound transitions due to H^- ions and other species, primarily in the near infrared spectra of stellar photospheres; and
iv) grain emission and extinction.

Among these, the grains are most amenable to chemical analysis. Grains have been discussed in other parts of this volume, in particular in the chapters by Greenberg and by Leger. Their spectral features generally are broader than atomic or molecular line emission and susceptible to anlayses through comparison of low resolution laboratory and astronomical spectra.

Discrete spectra can be produced by a variety of atomic, molecular and ionic transitions, both in absorption and in emission.

i) Electronic transitions can be seen in atoms as well as in ions. Most of these involve the absorption or emission in visible, ultraviolet or X-radiation. However, the transitions between high-lying levels close to the ionization continuum can be closely spaced, and give rise to infrared and even radio emission. The Rydberg transitions in hydrogen are particular examples. Predictions on the strengths of hydrogenic Rydberg transitions in the infrared have been derived by Petrosian (1970). He calculated that the grain continuum from HII region complexes would be sufficiently high at wavelengths around 20 μm and onward to the grain emission peak which generally is expected at ~100 μm, that spectral resolving powers in excess of $R \sim 10^5$ would be needed for detecting these lines. With ISO such lines would at best be detected with SWS and at sufficiently high signal to noise ratios to make these lines clearly apparent against a continuum background.

ii) Vibrational transitions in molecules typically appear in the near infrared, since binding energies of atoms in a molecule tend to be of the order of a few electron volts. Since several vibrational states normally lie between the vibrational ground state and the dissociation continuum, separations between levels tend to be of the order of a fraction of an electron Volt, corresponding to transition wavelengths of a few microns. Accordingly, vibrational transitions between adjacent vibrationally excited states of H_2 lie around 2 μm (cf Beckwith et al, 1983), and a water ice feature prominently absorbs near 3.0 μm in many interstellar clouds. This band apparently is also affected by NH_3 admixtures (cf van de Bult, Greenberg and Whittet, 1985); and CO vibrational transitions occur prominently around 4.7 μm.

For large molecules, attached atomic groups may exhibit similar vibrational transitions, potentially providing important indicators for partial recognition of the parent to which they are attached. CH stretching modes at 3.4 μm are typical recognition features in that sense.At high resolution, around $R \sim 10^4$ different isotopes of CO, such as $^{13}C^{16}O$, $^{12}C^{18}O$, and $^{12}C^{17}O$ can be discerned in the spectra of carbon stars (cf Rank et al, 1974).

Vibrational transitions generally are accompanied by changes in rotational quantum number, but pure rotational transitions also are possible, and normally can be excited at lower energies than vibrational states.

iii) Rotational transitions for diatomic atoms exhibits energies, E, which follow a particularly simple relationship as a function of rotational quantum number, J, moment of inertia, I, and Planck constant, h.

$$E=h^2J(J+1)/8\pi^2I$$

The higher the moment of inertia, the lower the energy of the low-lying energy states and the closer their separation. For CO, the transition J = 1 to 0 lies at a wavelength of 2.6 μm and the energy is so low that collisional excitations populate the J=1 level even in the coldest molecular clouds. At temperatures of 1000 K, transitions from levels as high as

J>30 have been observed at far infrared wavelengths. On the other hand, pure rotational states of H_2, which has a low moment of inertia, are observed in the mid- and near-infrared.

iv) Fine-structure transitions are exhibited by many atoms and ions in their ground states and are excited over a wide range of energies, permitting identification of density, temperature, chemical composition, and sometimes the clumpiness of gas clouds, over a wide range of temperatures. This is particularly true of species such as doubly-ionized oxygen whose ground state is split into three fine-structure levels. Transitions between the lowest two occur at 88.4 µm, while those between the two upper levels have a wavelength of 51.8 µm. Since the Einstein A coefficients and the collisional excitation cross sections for these states differ appreciably, the ratio of the states provides a measure of ambient density. That ratio is reasonably insensitive to temperature, so that an almost unambiguous density measure can be derived from the line ratio in most HII regions, particularly since the density affects the ratio most pronouncedly for $n_e \sim 10^2$ to 10^4 cm^{-3} which also happens to be the range most characteristic of HII domains. Since doubly ionized oxygen also can be excited to high-lying electronic states with successive emission of radiation at 4363 Å in a $^1S_0$ to $^1D_2$ transition, and a 5007 Å line from the $^1D_2$ to the ground $^3P_2$ state, the ratio of 4363 Å to 5007 Å line strength can serve as a temperature indicator for HII gases. Between these visible and the far-infrared lines a relatively reliable set of temperatures and densities can be established in that way (Dinnerstein, 1982).

The density-dependence derived from the fine-structure line ratios is linearly proportional to the electron density, n_e, while radio determinations based on free-free emission measurements provide an emission measure proportional to the integrated line-of-sight value of n_e^2. Comparison of radio and far-infrared data, therefore, permit an estimate of clumpiness $\langle n_e^2\rangle/\langle n_e\rangle^2$ (Melnick et al, 1979).

Recently, a number of fine-structure transitions of highly ionized atomic species that lie in the near infrared have been studied by Greenhouse et al, 1989. These are observed in novae and open the possibilities of providing insight into the temperatures and pressures found in a variety of coronal line regions exhibiting temperatures in the million degree range. This is a welcome possibility since, in contrast to the more traditional visible and ultraviolet line observations of such regions, infrared line measurements promise the possibility of observations on distant, highly dust-obscured domains at these high temperatures.

Table 1 shows a number of such spectral lines and indicates those lines already observed in astronomical sources. Between these lines one includes every known kind of astrophysical gaseous region, ranging from the coldest molecular clouds in which neutral carbon fine-structure lines are excited, to atomic oxygen and C^+ predominant in gases at ~10^2K, to O^{++} and O^{+++} lines, respectively seen in HII regions at 10^4K and planetary nebulae at higher temperatures, and finally Si^{8+} seen in novae and the million degree range. Table 2 shows many of the lines already observed in different astronomical sources. Between these a wide range of temperature regions are covered.

5. Discussion

With the ISO satellite a variety of applications of the above-named spectroscopic processes can be observed.

i) Grain chemistry: In analyzing the chemical structure of grains it is virtually essential to proceed phenomenologically. Whether one is interested in cometary dust, interstellar dust that potentially might be represented by refractory material alone or possibly by a refractory core embedded in an icy mantle, or alternatively if one considers the existence of polyaromatic hydrocarbons (PAHs) of different types, one needs to proceed by comparison to laboratory materials. In the laboratory one can synthesize varieties of simulated grain materials, respectively at high or at low temperatures, and to check their chemical and spectral characteristics. If identical spectra can then be obtained in astrophysical and laboratory measurements, a reasonably secure identification can be secured. Such an identificaiton is always facilitated by broad spectral coverage, and ISO will help us to obtain spectra in regions inaccesible through the atmosphere.

A great deal of effort will have to go into systematic studies of protostellar as well as highly evolved stellar sources, and interstellar clouds at different temperatures and densities in order to draw correlations between the relative strengths of the 3.3, 6.2, 7.7, 8.6 and 11.3 μm

Table 1. Some Isoelectronic Sequences Including Highly Ionized Species.**

$2s^22p$: $^2P_{3/2}-^2P_{1/2}$

Ion	λ
Si^{9+}	1.43 um
Al^{8+}	2.044
Mg^{7+}	3.028
Na^{6+}	4.65
Ne^{5+}	7.64
O^{3+}	25.87
N^{2+}	57.34
C^{+}	157.74
B	656.6

$2s^22p^2$: $^3P_1-^3P_0$, $^3P_2-^3P_1$

Ion	$^3P_1-^3P_0$	$^3P_2-^3P_1$
Si^{8+}	3.92 um,	2.58 um
Al^{7+}	5.85,	3.72
Mg^{6+}	9/03,	5.50
Na^{5+}	14.3,	8.61
Ne^{4+}	24.28,	14.32
F^{3+}	44.07,	25.83
O^{2+}	88.36	51.81
N^{+}	205.5,	121.9
C	609.1	370.4

$2s^22p_4$: $^3P_0-^3P_1$, $^3P_1-^3P_2$

Ion	$^3P_0-^3P_1$	$^3P_1-^3P_2$
Si^{6+}	6.5,	2.47 um
Al^{5+}		3.66
Mg^{4+}	13.5,	5.60
Ne^{2+}	36.0,	15.56
O	145.5	63.18

$2s^22p^5$: $^2P_{1/2}-^2P_{3/2}$

Ion	λ
P^{6+}	1.37 um
Si^{5+}	1.96,
Al^{4+}	2.88
Mg^{3+}	4.49
Na^{2+}	7.32
Ne^{+}	12.81
F	25.83

$3s^23p^1$: $^2P_{3/2}-^2P_{1/2}$

Ion	λ
Ca^{7+}	2.32
K^{6+}	3.19
Ar^{5+}	4.53
Cl^{4+}	6.70
S^{3+}	10.51
P^{2+}	17.9
Si^{+}	34.81
Al	89.2

$3s^23p^2$: $^3P_1-^3P_0$, $^3P_2-^3P_1$

Ion	$^3P_1-^3P_0$	$^3P_2-^3P_1$
Ar^{4+}	13.1,	7.9
Cl^{3+}	20.35,	11.74
S^{2+}	33.47,	18.71
Si	129.7,	68.5

$3s^23p^4$: $^3P_0-^3P_1$, $^3P_1-^3P_2$

Ion	$^3P_0-^3P_1$	$^3P_1-^3P_2$
Ar^{++}	21.8,	8.99
Cl^{+}	33.3,	14.4
S	56.3,	25.2

$3s^23p^5$: $^2P_{1/2}-^2P_{3/2}$

Ion	λ
Ar^{+}	6.99
Cl	11.3

(Underlined transitions have been astronomically observed by now)

**After Greenhouse et al, 1989 and Schmid-Burgk, 1982, with wavelengths from Kaufman and Sugar, 1976.

Table 2. Some Astronomically Detected Fine-Structure Transitions and Associated Ionization Energies**

Species	Wavelength		Ionization Energy Into and Out of Level	
Si^{8+}	3.92 um		401.4	476.0 eV
Al^{8+}	2.044		284.6	330.1
Si^{5+}	1.96		166.8	205.1
Ca^{7+}	2.32		127.7	143.3
Mg^{4+}	5.60		109.3	141.3
Ne^{4+}	24.28		97.1	126.2
Mg^{3+}	4.49		80.1	109.3
O^{3+}	25.87		54.9	77.4
Cl^{3+}	11.74		39.6	53.5
O^{2+}	51.81,	88.36	35.1	54.9
S^{3+}	10.51		34.8	47.3
N^{2+}	57.3		29.6	47.4
Ar^{2+}	8.99		27.6	40.7
S^{2+}	18.71,	33.47	23.3	34.8
Ne^{+}	12.81		21.6	41.0
Ar^{+}	6.99		15.8	27.6
C^{+}	157.74		11.3	24.4
Ni^{+}	6.62		7.6	18.2
O	63.18,	145.5	0	13.6
C	370.4,	609.1	0	11.3

**After Dinnerstein, 1982 and Greenhouse et al, 1989 with wavelengths from Kaufman and Sugar, 1986.

features currently associated with PAHs, and other astrophysical characteristics of the sources observed. Similar efforts will be needed to investigate what we think is ice at 3.1 or 45 µm; or silicates at 10 um, in order to study the conditions under which those spectral features appear. That should help clarify the nature and origins of those respective types of grains.

ii) Solar system Chemistry: In the atmosphere of planets one hopes to use ISO to determine D/H and H/He abundances, the existence of trace gases, and isotopic ratios of other species besides hydrogen. In comets, the parent molecules giving rise to dissociation products may be determined spectroscopically. On Mars, seasonal changes in atmospheric gas constituents might be observed. And many other types of promising chemical analyses can be contemplated (Encrenaz, 1988).

iii) Circumstellar and Interstellar Gases: We have already stated how atomic, ionic and molecular gases can be used to trace local densities and temperatures. Given these data, the abundance of rare atomic and molecular species that can be identified through their spectral lines can be determined. Such trace substances may represent intermediate stages of complex chemical reaction chains, and may therefore serve to tell us just how more complex molecules are formed in circumstellar or interstellar space. Such processes are discussed by A. Dalgarno in this volume of lectures.

iv) Novae and Coronal Gases: Extremely ionized gases can be observed through fine-structure transitions, as demonstrated by Greenhouse et al, 1989. Such gases may be observed also in extragalactic X-ray sources representing giant E galaxies embedded in haloes of million degree plasma. The ionic constituent of those haloes should be determinable in that fashion, provided the lines are sufficiently bright to be observed. Attempts might be made to search for them.

v) External Galaxies: We do not know how the abundance of interstellar chemical species changes in response to stellar evolution in different parts of galaxies. The ratio of N^{++} to O^{++} and other species should be readily traced in nearby extragalactic sources, as a function of galactocentric distance. Such studies should provide insight into the ways stellar evolution proceeds in different parts of a galaxy.

Other extragalactic studies might investigate the nature of ultraluminous far-infrared emitting sources. Spectral data should indicate whether these highly obscured sources, whose far-infrared luminosity can exceed the optical luminosity by one or two orders of magnitude, consist of luminous, dust shrouded, massive stars and HII regions, or else represent processes of a quite different nature (cf Harwit et al, 1987).

These are just samples of interesting studies that can be undertaken. In additon the chemical complexions of such interesting regions as protostellar domains, envelopes of planetary nebulae, supernova ejecta, potential protogalaxies and other key evolutionary stages of astrophysical sources can be and, if at all possible, will be studied by ISO.

It should be an exciting mission.

6. Other Space Observatories

The Infrared Space Observatory (ISO) is by no means the only observatory capable of providing chemical information, nor is it likely to be the last of the powerful observatories planned to perform cosmochemical analyses, over the coming decades.

We may expect partial chemical information on diffuse, widely distributed galactic regions from spectral data on the background radiation to be obtained with the Cosmic Background Explorer (COBE) to be launched late in 1989. Towards the late 1990s the Space Infrared Telescope Facility (SIRTF) will be launched with detectors which by that time will be considerably more sensitive than those on ISO, with arrays consisting of ten to a hundred times more pixels. The mapping of regions in far greater spatial detail with respect to their chemical properties should become possible.

About the time that ISO is launched, the Submillimeter Wavelength Satellite (SWAS) will be orbited as well. It will concentrate in the submillimeter band around half a millimeter, on water vapor (H_2O), molecular oxygen (O_2), neutral carbon (C) and ^{13}CO (J=5 to 4) emission, analyzing Galactic emission lines, with a velocity resolution of approximately 1 km sec^{-1}. The spatial resolution will be ~5 arc minutes, providing maps of regions that are comparable to panoramic CO maps of the Galaxy at 2.6 μm wavelengths.

In the more distant future, the Large Deployable Reflector (LDR), and the Far Infrared Space Telescope (FIRST), respectively planned by NASA and ESA, may fly either separately or as an amalagamated mission early in the 21st century. They will be dedicated in part to heterodyne spectroscopy capable of very powerful chemical analyses, and the large size of their collecting surfaces will also provide high spatial resolution of regions.

No doubt, other missions not mentioned here will also make contributions and our knowledge of chemical processes will grow.

7. Acknowledgements

The author's work on ISO is supported by NASA grant NAGW-1261. He wishes to thank the Ettore Majorana Centre for Scientific Culture for its hospitality at this meeting, as well as NATO, for travel support to attend this International School of Space Chemistry. Thanks also are due to Dr. Matthew Greenhouse for permission to cite unpublished work and for instructive discussions.

8. References

Beckwith, S., Evans, N.J.II, Gatley, I., Gull, G. and Russell, R.W., 1983 Ap.J. 264. 152.

Dinnerstein, H., 1982, NASA Technical Memorandum TM 84279.

Encrenaz, Therese, 1988, in Comets to Cosmology, A.Lawrence, ed., Lecture Notes in Physics, Springer Verlag, 297.

Greenhouse, M.A. Grasdalen, G.L., Woodward, C.E., Benson, J., Gehrz, R.D., Rosenthal, E. and Skrutskie, M.F., 1989, Submitted to Ap.J.

Harwit, M., Houck, J.R., Soifer, B.T. and Palumbo, G.G.C., 1987, Ap.J. 315, 28.

IRAS Science Team, 1986, A & A Supplement Series 65, 607 - 1065.

Kaufman, V. and Sugar, J., 1986, J. Phys. Chem. Ref. Data 15, 321.

Melnick, G., Gull, G.E. and Harwit, M., 1979, Ap.J. Lett. 227, L35.

Petrosian, V. 1970, Ap.J 159, 833.

Rank, D.M., Geballe, T.R. and Wollman, E.R., 1974, Ap.J Letters 187, L 111.

Schmid-Burgk, J. 1982 in Landolt-Bornstein Numerical Data and Functional Relationships in Science and Technology. New series. Ed. K.-H. Hellwege. Group VI: Astronomy, Astrophysics and Space Research, Vol 2, subvolume c. section 7.5, p.126.

van de Bult, C.E.P.M., Greenberg, J.M. and Whittet, D.C.B., 1985, M.N.R.A.S. 214, 289.

CHEMISTRY OF THE SOLAR NEBULA

J.S. LEWIS and M.L. HUTSON
Department of Planetary Science
University of Arizona
Tucson, AZ 85719
USA

ABSTRACT. The solar system is believed to have formed from a primordial gas and dust cloud, i.e., the primitive solar nebula (PSN). Constraints on the chemical composition and structure of the PSN are provided by ovserved densities and compositions of the planets, the asteroids and meteorites together with theoretical modelling. Many of these properties can be explained by equilibrium condensation from a solar composition gas. Other observations suggest that this model is too simple and additional processes, such as kinetic inhibition, heterogeneous accumulation, and volatile element transport must be considered.

1. INTRODUCTION

The sun, its planets, satellites, asteroids, and comets were created from a primordial gas and dust cloud commonly referred to as the primitive solar nebula (PSN). It is estimated that about 98% of the total mass of the PSN was a mixture of hydrogen and helium, slightly less than 2% was ice-forming materials, and about 0.5% was rock-forming solids. Elemental abundances for the PSN, the solar-system abundances, are inferred from meteorite and solar atmosphere compositions (see the recent review by Anders and Grevesse, 1989).

2. FORMATION AND STRUCTURE OF THE PSN

The solar system began as part of a dense interstellar gas and dust cloud. This interstellar cloud was compressed by some means, such as a passing shock wave from a stellar explosion or the radiation pressure of surrounding stars. The cloud became sufficiently dense for its gravitational potential energy to exceed its internal thermal energy, and began to collapse. The collapsing interstellar cloud's angular momentum caused fragmentation into dense cloudlets orbiting about a common center of mass. One of these cloudlets eventually became the PSN. This cloud continued to collapse by radiating off internal energy, and lost angular momentum by shedding mass from its outer edges and by further fragmentation.

As the PSN collapsed, conservation of angular momentum resulted in an increase of angular velocity. This spin-up inhibited further collapse toward the rotation axis, and the nebula collapsed parallel to

J. M. Greenberg and V. Pirronello (eds.), Chemistry in Space, 321–338.

the spin axis into a flattened, symmetrical, disklike shape. Material located in the inner part of the nebula heated up sufficiently to vaporize solids; farther out the heating was less, and dust grains inherited from the interstellar cloud survived. The disk was initially cooled by convection, and temperature decreased linearly with distance from the center of the nebula, following an adiabatic gradient. As temperatures decreased in the PSN, particles condensed. These particles were concentrated by gas drag in a thin layer in the center of the nebula, and began to agglomerate into larger objects.

At some point, temperatures and pressures in the center of the nebula became high enough that nuclear reactions could occur and the central star began its presumed T-Tauri phase. The young star was extremely luminous and emitted a stream of protons and electrons a million times stronger than our sun's present solar wind. Any gas and unaccreted dust was rapidly swept from the solar system; the PSN ceased to exist. Radiative transport replaced convective transport of heat. Temperatures no longer decreased linearly with distance, but rather as $r^{-1/2}$. Material in the inner solar system cooled when the nebular gas was removed, while material in the outer solar system heated up. Solid material continued to accrete, and the moons and planets formed.

3. CHEMICAL PROCESSES IN THE PSN

3.1 Equilibrium Condensation

The simplest model for the chemical processes that occurred in the PSN is that of equilibrium condensation. This model calculates the major element composition, mineralogy, and volatile element content of unaccreted solids in the PSN by chemical equilibrium methods as functions of temperature and pressure. Intimate contact between gas and condensed solids is required for strict thermodynamic equilibrium to be maintained.

The results of the equilibrium condensation model are path-independent. It is impossible to determine whether the presently observable chemical state was reached by heating or cooling. Although it is convenient to think in terms of a condensation sequence beginning at very high temperatures, it is neither certain or even likely that solid solar system materials were fully vaporized in the PSN. It is more likely that the condensation temperatures of solids in the PSN were actually equilibrium temperatures, reflecting the highest temperature at which gases and solids were intimately mixed.

Equilibrium condensation models have been used as tools to explain the varied mineralogies, oxidation states, and volatile element contents of the chondritic meteorites (e.g., Lord, 1965; Larimer, 1967; Larimer and Anders, 1967, 1970; Grossman, 1972, 1973; Grossman and Larimer, 1974). Major-element condensation calculations have also been used to interpret the available data on the compositions and structure of the planets, and on the chemical and isotopic compositions of planetary and satellite atmospheres (e.g., Lewis, 1972a,b, 1973a,b, 1974a,b; Barshay and Lewis, 1975, 1976, 1978; Fegley and Lewis, 1979, 1980). The results

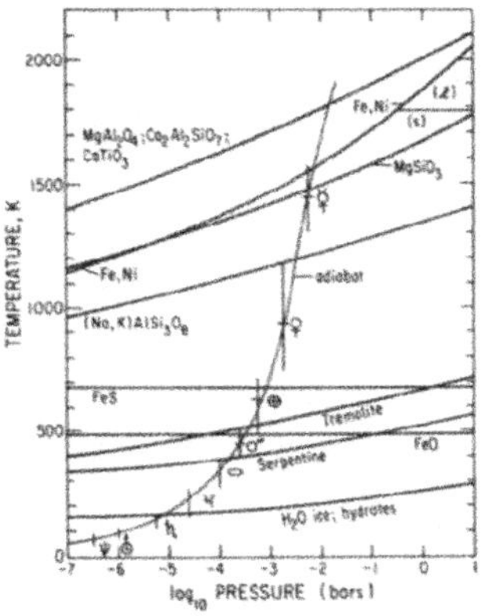

Figure 3.1. Equilibrium condensation sequence for a system of solar elemental composition. An adiabat for a representative model of the PSN is included for comparison. The symbols along the adiabat are, from high to low temperatures, Mercury, Venus, Earth, Mars, the asteroids, Jupiter, Saturn, Uranus, and Neptune.

of these calculations, starting with a solar composition gas, are given in Figure 3.1. Although the absolute temperatures at which a substance condenses varies with presure, with few exceptions the relative sequence of condensation is constant over the range of pressures expected for the PSN (10^{-2} to 10^{-6} atm).

For ease of calculation, condensation models usually begin at temperatures high enough that all of the nebular material is in the gas phase. The model thus describes the sequence of materials that form during equilibrium cooling of a solar mix of elements. The first condensates to form, those that condense at the highest temperatures, are the refractory siderophiles. These are elements of extremely low cosmic abundance, such as W, Os, Ir, etc., which condense as metals.

The second group of minerals to condense are the refractory oxides and silicates. These are defined as those minerals which condense at temperatures higher than that at which forsterite condenses. This group includes almost all of the highly electropositive elements that have valences of 2+ or higher, such as the important radioactive elements U and Th and the oxide moieties VO_2, Sc_2O_3, ZrO_2, rare earth element (REE) oxides, Al_2O_3, CaO, and TiO_2. The latter three combine to form many of the aluminum and calcium minerals observed in meteorites, including perovskite ($CaTiO_3$), gehlenite ($Ca_2Al_2SiO_7$), spinel ($MgAl_2O_4$), and anorthite ($CaAl_2Si_2O_8$). Despite the variety of elements included in this group, they comprise only about five percent of the total mass of the rock-forming materials available in the gas.

Iron-nickel metal is the next material to condense (although at very low pressures it will appear after the magnesium silicates). The metallic Fe-Ni alloy contains minor amounts of Co, Cu, Au, Pt, etc. as well as those nonmetals, such as P, C, and N, which have an appreciable solubility in iron at these temperatures and pressures.

The magnesium silicates, forsterite (Mg_2SiO_4) and enstatite ($MgSiO_3$), condense at temperatures slightly lower than that of Fe-Ni metal. Forsterite condenses first, but is quickly converted to enstatite with the addition of SiO_2 from the nebular gas. These minerals also contain certain other predominantly lithophile elements, including Mn, B, F, and some Cr.

At still lower temperatures, the alkali aluminosilicates ($(Na,K)AlSi_3O_8$) form by reaction of atomic alkali metal vapors with anorthite. Rubidium and cesium are incorporated into these aluminosilicates at slightly lower temperatures.

Below 680K, metallic iron reacts with nebular H_2S to form troilite (FeS). The moderately volatile chalcophiles, Zn, Pb, Ga, Ge, etc. are incorporated in troilite during and following FeS formation.

As the temperature decreases, the remaining Fe-metal is progressively oxidized to FeO, which enters into solid solution in olivine (as Fe_2SiO_4) and pyroxene (as $FeSiO_3$).

Tremolite ($Ca_2Mg_5Si_8O_{22}(OH)_2$) forms from enstatite, diopside ($CaMgSi_2O_6$), and water vapor, at a temperature either higher or lower than that at which Fe-metal is fully oxidized, depending on the pressure. Condensation of other hydroxyl silicates, by reaction of water vapor with ferromagnesian minerals, occurs slightly after tremolite formation.

All of the rock-forming minerals are condensed by approximately 400K. Below that temperature water ice, hydrates of ammonia and methane, and solid methane condense.

It must be emphasized that the model described above assumes strict thermodynamic equilibrium. The model predicts which minerals are present in a solar composition system at each point in the nebula; no turbulent transport of nebular gas is allowed. In addition to the effects of turbulent transport, the model neglects the effects of kinetic inhibition, which can inject a path-dependent character that is absent in equilibrium thermodynamic calculations.

3.2 Kinetic Inhibition

Strict thermodynamic equilibrium requires that, in addition to H_2O, CO and N_2 dominate the high-temperature regions of the PSN, as seen in Figure 3.2. At lower temperatures, CH_4 and NH_3 are more thermodynamically stable than CO and N_2. However, the kinetics of CO and N_2 reduction reactions catalyzed by metal grains were studied by Lewis and Prinn (1980). They found that only small amounts of NH_3 and CH_4 could be produced over the lifetime of the PSN. The gas composition is "frozen" at some temperature, referred to as the quench temperature (T_q). One interesting result is that two compounds are predicted to condense that do not appear in the standard equilibrium condensation sequence (Figure 3.3). These two compounds, NH_4HCO_3 (ammonium bicarbonate) and NH_4COONH_2 (ammonium carbamate) appear at temperatures near that at which water-ice condenses and may provide a source of organic material in the outer solar system.

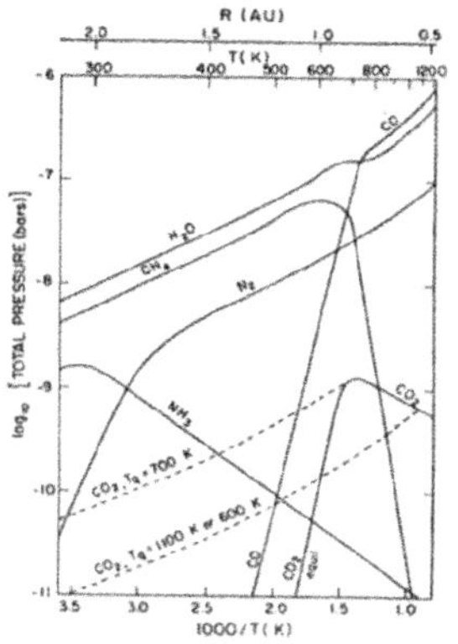

Figure 3.2. Partial pressures of major C, N, and O gases along a solar nebula adiabat. R is distance from the sun and T is temperature. The solid curves represent pressures calculated for strict thermodynamic equilibrium. The dotted lines illustrate the effects of quenching CO reduction reactions at 600, 700, and 1100 K on the CO_2 pressure profile (from Lewis and Prinn, 1980).

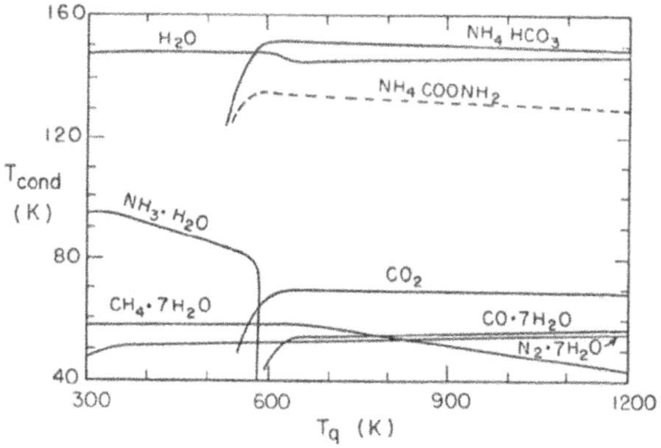

Figure 3.3. Condensation temperatures of carbon, nitrogen, and oxygen compounds as a function of the quench temperature for gas reactions (from Lewis and Prinn, 1980).

More recently, Fegley (1988) examined gas-grain chemical interactions in the PSN. Many of the low temperature reactions, such as hydration of preexisting silicates and the formation of FeO and FeS, require extensive solid state diffusion. Fegley (1988) calculated that diffusion times required to form the hydrated silicates from anhydrous crystalline silicates exceed the lifetime of the PSN. In addition, incorporation of FeO into silicates was found to be kinetically inhibited, although FeS formation was deemed feasible.

3.3 Heterogeneous Accumulation

Equilibrium condensation models require that condensed solids remain in intimate contact with the gas, i.e., that accretion of solids to form large bodies must be very slow compared to the rate of chemical reactions. However, as described in section 2, accretion of objects at least a few meters in diameter must have taken place before the dispersal of the PSN.

The effects of accretion on condensation can be examined with the help of two end-member models. The first model, equilibrium condensation, is discussed above. The other end-member model, heterogeneous accumulation, assumes that accretion occurs at such a rapid rate that any grain, once condensed, is instantly accreted and removed from contact with the nebular gas.

The results of heterogeneous accretion during cooling of a solar-composition gas have been discussed by Lewis (1974a) and Fegley and Lewis (1980). Unlike equilibrium condensation, the results are path-dependent.

The first objects to condense, upon cooling from high temperatures, are refractory siderophiles, essentially identical to those formed by equilibrium condensation. These are followed by the refractory oxides, (including U and Th) which show minor differences from the equilibrium condensation case.

As the temperature decreases, the refractory material will be surrounded by layers of Fe-Ni metal. The first metal nucleated is nickel-rich; subsequent layers are increasingly nickel-poor. As a result, there is a grandient of nickel content over the metal layer as a whole. It is interesting to note that this metal layer is accreted on top of a much less dense and highly radioactive refractory core.

The next material to condense is forsterite, followed quickly by enstatite. As forsterite is shielded from reaction with the gas, the gas phase will be left with an excess of Si, which condenses as quartz (SiO_2) with the last of the enstatite.

The alkali aluminosilicates are unable to form, since all of the aluminum compounds are buried in the refractory core. Sodium will condense as Na_2S and as NaF at lower temperatures. Potassium, rubidium, cesium, and possibly some magnesium will also condense, along with the moderately volatile chalcophiles, as sulfides. FeS, however, will not form, as all of the Fe metal is buried.

FeO does not form, for the same reason that FeS cannot; no Fe-metal is accessible. None of the hydrated silicates are created, as all of their precursor minerals are hidden beneath a layer of sulfides.

Below about 200K, ammonium salts, including NH_4Cl, NH_4Br, and $NH_4H_2PO_4$ condense. These are followed by a deep layer of water ice. Only a small surface of water ice remains in contact with the gas, so negligible amounts of $NH_3 . H_2O$ and no methane clathrate hydrate forms. Leftover H_2S reacts with NH_3 in the gas to precipitate ammonium hydrosulfide on top of the ice layer. At lower temperatures, ammonia and methane ices condense.

The final product of heterogeneous accumulation is a layered planetesimal. Each layer condensed at a different temperature, and is not in chemical equilibrium with other layers. The model assumes that only one planetesimal formed at any given location, and that this planetesimal was added to only by accretion of tiny condensates. Accretion of larger objects would erase the layering through impacts, and a more homogenous body would result. The mineralogy of heterogeneously accumulated planetesimals is very distinctive and corresponds poorly to the observed compositions of meteorites and planets.

3.4 Water Transport

All of the previously discussed nebular models begin with the assumption of a solar composition gas. That this assumption may not be valid is suggested by a recent study of eddy diffusion in the PSN (Stevenson and Lunine, 1988). As seen in Figure 3.4, the temperature in the outer part of the PSN was low enough that the stable phase of water was a solid;

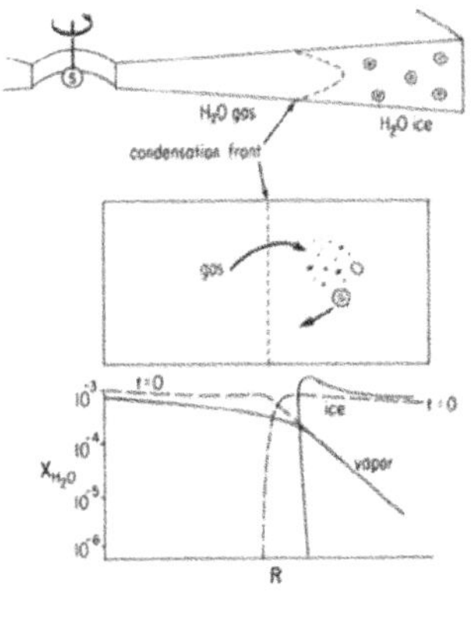

Figure 3.4. Water transport in the PSN. Water vapor is readily transported by turbulent mixing outward across the condensation threshold (dotted line), but solid ice particles are less efficiently transported. The dashed lines in the lower panel show the radial distribution of water vapor and ice at the outset of convective transport; the solid lines show the effect of continuing transport, including both depletion of water vapor sunward of the condensation front, and excess accumulation of ice just beyond the front.

nearer the sun, only water vapor was stable. A condensation front separated the two regions. In Stevenson and Lunine's (1988) model, eddy diffusion transports water vapor from the sunward side accross the condensation front, where it rapidly precipitates out as ice. The ice is assumed to mechanically decouple from the gas, i.e., it can no longer diffuse in the same sense as the gas. The net result is an accumulation of water ice just beyond the condensation front, and a broad zone of water-depleted gas interior to the condensation front. As the partial pressure of water vapor decreases in the inner zone, the condensation front, i.e., the region in which the partial pressure of the gas equals the vapor pressure of the solid, shifts outward (if temperature remains constant). However, if temperature decreases with time, the location of the condensation front will tend to move sunward, forming and excess water ice in a region previously depleted in water vapor. It is not known which of these two competing effects will dominate. The effect of water vapor depletions and water ice excesses on condensation models has not yet been determined.

4. SOURCES OF DATA ON THE PSN

4.1 Meteorites

The majority of meteorites that fall on the Earth are chondrites, named for the abundant spheroidal objects, i.e., chondrules, that they contain. The major rock-forming elements Fe, Mg, Si, S, Ca, Al, Na, Ni, Cr, Mn, P, K, etc., are found in chondritic meteorites in very nearly the same relative proportions as in the sun, with fractionations seldom larger than $\pm$10-20% from the mean. The textures of these meteorites clearly show that they are not of igneous origin. Some appear to be aggregates, with chondrules and mineral grains of incompatible compositions and diverse shock and thermal histories adjacent to one another. Because of the primitive, undifferentiated nature of these meteorites, they are often used as composition standards for discussions of the chemistry of solar system bodies. The interpretation of these compositions is, however, complicated by the fact that chondrites span a considerable range of oxidation state and volatile element content, and have experienced variable amounts of thermal and aqueous alteration.

Urey and Craig (1953) first demonstrated that chondritic meteorites consisted of several distinct and nonoverlapping chemical groups. These groups are usually portrayed on a diagram of total reduced iron (Fe metal plus FeS) versus total oxidized iron (FeO). Such a diagram, simplified from that given in Bild and Wasson (1977), is shown in Figure 4.1. Five groups are resolved on the basis of total iron content and oxidation state.

The most highly reduced chondrites, the E (enstatite) chondrites, are dominated by enstatite, metal, and troilite. Three groups, the progressively more oxidized and less iron-rich H (high-iron), L (low-iron), and LL (very low iron) groups are collectively called the ordinary chondrites. The most highly oxidized and volatile-rich chondrites belong to the C (carbonaceous) meteorites, subdivided into four groups (CI, CM, CO, CV), of which some contain sulfides and up to a few percent metal, whereas others have iron fully oxidized to ferrous and ferric silicates plus magnetite.

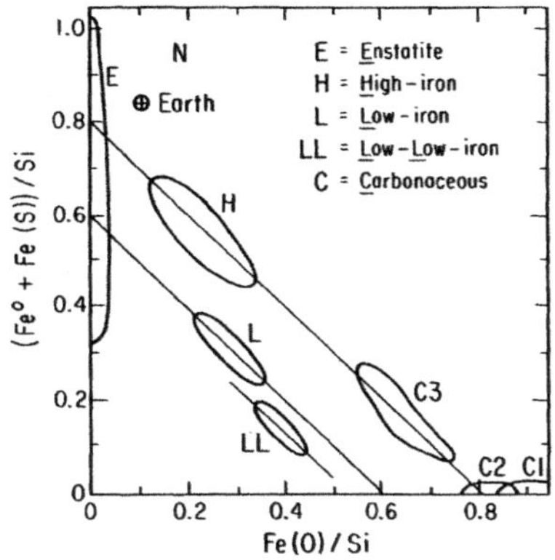

Figure 4.1. Urey-Craig chemical classification of chondrites. N denotes the Netschaevo chondrite described by Bild and Wasson (1977).

The iron variations observed by Urey and Craig (1953) have been attributed to a fractionation between metal and silicate components of unaccreted material in the PSN (Larimer and Anders, 1970). Two metal-silicate fractionations have been identified from examination of Fe and Ni variation in chondritic meteorites. The first metal-silicate fractionation produced carbonaceous and enstatite chondrites and an ordinary chondrite precursor from starting material similar to CI chondrites (Larimer and Anders, 1970); the second produced the H-, L- and LL-group chondrites from a L-like precursor (Dodd, 1981).

Larimer and Anders (1970) placed limits on the temperature at which the metal-silicate fractionation took place. The lower limit, 680 K (at 10^{-4} atm), is based on the fact that none of the chondrites show a simple correlation of Fe and S, implying that FeS had not yet condensed at the time of fractionation. The upper limit of 1050 K (at 10^{-4} atm) is derived from the observation that all the siderophile trace elements that condense above 1050 K are depleted by a constant factor in ordinary and enstatite chondrites, suggesting they were fully condensed when fractionation occurred.

The three classes of chondrites (ordinary, enstatite and carbonaceous) can also be distinguished by systematic variations in refractory lithophile elements (Al, Ca, Cr, Mg, Si, Ti), i.e., those elements which are among the first to condense from the PSN (Wasson, 1974). These variations are attributed to fractionations of early condensates.

The most refractory lithophile elements are fractionated, relative to Si, to roughly the same degree. Al, Ca and Ti are present in essentially constant ratios in chondrites, implying that a single, fixed set of minerals was involved and that fractionation occurred after Ca was fully condensed (Larimer and Anders, 1970; Kerridge, 1979; Larimer, 1979).

A few of the refractory lithophile elements (Mg, Cr, and Si) are fractionated to a lesser degree. Kerridge (1979) and Larimer (1979) reported that these elements represent two distinct fractionation trends, one for the carbonaceous chondrites, another for the ordinary and enstatite chondrites. The two trends intersect at a composition similar to that of CI chondrites (Kerridge, 1979), suggesting that the precursor material for all of the chondritic meteorites was similar in composition to CI chondrites. The Mg/Si ratio in this starting material must have been greater than one (Kerridge, 1979; Larimer, 1979), based on its current value in carbonaceous meteorites. This ratio occurs in nebular condensation models only after forsterite begins to condense. Apparently, carbonaceous chondrite precursor material experienced a fractionation event soon after forsterite began to condense; for the ordinary and enstatite chondrites, fractionation occurred after a large portion of the forsterite had condensed (Larimer, 1979).

Van Schmus and Wood (1967) used petrological criteria, including the degree of recrystallization of matrix material, the state of devitrification of chondritic glass, textural blurring of chondrule rims, degree of aqueous alteration, etc., to rank the members of each of the chemical groups in a sequence of six petrologic types.

Type 3 represents the least thermally or aqueously altered material. Types 4, 5, and 6 indicate increasing thermal re-equilibration and are sometimes considered to represent a metamorphic sequence. It is by no means clear that we are in fact seeing evidence of metamorphism in the usual sense, involving reheating from a colder early state. It may be simply that the different petrologic types were accreted and stored at somewhat different temperatures or with different cooling rates.

The carbonaceous chondrites are the only meteorites known of petrologic types 1 and 2. Meteorites of these two petrologic types show no evidence of thermal metamorphism, but have experienced varying degrees of aqueous alteration. The carbonaceous meteorites are sometimes grouped by petrologic type, as C1, C2, and C3. All of the type 1 meteorites are CI chondrites. All of the CM and some of the CV chondrites are of type 2. The remaining CV and the CO meteorites are mainly of type 3 (Dodd, 1981).

Chondritic meteorites provide evidence for isotopic inhomogeneities in the PSN. The variations of oxygen istopic compositions from class to class provide a fascinating example. As seen in Figure 4.2, the relative abundances of oxygen 16, 17, and 18 separate the chondritic meteorites into classes that cannot be interconverted by normal chemical or physical fractionation processes, which produce scatter along a line of slope approximately 1/2 in an oxygen isotope diagram (Clayton et al., 1976, 1977). High temperature mineral grains from the C2 and C3 chondrites (corresponding to the CM, CO, and CV groups) define a line of slope approximately 1.0 in Figure 4.2, due to mixing between two distinct nucleosynthetic components, one of which as extremely rich in

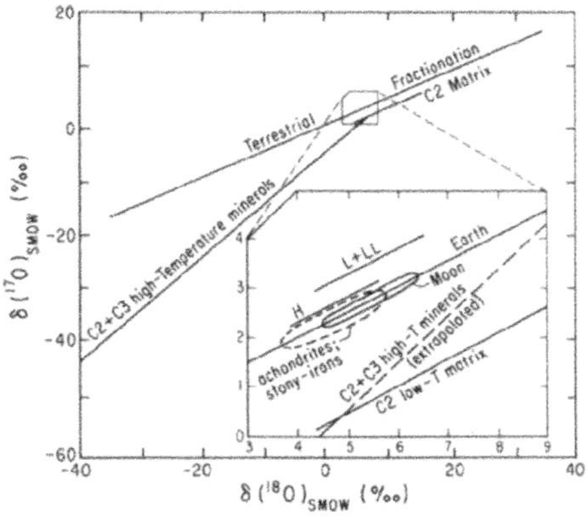

Figure 4.2. Oxygen isotope variations in the Solar System. (Data from Clayton, et al., 1976; 1977).

oxygen 16. These istotopic differences make it virtually certain that the substantial majority of preplanetary solids in the PSN were not fully vaporized, but were instead partially equilibrated with nebular gases.

4.2 Asteroids

The asteroids have been divided into a number of classes based on a combination of spectral and albedo data. A comprehensive review of asteroid classification is given by Tholen (1984). It is significant that different asteroid classes occupy discrete regions of the asteroid belt, as seen in Figure 4.3. The asteroid belt is dominated by two spectral classes, the inner belt by the S asteroids and the outer belt by the C asteroids.

The interpretation of the spectral properties of asteroid surfaces generally is limited to the detection and identification of major silicate phases (olivine, pyroxene and plagioclase), opaques (metal, carbonaceous material), and hydroxyl silicates (clays, etc.). Laboratory studies of the spectra of a large suite of meteorite and mineral samples then permits the tentative association of an asteroid class with a known meteorite class or a mixture of known minerals. Based on a comparison with meteorite spectra, the S,C, and D and P classes are assumed to represent volatile-poor stony or stony-iron material (S), carbonaceous chondrite-like material (C), and volatile-rich ultracarbonaceous material (D and P) (Gradie and Veverka, 1980).

The radial distribution of the various asteroid classes may reflect, in part, the nebular temperatures during the condensation and accretion of asteroidal material. However, two recent studies suggest

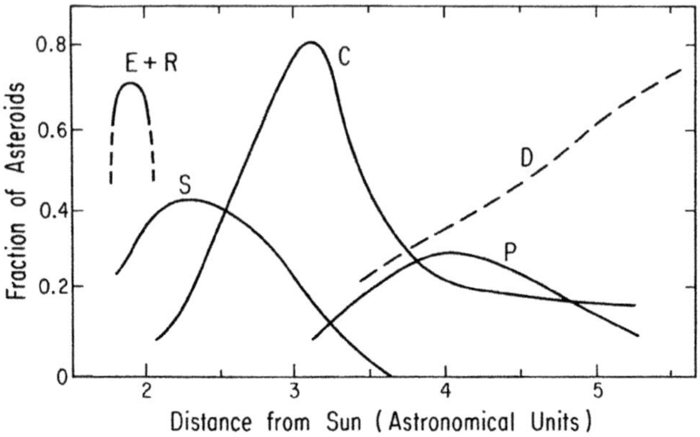

Figure 4.3. Distribution of asteroid spectral types versus distance from the sun. These smoothed and bias-corrected data are taken from Gradie and Tedesco (1982).

that some of the classes record evidence of an early radially varying post-accretional heating event.

Gaffey (1988) examined a group of inner belt asteroid classes, which include a range of compositions and mineralogies. Six of the classes studied consist of differentiated asteroids, i.e., asteroids which have apparently been igneously fractionated. Gaffey (1988) reported that the percentage of differentiated asteroids, regardless of class, decreased linearly with increasing distance from the sun.

Jones (1988) examined water distribution among the low albedo, outer belt asteroids. A majority of the C class contain hydrated silicates indicative of mild aqueous alteration, while the P and D asteroids appear anhydrous. In addition, the abundance of hydrated silicates in the C class gradually declines from 2.4 to 3.5 AU. Jones (1988) concluded that the original outer belt asteroids consisted of anhydrous silicates, water ice, and complex organics. Subsequent heating melted the water ice and produced the observed hydrated silicate abundance.

4.3 The Terrestrial Planets

Most of the solid material in the PSN was probably incorporated into the planets and their satellites. The bulk compositions of these objects reflect the chemistry and mineralogy of their precursor material, and thus provide constraints on the pressures and temperatures in the PSN before and during accretion.

The bulk densities of the terrestrial planets have been determined from spacecraft flybys coupled with ground-based observations. The uncompressed densities (corrected to zero pressure) of these planets are: Mercury approximately 5.3 g/cm^3, Venus 4.40 g/cm^3, Earth 4.45 g/cm^3, and Mars approximately 3.85 g/cm^3 (Lewis, 1972b). The variation of bulk density implies a different bulk chemistry for each of the terrestrial planets.

Lewis (1972b) compared the observed densities of the terrestrial planets with the densities of bulk condensates in equilibrium with a solar-composition gas. The temperature-pressure profile of the PSN was assumed to be adiabatic, as shown in Figure 3.1. Lewis (1972b) concluded that the observed planetary densities could be adequately predicted by equilibrium condensation theory, assuming that each planet sampled a narrow T,P region of the PSN. Subsequent modelling (Lewis, 1988) has determined that the observed density of Mercury is somewhat too large to be explained by equilibrium condensation and simple accretion alone. Some additional mechanism to provide an excess of dense core material is required.

The compositions calculated from equilibrium condensation models for the terrestrial planets are as follows (Lewis, 1972b):

Mercury is predicted to have a large Fe-Ni core surrounded by a mantle of refractory, Fe^{2+}-free magnesium silicates and oxides. Only trace amounts of FeO, alkali metals, and sulfur should be present.

Venus is predicted to have a massive mantle of Fe^{2+}-free magnesium silicates surrounding an Fe-Ni core. A silica-rich crust similar in composition to Earth's should be present. Only trace amounts of FeO and sulfur are expected.

Earth contains an inner Fe-Ni core and an outer core of Fe-FeS melt, surrounded by a mantle containing approximately 10% FeO. The observed increase in the density of Earth relative to Venus is explained by the incorporation in Earth of the relatively heavy element sulfur.

Mars should be essentially free of metallic iron. It is predicted to have a core composed primarily of FeS, surrounded by an FeO-rich mantle. Mars' volatile element content should be greater than that of Earth's.

The current state of knowledge regarding the compositions of the terrestrial planets remains consistent with the predictions of the equilibrium condensation model. Reflectance spectra constrain the FeO content of the surface of Mercury to at most a few percent (Vilas, 1988). However, as shown by Goettel (1988), the bulk composition of Mercury is essentially unconstrained at present.

X-ray fluorescence and gamma-ray results from the Veneras 8, 9, 13 and 14 and Vegas 1 and 2 suggest that the surface of Venus is covered with a variety of magmatic rocks similar in composition to those found on Earth's surface (Surkov, et al., 1986). The data have fairly large error bars, and thus the absolute abundance of FeO in Venus' crust in not well constrained.

That Venus' lithosphere has a far lower FeO content than Earth's is suggested by a model of sulfur chemistry in the atmosphere-lithosphere system based on Pioneer Venus data (Lewis and Kreimendahl, 1980). Lewis and Kreimendahl (1980) examined various sulfur-bearing gases, the abundances of which can only be achieved in a narrow range of oxygen fugacities. They concluded that the partial pressure of O_2 is buffered by a sulfide-sulfate buffer, consistent with the idea that Fe oxides are proportionately rarer relative to FeS in Venus' lithosphere than on Earth's.

The x-ray fluorescence results of the Viking landers indicate that the surface at these sites is covered with a fine material consisting largely of Fe-rich clays, a weathering product consistent with an Fe-rich crust (Carr, 1984).

The SNC meteorites (Shergottites, Nakhlites and Chassigny) are a small group of meteorites widely believed to originate from Mars. Evidence for this includes their young crystallization ages, the compositional similarity of Shergotty to Martian soil, and the presence in the shergottite EETA 79001 of a trapped gas component with element and isotope ratios matching those of the Martian atmosphere (Dreibus and Wanke, 1985).

On the assumption that these meteorites do indeed come from Mars, Dreibus and Wanke (1985) used the composition of the shergottites to estimate the bulk composition of Mars. Their model indicates that Mars has a mantle with a higher FeO content and a higher volatile element concentration than that of Earth.

Newsom and Drake (1987) examined the abundances of V, Cr and Mn in the mantle of the shergottite parent body (probably Mars). The abundances of these elements are consistent with segregation of a S-rich metallic liquid during core formation, and with a 30 wt% S-rich metallic core.

4.4 Other Data Sources

Additional constraints on the chemistry, temperature, oxidation states, etc., of the PSN have been obtained from studies of molecular clouds and comets. This information is excluded from this chapter, as extensive discussion of these objects is found elsewhere in this book.

5. CONCLUSIONS

Observed densities and compositions of the planets, the asteroids and meteorites together with theoretical modelling provides constraints on the chemical composition and structure of the PSN. Many of the observations can be explained by equilibrium condensation from a solar composition gas. Other observations suggest that this model is too simple. Additional theory is needed on kinetic inhibition and volatile element transport in the PSN, and on processes associated with accretion. More information on the densities and compositions of the planets, their satellites, comets and asteroids would also be useful. Some information will be forthcoming from missions such as CRAF (Comet Rendezvous Asteroid Flyby), Mars Observer (MO), Galileo and Cassini. However, no mission is scheduled to return to Mercury, and information on the chemistry of this planet is badly needed to constrain models of the inner PSN.

6. REFERENCES

Anders, E. and Grevesse, N. (1989) 'Abundances of the elements: meteoritic and solar', Geochim. Cosmochim. Acta 53, 197-214.

Barshay, S.S. and Lewis, J.S. (1975) 'Chemistry of solar material', in G.B. Field and A.G.W. Cameron (eds.), The Dusty Universe, Neale Watson Academic Publ. Inc., New York, pp. 33-40.

Barshay, S.S. and Lewis, J.S. (1976) 'Chemistry of primitive solar material', Ann. Rev. Astron. Astrophys. 14, 81-90.

Barshay, S.S. and Lewis, J.S. (1978) 'Chemical structure of the deep atmosphere of Jupiter', Icarus 33, 593-607.

Bild, R.W. and Wasson, J.T. (1977) 'Netschaevo: A new class of chondritic meteorite', Science 197, 58-62.

Carr, M.H. (1984) 'Mars', in M.H. Carr (ed.), The Geology of Terrestrial Planets, NASA SP-469, pp. 207-263.

Clayton, R.N., Onuma, N. and Mayeda, T.K. (1976) 'A classification of meteorites based on oxygen isotopes', Earth Plan. Sci. Lett. 30, 10-18.

Clayton, R.N., Onuma, N., Grossman, L. and Mayeda, T.K. (1977) 'Distribution of the pre-solar component in Allende and other carbonaceous chondrites', Earth Plan. Sci. Lett. 34, 209-224.

Dodd, R.T. (1981) Meteorites: A Petrologic-Chemical Synthesis, Cambridge Univ. Press, New York.

Dreibus, G. and Wanke, H. (1985) 'Mars, a volatile rich planet', Meteoritics 20, 367-381.

Fegley, B. (1988) 'Cosmochemical trends of volatile elements in the solar system', in Workshop on the Origins of Solar Systems, LPI Tech. Rept. #88-04, 51-60.

Fegley, B., Jr. and Lewis, J.S. (1979) 'Thermodynamics of selected trace elements in the Jovian atmosphere', Icarus 38, 166-179

Fegley, B., Jr. and Lewis,J.S. (1980) 'Volatile element chemistry in the solar nebula: Na, K, F, Cl, Br and P', Icarus 41, 439-455.

Gaffey, M.J. (1988) 'Thermal history of the asteroid belt: implications for accretion of the terrestrial planets', Lunar Plan. Sci. XIX, 369-370.

Goettel, K.A. (1988) 'Present bounds on the bulk composition of Mercury: Implications for planetary formation processes', in F. Vilas, C.R. Chapman, and M.S. Matthews (eds.), Mercury, Univ. of Arizona Press, Tucson, pp. 613-621.

Gradie, J. and Tedesco, E. (1982) 'Compositional structure of the asteroid belt' Science 216, 1405-1407.

Gradie, J.C. and Veverka, J. (1980) 'The composition of the Trojan asteroids', Nature 283, 840-842.

Grossman, L. (1972) 'Condensation in the primitive solar nebula', Geochim. Cosmochim. Acta 36, 597-619.

Grossman, L. (1973) 'Refractory trace elements in Ca-Al rich inclusions in the Allende meteorite', Geochim. Cosmochim. Acta 37, 1119-1140.

Grossman, L. and Larimer, J.W. (1974) 'Early chemical history of the solar system', Rev. Geophys. Space Phys. 12, 71-101.

Jones, T.D. (1988) 'An infrared reflectance study of water in outer belt asteroids: Clues to composition and origin', Ph.D. disseration, Univ. of Az.

Kerridge, J.F. (1979) 'Fractionation of refractory lithophile elements among chondritic meteorites', Proc. Lun. Plan. Sci. Conf. 10th, 989-996.

Larimer, J.W. (1967) 'Chemical fractionation in meteorites. I. Condensation of the elements', Geochim. Cosmochim. Acta 31, 1215-1238.

Larimer, J.W. (1979) 'The condensation and fractionation of refractory lithophile elements', Icarus 40, 446-454.

Larimer, J.W. and Anders, E. (1967) 'Chemical fractionation in meteorites. II. Abundance patterns and their interpretation', Geochim. Cosmochim. Acta 31, 1239-1270.

Larimer, J.W. and Anders, E. (1970) 'Chemical fractionations in meteorites III. Major element fractionations in chondrites', Geochim. Cosmochim. Acta 34, 367-387.

Lewis, J.S. (1972a) 'Low temperature condensation from the solar nebula', Icarus 16, 241-252.

Lewis, J.S. (1972b) 'Metal/silicate fractionation in the solar system', Earth Plan. Sci. Lett. 15, 286-290.

Lewis, J.S. (1973a) 'Chemistry of the outer solar system', Space Sci. Rev. 14, 401-410.

Lewis, J.S. (1973b) 'Origin and composition of the terrestrial planets and satellites of the outer planets', in H. Reeves (ed.), The Origin of the Solar System, CNRS, Paris, pp. 202-205.

Lewis, J.S. (1974a) 'The chemistry of the solar system', Sci. Amer. 230, 50-65.

Lewis, J.S. (1974b) 'The temperature gradient in the solar nebula', Science 186, 440-442.

Lewis, J.S. (1988) 'Origin and composition of Mercury', in F. Vilas, C.R. Chapman, and M.S. Matthews (eds.), Mercury, Univ. of Arizona Press, Tucson, pp. 651-666.

Lewis, J.S. and Kreimendahl, F.A. (1980) 'Oxidation state of the atmosphere and crust of Venus from Pioneer Venus results', Icarus 42, 330-337.

Lewis, J.S. and Prinn, R.G. (1980) 'Kinetic inhibition of CO and N_2 reduction in the solar nebula', Astrophys. J. 238, 357-364.

Lord, H.C., III. (1965) 'Molecular equilibria and condensation in a solar nebula and cool stellar atmospheres', Icarus 4, 279-288.

Newson, H.E. and Drake, M.J. (1987) 'Formation of the Moon and terrestrial planets: Constraints from V, Cr, and Mn abundances in planetary mantles and from new partitioning experiments', LPSC XVIII, 716-717.

Stevenson, D.J. and Lunine, J.I. (1988) 'Rapid formation of Jupiter by diffusive redistribution of water vapor in the solar nebula', Icarus 75, 146-155.

Surkov, Y.A., Moskalyova, L.P., Kharyukova, V.P., Dudin, A.D., Smirnov, G.G., and Zaitseva, S.Y. (1986) 'Venus rock composition at the Vega 2 landing site', Proc. 17th Lun. Plan. Sci. Conf., E215-E218.

Tholen, D.J. (1984) 'Asteroid taxonomy from cluster analysis of photometry', Ph.D dissertation, Univ. of Az.

Urey, H.C. and Craig, H. (1953) 'The composition of the stone meteorites and the origin of meteorites', Geochim. Cosmochim. Acta 4, 36-82.

Van Schmus, W.R. and Wood, J.A. (1967) 'A chemical petrologic classification for the chondritic meteorites', Geochim. Cosmochim. Acta 31, 747-766.

Vilas, F. (1988) 'Surface composition of Mercury from reflectance spectrophotometry', in F. Vilas, C.R. Chapman, and M.S. Matthews (eds.), Mercury, Univ. of Arizona Press, Tucson, pp. 59-76.

Wasson, J.T. (1974) Meteorites: Classification and Properties, Springer-Verlag, New York.

THE VOLATILE COMPOSITION OF COMETS DEDUCED FROM ULTRAVIOLET SPECTROSCOPY

PAUL D. FELDMAN
Center for Astrophysical Sciences
Department of Physics and Astronomy
The Johns Hopkins University
Baltimore, Maryland 21218, USA

ABSTRACT. The application of vacuum ultraviolet spectroscopy to the study of the composition of the volatile component of cometary coma is reviewed, particularly in the context of comparison with *in situ* measurements obtained during the recent apparition of comet P/Halley.

1. Introduction

Chemistry plays a major role in the life history of comets. The composition of the ice of the cometary nucleus, which accounts for most of the cometary mass, is determined by a variety of factors ranging from the nature of the cometesimal grains formed in the molecular cloud to the physical and chemical conditions existing in the primordial solar nebula at the time and location where the comets ultimately coalesce. As the ices are regarded to preserve, over the history of the solar system, the signature of this environment, much attention has focussed on the interpretation of recent cometary abundance measurements, particularly those made by instruments aboard the spacecraft which flew by periodic comet Halley (1986 III) during its 1985–86 apparition (Lunine 1989). At the other end of its life cycle, comets entering the inner solar system continuously lose mass as a result of solar heating, and it is this mass loss, which varies considerably with heliocentric distance, which not only makes a comet visible, but provides the means of determining the composition of the vaporized mass. It is in the coma, or cometary atmosphere, that primarily photochemical processes induced by solar ultraviolet radiation give rise to most of the species that are observed spectroscopically in the visible and ultraviolet regions of the electromagnetic spectrum.

The problem of composition reduces to determining, from observations of radicals, atoms and ions in the coma, the initial molecular progenitors (or "parent" molecules) present in the nucleus, presumably as ices. Some of the parent molecules can actually be detected directly, such as CO which fluoresces in the ultraviolet (Feldman and Brune 1976), or a number of species which fluoresce in vibrational transitions in the near-infrared such as H_2O, CO, CO_2, H_2CO, NH_3 and CH_4 (Crovisier 1988). However, except for measurements from the *Vega 1* and *2* IKS experiment (Combes *et al.* 1988), only H_2O has been unambiguously detected in the infrared (Larson *et al.* 1989). The atomic inventory is somewhat easier to derive as the resonance transitions of the cosmically abundant elements H, O, C, N and S all lie in the vacuum ultraviolet and, in principle, the total content of these species in the coma can be determined by an instrument with a suitably

J. M. Greenberg and V. Pirronello (eds.), Chemistry in Space, 339–361.

large field-of-view. Of course, a fraction of the atomic species of each element will be produced directly in ionic form, and will not be counted using this approach. In addition, another fraction of each element exists in the coma in the solid grains, and this component will also not be counted except for a small amount volatilized by evaporation or sputtering by energetic particles. The composition of the grains has been inferred from *in situ* measurements made by several instruments on the Halley encounter spacecraft (Kissel *et al.* 1986).

Although not all of the results from the *in situ* measurements are available (particularly from the *Giotto* neutral mass spectrometer experiment), it appears that there are no major inconsistencies between the *in situ* data and the compositional results deduced from remote astronomical observations. While the Halley missions offered a unique opportunity, the measurements so made represent a snapshot view of a complex, time-variable object, whose pattern could only be determined by remote observations over a long interval of time. Comparison of the Halley observations with those of other comets, both "new" and short-period comets, is necessary in order to place this large body of new information into the general context of cometary origin and evolution. For this purpose, ultraviolet spectroscopy is a very powerful tool, as will be seen below.

2. Photodestruction Processes in the Coma

As an example of the photolytic destruction processes occurring in the coma, we consider the dominant molecular species, water. Water vapor is assumed to leave the surface of the nucleus with some initial velocity, v_0, and flow radially outward, expanding into the vacuum of space, and increasing its velocity according to straightforward thermodynamics (Mendis *et al.* 1985). Even though collisions are important at distances typically up to 10^4 km (depending on the density and consequently, on the total gas production rate), the net flow of H_2O molecules is radially outward, such that the density varies as ρ^{-2} near the nucleus. This is the basis for spherically symmetric coma models (the number of particles flowing through a spherical surface is conserved), which assume isotropic gas production, but appears to hold equally well for the case of Halley, which clearly was not outgassing uniformly over its surface (Krankowsky *et al.* 1986). The photolysis of H_2O can proceed by:

$$
\begin{array}{llll}
H_2O + h\nu & \rightarrow & OH + H & (a) \\
 & \rightarrow & OH(A\,^2\Sigma) + H & (a') \\
 & \rightarrow & H_2 + O(^1D) & (b) \\
 & \rightarrow & H_2 + O(^1S) & (b') \\
 & \rightarrow & H_2O + e & (c) \\
 & \rightarrow & H + OH^+ + e & (d) \\
 & \rightarrow & H_2 + O^+ + e & (e) \\
 & \rightarrow & OH + H^+ + e & (f)
\end{array}
$$

The products are subsequently removed by

$$
\begin{array}{lcll}
OH + h\nu & \rightarrow & O + H & (g) \\
 & \rightarrow & OH^+ + e & (h) \\
H_2 + h\nu & \rightarrow & H + H & (i) \\
 & \rightarrow & H_2^+ + e & (j) \\
 & \rightarrow & H + H^+ + e & (k) \\
O + h\nu & \rightarrow & O^+ + e & (l) \\
H + h\nu & \rightarrow & H^+ + e & (m)
\end{array}
$$

Reactions *l* and *m* can also occur by resonant charge exchange with solar wind protons. The excited OH radicals produced in reaction *a'* and the metastable oxygen ^1D and ^1S atoms (reactions *b* and *b'*) spontaneously decay giving "prompt" OH emission with a "hot" rotational distribution (Bertaux 1986) and the auroral O I λλ6300, 6363 and λ5577 emissions (Festou and Feldman 1981) which can serve as useful tracers of the H_2O distribution in the inner coma (Spinrad 1982).

Table 1. Reaction thresholds and rate coefficients

Reaction	*Threshold* (Å)	*Rate* (10^{-6} s^{-1})	*Reference*
a	2424.6	10.2	1
a'			2
b	1450	1.35	1, 3
b'			3
c	984	0.334	1
d	684.4	5.54×10^{-2}	1
e	664.8	5.85×10^{-3}	1
f	662.3	1.31×10^{-2}	1
g	2823.0	18.0	1, 4, 5
h			
i	844.79	4.4×10^{-2}	1
j	803.67	5.4×10^{-2}	1
k	685.8	9.5×10^{-3}	1
l	910.44	0.21	1
m	911.75	0.073	1

References: 1. Huebner, Keady and Lyon (1982), 2. Bertaux (1986), 3. Festou and Feldman (1981), 4. Van Dishoeck and Dalgarno (1984), 5. Schleicher and A'Hearn (1988).

The energy thresholds (in wavelength units) and rate coefficients, for a quiet Sun at a heliocentric distance of 1 AU, are given in Table 1. From these numbers it is possible to determine a relationship between the total number of atoms of either oxygen or hydrogen in the coma and the total number of water molecules which were their parents, regardless of which reaction process was followed. However, it should be noted that it is quite difficult to measure the total content of these species because their lifetimes are long ($\sim 10^6$ s) and, in the case of H, the velocity acquired on dissociation is considerably greater than the canonical 1 km s^{-1} of the parent outflow. Also, both hydrogen and oxygen can be produced by the photodestruction of other parent molecules, so that while the atomic inventory may be fairly accurate, the initial molecular composition cannot be unambiguosuly determined. This is not quite so severe, as H_2O has been demonstrated to be the dominant constituent of the ice, both by direct measurement in Halley (Krankowsky *et al.* 1986), and by analysis of simultaneously observed emissions of all three neutral daughter products, O, H and OH in a number of comets (Feldman 1983). The strong hydroxyl emission at 3085 Å, which can be measured at high altitude observatories as well as from the *IUE*, is now routinely used as a monitor of the water production rate in comets (Feldman 1989).

Atomic carbon is somewhat more difficult to assess accurately as there are several possible molecular sources. The dominant ones are CO and CO_2, the former being variable in abundance relative to water in different comets (Feldman 1986). In comet Halley, the CO production rate, determined by long-slit ultraviolet spectroscopy (Woods *et al.* 1986), was ~ 0.17 that of water, but Eberhardt *et al.* (1987a), using data from the *Giotto* neutral mass spectrometer, showed that only one-third to one-half of that amount came directly from the nucleus. The remaining CO appeared to have a distributed source in the coma with a peak strength at about a distance of 10,000 km from the nucleus. The nature of this source remains one of the outstanding problems in comet chemistry. In Halley, CO_2 was less abundant than CO, $\sim$ 2–4% that of water (Krankowsky *et al.* 1986), while CH_4 was at most 1–2% that of water (Allen *et al.* 1987). The other observed carbon-bearing radicals, CN, CH, C_2 and CS do not contribute more than a fraction of one percent. In addition, carbon is a major constituent of the refractory grains which, for some comets, may comprise nearly half of the total cometary mass (again, this is not a very well determined number, even for Halley). As the variation in CO from comet to comet appears to be correlated with the dust-to-gas ratio (dustier comets have higher CO abundance) (Feldman *et al.* 1988), and while this does not appear to be the case for the other carbon containing radicals, it thus appears that the total carbon abundance, relative to hydrogen or oxygen, is variable from comet to comet.

3. Abundance determination from spectroscopic observations

We examine here the means whereby coma abundances can be determined from spectrophotometric measurements of either the total flux or the surface brightness in a given spectral feature. At the outset, it is important to note that the uncertainty in the derived abundances will include not only the measurement uncertainty, but also uncertainties in the atomic and molecular parameters and, in the case of surface brightness measurements, uncertainties in the model parameters used. For this reason, relative abundances, derived from observations of different comets with the same instrument and under similar geometrical conditions, are often more reliable (e.g. Roettger *et al.* 1989).

The cometary coma contains mostly neutral atoms and molecules which emit radiation primarily by means of resonance reradiation of solar photons. The few exceptions will be noted below. It is for this reason that ultraviolet spectroscopy has proved to be such an important tool as all of the cosmically abundant elements (H, C, N, O, S) and many simple molecules made up of these elements have their strongest resonance transitions in the ultraviolet. Assuming that the coma is optically thin in these transitions, the interpretation of the observed ultraviolet flux is straightforward. The total number of species i in the coma is

$$M_i = Q_i \tau_i(r) \tag{1}$$

where Q_i is the production rate (s^{-1}) of all sources of species i and $\tau_i(r)$ is the lifetime of this species at heliocentric distance r, $\tau_i(r) = \tau_i(1\ \text{AU})r^2$. The luminosity is then

$$L_i = M_i g_i(r), \tag{2}$$

where the fluorescence efficiency, $g_i(r) = g_i(1\ \text{AU})r^{-2}$, is given by

$$g_i(1\ \text{AU}) = \frac{\pi e^2}{mc^2}\lambda^2 f_\lambda \pi F_\odot \tilde{\omega}\ \text{photons s}^{-1}\ \text{atom}^{-1} \tag{3}$$

Here, e, m and c have their usual atomic values, λ is the transition wavelength, f_λ is the absorption oscillator strength, $\pi F_\odot$ is the solar flux per unit wavelength interval at 1 AU and $\tilde{\omega}$ is the albedo for single scattering. Thus, for a comet at a geocentric distance of Δ, the total flux from the coma is

$$F_i = \frac{L_i}{4\pi\Delta^2} = \frac{Q_i g_i(r)\tau_i(r)}{4\pi\Delta^2} \tag{4}$$

Note that the product $g_i(r)\tau_i(r)$ is independent of r.

Unfortunately, the scale lengths (the product of lifetime and outflow velocity) of almost all of the species of interest in the ultraviolet are of the order of 10^5—10^6 km at 1 AU. Thus, total flux measurements require fields-of-view ranging from several arc-minutes to a few degrees. As the emissions are well separated in wavelength, objective grating spectroscopy is well-suited to this requirement, as has been demonstrated by rocket observations of comet West (1976 VI) by Smith *et al.* (1980) and of P/Halley by Opal *et al.* (1987). Most of our information about the ultraviolet spectra of comets comes from observations made by the *International Ultraviolet Explorer* (*IUE*) satellite, which has been in geosynchronous orbit since 1978 and has observed ~ 30 comets to date (Festou and Feldman 1987). The *IUE* spectrographs have small apertures (nominally 10" × 20") and thus sample only a very small part of the total coma. In this case, again assuming an optically thin coma, the measured flux in the aperture, F', can be converted to an average surface brightness (in units of Rayleighs), B:

$$B = 4\pi 10^{-6} F' \Omega^{-1}, \tag{5}$$

where Ω is the solid angle subtended by the aperture. The brightness, in turn, is related to $\overline{N}_i$, the average column density of species i within the aperture, by

$$B = 10^{-6} g_i(r)\overline{N}_i \tag{6}$$

At this point the evaluation of $\overline{N}_i$ requires the use of a model of the density distribution of the species i. For parent molecules, produced directly by sublimation from the surface of the comet, a spherically symmetric radial outflow model is often adopted. Such a model assumes a steady-state gas production rate and a constant outflow velocity and gives rise to a density distribution which varies as ρ^{-2}, where ρ is the cometocentric distance. The basic validity of this model was demonstrated by the *Giotto* neutral mass spectrometer measurements of H_2O and CO_2 (Krankowsky *et al.* 1986), although detailed analysis revealed that the velocity of the water molecules increased from 0.8 to 1.1 km over a distance of 10,000 km (Lammerzahl *et al.* 1987). The dependence of outflow velocity on heliocentric distance remains uncertain, although Delsemme (1982) has suggested $r^{-1/2}$ based on thermodynamic arguments. Radio observations of the OH line width at 18 cm in P/Halley appear to show this behavior, but only pre-perihelion (Colom *et al.* 1989).

For the dissociation products (or "daughter molecules"), the modelling is more complex. This simplest model, due to Haser (1957) assumes continued radial outflow, although at a different velocity, such as to maintain a constant flux of the initial particle across an arbitrary spherical surface surrounding the nucleus. This model, which is valid only at distances equal to a few scalelengths of the product, is widely used as the densities can be easily expressed in analytical form. However, as surface brightness measurements are often (such as the case with *IUE*) made with small fields-of-view close to the nucleus, this model can lead to factor of two errors from the neglect of the dissociation kinematics. Since the solar photodissociation often leaves the product fragments with 1–2 eV of kinetic energy (see for example Okabe 1978 or Huebner and Carpenter 1979), the resultant motion (which is assumed to be isotropic in the parent molecule's rest frame), will contain a large non-radial component. Several approaches have been developed to properly account for the kinematics; notably the vectorial model of Festou (1981) and the average random walk model (a Monte Carlo method) of Combi and Delsemme (1980). The latter model has an advantage in the calculation of "granddaughter" densities, such as that of oxygen produced by reaction *(g)* above (Dymond *et al.* 1989).

In addition to the photodestruction chains, chemical reactions, particularly ion-molecule reactions, can alter the composition within a region of the coma known as the collision zone. The size of this zone is estimated to be $\sim$ 10,000 km for a comet with a total gas production rate of 1.0×10^{30} molecules s^{-1}, typical of P/Halley in the month following perihelion (Feldman 1989). While such reactions may produce numerous minor species (see eg. Huebner and Giguere 1980), they do not erase the signature of the original parent molecules as suggested by Oppenheimer and Downey (1980). In fact, detailed chemical models have clearly demonstrated the need for complex molecules to serve as the parents of the observed C_2 and C_3 radicals in the coma, strengthening the connection between comet formation and molecular cloud abundances (check suitable reference). Thus, the photochemical chains provide a valid means of relating the coma composition to that of the nucleus.

4. Observations

The $A^2\Sigma - X^2\Pi$ (0,0) band of the hydroxyl radical at ~3085 Å was well known from ground-based spectroscopic observations (Swings and Haser, 1957), but as this wavelength lies very close to the edge of the atmospheric transparency window, the strength of this feature (relative to that of other species) was not appreciated until 1970 when comet Bennett (1970 II) was observed from space by the *Orbiting Astronomical Observatory* (*OAO-2*) (Code *et al.* 1972). The spectrum is reproduced in Figure 1, and also shows a very strong, broadened H I Lyman-α emission from hydrogen, the other principal dissociation product of H_2O, which was independently discovered by measurements made with a photometer onboard the *Orbiting Geophysical Observatory* (*OGO-5*) designed to study the spatial distribution of geocoronal and interplanetary Lyman-α

radiation (Bertaux *et al.* 1973). The broad shape of Lyman-α seen in the objective grating spectrum of Figure 1 is due to the large spatial extent of the atomic hydrogen envelope, the result of a high velocity acquired in the photodissociation process and a long lifetime against ionization. Four years later, at the apparition of comet Kohoutek (1973 VII), atomic oxygen and carbon were identified in the spectra (Feldman *et al.* 1974) and direct UV images of the hydrogen coma, as well as of the O I and C I emissions, were obtained (Opal *et al.* 1974) from sounding rocket experiments. These were repeated in March 1976 for comet West (1976 VI) and produced the spectra in Figure 2 (Feldman and Brune 1976). It should be noted that these rocket observations were made at small solar elongation angles ($\sim 20°$) by using the limb of the Earth to occult the Sun and were thus able, taking advantage of the orbit geometry, to observe these two comets at small heliocentric distances ($\sim$ 0.4 AU) where they were very active and bright. This technique was later used to observe comet Halley at r $\sim$ 0.7 AU (McCoy *et al.* 1986, Woods *et al.* 1986).

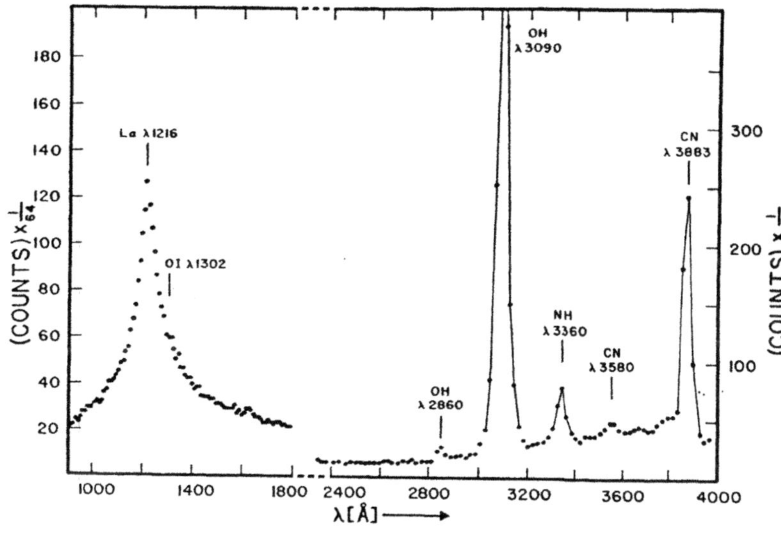

Figure 1. The ultraviolet spectrum of comet Bennett (1970 II) observed by the *OAO-2* satellite. From Code *et al.* (1972).

As noted above, since 1978, over 30 comets have been observed by the *International Ultraviolet Explorer* (*IUE*) satellite observatory. A summary of the highlights of these observations through 1986 has been given by Festou and Feldman (1987). For Halley alone, over 200 UV spectra were obtained over a period extending from September 1985 through July 1986 with the exception of a two month period around perihelion due to the 45° solar avoidance requirement of the satellite's telescope. Two examples are given here, composite spectra of a comet with a very low dust-to-gas ratio, Bradfield (1979 X) (Weaver *et al.* 1981) shown in Figure 3a, and the much dustier P/Halley (1986 III) shown in Figure 3b. For comparison with Figure 2, it should be noted that the field-of-view of

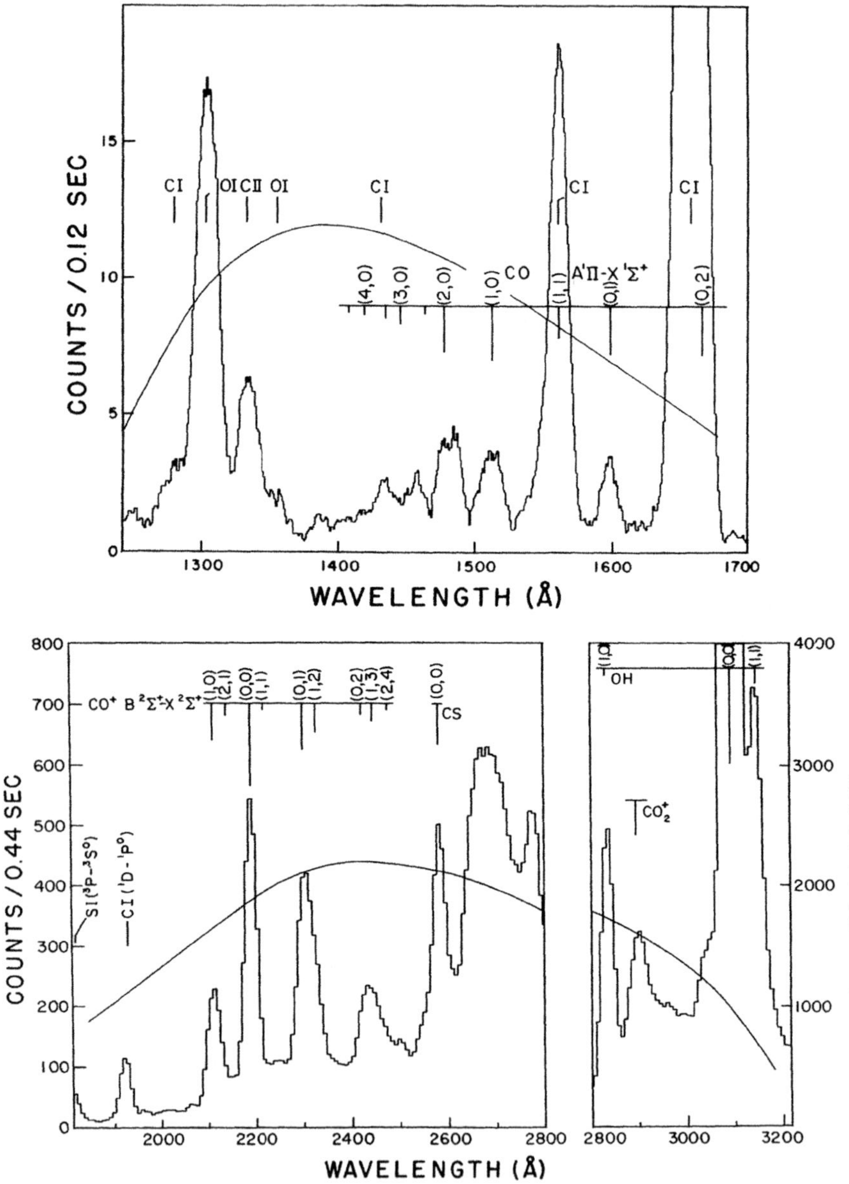

Figure 2. Ultraviolet spectra of comet West (1976 VII), recorded by a sounding rocket payload on 5 March 1976 (Feldman and Brune 1976). The solid line shows the relative instrumental response. In the upper spectrum, a CaF_2 filter was used to attenuate the strong H I Lyman-α emission. The spectral resolution was 15 and 22 Å for the upper and lower spectra, respectively.

the *IUE* aperture is ~ 10" × 20", while that of the rocket spectra of comet West was 35′×5′ (short wavelength) or 35′×10′ (long wavelength). Except for the CO^+ $B^2\Sigma^+ - X^2\Sigma^+$ bands, which have not been observed in any *IUE* spectrum and which may have been detected in comet West because of the long slit length employed, the gaseous emission features in the spectra of Figs. 2 and 3 are all the same, and this similarity of composition extends to the spectra of all comets observed in the UV to date. Differences in the relative intensities may reflect true differences in relative abundances, but are largely due to variations in the fluorescence efficiencies (g-factors) defined in Section 3 above with heliocentric velocity ("Swings effect") and, particularly in the short- wavelength spectra, with phase within the 11-year solar activity cycle. Note the relative weakness of the atomic oxygen and carbon emissions in the spectrum of comet Halley, taken at solar minimum in early 1986, compared to comet Bradfield (1979 X) observed in January 1980 at solar maximum. The H I Lyman-α emission, which is optically thick in the *IUE* aperture for most comets at r $\leq$ 1 AU, is a measure of the solar Lyman-α flux at line center, and is seen to vary by a factor of three, for a given OH column density, between 1980 and 1985 (Threadgill 1987).

Table 2. Observed species

Species	*Wavelength (Å)*	*Lifetime, g-factor*
H I	1216	1,2
O I	1304	1,3
C I	1561/1657	1,3
C I (^{1}D)	1931	4
S I	1814	5
C II	1335	4
CO	1478/1510	4,6
C_2	2313	7
CS	2575	8
S_2	2800-3100	9
OH	3085	10
CO^+	~2200	11
CO_2^+	2890	12

References: 1. Huebner, Keady and Lyon (1982), 2. Keller (1976), 3. Dymond *et al.* (1989), 4. Feldman and Brune (1976), 5. Roettger *et al* (1989), 6. Woods *et al.* (1986), 7. A'Hearn and Feldman (1980), 8. Jackson *et al.* (1982), 9. A'Hearn *et al.* (1983), 10. Schleicher and A'Hearn (1988), 11. Magnani and A'Hearn (1986), 12. Festou *et al.* (1982).

Table 2 lists the known cometary species which have been detected in the ultraviolet in at least one comet. In addition to the wavelengths, references are given for fluorescence efficiency and lifetime. As noted above, these are often variable with the comets's heliocentric velocity ("Swings effect") due to Fraunhofer structure in the solar continuum, as is the case for OH (Figure 4), or due to emission line profiles at shorter wavelengths such as O I λ1302 (Figure 5). The latter also illustrates the differential "Swings" or "Greenstein" effect in which atoms (or molecules) moving towards or away

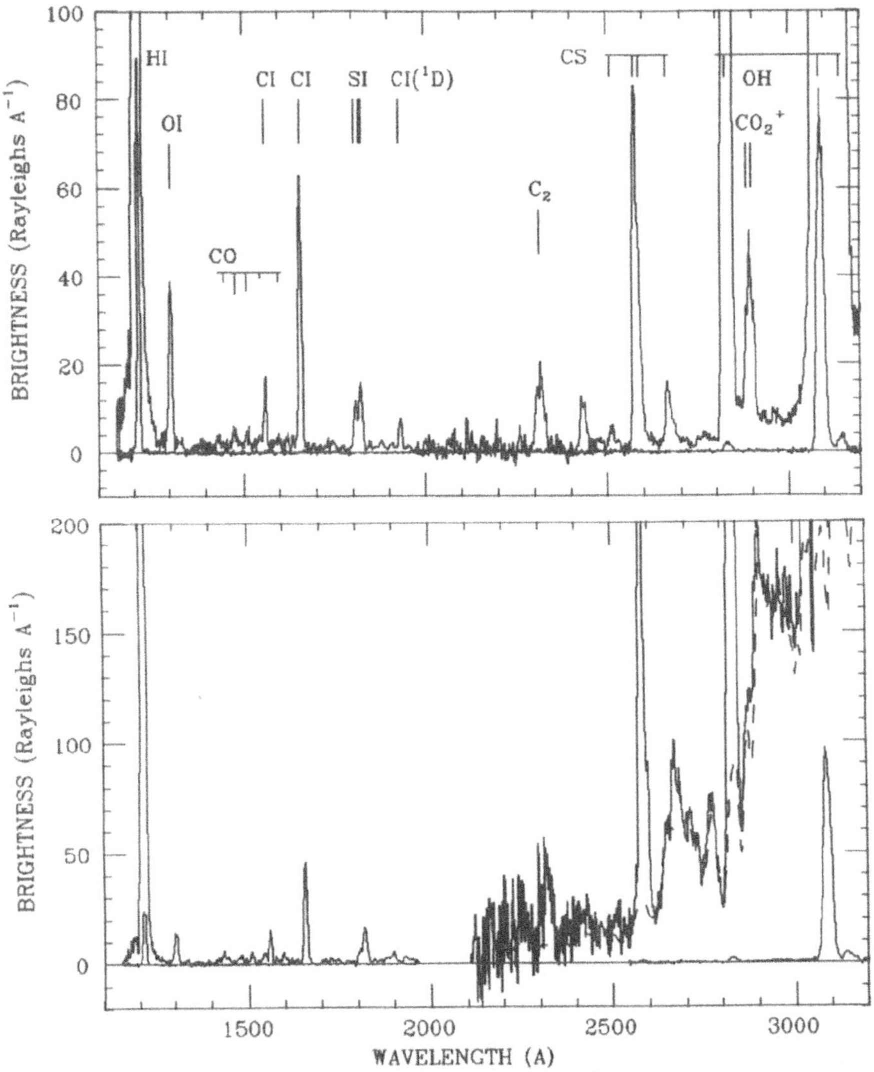

Figure 3. a) Composite *IUE* spectrum of comet Bradfield (1979 X) observed on 10 January 1980, when the comet was 0.71 AU from the sun (Weaver *et al.* 1981). The short exposures are reduced in scale by a factor of 400. b) A similar composite spectrum of comet Halley as obtained by *IUE* on 18 March 1986. The short exposures are reduced in scale by a factor of 200. The noise in the long-wavelength spectrum near 2000 Å (image number LWP 7824) results from limiting the exposure time to avoid saturating the camera with continuum emission near 3000 Å. Note the relative weakness of the atomic emission features due to the lower solar fluxes at solar minimum.

from the sun (in the center-of-mass of the comet) "see" different parts of the solar line profile and thus fluoresce at different rates leading to an asymmetric image of a symmetric coma (Dymond *et al.* 1989).

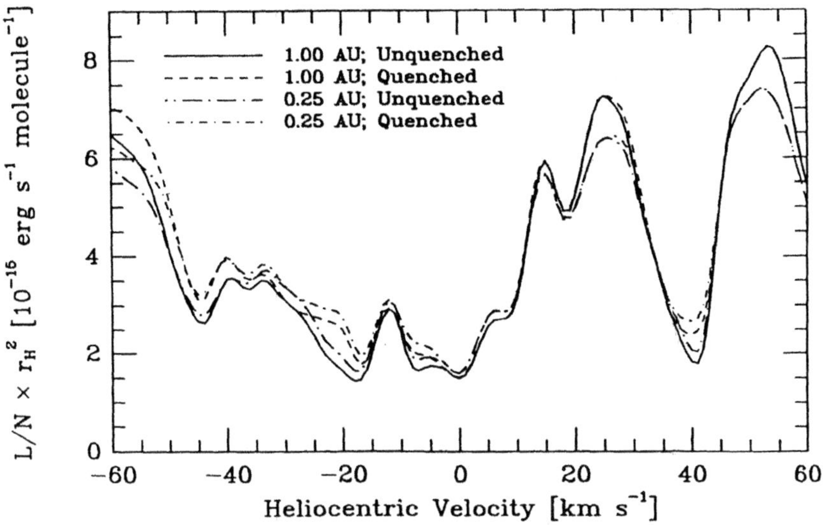

Figure 4. Fluorescence efficiency of the OH (0,0) band at 3085 Å as a function of heliocentric velocity at 0.25 and 1.0 AU (Schleicher and A'Hearn 1988).

Equally important is the list of species not yet detected given in Table 3. This list divides into two; those species whose resonance transitions occur at wavelengths shorter than 1150 Å (the short wavelength cut-off of *IUE*, and also of the *Hubble Space Telescope (HST)* to be launched in 1990), and those for which a combination of abundance and fluorescence efficiency yields a photon flux below the sensitivity threshold of the instruments used to date. For the former, it is necessary to await an astronomy mission dedicated to spectroscopy at $\lambda < 1200$ Å, such as *Lyman/FUSE*. *HST* offers the promise of significant improvement over *IUE*, particularly for D and OD. For Halley, Schleicher *et al.* (1986) report an upper limit of $D/H \leq 4 \times 10^{-4}$ based on OD/OH intensity measurements, which is consistent with the *Giotto* NMS measurement of Eberhardt *et al.* (1987b). Upper limits on H_2 and SO have been given by Dymond (1988) and Kim and A'Hearn (1989), respectively. For H_2, the limit is above the level expected from H_2O dissociation above (reaction *b*). Atomic nitrogen at 1200 Å is masked by the scattering wing of the intense H I Lyman-α emission at 1216 Å in *IUE* spectra.

In the following sections, examples of the application of UV spectroscopy to apparent compositional anomalies in some recently observed comets will be described.

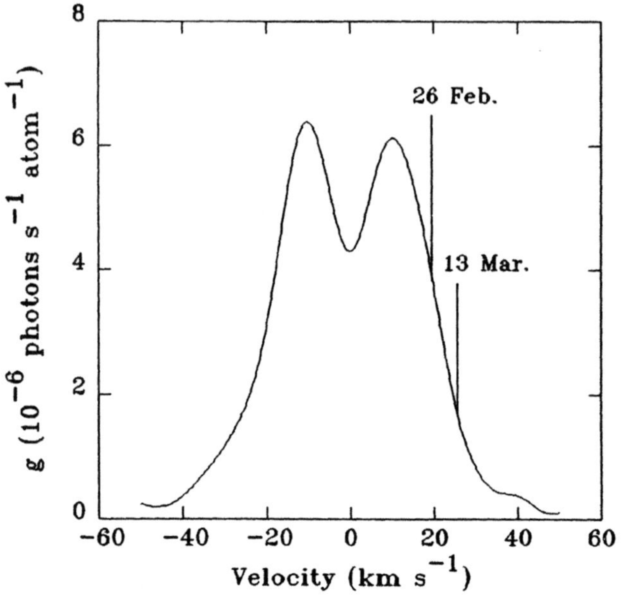

Figure 5. Fluorescence efficiency of the O I λ1302 multiplet as a function of heliocentric velocity (Dymond *et al.* 1989).

Table 3. Species not yet detected

Species	*Wavelength (Å)*
He I	584
He II	304
Ne I	736/744
Ar I	867/876
	1048/1067
O II	539, 834
N I	1134, 1200
D I	1215.3
H_2	1000-1610
N_2	~1400
CO_2	~2000
NO	2150
SO	~2500
OD	~3090

5. Temporal Variability of comet P/Halley

As noted above, the derived column abundances may be used, with suitable models when necessary, to derive the production rate of a given parent species. Thus, the entire suite of observations made during a comet's apparition (again, in those few cases for which it is possible) can be used to characterize the evolution of the gaseous component of the coma during the period of maximum activity of the nucleus. With *IUE*, it has been possible to do this for comets P/Halley (Feldman 1989), Wilson (1987 VII) (Roettger *et al.* 1989), Bradfield (1987s) and P/Tempel-2 (Roettger *et al.* 1990). Periodic comet Encke is not observable over an entire apparition, but as its period is only 3.3 years, it has been observed by *IUE* at three separate apparitions, so that a comparison of the gas production rates derivable from UV measurements with the visual light curve was possible. The light curve exhibits a pronounced asymmetry about perihelion, attributable to decreasing production of the parents of C_2 and CN post-perihelion. Similar behavior was found for the CS emission, but surprisingly, not for the OH, implying symmetric water production on either side of perihelion and an inhomogeneous composition of the cometary ice (A'Hearn *et al.* 1985).

Perhaps more diagnostic of the structure and composition of the cometary surface is the short-term variability, vividly illustrated by the visible light curves of comet P/Halley obtained by Millis and Schleicher (1986) over consecutive nights in March-April 1986. These authors derived a 7.4 day sidereal period for the light curve even though the data showed fine structure indicative of a shorter rotation period commensurate with the 2.2 day period claimed by other investigators. The light curve measures the quantity of gas and dust in the coma, not the nucleus, and its variation is attributed to changes in activity of various jets on the surface of the nucleus as these jets move in and out of sunlight due to the rotation dynamics of the nucleus. The exact nature of the rotation remains an open question (Belton 1990).

Light curve data on P/Halley were also obtained by the *IUE* Fine Error Sensor (FES), which is a slit-jaw camera used in the acquisition and tracking of the celestial targets being observed by the satellite. As a visible light photometer, the FES has an effective aperture of about 18", and because the *IUE* is in a geosynchronous orbit, it can uniquely provide continuous monitoring of the visible brightness of a comet over extended observation periods. During the months of March and April 1986 when there were frequent *IUE* observing shifts scheduled, large variations in the visible brightness were detected from day to day and also even within an 8 or 16-hour observing shift (Feldman *et al.* 1987). These data, while discontinuous, agree well, in both phase and amplitude, with the ground-based observations of Millis and Schleicher. Two features in these data bear noting.

Between 23 and 25 March 1986, *IUE* observations were made over a 40-hour period, all but eight of which were spent on comet Halley. The FES data, shown in the top panel of Figure 6, exhibit a smooth, nearly sinusoidal variation with an apparent period of about two days. The lower panel of Figure 6 shows that the ultraviolet emissions during the same period, from OH, CS, CO_2^+, and dust, all followed the visible light curve with the only difference being the amplitude of the variation (McFadden *et al.* 1987). In particular, OH, from the relatively long-lived water parent ($\tau \sim 1$ day), does not increase as rapidly as does CS (with a parent lifetime ~ 500 seconds) or the dust continuum. The large amplitude variations in both the visible light curve and the ultraviolet emissions continued through mid-April 1986 and then diminished during the remainder of the *IUE* observing dates.

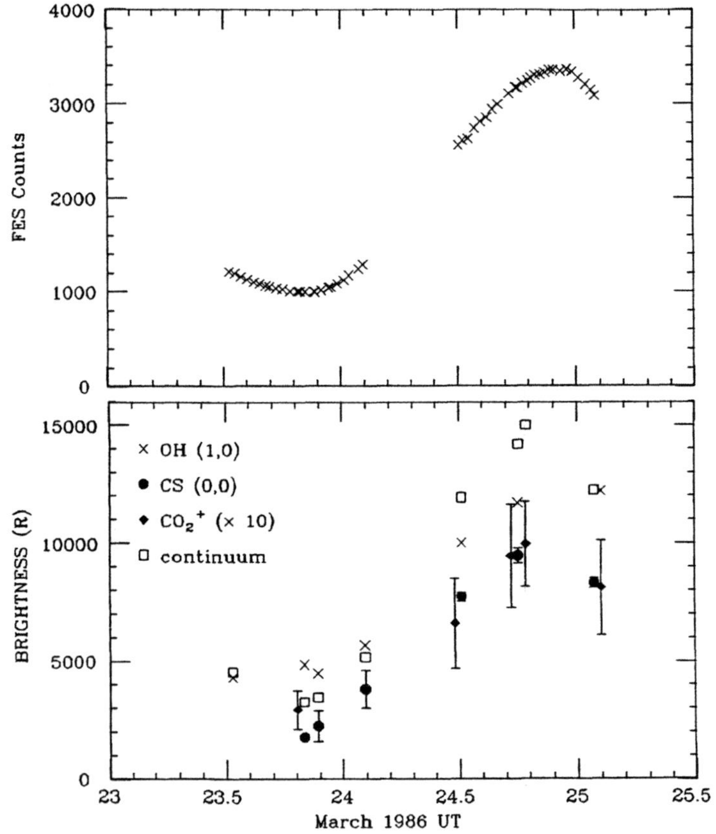

Figure 6. Comparison of the *IUE* FES light curve (top panel) with the variation in UV emsissions (bottom panel). From McFadden *et al.* (1987).

The other temporal event of note occurred on 18-19 March 1986 in the form of a transient outburst, of ~ 0.2 day duration, seen in Figure 7 in the FES count rate superimposed on the more gradual light curve of Millis and Schleicher. Accompanying this outburst was the detection of a large column abundance of CO_2^+ ions at a distance of 150,000 km from the nucleus (Figure 8), but without any increase in OH emission at the nucleus. This event has been interpreted by Feldman *et al.* (1986) as resulting from the eruption of a sub-surface pocket of volatile ices, which also produced an outward moving, extended clump of CO^+ ions detected from ground-based observations on 20 March 1986. Outbursts of this nature appear to be associated with many comets, most notably

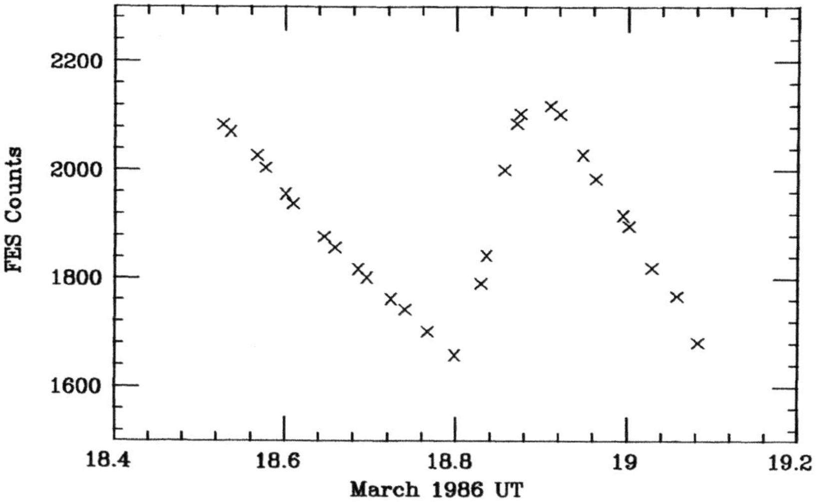

Figure 7. Visual light curve obtained with the *IUE* Fine Error Sensor, showing the outburst of comet Halley on 18 March 1986 (Feldman *et al.* 1986).

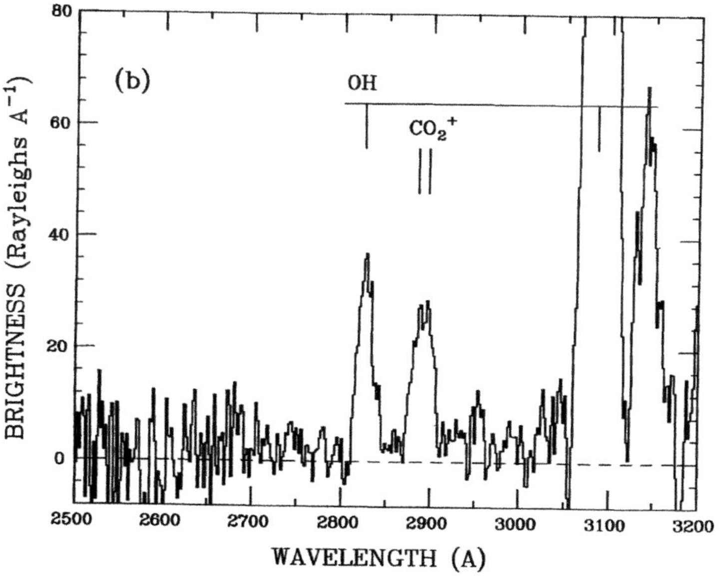

Figure 8. Long wavelength *IUE* spectrum obtained at the time of the peak of the outburst shown in Figure 7. The spectrograph aperture was offset 150,000 km from the nucleus in the tailward direction. The strong CO_2^+ emission is not seen in other offset exposures taken before or after the outburst.

P/Schwassmann-Wachmann 1, but the frequency of occurence on comet Halley must be relatively low.

6. The Distributed Source of CO in P/Halley

We noted above the apparent variation of CO from comet to comet and the tentative conclusion, based on UV detection of CO in only three comets, correlating the CO abundance with the amount of dust in the coma. The *in situ* measurements of CO in comet Halley by the *Giotto* NMS provided additional insight into the nature of this correlation. Unlike the H_2O and CO_2 densities measured simultaneously which showed the characteristic radial outflow ρ^{-2} decrease with distance near the nucleus, the CO distribution deviated markedly from ρ^{-2} for values of ρ less than 10,000 km. This is seen in Figure 9 (Eberhardt *et al.* 1987a) and the interpretation given is in terms of a distributed source of CO with a peak near 10,000 km (Figure 10), and which accounts for somewhat more than half of the carbon monoxide observed further out in the coma.

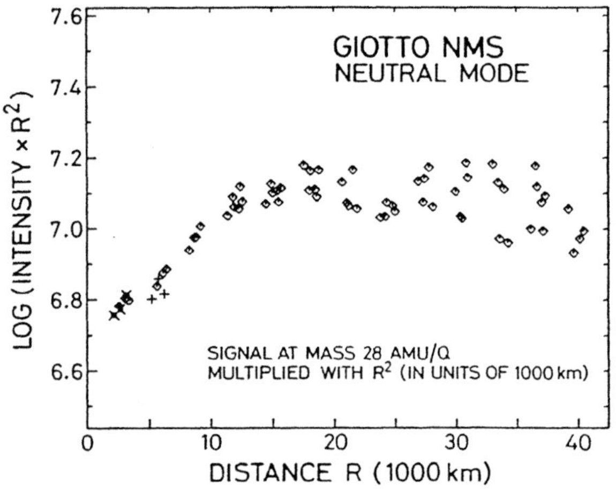

Figure 9. *Giotto* NMS measurements of the CO density in comet Halley (Eberhardt *et al.* 1987a). If the density followed a ρ^{-2} distribution, the data as plotted would lie on a straight line.

Additional information on the spatial distribution of CO in the coma of P/Halley is available from the long-slit ultraviolet spectra obtained from two sounding rocket experiments on 26 February and 13 March 1986 by Woods *et al.* (1986). The spatial profiles of the CO Fourth Positive emission bands near 1500 Å show the expected ρ^{-1} behavior (due to integrating the ρ^{-2} density distribution along a line-of-sight) except within ~ 20,000 km of the optical center of the comet, where they appear flattened. However, post-flight analysis has shown that the flattening is due primarily to rocket pointing error and jitter which masks any evidence of the distributed source in these data. Using the observed profile at 50,000 km, the derived production rate of CO relative to that of H_2O is found to be ~ 17%, in good agreement with the *Giotto* NMS measurement.

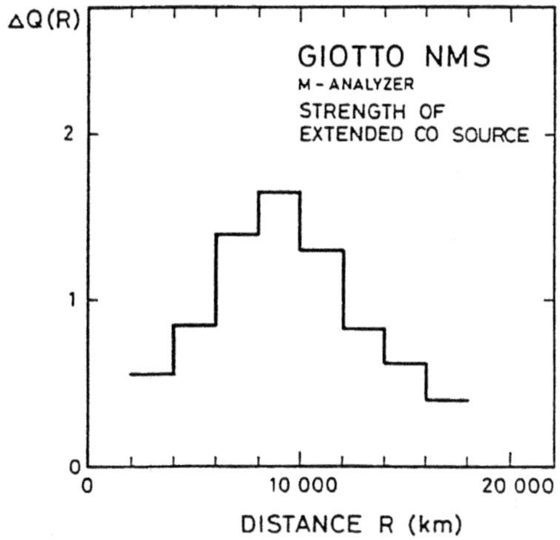

Figure 10. CO source distribution required to produce the data of Figure 9. From Eberhardt *et al.* (1987a).

Feldman *et al.* (1988) pointed out that the *IUE* observations of CO could be used to complement the rocket data as the small field-of-view of the *IUE* spectrograph apertures emphasizes the emission from gas present within 3,000-4,000 km of the nucleus. Furthermore, the comparison would provide a relative measure of the CO abundance near and far from the nucleus, independent of uncertainties in the CO fluorescence efficiencies, particularly at the time of the *Giotto* encounter on 13 March 1986 when rocket and *IUE* observations were made only hours apart. The results are shown in Figure 11 and again indicate a direct source from the nucleus contributing one-third to one-half the total CO in the coma. Moreover, it can be inferrred that the distributed source was not unique to the time of the *Giotto* encounter. The full details of this analysis will be presented elsewhere.

Several explanations have been advanced to account for the additional CO. These include the fragmentation of "CHON" particles and the dissociation of POMs, polyoxymethylene polymers released from dust grains, both hypotheses based on various other *in situ* data. The various merits of these proposals have been recently discussed by Krankowsky and Eberhardt (1989). It is likely that the same mechanism responsible for the CO is also the source of the observed CN "jets" of A'Hearn *et al.* (1986).

In connection with the CO observations, it should be noted that the observed spatial profile of the atomic carbon emissions is not consistent with photodissociation of CO as the only source of the carbon, confirming the apparent high carbon abundance (the "carbon puzzle") derived from *IUE* spectra of a number of comets (Festou 1984). Woods *et al.* (1987) have attempted to account for the high carbon content near the nucleus by evaluating various reaction paths initiated by electron impact of CO, the source of the electrons being photoionization, and the flux distribution based on *in situ Vega* measurements (Gringauz *et al.* 1986). While rough quantitative agreement can be found, the

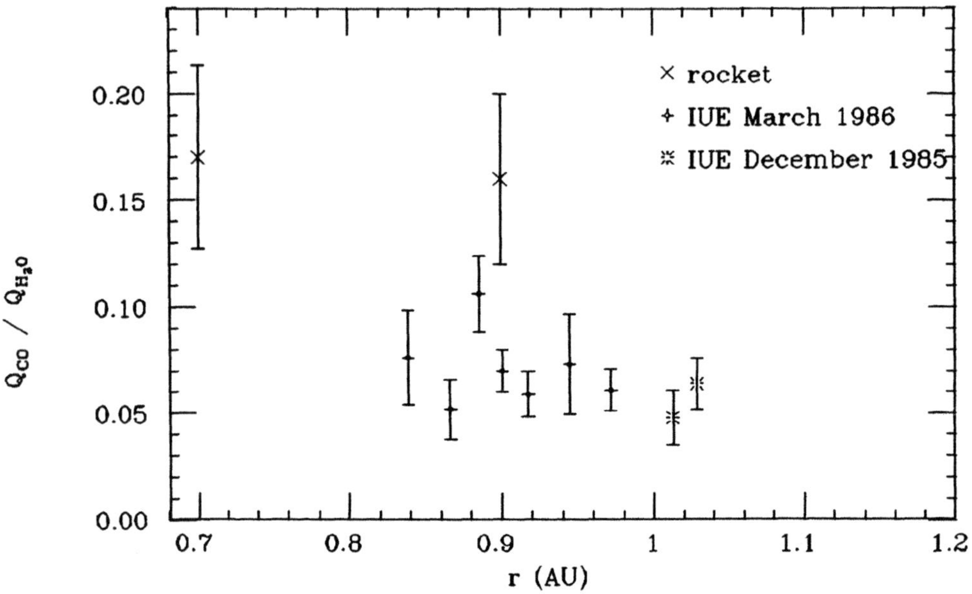

Figure 11. Apparent CO production rates derived from rocket and *IUE* data. The rocket data we evaluated at 50,000 km while the *IUE* spectra sample the region within 3,000-4,000 km of the nucleus.

detailed spatial distribution is less satisfactorily fit to the observations. The possibility of a grain source of C would not be consistent with the high carbon abundance seen in very dust-poor comets such as Bradfield (1979 X) or the recently observed P/Brorsen-Metcalf (1989o).

7. The S_2 Molecule in comet IRAS-Araki-Alcock

The presence of sulfur as a common constituent in comets has been established by the detection of carbon monosulfide (CS) in nearly all of the comets observed by the *IUE* with a production rate $\sim 10^{-3}$ that of water. Atomic sulfur is also detected in many, but is a weaker emission and is seen only in the brighter comets. However, if the presumed parent of CS, CS_2, is the only sulfur-bearing molecule present in cometary ice, then sulfur would be underabundant, by an order of magnitude, relative to oxygen, based on solar abundances, unless other molecules, such as H_2S or SO_2 were also present. With one exception, no other sulfur-containing molecule has been detected in the cometary coma (Kim and A'Hearn, 1989).

The exception is S_2, which was discovered serendipitously in comet IRAS–Araki–Alcock (1983 VII) by A'Hearn *et al.* (1983) on 11 May 1983, when the comet was at its closest approach to Earth, 0.032 AU. The spectrum, which is quite

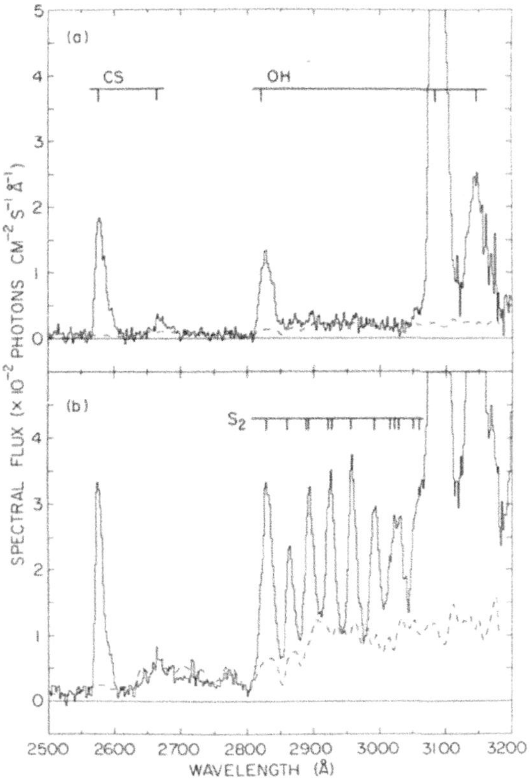

Figure 12. *IUE* spectra of comet IRAS–Araki–Alcock (1983 VII) obtained on 11 May 1983. The lower spectrum, showing S_2 emission, was taken with the aperture tracking the center-of-brightness of the comet. The upper spectrum had the aperture displaced 15 arc-seconds. From A'Hearn *et al.* (1983).

unambiguous, is shown in Figure 12. The deduced abundance of S_2, relative to H_2O, was $\sim 1 \times 10^{-3}$, comparable to the CS abundance, shortly after an outburst on 11 May, but had all but disappeared 28 hours later, despite the initial claim by Feldman *et al.* (1984) that the S_2 flux had decreased by a factor of 10. This is the only unambiguous reported detection of S_2 in any comet, or in any celestial object, and its discovery at the time prompted speculation that S_2 may serve as a unique indicator of the physical and chemical conditions at the site of comet formation in the primordial solar nebula. Grim and Greenberg (1987), on the basis of laboratory simulations, proposed that the S_2 would be formed by cosmic ray bombardment of cometesimal grains in the precursor molecular cloud of the solar nebula, and would therefore be present in the ices of all comets formed in the primordial nebula. Thus, although the conditions for observing S_2 were unique to comet IRAS-Araki-Alcock (i.e., very small Δ), it is surprising that no trace of it has appeared in any *IUE* spectrum of comets at moderately small Δ (~ 0.2 to 0.5 AU).

We have begun a systematic determination of S_2 upper limits from our existing *IUE* data, using a least-squares technique to determine the likelihood that a given spectrum matches the observed S_2 spectrum in comet IRAS-Araki-Alcock at some reduced level. The results of this analysis will be described in detail elsewhere, but we note here that for Halley at closest approach to the Earth on 9 April 1986 ($\Delta = 0.42$ AU), the derived upper limit to the abundance of S_2 relative to H_2O is four times lower than that observed in comet IRAS-Araki-Alcock at the time of the outburst. Further searches for S_2 in future comets, to a limiting sensitivity a factor of 100 lower, will be made with the *Hubble Space Telescope,* now scheduled for launch in 1990.

Acknowledgements

The preparation of this review was partially supported by NASA grants NAG 5-619 and NsG-5393.

References

A'Hearn, M. F., Birch, P. V., Feldman, P. D. and Millis, R. L. 1985, Comet Encke: Gas production and light curve, *Icarus* **64**, 1-10.

A'Hearn, M. F. and Feldman, P. D. 1980, Carbon in comet Bradfield 1979l, *Astrophys. J. (Letters)* **242**, L187-L190.

A'Hearn, M. F., Feldman, P. D., and Schleicher, D. G. 1983, The discovery of S_2 in comet IRAS-Araki-Alcock 1983d, *Astrophys. J. (Letters)* **274**, L99-L103.

A'Hearn, M. F., Hoban, S., Birch, P. V., Bowers, C., Martin, R., and Klinglesmith, D. A. 1986, Cyanogen jets in comet Halley, *Nature* **324**, 649-651.

Allen, M. *et al.* 1987, Evidence for methane and ammonia in the coma of comet Halley, *Astron. Astrophys.* **187**, 505-512.

Belton, M. J. S. 1990, Rationalization of Halley's periods: Evidence for an inhomogeneous nucleus?, *Icarus,* submitted.

Bertaux, J.-L. 1986, The UV bright spot of water vapor in comets, *Astron. Astrophys.* **160**, L7-L10.

Bertaux, J.-L., Blamont, J. E. and Festou, M. 1973, Interpretation of hydrogen Lyman-alpha observations of comets Bennett and Encke, *Astron. Astrophys* **25**, 415-430.

Code, A. D., Houck, T. E. and Lillie, C. F. 1972, Ultraviolet observations of comets, in *The Scientific Results from Orbiting Astronomical Observatory (OAO-2),* ed. A. D. Code, NASA SP-310, 109-114.

Colom, P., Gérard, E., Bockelée-Morvan, D. and Crovisier, J. 1989, The OH radio observations of comets at Nançay: new results and new questions, paper presented at *Comets in the Post-Halley Era,* Bamberg, FRG.

Combes, M. *et al.* 1988, The 2.5-12 μm spectrum of comet Halley from the IKS-*Vega* experiment, *Icarus* **71**, 404-436.

Combi, M. R. and Delsemme, A. H. 1980, Neutral cometary atmospheres 1. An average random walk model of photodissociation in comets, *Astrophys. J.* **237**, 633-640.

Crovisier, J. 1988, Infrared cometary spectroscopy, in *22nd ESLAB Symposium on Infrared Spectroscopy in Astronomy,* ESA SP-290.

Delsemme, A. H. 1982, Chemical composition of cometary nuclei, in *Comets,* ed. L. L. Wilkening, University of Arizona Press, Tucson, 85-130.

Dymond, K. F. 1988, Far-ultraviolet spectroscopy of comet Halley (1986 III), Ph.D. Dissertation, The Johns Hopkins University.

Dymond, K. F., Feldman, P. D., and Woods, T. N. 1989, Observations of a Greenstein effect in the O I λ1302 emission of comet Halley, *Astrophys. J.* **338**, 1115-1121.

Eberhardt, P. *et al.* 1987a, The CO and N_2 abundance in comet P/Halley, *Astron. Astrophys.* **187**, 481-484.

Eberhardt, P. *et al.* 1987b, The D/H ratio in water from comet P/Halley, *Astron. Astrophys.* **187**, 435-437.

Feldman, P. D. 1983, Ultraviolet spectroscopy and the composition of cometary ice, *Science* **219**, 347-354.

Feldman, P. D. 1986, Carbon monoxide in cometary ice, in *Asteroids, Comets, Meteors II,* eds. C. I. Lagerkvist, B. A. Lindblad, H. Lundsted and H. Rickman, Uppsala University, 263-267.

Feldman, P. D. 1989, Rocket and satellite observations of the ultraviolet emissions of comet Halley, in "Comet Halley 1986: World-Wide Investigations, Results and Interpretations", Ellis Horwood Ltd., Chichester, in press.

Feldman, P. D., A'Hearn, M. F., and Millis, R. L. 1984, Temporal and spatial behavior of the ultraviolet emissions of comet IRAS-Araki-Alcock (1983d), *Astrophys. J.* **282**, 799-802.

Feldman, P. D., A'Hearn, M. F., Festou, M. C., McFadden, L. A., Weaver, H. A. and Woods, T. N. 1986, Is CO_2 responsible for the outbursts of comet Halley?, *Nature* **324**, 433-436.

Feldman, P. D. and Brune, W. H. 1976, Carbon production in comet West (1975n), *Astrophys. J.* (Letters) **209**, L145-L148.

Feldman, P. D., Dymond, K. F., Roettger, E. E., and Weaver, H. A. 1988, The distribution of carbon monoxide in the coma of comet P/Halley, *B. A. A. S.* **20**, 827.

Feldman, P. D., Takacs, P. Z., Fastie, W. G. and Donn, B. 1974, Rocket ultraviolet spectrophotometry of comet Kohoutek (1973f), *Science* **185**, 705-707.

Feldman, P. D. *et al.* 1987, *IUE* observations of comet P/Halley: Evolution of the ultraviolet spectrum between September 1985 and July 1986, *Astron. Astrophys.* **187**, 325-328.

Festou, M. C. 1981, The density distribution of neutral compounds in cometary atmospheres: I. Models and equations, *Astron. Astrophys.* **95**, 69-79.

Festou, M. C. 1984, Aeronomical processes in cometary atmospheres: The carbon compounds' puzzle, *Adv. Space Res.* **4**, number 9, 165-175.

Festou, M. C. and Feldman, P. D. 1981, The forbidden oxygen lines in comets, *Astron. Astrophys.* **103**, 154-159.

Festou, M. C. and Feldman, P. D. 1987, Comets, in *Exploring the Universe With the IUE Satellite*, ed. Y. Kondo, Reidel, Dordrecht, 101-118.

Festou, M. C., Feldman, P. D. and Weaver, H. A. 1982, The ultraviolet bands of the CO_2^+ ion in comets, *Astrophys. J.* **256**, 331-338.

Grim, R. J. and Greenberg, J. M. 1987, Photoprocessing of H_2S in interstellar grain models as an explanation for S_2 in comets, *Astron. Astrophys.* **181**, 155-168.

Gringauz, K. I. *et al.* 1986, First *in situ* plasma and neutral gas measurements at comet Halley, *Nature* **321**, 282-285.

Haser, L. 1957, Distribution d'intensité dans la tête d'une comète, *Bull. Acad. Roy. Belgique*, Classe de Sciences **43**, 740-750.

Huebner, W. F. and Giguere, P. T. 1980, A model of cometary comae. II. Effects of solar photodissociative ionization, *Astrophys. J.* **238**, 753-762.

Huebner, W. F., Keady, J. J., and Lyon, S. P. 1982, Solar photo rates for planetary atmospheres and atmospheric pollutants, Los Alamos National Laboratory report.

Jackson, W. M., Halpern, J., Feldman, P. D. and Rahe, J. 1982, Production of CS and S in comet Bradfield (1979 X), *Astron. Astrophys.* **107**, 385-388.

Keller, H. U. 1976, The interpretations of ultraviolet observations of comets, *Space Science Rev.* **18**, 641-684.

Kim, S. J. and A'Hearn, M. F. 1989, Sulfur compounds in comets, paper presented at *Comets in the Post-Halley Era*, Bamberg, FRG.

Kissel, J. *et al.* 1986, Composition of comet Halley dust particles from *Giotto* observations, *Nature* **321**, 336-337.

Krankowsky, D. and Eberhardt, P. 1989, Evidence for the composition of ices in the nucleus of comet Halley, in "Comet Halley 1986: World-Wide Investigations, Results and Interpretations", Ellis Horwood Ltd., Chichester, in press.

Krankowsky, D. *et al.* 1986, *In situ* gas and ion measurements at comet Halley, *Nature* **321**, 326-329.

Lämmerzahl, P. *et al.* 1987, Expansion velocity and temperature of gas and ions measured in the coma of comet Halley, *Astron. Astrophys.* **187**, 169-173.

Larson, H. P. Weaver, H. A., Mumma, M. J., and Drapatz, S. 1989, Airborne infrared spectroscopy of comet Wilson (1986l) and comparisons with comet Halley, *Astrophys. J.* **338**, 1106-1114.

Lunine, J. I. 1989, Primitive bodies: Molecular abundances in comet Halley as probes of cometary formation environments, in *The Formation and Evolution of Planetary Systems*, ed. H. A. Weaver and L. Danly, Cambridge University Press, Cambridge.

Magnani, L. and A'Hearn, M. F. 1986, CO^+ fluorescence in comets, *Astrophys. J.* **302**, 477-487.

Mendis, D. A., Houpis, H. L. and Marconi, M. L. 1985, The physics of comets, *Fundamentals of Cosmic Physics* **10**, 1-380.

Millis, R. L. And Schleicher, D. G. 1986, Rotational period of comet Halley, *Nature* **324**, 646-649.

McCoy, R. P., Opal, C. B. and Carruthers, G. R. 1986, Far-ultraviolet spectral images of comet Halley from sounding rockets, *Nature* **324**, 439-441.

McFadden, L. A., A'Hearn, M. F., Feldman, P. D., Roettger, E. E., Edsall, D. M. and Butterworth, P. S. 1987, Activity of comet P/Halley 23-25 March 1986: *IUE* observations, *Astron. Astrophys.* **187**, 333-338.

Okabe, H. 1978, "Photochemistry of Small Molecules", Wiley, New York.

Opal, C. B., Carruthers, G. R., Prinz, D. K. and Meier, R. R. 1974, Comet Kohoutek: Ultraviolet images and spectrograms, *Science* **185**, 702-705.

Opal, C. B., McCoy, R. P., and Carruthers, G. R. 1987, Far-ultraviolet objective spectra of comet P/Halley from sounding rockets, *Astron. Astrophys.* **187**, 302-324.

Oppenheimer, M. and Downey, C. J. 1980, The effect of solar-cycle ultraviolet flux variations on cometary gas, *Astrophys. J. (Letters)* **241**, L123-L127.

Roettger, E. E., Feldman, P. D., A'Hearn, M. F., Festou, M. C., McFadden, L. A., and Gilmozzi, R. 1989, *IUE* observations of the evolution of comet Wilson (1986l), *Icarus* **80**, 303-314.

Roettger, E. E., Feldman, P. D., A'Hearn, M. F., and Festou, M. C. 1990, Comparison of ultraviolet and visual light curves for some recent comets, *Icarus*, submitted.

Schleicher, D. G. *et al.* 1986, Comets P/Giacobini-Zinner and P/Halley at high dispersion, in *New Insights in Astrophysics; 8 Years of UV astronomy with IUE,* ESA SP-263, 31-33.

Schleicher, D. G. and A'Hearn, M. F. 1988, The fluorescence of cometary OH, *Astrophys. J.* **331**, 1058-1077.

Smith, A. M., Stecher, T. P. and Casswell, L. 1980, Production of carbon, sulfur, and CS in comet West, *Astrophys. J.* **242**, 402-410.

Spinrad, H. 1982, Observations of the red auroral oxygen lines in nine comets, *Pub. A. S. P.* **94**, 1008-1016.

Swings, P. and Haser, L. 1957, "Atlas of Representative Cometary Spectra", AFCRC Technical Report 56-267.

Threadgill, M. 1987, A study of the Lyman-alpha brightness of eleven comets, M. A. Thesis, The Johns Hopkins University.

van Dishoeck, E. F. and Dalgarno, A., 1984, The dissociation of OH and OD in comets by solar radiation, *Icarus* **59**, 305-313.

Weaver, H. A. Feldman, P. D., Festou, M. C. and A'Hearn, M. F. 1981, Water production models for comet Bradfield (1979 X), *Astrophys. J.* **251**, 809-819.

Woods, T. N., Feldman, P. D., Dymond, K. F. and Sahnow, D. J. 1986, Rocket ultraviolet spectroscopy of comet Halley and abundance of carbon monoxide and carbon, *Nature* **324**, 436-438.

Woods, T. N., Feldman, P. D. and Dymond, K. F. 1987, The atomic carbon distribution in the coma of comet P/Halley, *Astron. Astrophys.* **187**, 380-384.

FROM INTERSTELLAR DUST TO COMETS TO COMET DUST : A TEST OF THE INTERSTELLAR DUST MODEL OF COMETS

J. M. GREENBERG and J.I. HAGE
Laboratory Astrophysics, University of Leiden
The Netherlands

ABSTRACT. The interstellar dust model of comets is used as a basis to simultaneously satisfy various observational constraints and to derive the porosity of comet dust. The observational constraints are : (1) the strengths of the 3.4 μm and 9.7 μm emission bands ; (2) the shape of the 9.7 μm band ; (3) the relative amount of silicates to organic materials ; (4) the mass distribution of the dust. The method used involves precise calculation of the temperatures and emission characteristics of porous aggregates of interstellar dust as a function of their mass, porosity and distance to the sun and the wavelength. The results indicate a high porosity of comet dust in the range $0.93 < P < 0.98$.

1 Introduction

The basic idea of the interstellar dust model of comets by Greenberg (1982, 1985) is that in the regions of the protosolar nebula where comets were born the temperature was so low that all components of the dust were preserved during aggregation into a comet. It has been proposed by Greenberg (1985) that during the formation of comets, aggregates of interstellar dust particles are made as illustrated schematically in figure 1. The individual interstellar dust grains contain a core of interstellar silicate and a mantle of organic refractory material. The aggregates are expected to have a packing factor of ≈ 0.2 (this corresponds to 80% vacuum, or porosity $P = 0.8$). The tangled structure of these aggregates, like a bird's nest, is suggested as a possible way to provide them with rigidity. The aggregates, in turn, coalesce to form much larger bodies and ultimately comets. When a comet comes close to the sun, the particles in the coma, which have been lifted from the comet surface, must initially consist of such fluffy aggregates of various sizes, with a porosity of about 0.8. The outer icy mantles (predominantly H_2O) evaporate quickly (close enough to the sun) once an individual aggregate is lifted from the comet and is fully exposed to the solar radiation. After evaporation of these volatile ices the

J. M. Greenberg and V. Pirronello (eds.), Chemistry in Space, 363–382.

aggregates consist of sub-micron particles ($m \approx 6 \times 10^{-14} g$) composed of organic refractory material and interstellar silicates. The predicted porosity of the coma dust is $P \approx 0.95$.

In this paper we use the interstellar dust model as a basis to quantitatively explain ground based infrared observations of the inner coma of comet Halley around $\lambda = 3.4 \mu m$ and $\lambda = 9.7 \mu m$ in terms of the particle size distributions measured in situ by the Giotto and Vega missions. At the same time we satisfy the relative amounts of organic and silicate material required by Giotto mass spectroscopy observations (Kissel and Krueger, 1987). When an earlier attempt was made (Greenberg, *et. al.*, 1989), it was found that the amount of small particles required to explain the infrared observations was between one and two orders of magnitude higher than then believed to exist (McDonnell *et. al.*, 1987). The present paper uses the results of the improved calibration of the PIA and DIDSY instruments (McDonnell *et. al.*, 1989), VEGA mass spectrometer data (Mazets *et. al.* 1987) and rigorous calculations of the temperatures and spectral properties of porous aggregates of interstellar dust to derive information about the porosity of the coma dust. The organisation of this paper is as follows. Section 2 is a brief recapitulation of the interstellar dust model of comets. In section 3a formulae are derived which show the relation between the observed size spectra in the coma and observations of the thermal emission. Section 3b lists our choice of observational results. In section 4 and 5 we derive the emission properties and temperatures of porous aggregates of interstellar dust as a function of their size, porosity and composition. In section 6 the results of sections 3, 4 and 5 are used to derive the porosity of the dust in the coma of comet Halley. Section 7 contains a discussion and our conclusions.

2 The interstellar dust model of comets

We briefly review the model of interstellar dust used here and derive the predicted properties of the comet nucleus and the coma dust according to that model.

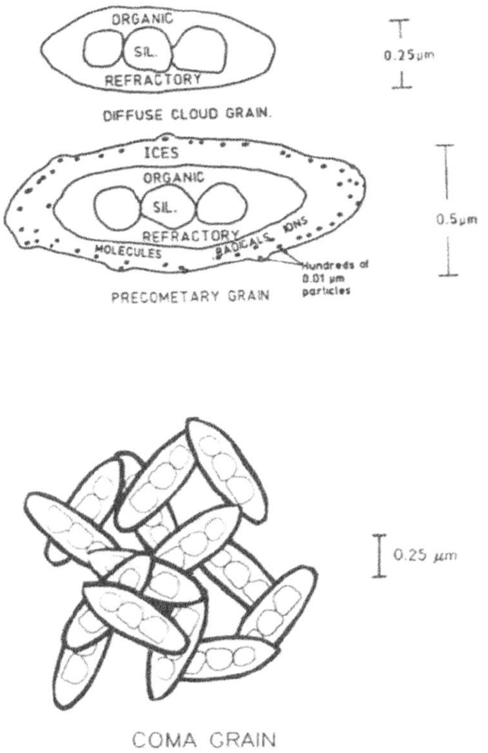

Figure 1. ***Top:*** ***Schematic of an interstellar dust grain, which contains a core of silicates and a mantle of organic refractory material. Middle: An interstellar grain as it would appear in the presolar dust cloud after accretion of gases on its surface. These grains make up a comet. Bottom: Schematic of a coma grain as it would appear according to the interstellar dust model. It would have a porosity of about 95%.***

2a INTERSTELLAR DUST IN THE PROTOSOLAR NEBULA

Protosolar nebula dust, out of which comets are formed, is a developed stage of diffuse cloud interstellar dust (Greenberg, 1985). The diffuse cloud interstellar dust consists of three major populations of particles as described by Greenberg, (1985). The major population, in terms of mass, consists of elongated particles with a silicate core and an "organic refractory" mantle, a few tenths of micrometers in size (i.e. "core-mantle particles", see figure 1. The organic refractory material originally started out as ices of simple chemical compounds, but has undergone up to billions of years of ultraviolet photoprocessing in interstellar space. The photoprocessing has changed the ice mixture into a carbon-rich and oxygen-poor refractory material containing many different organic molecules. The second population contains very small carbonaceous particles with radii smaller than $1/100\mu m$. Equally small silicate particles were originally proposed for the third population by Greenberg, but recent developments indicate that this population may consist of PAH molecules instead (see e.g. Leger and d'Hendecourt, this volume).

In the presolar molecular cloud the dust consists of the core-mantle particles with an additional outer mantle of volatile ices dominated by H_2O (see figure 1). The expected mass fractions of the three components of this dust are 0.20, 0.19 and 0.55, for the silicate (sil), organic refractory (or) and the volatile material (ice), respectively. These proportions are based on the model of interstellar dust (Greenberg, 1988) extrapolated to the protosolar nebular cloud stage and constrained by solar abundance. In a very dense region all the smaller particles could be imbedded in the icy mantles. The mass fraction of carbonaceous particles (c) is about 0.06.

2b THE COMET NUCLEUS AND COMA DUST

Regarding the formation of comets, it has been proposed by Greenberg that initially, aggregates of the protosolar nebula particles are made. Their tangled structure, like a bird's nest (Greenberg and Gustafson, 1981), is suggested as a possible way to provide them with rigidity. This structure may also be conjectured to be the result of random aggregation modified by slippage after particle collision at speeds of about $0.1 \mathrm{km\,s^{-1}}$. The aggregates, in turn, coalesce into larger bodies and ultimately comets. When a comet comes close to the sun, individual aggregates of various sizes are lifted from the comet and are exposed to the solar radiation. The ice mantles evaporate and the small carbonaceous particles are lost from the aggregates, as well. Some of the organic material may also evaporate. Hereafter the aggregates consist of core-mantle particles composed only of organic refractory material and interstellar silicates as illustrated schematically in figure 1. It is assumed that, under these conditions, this type of porous particle constitutes the coma dust in all but the most inner part of the coma. This assumption immediately implies that the 3.4 μm and 9.7 μm spectral features characteristic of interstellar dust, should, under appropriate conditions, show up in the infrared coma emission of a comet.

We now define the porosity, P, of an aggregate of interstellar dust as

$$P = 1 - \frac{V_{\mathrm{solid}}}{V}, \tag{2.1}$$

where V_{solid} is the volume of the solid material inside the aggregate, and V is the total volume of the aggregate, its volume being defined by an appropriate boundary. We define this boundary as the surface of the smallest convex volume which contains all the solid material of the aggregate (See Hage and Greenberg,

1990, for an example). We have $0 \leq P < 1$, where $P = 0$ corresponds to a solid aggregate and $P \approx 1$ corresponds to a cloud of independent interstellar dust particles. Note that a porosity P corresponds to a packing factor of $p = 1 - P$, and that if the porosity is high, a small change in the porosity corresponds to a large relative change in the packing factor. The interstellar dust model of comets predicts that the porosity of the comet could be as high as $P_{\text{nucleus}} = 0.83$, i.e. 83% of the total volume of a comet could be taken up by empty space (a packing factor of $p = 0.17$). The predicted porosity of the coma dust could be as high as $P_{\text{dust}} = 0.975$. These predictions are clarified in the remainder of this section.

If the aggregation into a comet were to lead to a fully compacted structure (i.e. P=0), the nucleus density would be

$$\rho_{\text{nucleus}} = \frac{M_{\text{sil}} + M_{\text{or}} + M_{\text{c}} + M_{\text{ice}}}{V_{\text{sil}} + V_{\text{or}} + V_{\text{c}} + V_{\text{ice}}}, \tag{2.2}$$

where M and V refer to the mass and volume fractions of the various materials in the protosolar nebula dust. Using the mass fractions of the protosolar nebula dust given in section 2[a] and densities of $\rho_{\text{sil}} = 3.5$, $\rho_{\text{or}} = 1.8$, $\rho_{\text{ice}} = 1.2$ and $\rho_{\text{c}} = 2$, in units of grams per cubic centimeter, one finds $\rho_{\text{nucleus}} = 1.54 \text{g cm}^{-3}$. This is about 15% higher than previously derived (Greenberg, 1986b), because of the present assumption of higher material densities. The present result assumes that the condensable elements in the comet have (proto-)solar abundance. The density and porosity of the coma dust may now be derived by assuming that the only difference between comet material and coma dust is that the volume filled originally by the ices and small carbonaceous particles in the nucleus material is taken up by an equal volume of vacuum inside the coma dust. This corresponds to taking $M_{\text{c}} = M_{\text{ice}} = 0$ in equation 2.2, resulting in $\rho_{\text{dust}} = 0.6 \text{g cm}^{-3}$. The porosity of the coma dust derived in this way would be

$$P_{\text{dust}} = 1 - \frac{V_{\text{sil}} + V_{\text{or}}}{V_{\text{sil}} + V_{\text{or}} + V_{\text{c}} + V_{\text{ice}}} = 0.75. \tag{2.3}$$

At this point we note that the mean density of meteors, some of which are clearly of cometary origin, is generally much less than 0.6 g cm^{-3}. According to Verniani (1969, 1973), those meteors whose mean aphelion distance is highest and which are therefore least affected by the sun have the lowest densities of about 0.2 g cm^{-3}. The interstellar dust model of comets assumes that the density of coma dust equals, or is less than, that of these meteors and this implies that the coma dust must have a higher porosity than derived above. The implied minimum density change by a factor of three now gives

$$P_{\text{dust}} = 1 - \frac{V_{\text{sil}} + V_{\text{or}}}{3 \times (V_{\text{sil}} + V_{\text{or}} + V_{\text{c}} + V_{\text{ice}})} = 0.92. \tag{2.4}$$

Consequently, by reconstitution of the ices and carbonaceous particles into the coma dust, the porosity and density of the comet itself may be derived:

$$P_{\text{nucleus}} = 1 - \frac{V_{\text{sil}} + V_{\text{or}} + V_{\text{ice}} + V_{\text{c}}}{3 \times (V_{\text{sil}} + V_{\text{or}} + V_{\text{c}} + V_{\text{ice}})} = 0.67 \tag{2.5}$$

$$\rho_{\text{nucleus}} = \frac{M_{\text{sil}} + M_{\text{or}} + M_{\text{c}} + M_{\text{ice}}}{3 \times (V_{\text{sil}} + V_{\text{or}} + V_{\text{c}} + V_{\text{ice}})} = 0.51. \tag{2.6}$$

Alternatively, one may assume a meteor density of 0.1 g cm^{-3} as given by Olsson-Steel (1989), which would give $P_{\text{dust}} = 0.96$, $P_{\text{nucleus}} = 0.83$, $\rho_{\text{nucleus}} = 0.26$. Another possibility is to take the comet density of 0.6 gcm^{-3} (as derived by Sagdeev *et al.* 1988) as a starting point and use the same volume and mass fractions as above, to derive the comet and coma dust porosities. These then turn out to be $P_{\text{dust}} = 0.91$ and $P_{\text{nucleus}} = 0.61$. Finally, one could start with a comet density of 0.25 gcm^{-3}, derived from the splitting of the comet Brooks 2 when it passed by Jupiter (Sekanina and Yeomans, 1985). This would give the same results as using the meteor density of Olsson-Steel (1989). A summary of these results is contained in table 1, where we have included the possibility for partial evaporation of the organic refractory component which actually is expected at high temperatures. Note that under the present assumptions, the lowest possible coma dust porosity (which corresponds to a fully compact comet nucleus), is $P = 0.75$, or $P = 0.83$ if half of the organics have evaporated.

TABLE 1

ρ_{nucleus} (g cm^{-3})	P_{nucleus}	$P^{1:1}_{\text{dust}}$	$P^{2:1}_{\text{dust}}$	$\rho^{1:1}_{\text{dust}}$ (g cm^{-3})	$\rho^{2:1}_{\text{dust}}$ (g cm^{-3})
1.54	0.	0.75	0.83	0.60	0.45
0.60	0.61	0.906	0.933	0.23	0.18
0.51	0.67	0.92	0.94	0.20	0.16
0.26	0.83	0.96	0.975	0.10	0.08

Table *1. Comet nucleus densities and porosities, with corresponding coma dust densities and porosities. The superscripts denote* $M_{\text{sil}} : M_{\text{or}}$ *in the dust. The case 1:1 is the original cometary ratio according to the model and the case 2:1 corresponds to the situation where half of the organic refractory has evaporated.*

3 Method.

3.a THEORY

For the purpose of our discussion, we consider the coma of comet Halley to be a spherically symmetric cloud of gas and dust, in which the particle density decreases as the inverse square of the distance to the comet nucleus. We shall assume that the outward flow of dust particles is directed radially from the nucleus and that the particle mass spectrum does not depend on the distance to the nucleus in the inner coma. Let the flux (per unit wavelength) observed at earth from the coma, seen within an aperture defined by a radius R_a at the comet, be denoted by $F(\lambda)$. Then the total power, W^o emitted in all directions from the central spherical region within the aperture is

$$W^o = \frac{2}{3} 4\pi\Delta^2 F(\lambda) \quad , \tag{3.1}$$

where Δ is the distance from the Earth to the comet at the time of observation. The factor 2/3 takes into account to a good approximation that the emission as seen from Earth comes from a cylindrical volume rather than from a sphere.

We define the mass absorption coefficient of a particle, $\kappa(\lambda)$, as

$$\kappa(\lambda) = \frac{C_{abs}(\lambda)}{M} \quad , \tag{3.2}$$

where C_{abs} is the cross section for absorption of the particle and M its mass. The emitted power by a spherical particle at a temperature, T, is then given by the product $4\pi M\kappa(\lambda)B(\lambda, T)$, where $B(\lambda, T)$ is the Planck function. In terms of the particles in the observed central spherical part of the coma, assuming that particles with the same mass are identical, the total emitted power per unit wavelength, W^e is

$$W^e = 4\pi(4\pi R_a^3) \int_0^\infty \kappa(\lambda)B(\lambda, T)\, M\, n(M)\mathrm{d}M \quad , \tag{3.3}$$

where $n(M)\mathrm{d}M$ is the number of particles of mass M per unit volume at a distance R_a from the comet, in the mass range $\mathrm{d}M$. A theoretical fit of the observed spectrum must therefore satisfy

$$4\pi R_a^3 \int_0^\infty \kappa(\lambda)B(\lambda, T)\, M\, n(M)\mathrm{d}M = \frac{2}{3}\Delta^2 F(\lambda) \quad , \tag{3.4}$$

for all wavelengths where the thermal emission dominates the reflected solar radiation.

Basic to this work is the consideration of the emission bands at 3.4 μm and 9.7 μm, which appear clearly above the continuum radiation from the coma. For either of these bands we define the emission above the continuum, F^{ex}, as

$$F^{ex}(\lambda) = F(\lambda) - \left[F(\lambda_1) + \frac{F(\lambda_2) - F(\lambda_1)}{\lambda_2 - \lambda_1}(\lambda - \lambda_1)\right] \quad . \tag{3.5}$$

Here λ_1 and λ_2 are the lower and upper wavelenghts of the bands, respectively. If we also define the effective mass absorption coefficient for a particle and for these bands, κ^{eff}, as

$$\kappa^{\text{eff}}(\lambda) = \frac{C_{\text{abs}}(\lambda) - \left[C_{\text{abs}}(\lambda_1) + \frac{C_{\text{abs}}(\lambda_2) - C_{\text{abs}}(\lambda_1)}{\lambda_2 - \lambda_1}(\lambda - \lambda_1)\right]}{M} \quad , \qquad (3.6)$$

then, if there is a well defined emission band in between λ_1 and λ_2, equation (3.4) becomes, to a good approximation :

$$4\pi R_{\text{a}}^3 \int_0^\infty \kappa^{\text{eff}}(\lambda_{\text{p}}) B(\lambda_{\text{p}}, T)\, M\, n(M) \text{d}M = \Delta^2 F^{\text{ex}}(\lambda_{\text{p}}) = \frac{2}{3} W^{\text{ex}}/4\pi \quad , \qquad (3.7)$$

for $\lambda_1 < \lambda_{\text{p}} < \lambda_2$. Here λ_{p} denotes the wavelength of the band peak. This approximation breaks down in case the Planck function has a sharper peak than κ^{eff}, which is, however, not the case we are interested in. Equation (3.7) shows how the relation between two observed quantities, $F^{\text{ex}}(\lambda)$ and $n(M)$, depends on the properties of the emitting dust, through the quantities κ^{eff} and T.

In this paper we shall assume that the coma dust has a certain porosity which is unknown, and which must therefore be taken as a free parameter in the calculations of the coma dust emission. The goal of this work is to calculate the left hand side of equation (3.7) as a function of porosity of the coma dust and then, by equating left and right hand sides, using equation (3.7) to determine the porosity of the coma dust. In particular, we shall satisfy equation (3.7) for the 3.4 μm and 9.7 μm bands. We shall calculate the quantities κ^{eff} and T of the coma dust particles exactly, based on the composition and structure of coma dust as derived from the interstellar dust model. The dependence of κ^{eff} and T on particle size (or mass M), particle porosity (P), wavelength (λ) and distance to the sun (r) will be taken fully into account. Furthermore, we use for $n(M)$ the dust mass spectra as measured in situ by the spacecraft missions to comet Halley, thus implicitly satisfying one observational constraint. The mean diameter of the individual grains in the coma particles follows from the interstellar dust model and is fixed at 0.13 μm, with the relative size of the core and mantle taken so that the mass ratio of the silicates to organic material is 2:1, thus satisfying another observational constraint (Kissel and Krueger, 1987). In sections 4 and 5, the calculation of $\kappa^{\text{eff}}(\lambda, P, M)$ and $T(P, M, r)$ is discussed in more detail.

3b OBSERVATIONS

Because of variability in the dust emission, it is necessary to choose observations of $n(M)$ and $F^{ex}(\lambda)$ as close together in time as possible. The Vega1, Vega2 and Giotto closest approaches occurred on March 6, 9 and 14 respectively. We have chosen to use the observations by Hanner *et. al.* (1987) on March 6.85, 12.8 and 13.75 . For use in equation (3.5), $\lambda_1 = 7.8\mu m$, and $\lambda_2 = 12.5$ for these observations. As far as we know no observations of the 3.4 μm band were done on these days. Therefore we have taken observations at a later date (March 28) to obtain the ratio of the 9.7 μm excess to the 3.4 μm excess and we use this ratio to estimate the value of $F^{ex}(3.4)$ on March 6.85, 12.8 and 13.75. The value deduced from observations on March 28.6 by Gehrz and Ney (1986) is $F^{ex}(9.7) = 1.2 \times 10^{-16}\ W\ cm^{-2} \mu m^{-1}$. The value for $F^{ex}(3.4)$ from Danks *et. al.* (1987) is $F^{ex}(3.4) = 2.6 \times 10^{-17}\ W\ cm^{-2} \mu m^{-1}$. After normalising these values to a single aperture, we obtain a ratio of about 10. However, we note that the measurements by Gehrz and Ney did not include the 7.8 μm filter, so that we used $\lambda_1 = 8.5\mu m$ in equation (3.5) instead, thereby underestimating the strength of the 9.7 μm band. We shall therefore assume $F^{ex}(9.7)/F^{ex}(3.4) > 10$ on March 6.85, 12.8 and 13.75.

The quantity $n(M)$ may be obtained from the cumulative particle fluxes measured by the spacecraft by differentiating the cumulative fluxes and dividing the result by the spacecraft velocity. We have used the results presented in figure 1 of McDonnell *et. al.* (1989) to derive $n(M)$ for the Giotto passage on March 14. We note that due to a recalibration of the instruments, the function $n(M)$ for $10^{-15} g < M < 10^{-10} g$ deduced from this are a factor of up to 25 times larger than the values that would be deduced from the results as presented in McDonnell *et. al.* (1987). We obtained $n(M)$ for the Vega missions from Mazets *et. al.* (1987) (their figures 5 and 8, top curves). We note that the Vega and Giotto differential particle mass distributions are similar up to masses of about $1 \times 10^{-9} g$, but that differences of up to two orders of magnitude occur in the differential amount of mass for the higher masses. Furthermore we note that the mass spectra we have used have been measured in the sunward hemisphere of the coma, where the particle densities are known to be higher than on the night side of the comet. If we take the density on the dark side a factor three lower (as indicated by Giotto post- and pre- encounter data) then the theoretical amount of emission as given by equation (3.7) must be multiplied by 2/3. Finally, whereas the Giotto spectrum we have used is an average over a pathlength of about 4100 km (McDonnell *et. al.*, (1989), the Vega data we have used are apparently influenced more by the jet-like structure of the inner coma. Judging from figure 10 of Mazets *et. al.* (1987), it seems likely that the flux curves at closest approach for the Vega spacecraft (which we have used) are representative of the particle density within a jet, and thus give an overestimate of the average density. From the same figure we deduce that the average particle density may be lower by a factor of 2 than the actual density measured by the Vega1 spacecraft. For the Vega2 measurements, a factor of 1.5 seems applicable.

4 Mass absorption coefficients.

4a OPTICAL CONSTANTS.

The mass absorption coefficients $\kappa(\lambda)$ and $\kappa^{\mathrm{eff}}(\lambda)$ depend on the dust material (through the optical constants), and the morphology and size parameter of the particles. The size parameter is defined as

$$x = \frac{2\pi a}{\lambda} \quad , \tag{4.1}$$

where a is a typical particle dimension. In the case of a sphere, for example, a would be the radius. For the silicates, we use the optical constants as presented by Draine and Lee (1984) for "astronomical silicates" . An important feature of this material is its strong 9.7 μm band, which we use to model the emission of comet Halley at this wavelength.

The optical constants of the organic refractories are adopted from the values given by Chlewicki and Greenberg (1990). These are based in part on laboratory measurements of the residues of ice mixtures irradiated with ultraviolet light. The visible properties of the organic refractory are consistent with the observed albedo, linear and the circular polarization of silicate core-organic refractory interstellar particles. There are two main features of this material with respect to the present paper : (1) the emission band at 3.4 μm, with which we represent the emission feature of comet Halley at around this wavelength, and (2) the absorptivity in the visual which is much higher than the absorptivity of the astronomical silicate in this wavelength region. This causes the dust in the coma of Halley which contains organic material to be hotter than purely silicate dust.

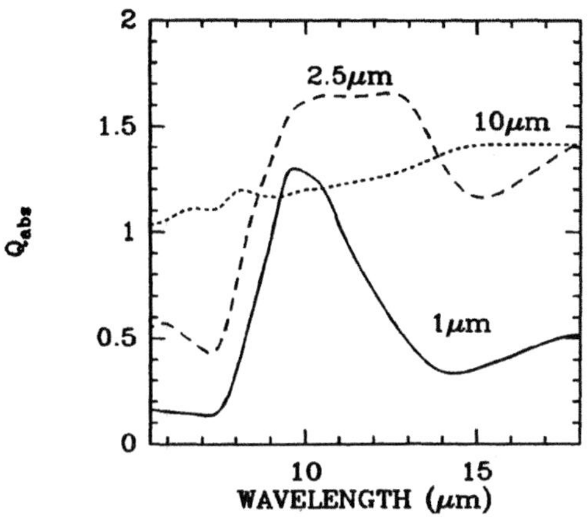

Figure 2. *The cross section of absorption of purely silicate spheres. The curves apply to spheres with radii of 0.1 μm, 2.5 μm and 10 μm. Clearly small particles emit the 9.7 μm better than large particles.*

4b EFFECT OF SIZE AND MORPHOLOGY

Figure 2 shows the shape of the 9.7 μm silicate band as it is emitted by purely silicate spheres of various sizes. From figure 2 it is clear that in order for a silicate sphere to emit (or absorb) effectively above the continuum at a wavelength $\lambda_p \approx 9.7\mu m$ it must satisfy approximately $a \leq 1\mu m$. For the larger spheres the absorption shape is distorted and κ^{eff} is low. Therefore, for $\lambda_p = 9.7\mu m$, only those *solid* silicate particles with a characteristic size smaller than about $1\mu m$ will contribute effectively to the $9.7\mu m$ *excess* to the comet emission.

To calculate κ^{eff} of porous aggregates we represent aggregates of interstellar dust, such as could have been present in the coma of comet Halley (c.f. figure 1), as a collection of identical particles with a fixed size, which may be either randomly situated and loosely bound to each other or more closely packed. The overall shape of the aggregate is taken as spherical. The parameters describing such an aggregate are its radius, R, the refractive index, m, size, a, and shape of its constituent particles, and its fluffiness, or porosity, P. The porosity P is defined by equation (2.1).

It is shown by Hage and Greenberg (1990) that if the inclusions are homogeneous and satisfy $x \ll 1$, the absorption cross sections of the aggregates described above can be determined by calculating an effective refractive index for the whole aggregate and subsequently using Mie scattering theory to obtain the various cross sections. The effective refractive index of the aggregate, m_{av}, is calculated according to the Maxwell-Garnett effective medium theory (Maxwell-Garnett, 1904, to be called Maxwell-Garnett EMT hereafter), with vacuum as the so-called "matrix" material and the individual particles as the "inclusions". The effective refractive index thus obtained depends on the porosity and is given by

$$m^2{}_{\mathrm{av}} = 1 + \frac{3(1-P)(m^2-1)/(m^2+2)}{1-(1-P)(m^2-1)/(m^2+2)}, \tag{4.2}$$

where m is the refractive index of the inclusions.

Since we assume that the inclusions are individual interstellar dust grains, which at $3.4\mu m$ and at $9.7\mu m$ satisfy $x \ll 1$ we may apply the above procedure. To apply the above method also to aggregates consisting of inhomogeneous particles such as interstellar dust grains, we describe these constituent particles as core-mantle spheres with an effective refractive index $m^{\mathrm{i}}_{\mathrm{av}}$, using the core as "inclusion" and the mantle as "matrix" material. The effective refractive index for the constituent particles thus obtained is

$$(m^{\mathrm{i}}_{\mathrm{av}})^2 = m_1^2\left[1+3q^3\left(\frac{m_2^2-m_1^2}{m_2^2+2m_1^2}\right)\left\{1-q^3\left(\frac{m_2^2-m_1^2}{m_2^2+2m_1^2}\right)\right\}^{-1}\right] \quad , \tag{4.3}$$

where a is the radius of the constituent particle, q and m_2 the fractional radius and refractive index of its core, and m_1 the refractive index of its mantle. This approach is justified because the polarizability of the core-mantle spheres (which completely determines their scattering properties) implied by equation (4.3) is exactly correct for concentric core-mantle spheres in the limit $x \to 0$. The effective refractive index given by equation (4.3) is to be used in the right hand side of equation (4.2) to obtain the effective refractive index of the aggregate as a whole.

Figure 3 shows κ^{eff} (9.7) of the coma dust particles as a function of their porosity and size. The labels correspond to the aggregate radii at $P = 1$. For example, an aggregate with porosity P containing the same amount of material as a solid aggregate with a radius of R, has a radius of $a = R/(1-P)^{1/3} = 32\mu m$ for $P = 0.97$ and $R = 10\mu m$.

Figure 3 shows that, at a certain porosity, κ^{eff} (9.7) decreases as the particle size increases. Note how rapidly the effective absorption drops for solid particles ($P = 0$): from 0.1 to 2.5 μm the reduction is about 1/2 ,and for 10 μm $\kappa^{\mathrm{eff}} \approx 0$. At a given size (but greater than approximately 2 μm), κ^{eff} (9.7) increases as the porosity increases. In the limit $P \to 1$, all aggregates have an effective mass absorption coefficient as high as that of a small particle. Furthermore, for sufficiently high porosities, the shape of the absorption as a function of wavelength is like that of a small particle and not distorted as in figure 2 (see Hage and Greenberg, 1990).

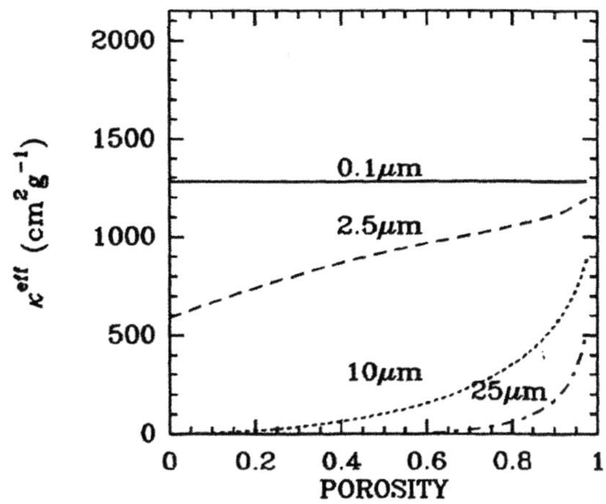

Figure 3. *The mass absorption coefficient at $\lambda = 9.7\mu m$ for coma dust as a function of the dust porosity and size. The labels indicate the radius of the aggregate at zero porosity. A single curve applies to a particle of constant mass.*

5 Dust temperatures

We have used the following standard equation to calculate the temperature of a spherical body in a radiation field

$$\int_0^\infty C_{abs}(\lambda)F_r(\lambda)d\lambda = 4\pi \int_0^\infty C_{abs}(\lambda)B(\lambda,T)d\lambda \quad . \tag{5.1}$$

Here F_r denotes the ambient radiation field, C_{abs} and T are the absorption cross section and temperature of the dust particle respectively. Equation (5.1) expresses the fact that in equilibrium with the radiation field, a body emits the same amount of energy as thermal emission as it absorbs at other wavelengths. Let us first consider a few special limiting cases: (1) absorbing particles which are relatively large, will have $Q_{abs}(\lambda) \approx 1$ for a large range in λ, so that they will approximately have the temperature of a black body, $T \approx 279r^{-0.5}$, where r is the distance to the sun, in astronomical units; (2) individual core- mantle dust particles with a typical size of e.g. a few tenths of a micrometer will be much hotter, because they emit relatively poorly in the infrared (since $C_{abs} \sim 2\pi/\lambda$), but absorb well in the visual and ultraviolet; (3) lastly, the temperature attained by an aggregate in the limit $P \rightarrow 1$ is equal to the temperature of a single inclusion if it were exposed to the radiation field on its own. This is because the inclusions in very porous aggregates do not shadow each other much and absorb and emit radiation nearly as if independent. In general, the temperature of an aggregate depends on its distance to the sun, on its size and fluffiness and on the properties of the individual constituents.

To satisfy equation (5.1) for a certain particle in practice, one has to calculate the emitted power for a range of temperatures, calculate the absorbed power and then interpolate to match emission and absorption and find the correct particle temperature (see, e.g. Greenberg, 1971). To calculate the absorption cross sections for the solid particles, we have used the standard Mie theory. For solid core-mantle particles, we have used another appropriate version of the Mie-theory. As far as the aggregates are concerned, we may only use the combined Mie theory–Maxwell-Garnett EMT approach to calculate their absorption cross sections if $x \ll 1$ for the individual constituents. Since the individual constituents are interstellar dust grains with a radius of about 0.1 μm, the above approach is valid up to $\lambda \approx 1\mu m$. For shorter wavelengths, we consider these aggregates to have the optical characteristics of a cloud of independent particles. This approximation works well if $X > 10$. The mean optical depth, τ, of a spherical cloud of identical small ($x \ll 1$), absorbing particles is given by

$$\tau \approx \frac{4}{3} N R C^{\mathrm{i}}_{\mathrm{abs}} \tag{5.2}$$

where N is the number of particles per unit volume, R the radius of the cloud and $C^{\mathrm{i}}_{\mathrm{abs}}$ the cross section for absorption of an individual particle. To a first approximation, the flux of a plane wave incident on such a cloud is reduced by a factor $e^{-\tau}$. In this approximation, the absorption cross section of the aggregate is

$$C_{\mathrm{abs}} = G(1 - e^{-\tau}), \tag{5.3}$$

where G is the geometrical cross section of the cloud ($G = \pi R^2$).

Results for the temperatures of the aggregates as a function of porosity and size are shown in figure 4. The temperatures shown are computed for a solar distance of $0.9AU$. In general, the trends discussed above are borne out by the results of the exact calculations used to draw figure 4. Note that the rise in temperature to the limiting value does not become rapid until quite high values of P.

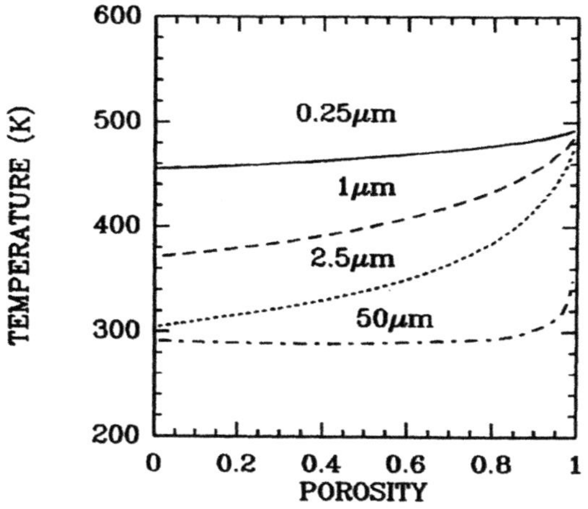

Figure 4. *The temperature of coma dust as a function of the dust porosity and size. The labels indicate the radius of the aggregate at zero porosity. A single curve applies to a particle of constant mass.*

6 Results

Figure 5 shows the theoretical emission from comet Halley's coma at $9.7\mu m$ in Watts per centimeter, as a function of the porosity of the coma particles, calculated using equation (3.7) and the methods described in sections 4 and 5. The continuous line shows the emission expected from the coma in an aperture corresponding to 5620 km at the comet nucleus on March 6.85, using the particle mass spectrum measured by Vega1. The long-dashed and short-dashed lines show the emission from the coma in an aperture corresponding to 4900 km at the nucleus on March 12.8, using mass spectra measured by Vega2 and Giotto, respectively. Figure 6 shows the same set of curves, but now for the emission at $3.4\mu m$.

We may use the observed values of $F^{ex}(9.7)$ on the corresponding dates, and the observed value of the ratio $F^{ex}(9.7)/F^{ex}(3.4) > 10$ on March 28 to read the required porosities from these figures, in order to satisfy equation (3.7). Table 2 shows the results obtained.

After having thus found the porosity necessary to fit the amount of excess emission, it is possible to compute the shape of the $9.7\mu m$ band also, using the values for the porosity obtained from figure 5. The curve in figure 7 shows the result for the band shape on March 12.8, using a porosity of about 0.95 and the Giotto size distribution. The crosses in this figure represent ground based observations

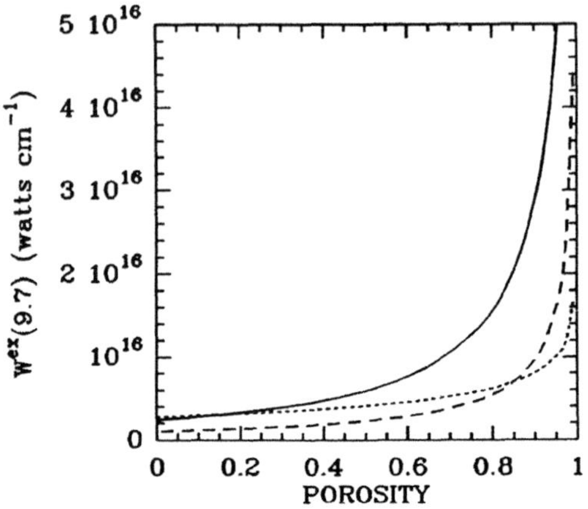

Figure 5. *The power per unit wavelength at 9.7 μm emitted from coma of Halley calculated using equation (7), as a function of the porosity of the coma dust. Continuous line : the power emitted from the spherical region around the nucleus with a radius of 5620 km on March 6.85, using the particle size distribution measured by Vega1. The observed power was about* $1.6 \times 10^{16} W\ cm^{-1}$ *Long dashes : the power emitted from the spherical region around the nucleus with a radius of 4900 km on March 12.8, using the particle size distribution measured by Vega2. Short dashes : the power emitted from the spherical region around the nucleus with a radius of 4900 km on March 12.8, using the particle size distribution measured by Giotto. The observed power on March 12.8 was about* $1 \times 10^{16} W\ cm^{-1}$

observations on this date, by Hanner *et. al.*, (1987). Figure 7 shows that a reasonable match is obtained. However, using $P \approx 0.8$ and the Vega1 mass spectrum, the resulting bandwidth is about 50% too wide compared to the observations by Hanner *et. al.* on March 6.85 (see figure 8). The Vega2 mass spectrum gives a bandwidth which is about 30% too wide on March 12.8 .

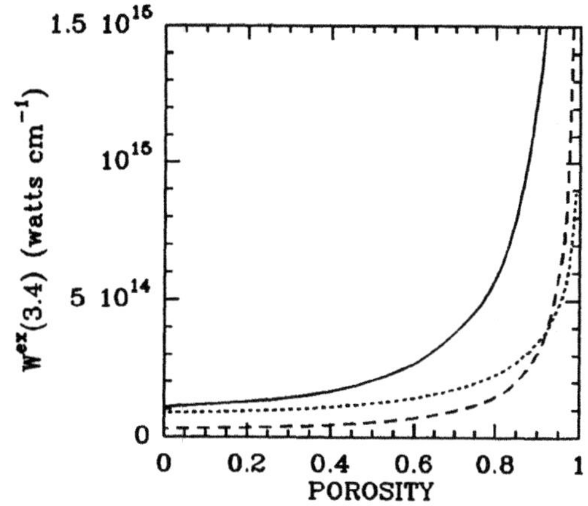

Figure 6. *The power per unit wavelength at 3.4 μm emitted from coma of Halley calculated using equation (7), as a function of the porosity of the coma dust. Continuous line: the power emitted from the spherical region around the nucleus with a radius of 5620 km on March 6.85, using the particle size distribution measured by Vega1. The observed power was about $1.6 \times 10^{15} W\ cm^{-1}$ Long dashes: the power emitted from the spherical region around the nucleus with a radius of 4900 km on March 12.8, using the particle size distribution measured by Vega2. Short dashes: the power emitted from the spherical region around the nucleus with a radius of 4900 km on March 12.8, using the particle size distribution measured by Giotto. The observed power on March 12.8 in the 3.4μm band was about $1 \times 10^{15} W\ cm^{-1}$*

7 Discussion and conclusions

Table 2 shows that the interstellar dust model of comets is able to provide the basis to quantitatively and simultaneously satisfy the observational constraints of: (1) the particle mass spectra; (2) the mass ratio of organic to silicate material in the coma dust; (3) the absolute amount of excess emission in the 9.7μm band; (4) the absolute amount of emission in the 3.4μm band. The fact that the porosities required to match the 3.4μm and the 9.7μm bands on the same date are not too far apart gives confidence in the consistency of the model. However, although $P > 0$ is always consistently required, the result $P = 0.8$ obtained for March 6.85 and the Vega1 data is rather low compared with the predictions of the interstellar dust model. We may also include the band shape at 9.7μm as another constraint in the discussion. Although the Giotto mass spectrum gives the correct band shape, the

TABLE 2

date	$n(M)$	$P(9.7)$	$P(3.4)$
March 6.85	Vega1	0.8	> 0.93
March 12.8	Vega2	0.91	> 0.97
March 12.8	Giotto	0.95	> 0.99

Table 2. *The porosity of the coma dust derived by application of equation (3.7). $P(9.7)$ is the porosity required to match the strength of the 9.7μm band, and $P(3.4)$ is the result for the 3.4μm band.*

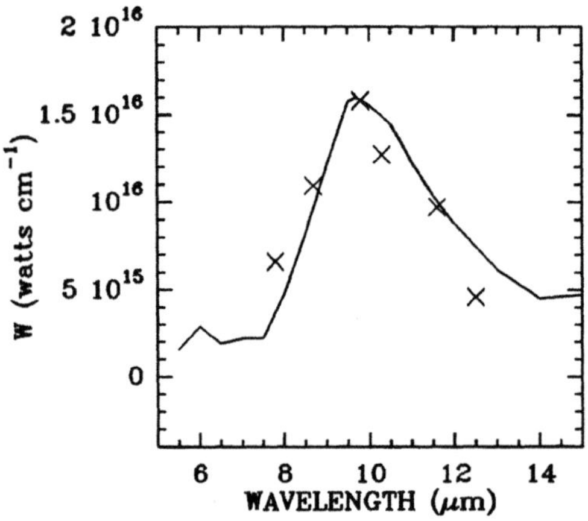

Figure 7. *The shape of the 9.7 μm band calculated using equation (3.3), using the Giotto dust size distribution and a porosity of $P \approx 0.97$. The crosses show the observations by Hanner et. al. on March 12.8*

Vega mass spectra result in band shapes that are too wide. The 9.7*μm* band shape is distorted only by the relatively large, low porosity particles (c.f. figure 2). If the porosities as deduced from the calculations using the Vega mass spectra were significantly higher, then these calculations would have given narrower bandshapes, but would also produce more emission in the 9.7 *μm* band than observed. It is important to remark that if the constituent grains are taken much smaller or larger than the interstellar grains, the porosity requirement becomes unrealistically large.

Assuming that the numbers in table 2 are correct we conclude that the coma

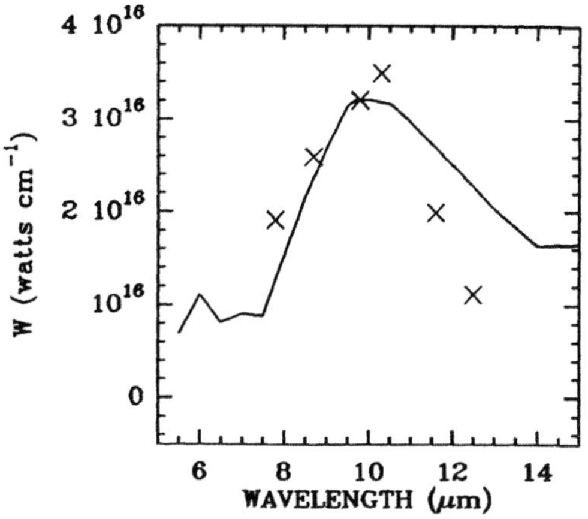

Figure 8. *The shape of the 9.7 μm band calculated using equation (3), using the Vega1 dust size distribution and a porosity of $P \approx 0.8$. The crosses show the observations by Hanner et. al. on March 6.85*

dust has a porosity of at least 0.8, but more likely in the range $P > 0.95$. These higher values improve the 9.7 μm band shapes and are consistent with the interstellar dust model. Finally, if the value of 0.95 applies to the coma dust we may reconstitute the volatiles into the dust to find the porosity and density of the comet itself. Based on the interstellar dust model this results in a comet porosity of 0.67 and a comet density of 0.51 gcm^{-3}. We note that this density is close to the value of 0.6 gcm^{-3} found by Sagdeev *et. al.*, (1988).

Acknowledgements

We would like to acknowledge partial support for this research from NASA grant # NGR33018148.

References

Chlewicki, G. and Greenberg, J.M., 1990, *Astrophys. J.*, in press.

Danks, A.C., Encrenaz, T., Bouchet, P., LeBertre, T. and Chalabaeu, A., 1987, *Astron. Astrophys.* **184**, 329-332.

Draine, B.T. and Lee, H.M., 1984, *Astrophys. J.* **285** 89-108.

Gehrz, R.D. and Ney, E.P., 1986, in Proc. *20th ESLAB Symposium on the Exploration of Halley's Comet*, **ESA SP-250**, 1010-105.

Greenberg, J. M., 1971, Astron. Astrophys. 12, 204-249.

Greenberg, J.M., 1982, in "Comets", ed. L.L. Wilkening, 131-163 (U. of Arizona Press, Tucson).

Greenberg, J. M., 1985, *Physica Scripta*, **11**, 14.

Greenberg, J.M., 1986, *Nature*, **321**, 385.

Greenberg, J. M., 1986b, in *Asteroids, Comets and Meteors II*, eds. Lagerkvist, C. I., Lindblad, B. A., Lundstedt, H. and Rickman, H. (Uppsala University Press), 221.

Greenberg, J.M., 1988, in *Dust in the Universe*, eds. Bailey, M. and Williams, D. A., (Cambridge University Press), 121.

Greenberg, J. M. and Gustafson, B., 1981, *Astron. Astrophys.*, **93**, 35.

Greenberg, J. M., Zhao N. S. and Hage, J. I., 1989 *Adv. Space Res.* **9**, no. 3, 3.

Hage, J. I. and Greenberg J. M., 1990, *Astrophys. J.* in press.

Hanner, M.S. Tokunaga, A.T., Golisch, W.E., Gripe, D.M. and Kaminski, C.D., 1987, *Astron. Astrophys.* **187**, 653-660.

Kissel, J. and Krueger, F.R., 1987, *Nature*, **326**, 755-760.

Maxwell-Garnett, J. C., 1904, *Phil. Trans. R. Soc. Lond.*, **A203**, 835.

Mazets E. P. *et. al.*, 1987, *Astron. Astrophys.* **187**, 699-706.

McDonnell, J.A.M. *et. al.*, 1987, *Astron. Astrophys.* **182**, 719-714.

McDonnell, J.A.M. *et. al.*, 1989, in: Workshop on Analysis of Returned Comet Nucleus Samples, Milpitas, California.

Olsson-Steel, 1989, in *Highlights of Astronomy*, **8**, ed. P. McNally.

Sagdeev, R. Z. *et. al.*, 1988, *Nature*, **331**, p240-242.

Sekanina, Z. and Yeomans, D. K., 1985, *Astron. J.*, **90**, 2331.

Verniani F., 1969, *Space Science Rev.*, **10**, 230.

Verniani F., 1973, *J. Geophys. Res.*, **78**, 8429.

FEATURES OF EXPERIMENTAL STUDYING OF COMET HALLEY DUST PARTICLES ELEMENTAL COMPOSITION

E.N.EVLANOV, M.N.FOMENKOVA, V.N.KHROMOV, L.M.MUKHIN,
O.F.PRILUTSKY, R.Z.SAGDEEV, B.V.ZUBKOV
Space Research Institute, USSR Academy of Sciences
117810, Profsojusnaja st., 84/32
Moscow
USSR

ABSTRACT. Time-of-flight dust-impact mass-analysers PUMA-1,2 aboard VEGA spacecrafts were intended to investigate the properties of Halley comet dust component by means of direct measurements in the nucleus vicinity. More then 2000 mass-spectra were obtained by PUMA-1 and more then 500 ones were obtained by PUMA-2. Masses of studied particles are in the range 5 10^{-17} - 10^{-12} g. The spectra measured in time analyzer mode (zero mode) were the most convinient for data treatment; the number of these spectra was about 4-5% from the total number of PUMA spectra. The principal elements (H, C, N, O, Na, Mg, Al, Si, S, Cl, Ca and Fe) and their most abundant isotopes were found. The isotopic composition of C, Mg, Si, S, Cl and Fe is similar to mean cosmic isotopic abundance.The cometary dust particles contain a significant amount of light elements (H, C, N and O). Their presence can be easily explained by the presence of organic compounds similar to observed in carbonaceous chondrites. The results of cometary dust composition measurements shows to a possible similarity of interstellar and cometary dust and can be explained by a model of cometary nuclei as aggregates of interstellar dust particles with no substancial chemical evolution.

1. Introduction

The spacecraft flyby near comet Halley's nucleus offered qualitatively new possibilities to investigate cometary dust particles. To measure their composition the PUMA dust impact mass-analyzers [1] were installed on board the VEGA-1 and -2 spacecrafts as well as the PIA instrument on board the GIOTTO ESA space probe [2]. Scientists from France, the FRG, and the USSR participated in the PUMA experiment whose scentific systems were developed and manufactured at the Space Research Institute, USSR Academy of Sciences and at Max-Planck Institute for Nuclear Physics (Heidelberg,FRG). The scientific objectives of PUMA experiment were:

- the elemental composition of individual dust-particles;
- physical parameters of dust-particles: density, size, structure;
- abundance of individual isotopes;

J. M. Greenberg and V. Pirronello (eds.), Chemistry in Space, 383–387.

- the elemental composition dependence on a distance from the cometary nucleus;
- the molecular composition of dust particles;
- the search for the organic component of dust particles;
- the mass distribution function for particles;
- the spatial distribution of dust near the Halley Comet's nucleus.

These data are necessary not only to study the comet itself but also to set up the evolution relationship between the interstellar, cometary, and interplanetary dust.
Both PUMA instruments operated satisfactorily during flybys near Halley's nucleus on March 6 and 9, 1986. First results of the experiment were published in the Nature magazine [3]. This paper discusses instruments, their specific features and experimental conditions.

2. PUMA dust-impact mass-analyzer

The VEGA spacecraft velocity with respect to the Halley comet's nucleus was very high at the encounter (79.2 km/s for VEGA-1; 76.8 km/s for VEGA-2, and 68.4 km/s for GIOTTO). Such a velocity, on the one hand, made it difficult to perform in-flight measurements because the instruments were near the cometary nucleus for a very short period of time; on the other hand, this velocity happened to be rather useful to study dust particles and their composition, in particular. The gaseous cloud is produced during the high-velocity impact of a dust particle with a solid target. To measure the ion composition of the plasmacloud caused by the high-velocity impact the PUMA instrument used the time-of-flight method. This method helps to determine an ion mass (an ion mass to ion charge ratio) from the time the ion needs to cover a fixed distance. Fig.1 shows schematically the PUMA analyzer system. Due to the high relative velocity cometary dust-envelope particles are falling onto the spacecraft as an almost parallel beam with a divergence angle less than one degree. Prior to impact with the target dust particles pass through a blende which protects the instrument from the background light.
The rather high requirements were imposed upon the target surface cleannes and purity. That was dictated by the fact that the number of target ions in spectrum is comparable with the number of ions formed from the dust particle matter. Targets were made of silver subjected to cleaning by ion beams. The target's effective area was equal to 5 cm^2.
At the stage of the PUMA experiment preparation certain problems were theoretically analyzed, they are: a high-velocity impact, processes in the plasma cloud caused by impact, and characteristics of ion optics used in the instrument [4,5].
The numerical simulation of an oblique high-velocity impact [5] showed that the most probable direction of particle movements in the plasma cluster corresponds to the mirror reflection of the falling dust-particle trajectory from the target plane. For the flat target case the real geometry of the instrument (see Fig.1.B1) is so that the ion escape direction does not coincide with the ion-optics axis; this leads to decreasing transmission and sensitivity of the instrument. To optimize transmission a corrugated target was used (Fig.1.B2) which assures the

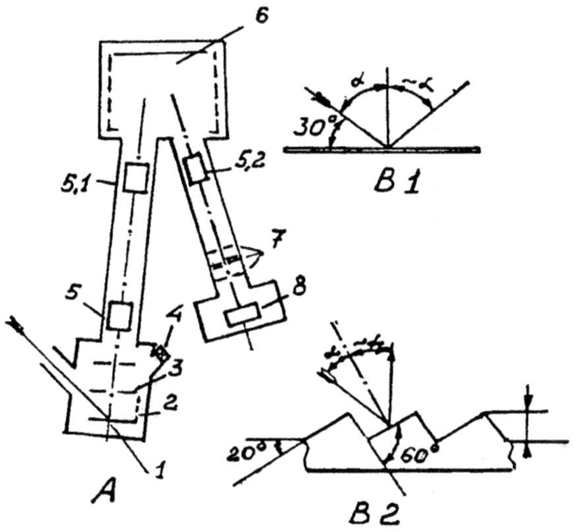

Figure 1. PUMA mass-analyser: 1 - target, 2 - catcher, 3 - accelerating grid, 4 - photomultiplier, 5 - ion lenses, 6 - reflector, 7 - retarding grid, 8 - secondary electron multiplier (SEM).

coincidence of the most probable direction of ion escape with the ion optics axis. A specially formed target (a corrugating angle 25°, a structure step of 0.5 mm) was assembled within PUMA-1 whereas PUMA-2 used a flat target. The indirect data indicated [6] that the PUMA-1 structural target indeed increased the sensitivity, the instrument recorded dust-particles with a mass up to 10^{-17} g. Unfortunately, the direct comparison of the data obtained with structural and flat forgets turned out to be impossible due to unsanctioned decrease in voltage of the PUMA-2 electron multipliers during the flyby.

A dust-particle impact with the target is followed by the short-time burst of the electromagnetic emission, the ejection of ions, electrons, neutral atoms, and fragments of the target material, and the microcrater formation. The instrument photomultiplies records the emission burst in UV spectral bands (310 to 350 nm).

Ions thus generated are pulled out of the plasma cloud by the accelerating grid (Figure 1); the target potential is +1000 V; the accelerating grid potential, -2000 V; and the potential of the output grid and drift tubes is equal to zero. A total energy with which a particle leaves the accelerating system depends on an initial energy a particle gains during the plasma cloud scattering.

The PUMA instrument should resolve individual isotopes in the whole mass range, from hydrogen to silver. This means that the mass resolution M/dM should be equal to 150 - 200. To achieve this value the instrument employed an "energy focusing" method which assumes the simultaneous arrival of ions with the same mass but with different energies at a detector [7]. After passing the accelerating system ions enter the analyzer's drift tube and are focused by electrostatic lenses. Then, the ion beam falls to the reflector section were correcting electrodes and the reflecting plate create special electric field distribution that can provide for "energy focusing". After the reflector ions enter the second drift tube at the end of which there are the retarding grid with a potential of +1000V and the ion detector. The ion energy interval at this detector is determined by the potential difference between the retarding grid and the reflecting plate.

The use of both a wide and narrow energy window permitted to improve the experiment characteristics and yielded an optimized inflight functioning of the experiment over the full mass range. In the PUMA instrument the energy window is defined by the potential difference between the reflector (upper limit for the ion energy) and the retarding grid (lower limit). The retarding grid is located near the detector where the plasma effects are negligible. The interest of a narrow energy window mode for ion detection in the PUMA experiment was demonstrated in laboratory laser simulation of high velocity impact [8]. Plasma clouds were generated in this case with a YAG-laser (532 nm, 20 ns, 3 mj). The dependence of the mass resolution for silver doublet (107 and 109 a.m.u.) on ion number in the cloud (an induced charge on the accelerating grid of PUMA) is shown in Fig.2; the resolution was decreased from 300 ($N = 10^5$ ions) to 30 (N = 106). The decrease of resolution was mainly due to plasma screening of the acceleration electric field, which the spread of the ion start times.

This was confirmed by a hydrodynamical model [8] of the ion cloud expansion in the external electric field ($T = 10$eV, $N = 10^8 - 10^{10}$). An incomplete electric field penetration into the expanding plasma cloud allowed acceleration only for a small fraction of the ions in the boundary region of the cloud. When the particle concentration in the plasma cloud decreases below a critical level, the electric field penetrates into the cloud and the screening can be neglected. The main consequence of the plasma screening is an increase of the duration of the ion pulse.

Due to this effect the resolution $R=T/2dT$ is about 20 for $N = 10^{10}$. This value approximately the same (R = 30 for N = 108) as that found in the laser simulation. There is a time dependence of energy of the ions in the current pulse. Ions at the front edge of the current pulse have more energy than the acceleration potential difference, while ions in the tail of the pulse have less energy. Therefore a narrow energy acceptance window reduces the spread of effective ion start times. It is necessary to choose the optimum energy and the energy spectrum width for the detected ions. The median energy of the window was selected (corresponding to the target potential) to make sure that it lies in the energy range of ions coming from the target. The energy window width was selected to optimize transmission and resolution.

The implementation of a structured target can strongly increase the transmission. The optimum result (according to calculations for the transmission by Sagdeev et al. [9]) is achieved for a structured target with an angle near 30 degrees (see Fig.1). For silicon the transmission is increased to about 10%. The geometrical parameters of the target for PUMA-1 are shown in Fig.1 (the structure angle is close to 20 degrees and the structure height is 0.5 mm). Unfortunately a comparison of the efficiency of the PUMA-1 and PUMA-2 targets was not possible due to the

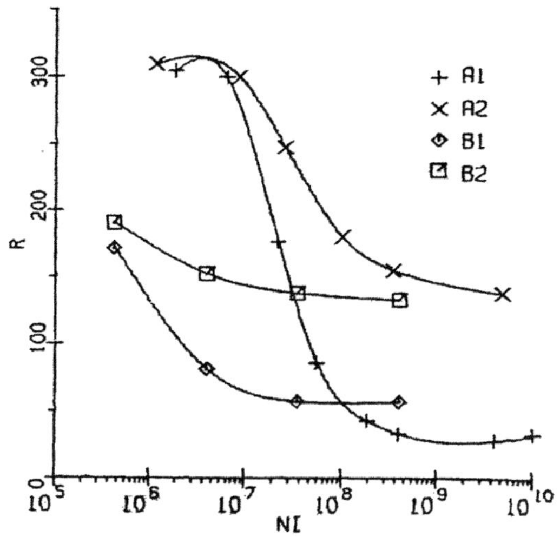

Figure 2. Mass resolution for silver ions in calibration and cometary spectra (N: target ion yield, A: laser calibration, B: comet measurements, 1: wide and 2: narrow window).

decrease of the high voltage at the multiplier to 2400 V (instead of 3300 V).
From Fig.2 (A-laser simulation) it is clear that in the narrow energy window the mass resolution is high enough (R = 200 for N = 10^{10} ions) to separate individual mass lines from the major elements H, C, N, O, Mg, Si, S, Ca, Fe, but not high enough to separate isotopes. There is a drop of sensitivity for the narrow window (about 10 times).
Based on the result of the laser simulation of plasma effects the following potential values were chosen:

a)Wide energy window:
U(target)=1020V
U(reflector)=1100V
U(retarding grid)<1000V

b)Narrow energy window:
U(target)=1020V
U(reflector)=1020V
U(retarding grid)=1000V

The instrument design realized the change in the reflecting plate potential from +1100 V to +1010 V and inversely which occurred every 30 s. Thus, in the wide energy window mode (the reflector plate potential is +1100V) ions with initial energies from 0 to 100 eV are measured; and in the narrow energy window mode (this potential is +1010V), ions with initial energies from 0 to 17 eV (taking into account the 5° angle between the velocity vector and the normal to the reflector surface) are registered.
Passing through the analyzer ions are groupped into clusters with the same mass-to-charge ratio (with the same velocity). The so-formed time of-flight mass spectrum is recorded by the secondary electronic multi-

plyier (SEM) of the MM-1 type [10]. Ions of an energy of about 4 keV (with due account of SEM voltage) falling onto the SEM cathode knock out secondary electrons with the efficiency close to unit which depends weakly on the mass number and atomic charge of an ion [10]. The electronic multiplying process takes place in 20 sequential sections.

3. The PUMA experiment on board the VEGA spacecraft [3].

VEGA-1 passed near the nucleus of Halley Comet on March 6, 1986. PUMA-1 was switched on at a distance of 5.5 10^5 km from the cometary nucleus and switched off at a distance of 3.0 10^5 km after the encounter at a distance of 8889 km. The instrument operated in its normal mode during the whole period (2 hours 57 minutes 52 seconds); the high voltage of the multiplier was equal to its nominal value, 3328 V. For this interval of time 6400 frames with physical events and 3945 frames with test pulses were transmitted to the Earth.
VEGA-2 passed near Halley Comet on March 9,1986. PUMA-2 was switched on at a distance of 5.2 10^5 km from the cometary nucleus and operated in its normal mode during 1 hour 3 minutes 44 seconds with the high multiplier voltage equal to 3328 V. 10559 frames with physical events and 1241 frames with test pulses were dumped to the Earth. PUMA-2 malfunction occurred at a distance of 2.3 10^5 km from the nucleus. During 131 seconds the instrument did not function, then it again began its operations, however, the multiplier voltage decreased down to 2400V. This led to the lower gain coefficient of the secondary electronic multiplier (SEM), the shorter dynamic range, and to the lower sensitivity of PUMA-2. With such a multiplier voltage the instrument operated during 2 hours 8 minutes 52 seconds (except for one more 43 sec malfunction at a distance of 6 10^4 km from the cometary nucleus), passed the nucleus at a minimum distance of 8030 km from the nucleus, and was switched off at 3.7 10^5 km from the nucleus. 7415 frames with physical events and 975 frames with test pulses were dumped to the Earth. A great number of spectra, in particular at the initial and final portions of the encounter trajectory, have no those peaks which could be identified as peaks of basic elements (such spectra can be recorded only if the multiplier voltage is normal, this was not the case far PUMA-2). Figure 3 shows examples of these spectra. These events are identified with impacts of very small dust-particles [11].

4. Basic experimental results.

The first stage of the PUMA data processing was the analysis of zero-mode spectra. In this mode the output data are not compressed, therefore, the telemetry information corresponds most closely to real processes in the instrument. A signal from the spectrum recording system output is interrogated with a frequency of 15 MHz (every 66.7 ns), is numbered, and the result is stored in the fast memory. The interrogation begins 0.0035 ms after the event recording system operates, and continues 0.045 ms. Each spectrum contains 660 to 670 values of the amplitude.

Zero-mode spectra show distinctly mass lines of basic elements; this forms the basis for correct interpretation of the data acquired. The detailed information about the shape of mass lines makes it possible to

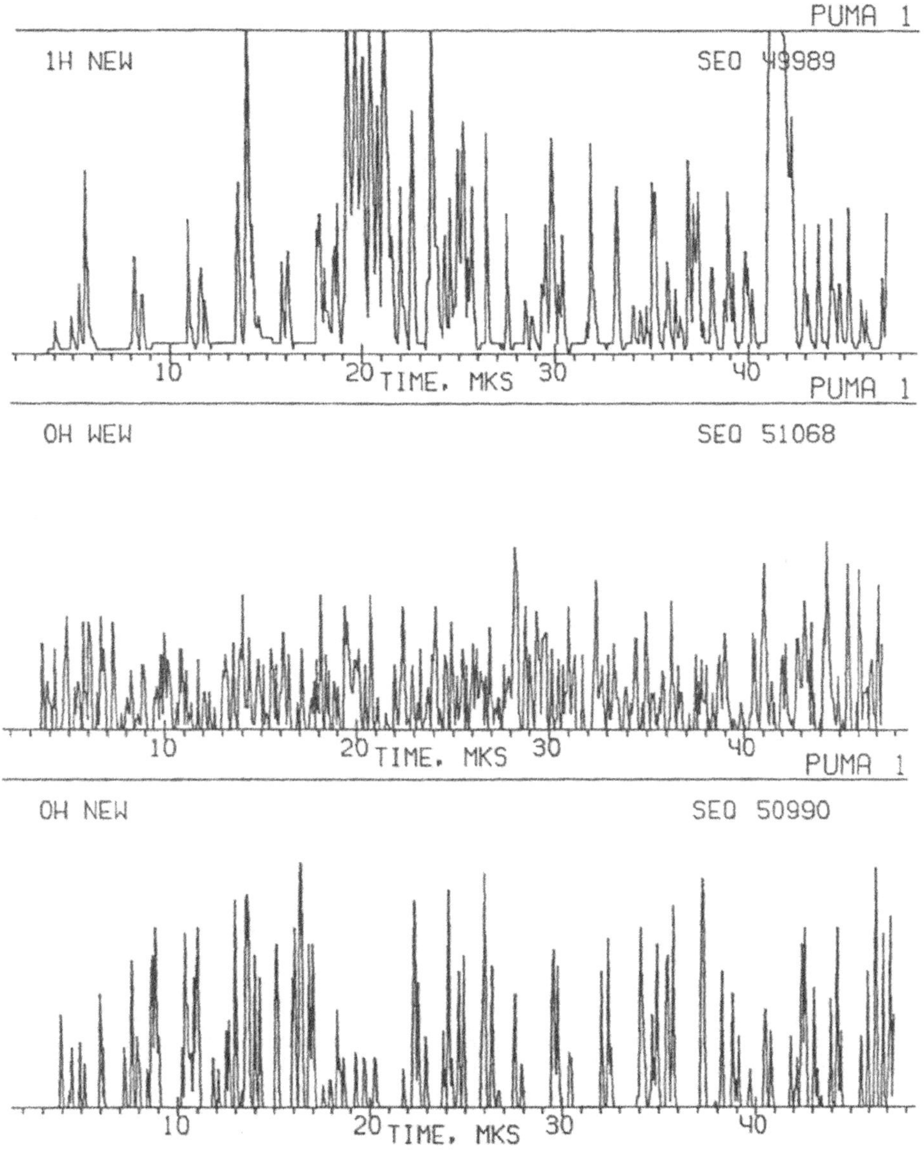

Figure 3. Examples of PUMA-1 spectra:
1. a dust particle spectrum with mass 10^{-16} g in narrow energy window;
2,3. a quasi-noise spectra in wide and narrow energy windows.

study the spread function and the mass-spectrometer resolution, to estimate the effective width of peaks and its dependence on a dust-particle size and impact conditions. A zero-level signal can be obtained from those spectra which have no clearly-expressed lines. Thus, regardless of a comparatively small number of zero-mode spectra (about 10% of the total amount) they are most suitable for understanding the processes occuring in the instrument during measurements and for developing models of spectrum formation.

85 and 20 spectra of PUMA-1 and PUMA-2, respectively, were selected for analysis. They are the spectra which have clearly-expressed peaks identified with ions of target silver or, if only, with one of basic elements: H, C, O, Mg, Si, Fe.

The peak maximum recording time (an ion with the mass M) taken from the beginning of the frame format was evaluated from the equation:

$$t = AM^{1/2} + B \qquad (1)$$

The factor A depends on the energy and effective length of the ion flight in the reflector [7]; the factor B does on the delay of the time scan beginning relative to an event. In the wide "window" mode the effective flight length and, accordingly the factor A is lower than that in the "narrow" window mode. Table 1 gives the mean coefficients A and B for PUMA-1 and PUMA-2, respectively.

TABLE 1. Mean Approximation Coefficients A and B.

	Narrow window			Wide window		
	A	B	Number of spectra	A	B	Number of spectra
PUMA-1	3.951	-2.891	31	3.749	-3.178	37
PUMA-2	3.841	-3.148	15	3.725	-3.366	10

The qualitative quick-look analysis [3] of mass-spectra revealed the following features. PUMA-2 spectra because of the SEM low gain factor and have the peak amplitude considerably lower than those recorded by PUMA-1 with the same input changes. For this reason, the PUMA-2 data show a greater number of large dust-particles, since signals iniciated by fine particle impacts lie below the zero level. PUMA-2 spectra include much less lines than PUMA-1 spectra. As a rule they involve only basic elements, no minor components are observed as well as noises.

The instrument resolution is higher in the narrow window mode that is consistent with theoretical predictions [8], however, discrimination of light elements takes place hydrogen suppresion is particularly observed. The mass resolution for the silver doublet as a function of the total number of ions (charge on the acceleration grid) entering the analyser is shown in Fig.3. With the wide energy window, the resolution for silver ions decreases strongly for N = 10^7 (R<50). The same effect was observed for N = 10^8 with the laser simulation. Resolution does not change significantly (from 180 to 140) for a range of N from 10^6 to 10^{10} (see Fig.2B for the narrow window).

The average resolution as a function of element mass number is represented in Fig.4. For a small dust particle ($N<10^5$) we have an increase of resolution with the mass number in both regimes. This suggests a fixed spread of ion start times in that case. For the wide energy window and $N<10^6$ there is no actual dependence of the resolution on the mass number of the element. This results from plasma effects, since the spread of the start times will increase with mass number [5].

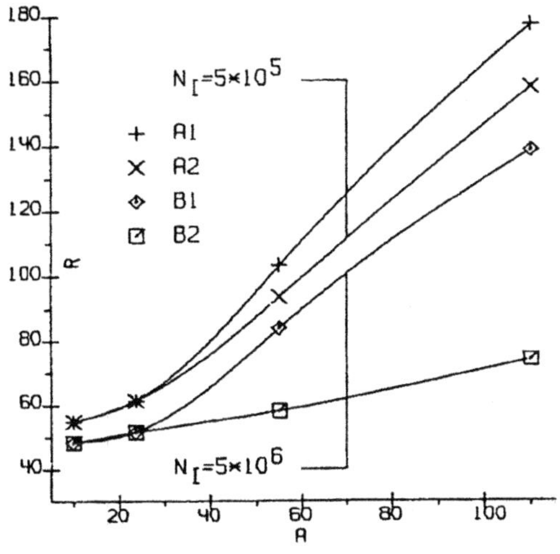

Figure 4. Dependence of mass resolution on the mass number for the wide (A2,B2) and narrow (A1,B1) windows for two levels of the accelerating grid signal (N = 5 10^5 ions and N = 5 10^6 ions).

It is difficult to make an exact comparison for the transmission of the narrow and wide windows since we have no spectra of two identical dust particles. To estimate the relative transmission of the wide and narrow energy windows, we take the ratio j of average charges Q of mass lines for the wide and narrow windows).

The lower transmission in the wide window mode for large particles compared to the narrow window mode can possibly be explained by signal saturation effect in the multiplier. An indication of such a saturation effect in the wide window is the dependence of the count rate of noise pulses and their average amplitude on the acceleration grid signal. For the narrow window the count rate and the average amplitude depend only

slightly on the acceleration grid signal. In the wide window mode, a sharp decrease of the count rate and the average amplitude can be seen even if the signal on the accelerating grid is increased.
Another indication of a multiplier saturation is the dependence of the number of silver ions detected by the multiplier on the accelerating grid signal (see Fig.5). A decrease of the average number of the silver ions was observed with the increase of the AC signal in the wide window of PUMA-1 (mode 0); for the narrow window this value is constant up to $Q=10^{-12}$ C. For PUMA-2 the number of the silver ions is almost proportional to the AC signal (the gain coefficient is about 10^5) and the saturation of the multiplier could be negligible except for very large particles.
At the next stage of processing of 68 PUMA-1 and 25 PUMA-2 spectra the transition from the time to mass scale was made and ion concentrations were calculated. Spectra from both instruments reliably display major elements H, C, N, O, Na, Mg, Al, Si, S, Cl, Ca, Fe (in many cases with isotopes). Ion concentrations may be directly used in studying isotopic ratios, since isotopes of one element (close in mass) behave actually alike in the process of ion cloud formation and during their flight through the ion-optical system. Individual spectra show a considerable spread of isotopic ratio values.
Mass spectrometry of cometary dust particles allows the determination of isotope ratios. Meteoritic studies have revealed important information on the origin, formation and evolution of the early solar system material. The study of isotopes in the cometary dust should lead to equally important results. Cometary material is probably least affected by planetary processes which tend to dilute isotopic peculiarities.
PUMA-1 spectra have a great number of pulses which are not identified with basic elements and are not repeated from one to another. To determine the mean composition of dust particles and to search for minor components (elements and isotopes with low abundance, of molecular ions) PUMA-1 and PUMA-2 zero-mode sum spectra were analyzed.
Sum histograms have been built for the distribution of the number of peaks observed and integral charges over masses. Fig.6 and Fig.7 summarize the histograms thus obtained. In both histograms peaks H, C, N, O, Ne, Mg, Al, Si, S, Cl, K, Ca, Fe, Ag, are obviously identified against the background of uniform (in terms of recording time) of quasi-noise pulses. The isotope ratios of the above listed elements agree well enough with their mean cosmic values (Table 2).
To determine the elemental composition corrections should be necesserily introduced which permit transition from ion to atomic number densities, taking into account the properties of a plasma cloud formed as a result of high-speed impact of a dust particle against the target [5]. Unfortunatelly, the available methods of this problem solutions [4,5,12] yield contradictory results and cannot be employed for interpretation. We hope, in visible future the laboratory calibration for 80 km s velocities and the development of high-velocity impact theory will
Along with the peaks of main elements and their isotopes, a quasi-noise component appears on integral spectra. The amplitude and frequency of noise pulses in PUMA-1 measurements considerably exceed those measured by PUMA-2. The statistical stochastic analysis of time distribution

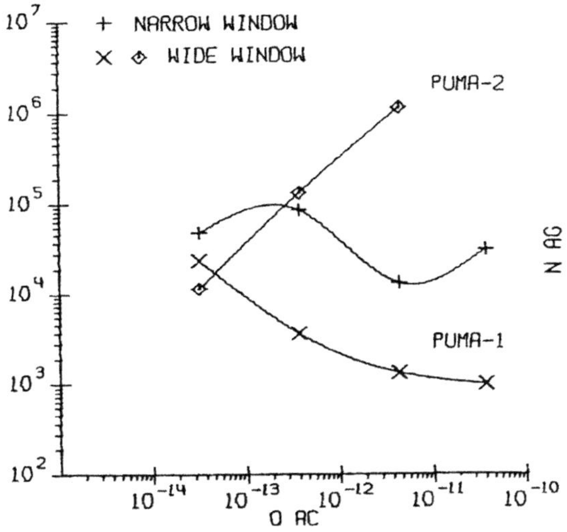

Figure 5. Yield of silver ions as a function of accelerating grid signal for PUMA-1 and PUMA-2.

TABLE 2. Comparison of isotope ratios from the PUMA-1 and PUMA-2 experiments with their mean cosmic values

Isotopes	Experimental results PUMA-1	PUMA-2	Mean cosmic value
^{12}C, ^{13}C	98,0	-	90,0
^{24}Mg, ^{25}Mg, ^{26}Mg	9,8; 11,2	5,2; 8,9	7,8; 7,0
^{28}Si, ^{29}Si, ^{30}Si	12,6; 30,9	-	19,6; 29,7
^{32}S, ^{34}S	16,6	25,1	22,4
^{35}Cl, ^{37}Cl	3,5	-	3,1
^{54}Fe, ^{56}Fe	28,8	15,5	15,5
^{107}Ag, ^{109}Ag	1,2	1,5	1,1

shows that those pulses and randomly distributed over the entire timebase (including the mass range from 2 to 11 where hardly could the peaks of molecular ions appear). This circumstance puts to doubt the identifi-

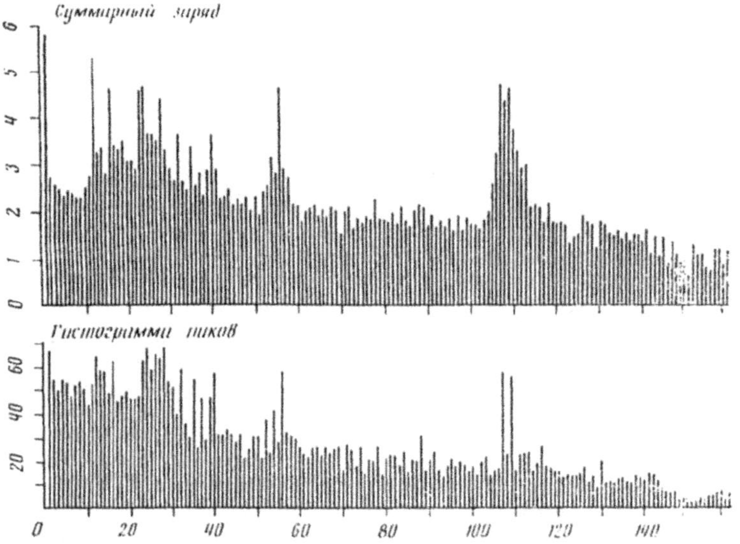

Figure 6. Summary histogram of zero mode spectra of PUMA-1.

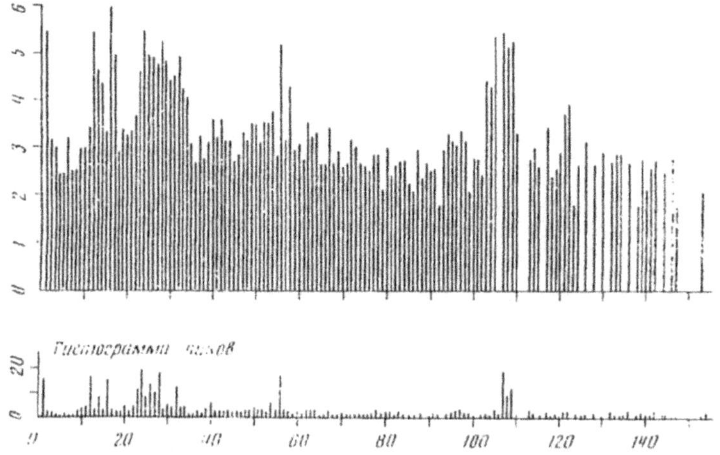

Figure 7. Summary histogram of zero mode spectra of PUMA-2.

cation of noise pulses with complex molecular organic ions in [13]. Note also that the characteristics of noise in the spectra with both high and low carbon content are about the same. Halley Comet's dust particles contain a large amount of light elements. It is most natural to explain their presence by the presence of organic compounds, they may be similar to those found in carbonaceous chondrites (kerogens, amino-acids). The composition of dust particles in the range of elements from Mg to Fe can be attributed to the presence of silicates or, sometimes, troilite. The abundance ratio for main elements fairly correspond to those typical of similar formations within the Solar system.

5. Conclusions

The cometary dust which composition was studied by means of PUMA instruments can be genetically connected with two space dust components:

- the interplanetary dust, one of the sources of which (maybe, one of the main sources) are comets;
- the interstellar dust that served the source material not only for comets but also for other small and large Solar-system bodies.

The results of comparison of measured cometary dust particles composition with data on interstellar dust-particles composition correspond to the assumption that the comets consist of interstellar dust particles, that didn't undergo any serious chemical changes in the process of the comet's formation and evolution. The summarized characteristics of interstellar and cometary dust are the same. The only exception is the lack of the cometary dust particles volatile shell, that can not be stable under the conditions of inner area of the Solar system, where the temperature of the dust particles reaches several hundred degrees Kelvin. But since there are no detailed data on the mineralogical composition of the silicate nuclei and the chemical composiution of the non-volatile organic shels of interstellar dust particles therefore the more accurate evalution of PUMA results in the aspect in questions is impossible.

At present there are performed the detailed analytical researches of interplanetary dust particles [14]. The following results are obtained in the analysis of these dust particles:

- the mineralogical composition of dust particles is different from particle to particle;
- they can't be united in one common class and must be considered separately;
- the average content of non-volatile elements in dust particles is close to the solar one;
- a considerable amount of carbon is in the content of the dust particles.

The results of detailed analytical researches of interplanetary dust particles are very similar to the main results of data analysis of PUMA, that proves the genetic bond between interplanetary and cometary dust. Nevertheless, there should be certain differences between these substances, connected, in particular, with the facts, that:

- interplanetary dust particles can undergo additional

transformations while moving in near Solar space and braking in terrestrial atmosphere;
- the cometary dust is not the only possible source of interplanetary dust that can be formed, for instance, in the collision of the bigger bodies (meteorites or asteroids).

The average composition of the "silicate" component of the cometary dust within the limits of the experiment's accuracy coincides with the average composition of carbonaceous chondrites.

On the whole, it can be assumed that the results of PUMA form the following picture of the space dust evolution:

a) the dust formation in the cool stars neighbourhood, its ejection into space and forming of interstellar dust and gas clouds;
b) interstellar clouds collapse and dust condensation into macroscopic formations, partially preserved till the present time almost without any changes (the comet's nuclei), or undergone certain transformations (carbonaceous chondrites).

Thus, PUMA experiment of VEGA project became one of the stages of the Solar system primordial matter reserach. This experiment must be followed by other experiments on studying of the Solar system small bodies and, later, the laboratory researches of the matters's samples returned from the comet's nuclei).

REFERENCES

1. Balebanov, V.M., Vorontzova, E.N., Scuridin, G.A. (eds.) (1984) Venus-Halley project, Lui-Jean, Paris.
2. Kissel J. (1986) 'The Giotto particulate impact analyser' ESA SP-1077, 67-83.
3. Kissel, J., Sagdeev, R.Z., Bertaux, J.L., et al. (1986) 'Composition of comet Halley dust particles from Vega observations', Nature 321, 280-282.
4. Inogamov N.A. (1986) Evoluciya plazmennogo oblaka v electricheskom pole', Dokl. AN SSSR 277, 1375-1379.
5. Sagdeev R.Z., Churaev R.S., Shapiro V.D., Shevchenko V.I. (1983) 'Uchet plazmennikh effektov v udarnom vremyproletnom mass spectrometre ("PUMA")', Preprint 856,IKI AN SSSR.
6. Sagdeev R.Z., Kissel J., Evlanov E.N., et al. (1987) 'Dependence of mass resolution and sensitivity of PUMA instrument on the energy spread of ions produced by hypervelocity impacts', Astron. Astrophys., 179-182.
7. Mamirin B.A., Karasev V.I., Shmikk D.V., Zagulin V.A. (1983) Mass-reflektron. Novij bezmagnitnij vremyproletnij mass-spectrometr s visokoj razreshaucjej sposobnostju', Zhurn. eksperim. i teoret. fiz., 64, 82-89.
8. Evlanov E.N., Zubkov B.V., Krasovskij G.N. et al. (1984) 'Rezultati ispitanij poljotnogo cetochnogo bloka', Otchet IKI AN SSSR.
9. Sagdeev R.Z., Anisimov S.I., Zhitenev S.B. et al. (1984) 'O propusknoj sposobnosti vremiaproletnogo mass-spectrometra s

pyleudarnim istochnikom ionov', Dokl. AN SSSR, 279, 613-616.
10. Secondary electron multiplier MM-1. Technical specifications. (1982) Johnson Labs.
11. Sagdeev R.Z., Evlanov E.N., Fomenkova M.N. et al. (1988) 'Tiny dust particles near Halley's comet', Proc. of XXVII COSPAR Meeting, 5-3-10.
12. Kissel J., Krueger F.R. (1984) 'Experimental investigations on ion emission with dust impact on solid surfaces', ESA SP-224, 43-51.
13. Kissel J., Krueger F.R. (1987) 'The organic component in dust comet Halley as measured by the PUMA mass spectrometer on board VEGA, Nature, 326, 755-760.
14. Zinner E., McKeegan K.D., Walker R.M. (1983) 'Laboratory measurements of D/H ratios in interplanetary dust', Nature, 305, 119-121.

CHEMICAL COMPOSITION OF HALLEY'S DUST COMPONENT FROM THE PUMA-2 DATA

L.M. MUKHIN*, Yu.P. DIKOV**, E.N. EVLANOV*, M.N.FOMENKOVA*, A.D. GRECHINSKIY*, M.A. NAZAROV***, O.F. PRILUTSKIY*, T.V. RUZMAIKINA[x], R.Z. SAGDEEV*, B.V. ZUBKOV*.

*Space Research Institute, Profsoyuznaya 84/32, Moscow, 117810, USSR
**Geology and Mineralogy Institute, Staromonetniy 35, Moscow, 109017, USSR
***Geochemistry and Analytical Chemistry Institute, Kosygina 19, 117975, Moscow, USSR
[x]Earth's Physics Institute, B. Gruzinskaya 10, Moscow, 123810, USSR

ABSTRACT. The survey of elemental composition of P/Halley dust component according to the data acquired by the PUMA-2 instrument installed on the board of the VEGA-2 spacecraft is presented. Possible mineralogical composition of dust particles is considered. Some reasons for possible presence in comet's dust particles of virgin organic compounds are presented.

INTRODUCTION

Among the Solar system's small bodies comets occupy a special place. There are solid grounds to consider them relic objects that didn't undergo any serious changes since the time of their formation. There is no common point of view on the time and place of comets formation. The existing hypotheses can be divided into three groups. According to the most popular one, comets were formed in Uranus-Neptune zone, and later were ejected by gravitational interaction into the outer regions of the Solar system, where they formed the so-called Oort cloud.

The hypothesis on comet's interstellar origin has been greatly developed in recent time by Clube and Napier /1/, /2/. This hypothesis faces certain difficulties, connected mainly with isotopic ratios $^{12}C/^{13}C$ and D/H, that happend to be more closer, in case of Halley's Comet, to Solar rather than to interstellar ratios.

The third group of hypotheses studies the comet formation in situ, at a considerable distance of the order of 10^3 - 10^4 AU from the Sun /3, 4, 5, 6/. There certainly exists a number of other models /7, 8, 9, 10/.

It seems evident that to solve the principal cosmogonic problem on

J. M. Greenberg and V. Pirronello (eds.), Chemistry in Space, 399–414.

the time and place of formation of comets the study of the chemical composition of cometary nuclei can be of particular concern. For VEGA and GIOTTO Missions to Halley Comet only spectroscopic measurements in cometary atmospheres were possible. These experiments allowed a great amount of different molecules and elements to be determined in cometary comas. Tab. 1 summarized these data.

TABLE 1. Molecules and atoms in cometary spectra

Molecule or atom	Wave lenght A	Molecule or atom	Wave lenght A
Al	3961	Cu	
C	1561, 1657	Cr	
C^+		Fe	
C_2	4737, 5165,	H	1216
	5635, 6195	H_2O	1.35 cm
C_3	3880-4100	H_2O^+	5490-7500
	3970-4070	HCN	3.4 mm
	4020-4045	He	
	3980-4010	K	
CH	3871, 3889,	Mg	
	4314, 9 cm	N_2^+	3540-4280
CH^+	3950-4260	NH	3360, 3370
CH_3CN	2.7 mm	NH_2	5575 -7534
CN	3590, 3883,	Na	5890, 5896
	4216, 7876,	Ni	
	9148, 10833	O	1302-1306
CO^+	3416-6200	OI	6300-6364
	3500-4240	OH	3064, 3428
CO_2^+	3378-3839		1.08 mcm, 18 cm
CS		OH^+	3594-3616,
Ca	4227		3904
Ca^+	3934-3968	NH_3	23.9 HHz
Co		S_2	2800-3100

For clear reasons, the spectoscopic data are rather difficult to use in solving the key problem on so-called parent molecules in the comet and, moreover, in determining the mineralogical composition of dust component. It is just the case why the VEGA-1 and 2 and GIOTTO data are of principal significance. It should be noted that it is these missions during which, for the first time, the experimets were performed on direct determination of the elementary composition of Halley Comet's dust component.In the present review we will discuss the results and preliminary interpretation of the data acquired by the PUMA-2 dust-impact mass--analyzer installed on the board of the VEGA-2 spacecraft.

We shall remind you briefly the main concepts of the design of the experiment for studying elementary composition of Halley's Comet dust component.

The Halley Comet dust envelope particles shock against the silver tracted groups.

Combinations of elements and their relationship have been studied for 511 cometary particles, their composition being calculated according to non-zero mode spectra, recorded both in the "narrow" and "wide" windows.

Na, Ca, C, H, N, S, Si, Mg, Fe, Cr and Al were taken into consideration. Unfortunately, the data on oxygen are not reliable and they cannot be used, but the presence of oxygen may be supposed in all the particles. During the systematization it was assumed that if element-to-Mg ratio in the particle is less than 0.1, than such a particle does not contain this element; and if this ratio in the particle for element with maximum concentration is above 10, then it was assumed that this particle contains neither Mg not any other elements for which this ratio was also above 10. Certainly, these limits are conventional and they are defined only by the fact that with the given concentrations the impurity element does not affect the main elements ratio to a large extent.

Theoretically possible number of combinations of elements mentioned above is 1023 if (Mg, Fe) association is assumed as an "element". Only 138 combinations are realized in cometary particles out of this number. That indicates the presence of certain mineral phases and their paragenesises in the cometary dust. A special article will be dedicated to the detailed characterization of the discovered elemental association.

PARTICLES WITH CHON ELEMENTS

Probably the most interesting result obtained with the help of the PUMA is the discovery of anomalous concentrations of light elements (H, C, N, O) in cometary dust particles. It can be seen from histogram which show the distribution of the carbon-to-total rockforming elements ratio that in a number of cases it exceeds 100. In 95% cases in WEW dust particles contain a set of light elements in various combinations (see Tab. 2).

Extremely important is the fact that micron and submicron particles analyzed by PUMA instruments, keeps their temperatures within 400 to 500 K during about 10^5 s. This is also the reason why easily-boiling constituents which the organic component of dust could potentially incorporate, must have disappeared by the moment when the mass-spectrometer was counting dust particles. Thus, the organic fraction of dust is being modified during the time particles need to move from the nucleus to the spacecraft.

TABLE 2. Number of particles, containing light elements

Light elements	Wide energy window		Narrow energy window	
	with O	without O	with O	without O
H, C, N (O)	71	7	27	1
H, C (O)	109	8	55	10
H, N (O)	7	0	0	1
C, N (O)	3	0	3	1
H (O)	25	8	2	16
C (O)	13	0	38	15
N (O)	2	0	2	0

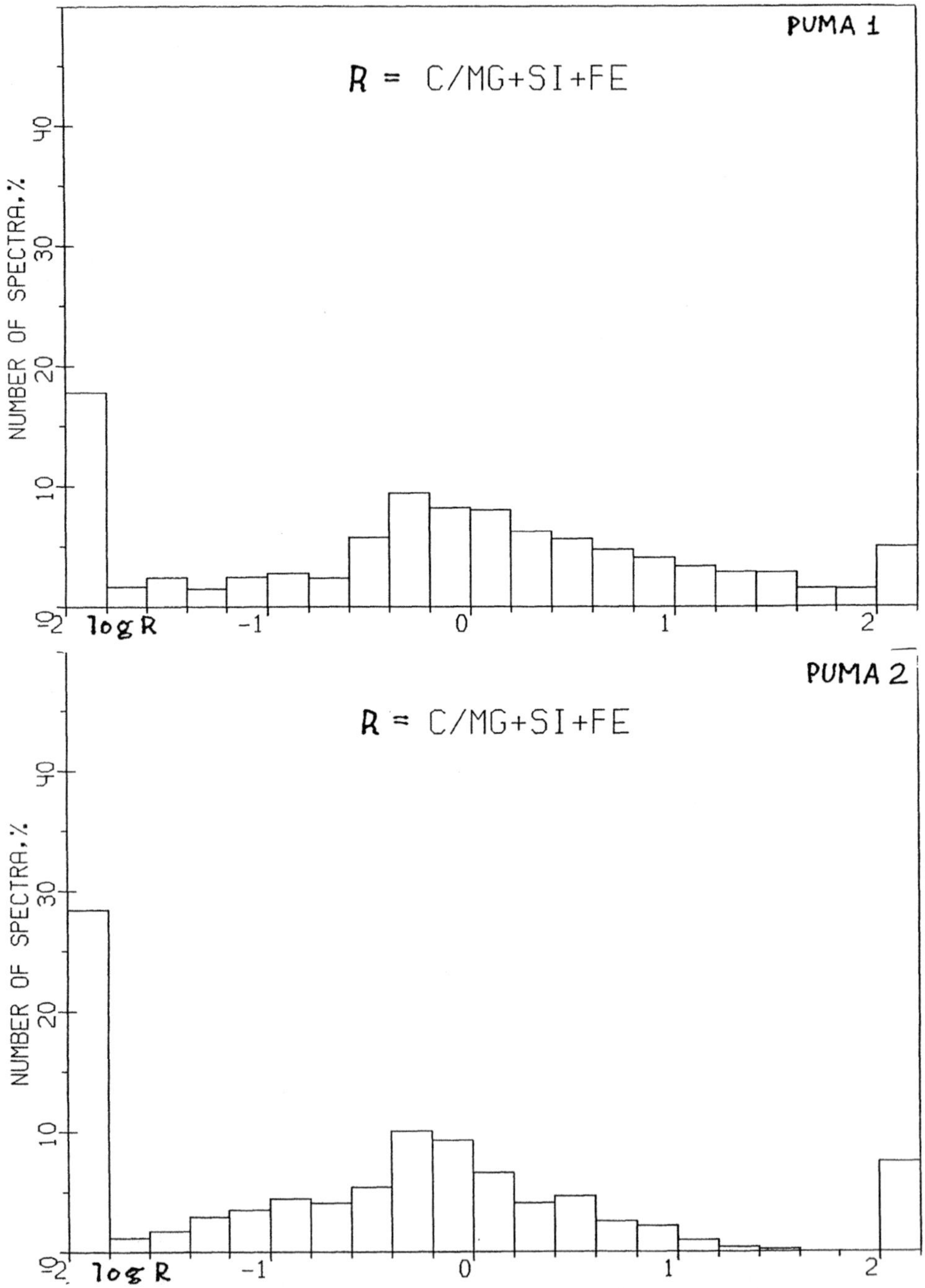

FIG. 1. The distribution of carbon-to total rockforming elements.

target of the instrument with the speed of about 80 km/s. At the initial moment of the shock the particle's matter and the adjacent areas of the target are heated upto very high temperatures (10^5 K); in the impact area pressures grow upto 10 Mbar and most of plasma cloud ions get ionized many times. Plasma cloud expands, temperature and density decrease, ions are recombined and ionization gets "frozen". As the results of measurements show, most of the ions analyzed by the instrument are ionized once. Time-of-flight technics was used in dust particle impact mass-analyzer PUMA.

Passing through the TOF analyzer ions gather into the "chain of packages", i.e. clusters of ions with the same mass-to-charge ratio. The obtained spectrum is recorded by secondary electron multiplier. The detailed description of the PUMA operation and peculiarities - see /11/.

There were two modes of recording in PUMA instruments differing in recorded ions energy range: "narrow" energy window (ΔE = 10 eV) and "wide" energy window (ΔE = 100 eV), average ion energy being equal to 1020 eV. The modes were switched over every 30 sec. In the "narrow" window mode less H, C - containing particles were recorded due to specific formation of the ion cloud in the process of high-speed impact. Under determination of light elements and sulfur should be taken into consideration while interpreting the spectra, recorded in the "narrow" window. As calibration of the instruments on particle accelerator with the collision speed of 80 km/s is not feesible, all the following conclusion were made basing on the analysis of element ion concentrations, assuming that ion composition is adequate to elemental composition of dust particles. This assumption is justified only in case the temperatures of ion quenching in plasma cloud are high enough. It is possible that such condition is not realized completely, and transfer of ion concentrations into the atomic ones needs corrections. Unfortunately, detailed plasma-chemical calculations have not been completed yet ., and the character of these corrections remains unclear. Corrections used by the authors /12/, as it will be shown below, hinder greatly PUMA-2 data interpretation, at least for some elements.

In the present paper dust particles are analyzed on the basis of non-zero mode spectra, obtained with data compression aboard the spacecraft. Only the amplitude and position in the spectrum are known for each recorded peak. That also brings certain ambiguity into ion composition due to unknown width of the peaks. According to theoretical assessment /13/ the width of peaks increases with the growth of the particles size and the mass of ions, forming a peak.

Semiquantitative character of analitical data and complexity of cometary particles composition create a serious problem, how to identify the origin of their mineral part. To solve this problem it is necessary to analyze element combination patterns in the cometary dust, and on this basis to reveal the simplest monomineral particles, which makes it possible to acquire certain limitations on the phase composition of more complicated cometary particles, because the latter may contain the same phases which make monomineral dust particles up. Such an approach requires the systematization of cometary particles according to their qualitative chemical composition, i.e. to the combinations of elements, to be followed by quantitative analysis of elemental relations within the ex-

The results of mass-spectrometry experiment do not yield explicit information about the specific composition of organic compounds. However, if we use the results of numerical calculations of cometary dust temperature /14/, we may assume that dust particles contain compounds with the decomposition temperature $T \sim 400$ K. Data on thermostability of vareous organic compounds do not disagree with the presence in dust particles of compounds of kerogen type /16/, yellow organic stuff Greenberg obtained in his experiments /15/. It is also possible that dust contains such biologically important compounds as amino acids. (Note that the characteristic time of the thermal destruction of some amino acids is more than 10^5 s /17/, which is more than the characteristic time-of-flight of particles $\lesssim 10^4$ s from the comet nucleus to the spacecraft).

On the whole, the results of mass-spectrometry experiments are explained well in the model, according to which the comet contains a great number of dust particles consisting of a silicate nucleus and a mantle made of organic compounds and volatiles. According to Greenberg the particles with such a morphology form in the interstellar clouds /15/. However, Greenberg argues in favor of the possibility of organic compounds formation in the diffusive and molecular clouds. The possibility itself cannot serve as a proof of interstellar genesis of dust particles. The interstellar origin of cometary dust can be imediately evident from studying the cometary gas isotopic composition. These data are not available now. Nonetheless the up-to-date inderect data make it possible to state that cometary dust probably has an interstellar nature.

In the given paper we discuss these data: radical difference between the mineral content of cometary gas and the meteorite matter, the possibility of survival of dust particle organic mantle during the formation of the protoplanetary disk, and the diffuculty of formation of dust particles with a silicate nucleus and an organic mantle in the protoplanetary disk itself. These facts will be analyzed below.

There is sufficient difference in Mg/Fe distribution in the submicron samples of carbonaceous chondrites Orguell and Murchison and cometary dust /18/. This difference may be interpreted by the MgO and FeO presence in cometary dust. And according to the results /23/ it is the diatomic metal oxides, exist in the interstellar medium. The fact that cometary dust particles contain the complicated organic compounds is an argument in favor of their interstellar origin.

Irradiation of the ice mixture CH_4, H_2O, NH_3, HCN and so on by ultraviolet, X-rays or cosmic rays is essential to formation of complicated organic compounds on dust particles. And the radiation dose, should be considerable, not less than 10^{16} rad = 10^{18} erg/g. It is unlikely that the cometary matter in the Oort cloud was exposed to this radiation dose during 4.5 bill years. According to /25/ the comet layer $\sim$ 100 m thick is exposed to the above mentioned ionizing radiation dose. Since, according to the available estimates Halley Comet has made about 1000 passages near the Sun and lost in each passage the matter layer about 1m thick, it becomes evident that during the latter passage those particles were analyzed which were exposed to the radiation dose in the Oort cloud much lower than it is necessary for organic molecules formation.

The necessity of exposure to great radiation doses heavily restrict organic compounds formation on dust particles in the protoplanetary disk.

The disk is opaque for the ultraviolet solar radiation. Indeed, the optical depth in the direction perpendicular to the disk plane is

$$\tau = \frac{\pi \cdot a^2}{\mu \cdot m_H} \cdot \frac{n_d}{n} \sigma \quad , \tag{1}$$

where a is the dust particle radius, n_d is the number of dust particles in the unit volume, σ is the disk surface density. Assuming that the characteristic value a is the same as in interstellar medium $a \sim 10^{-5}$ cm and $n_d/n \sim 10^{-13}$, we find

$$\tau \simeq 10 \cdot (\sigma / 1 \text{ g cm}^{-2}). \tag{2}$$

From this expression it follows that whether the dust concentration is normal, the disk becomes transparent along vertical for radiation, i.e. $\tau < 1$ only if $\sigma < \sigma_0 \simeq 0.1$ g cm^{-2}. In the protoplanetary disk even with the minimum mass $M \sim 0.01\ M_\odot$, the surface density is at least two orders higher σ_0 up to the region of outer planets Uranus and Neptune (30 AU). These density restrictions in the region of outer planets were obtained by addition of H and He volatiles, which have been probably lost in the process of planets accumulation prior to the solar composition restoration in the zone of their formation. Thus, obtained surface density dependence on radius, proves to be proportional to $R^{-3/2}$.

According to Yamamoto /19/, the region where Halley Comet formed, was confined between 14 and 115 AU. Extrapolating the surface density distribution $\sigma \propto R^{-3/2}$ up to the boundary of the region, where Halley Comet formed, we find that $\sigma > 1$ g cm^{-2}. Therefore, the disk optical depth for UV radiation is great: $\tau > 10$. So, UV radiation is effectively absorbed in the high rarefied disk layers which contain a relateveIy small portion of dust matter.

At the stage of matter accretion from the molecular cloud there is a developed turbulence in the disk that results in dust and gas mixing. Participating in the turbulent motion dust transports in the vertical direction. During the accretion time $t_{acc} \sim 10^5$ years a particle, being at R distance from the Sun exists in the region, transparent for the shortwave radiation ($\tau < 1$) during

$$t_0(R) = t_{acc} \cdot \frac{2\int_{z_0}^{\infty} \rho\,(R,z)\, dz}{\sigma\,(R)} \simeq \frac{t_{acc}}{10 \cdot (\sigma(R)/1 \text{ g cm}^{-2})} , \tag{3}$$

here $z_0(R)$ is determined from the condition $\tau\ (R, z_0) = 1$. During this time it will be exposed to the radiation dose

$$E = \int_{\nu_c}^{\infty} B(\nu, T_\odot)\, d\nu\, (R_\odot/R)^2 \frac{\pi \cdot a^2}{m} t_0(R) \sim$$

$$\sim 10^{14} \cdot (1 \text{ AU}/R)^{1/2} \quad \text{rad} \tag{4}$$

Here $\nu_c = c/\lambda_c$, $\lambda_c \simeq 200$ nm, m is the dust particle mass.

Thus, in the zone, where Halley Comet forms dust particles were exposed to the dose about two orders less than that necessary for organic matter formation from the ice mixture. Dust transport from the near solar regions cannot be an explanation to the great amount of organics, since in these regions ice sublimation occurs.

As the Sun passed through the T TAU stage the intensity of UV radiation increases in 10^2 - 10^4 times but ices sublimation due to increasing particles temperature (up to 160 K at a distance 100 AU) prevents organic formation.

After accretion is completed and turbulence damped in the disk,dust rapidly settles down in the central plane, that does not allow its long exposure in the outer layer.

The X-rays radiation with the quantum energy $\gtrsim 20$ keV can play great role in the irradiation of dust particles.

The exposure time necessary for the radiation dose is sufficient to form organic compounds made of ices which surround a silicate nucleus $A > A_0 = 10^{18}$ erg/g, is

$$t > \frac{A_0 \cdot a \cdot \rho_p}{\alpha \int I(E)\, dE}\,, \qquad (5)$$

where I(E) is the radiation spectral density /24/, α is the portion of X-ray radiation, absorbed by dust particles.

When the heavy component density z = 0.02 is standard for the interstellar matter and in all parts of the disk should be fully concentrated in dust, $\alpha \lesssim z = 0.02$. With the assumption that $a \sim 10^{-5}$ cm and $\rho_p \sim 1$ g cm^3, obtain $t \gtrsim 10^{20}$ s$\sim 3 \cdot 10^{10}$ years, that is longer than the Solar system lifetime. When particles become larger the necessary exposure time grows. Note that the real time of dust exposure in the protoplanetary disk is shorter than tha Solar systam lifetime. It is restricted by the time of dust sedimentation in the central disk plane and of solid body formation. For the disk with the normal interstellar gas-to-dust ratio the characteristic time of dust particle sedimentation is equal to /26/

$$t_{sed} \sim 10^3 \cdot (R/1\ AU)^{3/2} \ \text{years}. \qquad (6)$$

Let us consider now the possibility of "survival" of organic compounds in the accretion process as the disk forms. According to Ruzmaikina and Maeva /27/ the matter flows from the edge to the center ($u_R < 0$) in the external part of the disk within the range $0.6\ R_d \lesssim R \lesssim 0.9\ R_d$. This region is surrounded by the parts with $u_R > 0$. Such a velocity distribution in the disk means that the element joined first near the disk edge, approaches the center, and reaching some radius $\simeq 0.6\ R_d$ reverses its direction of motion and starts to remove. Due to this velocity distribution the matter in the external part of the disk never approached the the center to $R \lesssim 0.6\ R_d$, and hence did not occur in the high temperature region.

Taking into account of dust particles turbulent diffusion leaves this pattern unchanged. Solving the dust transport equation, we can show that about 1/3 (mass) of dust accreted to the disk never occur in the near-solar region and be heated to temperatures exceeding the organic destruction temperature.

The turbulence dissipation is the source of heating in the disk. Cooling takes place due to radiation from the disk surface. Since during the accretion stage the outer parts of the disk are most of the time shielded from the nucleus radiation, the contribution due to heating is is neglected. Then, as follows from the stationary equation of power balance the temperature at the disk photosphere level is determined by

$$T = (9 G M \sigma \nu R^{1/2} / 8 \sigma_{SB}) \cdot R^{-7/8} , \qquad (7)$$

where $\sigma_{SB} = 5.67 \cdot 10^{-5}$ erg cm^{-2} grad^{-4} is the Stephan-Boltzmann constant, σ is the disk surface density. In the majority of the disk the density distribution is close to stationary $\sigma \nu\ R^{1/2} \simeq$ const, and since for R = = 1 AU, $\sigma \sim (10^3 \div 10^4)$ g cm^{-2}, then for the limited values ν the temperature

$$T \simeq (400 \div 500) \cdot (1\ AU/\ R)^{7/8}\ \ K. \qquad (8)$$

The accretion matter fall velocity is

$$V = (GM\ /\ R)^{1/2} \sim 3 \cdot 10^6\ (1\ AU\ /\ R)^{1/2}\ \ cm/s, \qquad (9)$$

that is much higher than the sound velocity in the disk:

$$c_s = (k_B T\ / \mu\ m_H)^{1/2} = 10^5\ (1\ AU\ /\ R)^{7/16}\ cm/s. \qquad (10)$$

Therefore, on the disk equator, where the matter falls down almost perpendicularly to the disk surface, there should be a shock wave. Heating in the accretion shock wave is most dangerous for survival of the organic mantle on dust particles. So we consider this aspect in greater detail. Heating of dust particles in the shock wave is discussed in papers /21/ and /22/. The temperature is determined by the balance of energy a dust particle acquires and radiates in the IR diapason. Behind the shock front the particle heating sources are: UV radiation quantum absorption and particle collision with gas molecules. The energy acquired by a particle is equal to

$$L_+ = 3 \pi a^2 n_0 \mu V^3 m_H, \qquad (11)$$

where n_0 is the particle density of the accreting gas, μ is the its molecular weight, a is the dust particle radius, V is the accreting matter velocity. At the distances $R \sim R_d$ the envelope rotation is insignificant. Gas falls down spherically-symmetrically and its density may be expressed as

$$n_0 = \dot{M}\ /\ 4 \pi V R^2 \mu m_H = \qquad (12)$$
$$= 10^{11}\ (M_\odot /\ M)^{1/2} \cdot (1\ AU/\ R)^{3/2} \cdot (10^5\ yrs\ /\ t_{acc})\ \ cm^{-3},$$

where t_{acc} is the full accretion time. The energy lost by a particle depends on emissivity. According to Mie theory the emissivity of a particle in the IR range is determined by the ratio

$$Q(a,\lambda) = c_1 \cdot (a/\lambda) \cdot \lambda^{-\beta}, \tag{13}$$

where c_1 and β are the dimensionless coefficients which depend on a sort of particle material, λ is the wavelenght. For graphite, silicate and organic compounds the coefficients are β = 1; 0; 0 and c_1 = 21; 1; 2, respectively. For this $Q(a,\lambda)$ the energy radiated by a particle of a radius, is determined by

$$L_- = \frac{8\pi^2 a^3 h\, 10^{-4\beta} c_1}{c^{3+\beta}} \cdot (\beta+4)! \cdot (k_B T / h)^{5+\beta}, \tag{14}$$

where c is the speed of light, h is the Planck constant. From $L_+ = L_-$ it follows that the temperature of graphite, silicate and organic compounds particles is approximately equal to

$$T \simeq 400\ ((M/M_\odot)\cdot(10^{-4}\ cm/a)\cdot(10^5\ yrs/t_{acc}))^{1/5}\cdot(1\ AU/R)^{3/5}\ K. \tag{15}$$

The given formulas suggest that in the shock wave at the disk edge the temperature of micron dust particles does not reach 400 K when the disk radius is slightly more than 1 AU. For 0.1 micron particles this is fulfilled if the disk radius is more than 3 AU.

It is essential that in the region at hand $R>15$ AU of planets-giants and outer planets this condition will be certainly met. Hence in these regions the comets can form which involve interstellar dust, containing the organic matter.

POSSIBLE MINERAL COMPOSITION OF HALLEY COMET DUST PARTICLES

1. Si poor particles.

170 particles were discovered, where Si is either absent or is present in a very low concentration. 67 particles of this type contain only sulphur and light elements. In other particles in 92.2% cases (Mg, Fe) are present, in association with light elements, as a rule. Chromium was discovered in 28 particles, sodium was discovered in 15 ones.

The particles, comprised of light elements in this group, have the following qualitative composition: (H, C, ⟨O⟩) - 38, (H, C, N, ⟨O⟩) - 8, (H, C, S, ⟨O⟩) - 4, and (H, C, N, S, ⟨O⟩) - 2. This allows to assume that these particles consist mostly of organic matter. Besides, the particles containing only (C, ⟨O⟩) - 5 (carbon matter) and (H,⟨O⟩) - 8 (water ice?) were identified.

The particles, containing Mg and Fe, contain sulfur and light elements in 70 % cases. Most usual compositions of this type of particles are as follows (Mg, Fe)- 16, (Mg, Fe, H, C) - 15, (Mg, Fe, H) - 7, (Mg, Fe, C) - 6. Other combinations were observed in isolated cases. (Mg, Fe) particles may be interpreted as oxide particles, fused magnesia mainly, because only two oxide particles with high Fe content were discovered, i.e. magnetite and magnesioferrite. But it should be noted that the discovered oxide particles were usually very small and the majority of their spectra were obtained in a narrow angle. That is why presence of light elements should not be excluded and such dust particles may actually consist of more complicated phases, as carbonates and hydroxides.

Presence of such phases in cometary dust is confirmed by the analysis of elemental association in poor Si particles. Thus the association (Mg, Fe, H) indicates definitely the existance of brucite. Carbonates, mainly magnesite or hydromagnesite may occur in the (Mg, Fe, C)- or (Mg, Fe, C, H)-particles. Certainly, carbonaceous and organic matter may exist in the particles of this type, but it is unlikely that their composition is defined only by organic compounds mixed with fused magnesia. Judging from the prevalence of (Mg, Fe, C, H)- , (Mg, Fe, C)- , (Mg, Fe, H)-particles, brucite and carbonates may be basic non-organic components in Si-poor cometary particles.

Fe and Mg sulphides, sulphates, nitrates or their hydrated analogues are possible accessory phases in Si-poor particles. Their presence is indicated by (Mg, Fe, S), (Mg, Fe, H, S), (Mg, Fe, H, N) associations, observed in some particles. It is typical that Fe content is increase in the particles, containing sulphur. This correlation is an evidence of sulphide and sulphate Fe in cometary matter. High Cr of some Si-poor particles can be explained by the presence of magnesiochromite, as (Mg, Cr) association is typical for such particles. However, some particles were discovered where only Cr in association with light elements were observed. It is quite possible that such particles may contain escolaite. Na carbonates, sulphates, nitrates, cianides or rhodanides may be present in Si-poor particles as accessory phases, as this element is usually associated with C, H, N and S. Na compounds content of Si-poor particles is not high, because in all cases they are associated with Mg-Fe phases.

2. Si-rich particles.

This group incorporate 341 particles, 94.4% of these particles containing Mg and Fe, 80.9% - S and light elements, 50% - Na. The most usual associations of metal ions, observed in this group are as follows: (Mg, Fe) - 127 particles, (Mg, Fe, Na) - 106 particles, (Mg, Fe, Na, Al) - 21 particles, (Mg, Fe, Na, Ca) - 15 particles and (Mg, Fe, Na, Ca, Al) - 9 particles. In the majority of cases these associations include sulphur and light elements. Chromium is found in 28 particles, contrary to Si-poor particles where association (Mg, Fe, Cr) is typical, in Si-rich particles chromium is usually found not only together with Mg and Fe but also with Ca and Na. Among Si-rich particles without Mg and Fe there are dust particles which in addition to light elements contain only Si - 11 particles, only Al and Si - 3 particles and only Na and Si - 5 particles.

The association of elements observed will enable us to assume that the composition of Si-rich particles together with organic matter should include Mg-Fe silicates and hydrosilicates as dominating phase. The latter could be explained by presence of the particles of the (Mg, Fe, Si, H, ⟨O⟩) composition (8 particles) and hydrated phases in Si-poor particles. Silicates of Na is probably a secondary but important phase in Si-

-rich dust particles, because there is some correlation between Si and Na and one particle containing (Na, Si) was discovered. Na (sodium) could be present in other silicate phases. Thus coexistance of Na, Ca and Al assumes the presence of plagioclases, feldspathoides for example sodalite, nepheline, scapolite and at last zeolites.

Beside one cannot exclude unsructural position of Na into surface layer silicates. Cr-containing pyroxenes (ureite), hydrosilicates of Al (caolinite), silicate phases (most likely opal) will be present as other accessory phases in Si-rich particles.

Silicate phases and organic matter probably are not the only components of the Si-rich particles. Thus the analysis shows, that Na, Ca and Al have a definite tendency to associate with S and N. These elements are observed with Na, Ca and Al in 60 - 70% cases, whereas Mg and Fe associate with S and N only in 20% particles. It might mean that Na, Ca and Al can be present in rich Si particles in a form of mineral salts such as sulphates, nitrates, cianides or rhodanides. Possible existance of such compounds as well as Na carbonates in a comet dust are supported by data on Si-poor particles. It is very likely that mineral salts Na, Ca and Al in rich Si particles are more spread than their silicates, because the presence of these elements do not effect significantly the ratio of Si, Mg and Fe in the Si-rich particles.

The latter is very important, because it will help to define the nature of comet Mg, Fe-phases based on the Si, Mg and Fe ratious and to use the whole set of analytic data available.

The observed Si/(Mg + Fe) distribution displays that the composition of the Si-rich particles are not based on the mixture of the non-silicate phases with talk or sepiolite because there are modes typical for the phases of an intermediate composition. Actually on all cases the mode of poor Ca pyroxene, in basic enstatite, is clearly observed. Olivine, usually forsterite, is dominating only in the particles, where hydrogen is absent. In the particles containing hydrogen, as expected, the olivinous cluster is suppressed and the serpentine composition is more clearly observed. A comparison of the Si/Mg and Si/(Mg + Fe) distributions enables us to estimate the form of the Fe in rich Si particles. A clearer grouping of compositions on the Si/(Mg + Fe) distribution with a better correspondance of modal values to the compositions of known mineral phases, means that a considerable part of Fe is in silicates. It indicates the oxydized nature of the comet matter.

Therefore, the composition of the Si-rich particles should include serpentine, olivine and poor Ca pyroxene as basic silicate phases. Mixing of these phases with hydroxides or with Fe and Mg carbonates as well as with talk or sepiolite could explain the whole dispersion of Si/(Mg + Fe) ratios observed in these particles. Besides, sulphates or sulphides of Fe could be found in dust particles containing sulphur because the Fe/Mg ratio in such particles is much higher. It is also possible that (Fe, S) component could be present in the composition of mixed layers silicates of the tochilinite type because this component is obviously associated with serpentine.

Modal values of Si/(Mg + Fe) distribution are in an exelent agreement with the phase composition observed in a meteorite matter. It indicates the validity of data on Si, Mg and Fe obtained by PUMA-2 instrument

An identification of olivine based on these data corresponds to the appearence of the absorbtion line of this mineral in according to the results obtained by the IKS method during the VEGA mission /28/. Therefore the PUMA data for Si, Mg and Fe do not need any corrections suggested by /12/, because if the corrections are used, it will cause a shift of modal values (including the olivine mode) in the Si/(Mg + Fe) distribution to the region of unreal silicate compositions with the Si/(Mg + Fe) ratio more than 1.5.

THE DISCUSSION OF RESULTS

The spectra analysis shows that in the mineral composition of the comet dust, next components are important:
1.Silicates, where olivine, poor Ca pyroxene and serpentine dominate . Silicate of Na is also definitely present. The presence of alumosilicates of Na and Ca, kaolinite, and rich Cr pyroxene is also possible.
2. Oxides and hydroxides. The most important phases are periclase and/or brucite, accessory phases - magnesiochromite, opal. Possible are also: magnetite, magnesioferrite, eskolaite, goethite.
3. Carbonates and hydrocarbonates. Among them carbonates of Fe, Mg and Na are the most popular.
4. Sulphates and sulphides. Probably, both types of such compounds are present. Sulphates are more typical for Na, Ca and Al, sulphides - for Fe.

In comet particles these mineral components produce various mixtures with an organic matter, Mg and Fe are basically included in the composition of silicates, hydroxides and carbonates, and Na, Ca and Al are included in the composition of mineral salts having a rather accessory meaning. The most typical is the association of organic matter with the layer silicates. The more complicated the composition of organic matter, the bigger the Si/Mg ratio in comet particles upto space proportion. The Fe/Mg ratio also increases and in associations containing the organic matter with sulphur it achieves values observed in silicate phases of ordinary chondrites, although on the whole the Fe/Mg ratio in the comet particles is much lower than in chondrites.These tendencies are followed by the increase of the amount of mineral salts, as the latters basically associate with sulphur and nitrogen.

When comparing the mineral composition of comet particles with that of meteorites of different types, it appears that they are similar to C chondrites, which contain a complex of layer silicates, sulphates, carbonates and organic matter. However, in the comet particles a phases and organic matters are more propagated. Besides, in CI chondrites olivine and pyroxene are not typical. These minerals are more propagated in C3 chondrites, but they have very small amount of the organic matter, mineral salts and layer silicates. At last it is important, that comet particles in the majority of cases are noticeably more magnesiable than the chondrite matter /18/. Such prevalence of rich Mg phase (forsterite and enstatite)in meteorites are not noticed, except for enstatite chondrites and achondrites, but the latters have a reduction character, they do not contain the organic matter and layer silicates. The particles of the interplanetary dust /29/ at least some of them are closer to the comet particles. They also can be enriched by Na, organic matter and they have a similar distribution of the Fe/Mg ratio. They also contain layer silicates, olivine, pyroxene, opal. However, on the average the interplanetary dust has less volatile substances than the comet particles. One can assume that the formation of the interplanetary dust could be connected with the destruction of cometary bodies, and the interplanetary particles just lost the volatile substances while being in space or when entering the atmosphere of Earth.

The presence of very high concentration of organic matter (80% of studied particles contain light elements) shows a great difference by the chemical composition between the cometary dust and any other matter of the Solar system known in cosmochemistry. As indicated in /30/ this fact allows to conclude that comet dust is of interstellar origin. If this model is valid, the interstellar dust should contain high temperature condensates such as olivine and pyroxene. The presence of low temperature phases (serpentine, hydroxides, carbonates etc.) in cometary particles may show a primary inhomogenety of the interstellar dust which consists of the mater formed at different conditions and from different sources. However, the formation of low-temperature minerals could be considered as a result of processes of a secondary change of the primary condensates (olivine and pyroxene). Such processes could have occured in the upper layers of the comet nuclei (it was discussed in /31/), as possibly in the interstellar medium due to dehydration and decarboxylision reactions of the organic matter coexisting with olivine in the interstellar dust. In these reactions olivine can be transformed into serpentine. The association of the organic matter and serpentine observed in the comet particles is compatible with such scheme of the process. Unfortunately, at present it is difficult to make certain conclusions on genesis of the low-temperature phases in the comet dust component. It is clear that the cometary matter has traces of quite complicated and various cosmochemical processes.

REFERENCES

1. Clube, S.V.M. and Napier, W.M. (1982) 'Spiral arms, comets and terrestrial catastrophism', Quart. J. Roy. Astron. Soc. 23, 45-66.
2. Clube, S.V.M. and Napier, W.M. (1984) 'Comet capture from molecular clouds: A dynamical constraint on star and planet formation', Monthly Not. Roy. Astron. Soc. 208, 575-588.
3. Cameron, A.G.W. (1973) 'Accumulation processes in the primitive solar nebula', Icarus 18, 407-450.
4. Cameron, A.G.W. (1978) 'The primitive solar accretion disk and the formation of the planets', in S.F. Dermott (ed.), The Origin of the Solar System, New York, pp. 49-75.
5. Donn, B. (1976) 'Comets, interstellar clouds, and star clusters', in The Study of Comets, NASA SP-393, Washington, pp. 663-672.
6. Biermann, L. and Michel, K.W. (1978) 'On the origin of cometary nuclei in the presolar nebula', Moon and Planets 18, 447-464.
7. Lyttleton, R.A. (1948) 'On the origin of comets', Monthly Not. Roy. Astron. Soc. 108, 465-475.

8. Vsekhsvyatskii, S.K. (1967) The Nature and Origin of Comets and Meteors, Prosveschchenie Press, Moscow.
9. McCrea, W.H. (1975) 'Solar system as space probe', Observatory 95, 239-255.
10. Van Flandern, T.C. (1978) 'A former asteroidal planet as the origin of comets', Icarus 36, 51-74.
11. Venus-Halley Project. (1985) V.M. Balebanov, E.N. Vorontzova, G.A. Scuridin (eds.), Lui-Jean, Paris.
12. Jesberger, E.K. et al. (1988) 'Aspects of the major element composition of Halley's dust', Nature 332, 691-695.
13. Inogamov, N.A. (1984) 'Plasma cloud evolution in electric field', Doklady Academii Nauk SSSR 227, 1375-1379.
14. Hunner, M.S. (1985) 'A preliminary look at the dust in Comet Halley', Adv. in Space Res. 5, N 12, 325-334.
15. Greenberg, J.M. (1987) 'Evidence for the pristine nature of Comet Halley', in The Comet Nucleus Sample Return Mission, Proc. Workshop, Canterbury, pp. 47-57.
16. Wallis, M.K. et al. (1986) 'Evaporating grains in Halley's coma',in B.Battrick, E.J. Rolfe, R. Reinhard (eds.), Proc. Int. Symp. on the Exploration of Halley's Comet, Heidelberg, v.", pp. 251-254.
17. Vallentyne, J.R. (1966) 'Thermal stability of threonine in the presence of a marine polyphenolic material', Science 151, 214-215.
18. Brownlee,D.E. et al. (1987) 'A quantitative comparison of Comet Halley and carbonaceous chondrites at the submicron level', in Lunar and Planetary Sciences, XVIII, Houston, Pt. 1, pp. 133-134.
19. Yamamoto, T. (1985) 'Formation history and enviroment of cometary nuclei', in S. Klinger et al. (eds.), Ices in Solar System, Dordrecht, pp. 205-219.
20. Feldman, P.D. (1983) 'Ultraviolet spectroscopy and the composition of cometary ices', Science 219, 347-354.
21. Hollenbach, D. and McKee, C.F. (1979) 'Molecule formation and infrared emission in fast interstellar shocks. 1. Physical processes', Astrophys. J. Suppl. 41, 555-592.
22. Ruzmaikina, T.V. and Safronov, V.S. (1985) 'Premature particles in the solar nebula', in Lunar and Planetary Sciences, XVI, Houston, p. 720.
23. Dule, W.W. et al. (1979) Appl. Space Science 65, p.69.
24. Schwartz, D. and Gursky,H. (1974) 'The cosmic X-ray background', in R. Giacconi and H. Gursky (eds.), X-Ray Astronomy, Dordrecht, pp. 359-388.
25. Strazzulla, G. (1986) '"Primitive" galactic dust in Solar system?', Icarus 67, 63-70.
26. Safronov, V.S. (1969) The Evolution of Protoplanetary Nebula and Formation of Earth and Planets, Nauka, Moscow.
27. Ruzmaikina, T.V. and Maeva, S.V. (1986) 'Study the process of formation of protoplanetary disk', Astronomicheskii Vestnic 20, 212-227
28. Combes, M. et al. (1988) 'The 2.5 to 12 μm spectrum of Comet Halley from the IKS-VEGA experiment', Icarus,
29. Mackinnon, I.D.R. and Rietmeijer, F.J.M. (1987) 'Mineralogy at chondritic interplanetary dust particles', Rew. of Geophisics 25, 1527-1553.

30. Rietmeijer, F.J.M. (1988) 'A quantitative comparison of fine--grained chondritic interplanetary dust and Comet Halley dust' ,in Lunar and Planetary Sciences, XIX, Houston, pp. 980-981.
31. Rietmeijer, F.J.M. (1985) 'A model for diagenesis in protoplanetary bodies', Nature 313, 293-294.

Contributed Papers

HEAVY METAL ABUNDANCES IN NOVA EJECTA REVEALED BY INFRARED CORONAL EMISSION LINES

MATTHEW A. GREENHOUSE
Laboratory for Astrophysics
National Air and Space Museum
Smithsonian Institution
Washington, D.C. 20560

ABSTRACT. This article discusses the recent discovery of infrared coronal emission lines in the ejecta of Nova QU Vul, V1819 Cyg, and Nova Herculis 1987. These data suggest that infrared coronal line nova shells are typically highly overabundant in Al, Mg, and Ne, and may contribute significantly to the gas phase abundance of these elements in the interstellar medium of galaxies.

Discussion

The term "coronal line" refers to forbidden ground state fine structure transitions in species with ionization potential $\geq$ 100 ev. Coronal lines of heavy elements were first observed in optical spectra of the Solar corona; however, highly ionized elements are also found in lower density regions such as planetary nebulae, nova shells, supernova remnants, and throughout the hot phase of the interstellar medium (ISM) of galaxies. Many of these latter sources occur in locations where heavy dust obscuration precludes their study via ultraviolet or optical lines.

Infrared coronal lines of several abundant elements have now been observed in one planetary nebula and several classical novae (see Grasdalen and Joyce 1976; Ashley and Hyland 1988; Greenhouse *et al.* 1988, 1989). Infrared coronal lines observed to date are listed in Table 1. This emission not only reveals the presence of coronal temperature gas in obscured regions, but can also yield the electron temperature and elemental abundances in the gas. The results discussed below have been described in detail by Greenhouse *et al.* (1989).

Three of the 4 infrared coronal line novae observed to date were discovered since 1986. A typical spectrum of one of them, Nova QU Vul, is shown in Figure 1. Intense, high contrast emission lines of several highly ionized heavy elements dominate the emission in the wavelength region shown. In novae one often finds that the ions are in collisional ionization equilibrium and that the transitions are not thermalized. The electron temperature (T_e) and elemental abundances can be determined from the spectrum shown in Figure 1 in a simple way.

Lines due to adjacent ionization states of Si are apparent in the spectrum near 2 μm. The emission is optically thin, so the [Si VI]/[Si VII] line intensity ratio is proportional to the ratio of the density of emitters $n(Si^{+5})/n(Si^{+6})$. For a gas in collisional ionization equilibrium, this latter quantity is determined only by T_e. Hence, one can find T_e by comparing the measured line ratio with that predicted by model collisionally ionized plasmas (see Shull and Van Steenberg [1982] for a recent model). One can then find relative abundances among the coronal species.

For example, the measured relative density $n(Al^{+5})/n(Si^{+6})$ can be used with the ionization fractions $n(Al^{+5})/n(Al^0)$ and $n(Si^{+6})/n(Si^0)$ predicted at T_e by the model plasma to find the relative abundance $n(Al^0)/n(Si^0)$ of the neutral elements. This relative abundance can then be compared to that expected in a solar abundance pattern. Results obtained on infrared coronal line novae observed to date are shown in Table 2. One can see that both

J. M. Greenberg and V. Pirronello (eds.), Chemistry in Space, 417–418.

V1500 Cyg and QU Vul exhibit highly non-solar abundance patterns in Al and Mg. The latter nova is also extremely overabundant in Ne.

This heavy metal enrichment of nova shells primarily results from diffusion and/or turbulent mixing of white dwarf core material into the accreted H rich envelope prior to the nova outburst (see Shara [1989] and references therein). Novae provide a mechanism that returns enriched white dwarf core material to the ISM. The close binary systems that give rise to classical novae may under go $\sim 10^3$ outbursts during a Hubble time (Ford 1978). Hence, these new abundances, revealed by infrared coronal line spectra, suggest that novae can contribute significantly to heavy metal gas phase abundances in the interstellar medium of galaxies.

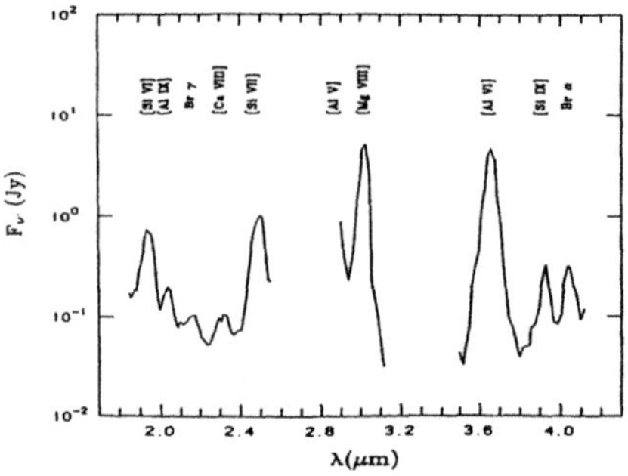

Figure 1. Nova QU Vul on 30 Jul 1986

TABLE 1. Coronal lines of recent novae

Ion	Transition	λ (μm)
Ca VIII	$^2P^0(3/2,1/2)$	2.32
Al V	$^2P(1/2,3/2)$	2.88
Ne VI	$^2P(1/2,3/2)$	7.60
Al VI	$^3P(1,2)$	3.66
Si VI	$^2P(1/2,3/2)$	1.96
Si VII	$^3P^0(1,2)$	2.47
Mg VIII	$^2P^0(3/2,1/2)$	3.02
Al IX	$^2P^0(3/2,1/2)$	2.04
Si IX	$^3P(1,0)$	3.92

TABLE 2†. Electron temperature and elemental abundances

Object	Log T_e (K)	Al/Si	Mg/Si	Ne/Si
Sun	–	0.1	0.8	2.5
V1500 Cyg	5.53	4.9	2.3	–
QU Vul	5.57	12.8	2.2	335.8
V1819 Cyg	5.46	–	–	–
Herculis 1987	5.50	–	–	–

† see Greenhouse *et al.* (1989) for further details

References

Ashley, M. C. B., Hyland, A. R. 1988, *Ap. J.*, **331**, 532.

Ford, H. C. 1978, *Ap. J.*, **219**, 595.

Grasdalen, G. L., and Joyce, R. R. 1976, *Nature*, **259**, 187.

Greenhouse, M. A., Grasdalen, G. L., Hayward, T. L., Gehrz, R. D., and Jones, T. J. 1988, *A. J.*, **95**, 172, erratum 1989, *A. J.*, **98**, 738.

Greenhouse, M. A., Grasdalen, G. L., Woodward, C. E., Benson, J., Gehrz, R. D., Rosenthal, E., and Skrutskie, M. F. 1989, *Ap. J.*, in press.

Shara, M. M. 1989, *Pub. A. S. P.*, **101**, 5.

Shull, M., and Van Steenberg 1982, *Ap. J. Suppl.*, **48**, 95, erratum 1982, *Ap. J.*, **49**, 351.

IMPULSIVE MODEL OF COLLISIONS

N.PETROVIĆ AND S.D.BOSANAC
Ruđer Bošković Institute
Department of Physical Chemistry
Bijenička 54
P.O.Box 1016
41000 Zagreb, Croatia
Yugoslavia

ABSTRACT. Quantum theory of impulsive collisions is developed. It is suited for the study of nonreactive and dissociative collisions.

Energy transfer in nonreactive collisions of atoms and molecules is mainly determined by the short range repulsive part of the potential. This fact can be used for the modelling of these collisions and in the simplest case the repulsive part is replaced by a hard wall while the remaining potential is neglected. The model works well in a number of cases, primarily when single impact collisions are considered (Bosanac [1]). We are developing the quantum theory of this model in a more general case when potential is spherically symmetric outside the repulsive wall. The deviation from the spherical symmetry is treated as perturbation. For atom-molecule system we write the potential as

$$V(\underline{R}, \underline{r}) \sim V(R)\,\eta(\underline{R}, \underline{r}) \qquad (1)$$

where $\underline{R}$ is the separation vector of the atom with respect to the centre of mass of the molecule and $\underline{r}$ is the separation vector of the two atoms in the molecule. The funcition η is equal to unity when the atom is beyond the reach of the hard wall but otherwise it has infinite value. Hence this function defines boundary of the molecule

$$R = g(\theta, \phi\,;\underline{r}) \qquad (2)$$

where θ and ϕ are spherical angles of the vector $\underline{R}$.

A particular solution of the Shrödinger equation ,when the atom is outside the boundary, is given as a product of the wave functions for the molecule and the relative translational motion of the atom and molecule. However, the most general solution is a linear combination of the particular solutions, but we seek the one which obey the following conditions: a) all the particular solutions have the same total energy, b) at the boundary the solution must be zero and c) for large atom-molecule separation the solution must have proper scattering boundary condition. We can write this solution as

J. M. Greenberg and V. Pirronello (eds.), Chemistry in Space, 419–420.

$$\Psi = \rho_{N,J,M}(\underline{r})\, \omega^{z}_{N,J}(\underline{R}) + \sum_{n,j,m} \rho_{n,j,m}(\underline{r})\, \omega^{+}_{n,j,m}(\underline{R}) \tag{3}$$

where $\rho_{N,J,M}$ is eigenfunction of the molecule in the vibrational state N and rotational state J,M. $\omega^{+}(\underline{R})$ represents the incident wave which has asymptotic form of a plane wave propagating along the z-axis. The wave function $\omega^{+}(\underline{R})$ has the asymptotic form

$$\omega^{+}_{n,j,m} \sim \frac{\exp(iRk_{n,j})}{R}\, f_{n,j,m;N,J,m}(\theta,\phi) \tag{4}$$

where $f(\theta,\phi)$ is the scattering amplitude.

At the boundary (2) the solution (3) must be zero, for all θ, ϕ and $\underline{r}$. This condition defines a set of linear equations for the unknow scattering amplitide, which can be solved by simple inversion of appropriate complex matrix. In the impulsive model, therefore, it is not required to solve differential equations as it is normally the case. The model has been used in molecule-surface studies of energy transfer (Bačić and Bosanac [2]) but we are presently investigating its validity for simple cases of atom-molecule collisions (Petrović and Bosanac [3]).Our intension is to apply the model to the tree dimensional nonreactive collisions of atoms and molecules, and also to the collision induced dissociation processes.

REFERENCES

1. Bosanac, S. D. (1982) ' Origin of the left-right asymmetry in rotationally inelastic differential cross section ', Phys. Rev. A 26, 816-824.
2. Bačić, Z. and Bosanac, S. D. (1984) ' Two-dimensional model of molecule-surface scattering ', Chem.Phys. Lett. 105, 518-522.
3. Petrović , N. and Bosanac, S. D. ' Theory of impulsive collisions ', work in preparation.

ADIABATIC APPROXIMATIONS FOR MOLECULES IN MAGNETIC FIELDS

P.Schmelcher, L.S.Cederbaum and H.D.Meyer
Theoretical Chemistry, Institute for Physical Chemistry
Im Neuenheimer Feld 253
D–6900 Heidelberg
Federal Republic of Germany

The astrophysical discovery of strong magnetic fields on white dwarfs ($\approx 10^8$G) and neutron stars ($\approx 10^{12}$G) motivated many studies on the behavior of atoms and molecules in strong magnetic fields[1-2]. In contrast to the numerous investigations on the hydrogen atom, the information about the behavior of many electron atoms and, in particular, molecules in strong magnetic fields is very scarce. Nevertheless, the existing knowledge about molecules indicates a richness of new phenomena and shows drastic changes of the molecular properties in comparison with the zero–field case. It is therefore of relevance to investigate the validity of the foundations of molecular physics and in particular of the Born–Oppenheimer approximation in the presence of a magnetic field[3-4]. The starting point for such an investigation is the total molecular Hamiltonian in Cartesian coordinates. For the sake of simplicity we will specialize to the case of a homonuclear diatomic molecule. As a first step we perform the center of mass pseudoseparation of the neutral molecule. The only consequence of the center of mass motion is the appearance of a Stark effect due to a motional electric field[3-5]. From the resulting molecular Hamiltonian we obtain the corresponding electronic Hamiltonian by setting the nuclear relative velocity equal to zero. To solve the Schrödinger equation we expand as usual the total wave function in a series of products of electronic and nuclear wave functions. Projecting on one electronic eigenfunction finally yields the following set of coupled nuclear equations of motion[3-4]

$$\left[\frac{2}{M_0}\cdot\left[\vec{P} + \frac{eZ}{4}\cdot\vec{B}\times\vec{R}\right]^2 + \epsilon_m(\vec{K};\vec{R}) - E\right]\chi_m(\vec{R}) = \sum_n \Lambda_{mn}\chi_n(\vec{R}) \qquad (1)$$

where M_0 is the total nuclear mass, Z the nuclear charge number and $\vec{R}$ the relative vector of the two nuclei. The vector potential is chosen in the symmetric gauge. ϵ_m is the electronic potential surface which depends parametrically on the constant pseudomomentum k. The nonadiabatic coupling elements Λ_{mn} take on the following appearance

J. M. Greenberg and V. Pirronello (eds.), Chemistry in Space, 421–422.

$$\Lambda_{mn} = -\frac{2}{M_0}\cdot\left[<\phi_m|\vec{P}^2|\phi_n> + 2\cdot<\phi_m|\vec{P}|\phi_n>\cdot\vec{P} + \frac{e}{2}Z\cdot\vec{B}x\vec{R}\cdot<\phi_m|\vec{P}|\phi_n>\right] \quad (2)$$

In addition to the nonadiabatic coupling elements present also in the zero–field case, we have now additional explicit magnetic field–dependent couplings. The Born–Oppenheimer approximation is obtained by setting the right hand side of eq.(1) equal to zero. The resulting nuclear equation of motion shows no screening of the nuclear charge against the magnetic field, i.e. it contains, independently of the internuclear distance R, the para– and diamagnetic terms of the nuclear relative motion. This is in contrast to the fact that we expect a free nuclear relative equation of motion for infinitely separated neutral atoms. To investigate this problem we consider the dissociation limit of the molecule into neutral atoms. For this limit we have calculated[3-4] the electronic energy ϵ and the diagonal terms Λ_d of the nonadiabatic couplings (2). The result is

$$\epsilon = 2E_a + \mathscr{C}$$

$$\Lambda_d = \frac{1}{M_0}\cdot Ze\cdot(\vec{B}x\vec{R})\vec{P} + \frac{1}{8M_0}\cdot Z^2e^2\cdot(\vec{B}x\vec{R})^2 + \mathscr{C} \quad (3)$$

where E_a is the energy of the atom in the presence of a magnetic field. $\mathscr{C}$ is an irrelevant constant. From Λ_d in eqs.(3) one immediately realizes that the screening of the nuclear charge is contained in the diagonal term of the nonadiabatic coupling elements. If we therefore add Λ_d to the nuclear equation of motion (1) in the Born–Oppenheimer approximation we arrive at a free nuclear relative equation of motion for the dissociation limit of the molecule into atoms. The diagonal term of the nonadiabatic coupling elements must therefore be included in the nuclear equation of motion in the Born–Oppenheimer approximation in order to obtain a reasonable description of nuclear dynamics in the presence of a magnetic field.[3-4] In contrast to the zero–field case it is therefore not sufficient to perform a Born–Oppenheimer or Born–Huang adiabatic approximation, instead one has to include the complete diagonal term of the nonadiabatic couplings in the adiabatic approximation scheme. We finally refer the reader to an alternative approach, the so–called screened Born–Oppenheimer approximation, which has been discussed in reference 4.

REFERENCES

[1] D.M.Larsen, Phys.Rev.A 25, 1295 (1982)
[2] U.Wille, Phys.Rev.A 38, 3210 (1988)
[3] P.Schmelcher, L.S.Cederbaum and H.D.Meyer, J.Phys.B 21, L445 (1988)
[4] P.Schmelcher, L.S.Cederbaum and H.D.Meyer, Phys.Rev.A 38, 6066 (1988)
[5] M.Vincke and D.Baye, J.Phys.B 21, 2407 (1988)

THE CALCULATION OF RO-VIBRATIONAL MOLECULAR SPECTRA FROM FIRST PRINCIPLES.

JAMES R. HENDERSON, JONATHAN TENNYSON and STEVEN MILLER
Department of Physics and Astronomy,
University College London,
Gower Street, LONDON WC1E 6BT, U.K.

ABSTRACT. A discussion of our recent developments in the ab initio calculation of molecular spectra and our involvement with astro-chemistry is presented. Particular attention is paid to the problem of dealing with high-lying and disordered states, with specific reference to the important H_3^+ molecular ion.

Quantum chemists are now beginning to be able to perform large and accurate calculations to realistically investigate the behaviour of small molecules and to address challenging spectroscopic problems. Obviously, this is of great use and importance to the world of astro-physics and interstellar spectroscopy.

When attempting to perform ab initio calculations to predict the spectra of polyatomic molecules, a major difficulty is the accurate datermination of the bound-state energies and wavefunctions. This is done by solving the nuclear-motion quantum mechanics on a particular potential energy hypersurface. Using conventional basis set expansion methods it has not generally been possible to investigate regions of high energy - particularly in molecules which undergo large amplitude vibrational motion.

We have recently developed a new approach to deal with these large amplitude motions in triatomic molecules that can support as many as several hundred bound states.

The method, which we term a Discrete Variable Representation or DVR, was first suggested by Harris et al. (1) but has not been seriously investigated until recent years by Light et al. (2). The authors have now developed and extended the theory to operate more efficiently and to enable rotational excitation in the DVR to be implemented for the first time ever (3).

The main idea is to treat one or more of the internal co-ordinates in a finite element manner. In this way, Hamiltonian matrices are constructed for a reduced dimensional problem, i.e. 1-D and 2-D, and diagonalised to yield intermediate eigenfunctions which they serve as an effective basis to proceed and solve the next, higher-dimension matrices. For a triatomic system, the procedure will stop when the 3D

J. M. Greenberg and V. Pirronello (eds.), Chemistry in Space, 423–424.

Hamiltonian has been solved. The major advantage of the DVR is that not all of the intermediate eigenfunctions are needed for the next step and so only a subset need be selected to form a very accurate and compact basis. The result is that the final matrix to be diagonalised is of a much reduced secular size. Hence one can investigate much higher ro-vibrational states than has previously been the case.

Our recent work on the H_3^+ molecular ion (3) has produced accurately some 180 vibrational states - two-thirds of the way to dissociation, and J = 1←0 excitation energies for the lowest 41 vibrational states, substantially more than has been achieved before. We hope in the very near future to be able to seriously address the challenging problem of the 27,000 near-dissociation laboratory lines observed by Carrington et al. (4) in 1982.

The molecules which interest us most are indeed those which are of great importance in the interstellar medium - species such as HCN/HNC and H_3^+ along with its deuterated isotopomers. In fact, recent calculations by Miller and Tennyson (5) have enabled H_3^+ to be detected extraterrestrially for the first time ever. The calculated results are in excellent agreement with laboratory data as well as the observed spectra and have enabled 24 out of 32 observed lines to be assigned in a K-window overtone spectrum of H_3^+ in the magnetosphere of Jupiter at a temperature of around 1000 K (6). It is hoped that this recent breakthrough will assist in the much awaited interstellar and extragalactic detection of H_3^+ and exemplifies the rôle of the theoretical molecular physicist in the world of astrophysics and interstellar chemistry.

REFERENCES

(1) Harris, D.O., Engerholm, G.G. and Gwin, W.D. (1965) 'Calculation of matrix elements for one-dimensional quantum-mechanical problems and the application to anharmonic oscillators', J.Chem.Phys. 43, 1515-7.

(2) See for example, Bačić Z. and Light, J.C. (1987) 'Accurate localised and de-localised vibrational states of HCN/HNC', J.Chem.Phys. 86, 3065-77.

(3) Tennyson, J. and Henderson, J.R. (1989) 'Highly-excited ro-vibrational states using a discrete variable representation : the H_3^+ molecular ion', J.Chem.Phys. (in press).

(4) Carrington, A. and Kennedy, R.A. (1984) 'Infrared predissociation spectrum of the H_3^+ ion', J.Chem.Phys. 81, 91-112.

(5) Miller, S. and Tennyson, J. (1987) 'First principles calculation of the molecular constants of H_3^+, H_2D^+, D_2H^+ and D_3^+', J.Mol.Spectr. 128, 183-92; (1988) 'Overtone bands of H_3^+: first principles calculation', 128, 530-9, and (in press) 'Hot band transition frequencies and line strengths in H_3^+'.

(6) Drossart, P., Maillard, J.-P., Caldwell, J., Kim, S.J., Watson, J.K.G., Majewski, M.A., Tennyson, J., Miller, S., Atreya, S., Clarke, J., Waite, J.H.Jr. and Wagener, R. (1989) 'Detection of H_3^+ on Jupiter', Nature (in press).

IRRADIATION PARAMETERS IN COSMIC CHEMISTRY

N. J. Sack
Department of Nuclear Engineering and Engineering Physics
University of Virginia
Charlottesville, VA 22901

ABSTRACT. The radiation chemistry of various dense astrophysical environments has been subject to many laboratory investigations which dealt with the radiation of simple gas or ice mixtures with UV radiation or high energy particle irradiation. Certainly the experimental irradiation time is always by many orders of magnitude shorter than the time scales of the simulated astrophysical system. To obtain comparable total absorbed doses of energy, high dose rates, respectively current densities of radiation, have been applied in the simulations. This paper deals with problems in the application of experimental data on the chemistry of the astrophysical environments arising by the difference in dose rate.

The chemistry in various dense (density $> 10^{18}$ cm^{-3}) astrophysical environments containing the elements C, H, O, and N like atmospheres of planets and moons or surfaces of comets and outer planets has been simulated in numerous laboratory experiments (for references, see, e.g., Sack 1989). This chemistry is mainly a radiation chemistry driven by UV-radiation or high energy particle radiations like galactic cosmic rays or solar wind.

In primary processes like ionizations or excitations, which may be followed by dissociations, various reactive species like ions, radicals, or electrons are produced. The kinetic energy of these particles ranges from meV to some eV.

Then these reactive species undergo many chemical reactions, as, ion-molecule reactions, neutralizations, or neutral reactions. This leads finally to the formation of higher organic compounds like alcohols, amines, cyanides, or aldehydes.

Besides the classification of the reactions by, e.g., ion-molecule, neutral reactions, or neutralizations they can also be classified in another way: there are reactions between reactive species and stable "target molecules", e.g.,

$$H + CH_4 \longrightarrow H_2 + CH_3 \qquad (1)$$

J. M. Greenberg and V. Pirronello (eds.), Chemistry in Space, 425–426.

and reactions among two reactive species, e.g.,

$$H + C_2H_5 \longrightarrow C_2H_6 \quad . \qquad (2)$$

As the C_2H_5 is not a stable molecule but could also react with many other species, reactions of type (2) depend much more on the density of the reactive species than reactions of type (1). Hence the relative amount of reactions among reactive species (type 2) will increase with increasing density of reactive species.

The density of the reactive species in gaseous systems is regulated by the density of the primary events, which itself is dependent on the power density of the radiation. It therefore can be stated that in gaseous systems reactions among reactive species are enhanced at high power densities of the radiation.

This is in agreement with the measurements of Sack et al. (1989), Meisels (1965), and Woodward and Back (1963), who have studied the influence of these irradiation parameters, namely power density, on the radiolysis of various gas mixtures with different radiations.

In frozen systems like the surfaces of comets, radicals produced by radiation induced processes can be stored for a considerable period of time (Johnson et al. 1986). Therefore it seems possible that here the total absorbed dose of radiation regulates the density of reactive species, which may react when the surface of the comet is warmed up during the approach of the comet and the sun. Nevertheless the importance of very slow low temperature reactions is not quite clear; it has to be taken into account that the astrophysical time scales are very long so that even reactions with very low rate coefficients may contribute to the chemistry of these systems.

Hence it has been shown that radiolytical processes in dense astrophysical environments depend strongly on irradiation parameters; high power densities (respectively high total absorbed doses) will not only lead to high radiolytical yields but will influence the whole chemical network by enhancing reactions among reactive species compared to reactions of reactive species with stable target molecules. This has to be taken into account for simulations of radiolytical processes in astrophysical environments.

References

Johnson, R. E., J. F. Cooper, and L. J. Lanzerotti 1986, in Proc. 20th ESLAB Symposium on the Exploration of Halley's Comet, Heidelberg, 27 - 31 October 1986, pp. 269 - 271. ESA SP-250 (December 1986).

Meisels, G. G. 1965, J. Amer. Chem. Soc. 87, 950 - 957.

Sack, N. J. 1989 'On the importance of radical density in cosmic chemistry', submitted to Earth, Moon, and Planets.

Sack, N. J., R. Schuster, and A. Hofmann 1989 'Current density effects and the importance of charge transfer processes in the radiolysis of methane - astrophysical implications', submitted to The Astrophys. J.

Woodward, T. W., and R. A. Back 1963, in 'the γ- radiolysis of hydrocarbon gases', Can. J. Chem. 41, 1463 - 1468.

COSMIC-RAY-INDUCED PHOTONS: ARE THEY REALLY IMPORTANT?

S. AIELLO - C. CECCHI-PESTELLINI
Dipartimento di Fisica/Spazio
Università degli Studi di Firenze
via Leone Pancaldo, 3/45
I-50127 Firenze, Italy

ABSTRACT. In this paper we analyze the production and transfer of U.V. photons inside the core of a spherical core-halo molecular cloud, using various assumptions on extinction and scattering dust properties. In our computations we take into account the contributions to the excitation of the $B^1\Sigma_u^+, C^1\Pi_u$ states of molecular hydrogen by both primary protons and secondary electrons. The computations were performed assuming that the core has a uniform gas density of 10^4 cm^{-3} and a visual optical depth (edge to edge) of 10. We found the intensity of cosmic-ray-induced UV photons inside the core to be of a higher order of magnitude than that produced by the interstellar radiation field. Finally, we noted the effects of this UV radiation field on the evolution of grain mantles.

1. INTRODUCTION

The physical conditions and chemistry in diffuse clouds are driven by interstellar UV photons (through photodissociation and photoionization processes). Interstellar radiation is the most important source of energy in these clouds. The conditions in dense clouds are quite different since, due to screening by dust, interstellar UV radiation undergoes a strong attenuation. Consequently, cosmic-ray ionization was in general believed to be the sole driver of gas-phase chemistry, and the photo-effects were neglected. However, in 1983 Prasad and Tarafdar suggested that a part of the energy deposited by C.R. inside dense clouds could be transformed into a chemically-significant UV photon flux through the excitation of the Lyman and Werner system of H_2 molecules by energetic secondary electrons. Their suggestions was later investigated by Lepp et al. (1987), Sternberg et al. (1987) and Gredel et al. (1987), all of whom found that cosmic-ray-induced photons really

J. M. Greenberg and V. Pirronello (eds.), Chemistry in Space, 427–431.

significantly affect the chemistry of dense clouds.

The flux obtained by Prasad and Tarafdar appears to be underestimated. Indeed, the evaluation of any radiation field within a gas-dust cloud implies the treatment of radiation transfer, a task which can present many difficulties. In fact, the transfer of radiation within clouds is dominated by its interaction with dust grains, the optical properties of which strongly depend on the wavelength. Particular attention must be paid to the role of multiple scattering on dust particles. Ignoring it may lead to large errors in computing the transmissivity. Moreover, the band systems may also be directly excited by primary cosmic protons. Thus, the contribution of the latter to the production of UV photons must be considered in computations. It therefore seemed worthwhile to reconsider the production of UV photons by cosmic rays and their transfer in a dense cloud, taking into account both the contribution by primary protons and secondary electrons, and investigating the effects of dust-model dependence of the extinction and scattering properties of dust particles on the resulting UV spectrum.

2. ENERGY DEPOSITION BY C.R. PROTONS

The crucial parameter in the production of UV photons is the collisional excitation coefficient at each individual level of the $B^1\Sigma_u^+, C^1\Pi_u$ states of the hydrogen molecule. The calculation was made in the framework of a comprehensive analysis of the energy deposition by C.R. protons in a cloud of H_2 molecules.

When traversing the interstellar medium, cosmic rays lose their energy by ionizing, exciting and dissociating the atoms and molecules of the interstellar gas. The ionization, dissociation and excitation efficiencies are enhanced by the secondary energetic electrons produced in the primary ionization.

The spectrum of the ejected electron energies was computed up to the third generation. The energy spectrum of the electrons produced by proton-impact ionization was computed by taking into account that the energy degradation of the protons within interstellar clouds is far from being complete. For this reason, we computed the volume production of secondary electrons. On the contrary, in comparison with the size of the clouds, the range of electrons is small enough to cause their entire energy degradation to be considered "local". We computed the spectra to the second and third generations. We obtained for ς, the c.r. ionization rate, in the center of a dense cloud ($\tau_v = 10$, $n(H_2) = 10^4$ cm^{-3}) $\approx 3\times10^{-17}$ s^{-1}, in agreement with the values obtained with other methods.

Finally, we computed the excitation rate as a function of the quantum number.

3. PHOTON PRODUCTION AND TRANSFER

The transitions involved are:

$$H_2\,(X\,^1\Sigma_g^+, v''=o) + p \longrightarrow H_2(B\,^1\Sigma_u^+, v') + p$$
$$\longrightarrow H_2(C\,^1\Pi_u, v') + p$$

In solving the equation of statistical equilibrium, we assumed that the interstellar UV radiation does not enter the cloud. Such a situation may occur in the core of dense clouds with a core-halo structure. The extinction curve plays a major role in determining the transfer of radiation. Since in dense parts of molecular clouds the properties of dust may differ from those derived for the diffuse interstellar medium, adopting the mean interstellar extinction curve in computations may be inappropriate.

We therefore carried out the computations in two cases:

I. The extinction curve towards σ Sco (Snow & York, 1975) is adopted as a representative curve for dense clouds. In this case for ω and g we adopted the values measured in Orion, a region which also exhibits a flat extinction in UV (Mathis et al., 1981).

II. The mean interstellar extinction curve extrapolated to 900 A is adopted. The albedo (ω) and the phase parameter (g) are those computed by Draine (1987) for a graphite silicate mixture.

All computations were performed for a core optical depth of 10 and uniform density for the H_2 molecules of 10^4 cm^{-3}. Multiple scattering doubles UV penetration of photons by a factor of two. The total flux in the 900-1800 A range is about 9000 ph cm^{-2} s^{-1} In Case II, the total flux is lower (5000 ph cm^{-2} s^{-1}). In Case I, an almost two-fold number of photons is found in comparison with the case of "normal" extinction curves. This result appears to be especially significant since the regions most screened by dust from interstellar UV radiation (that is, those regions in which cosmic-ray-induced photons may play an important role in cloud chemistry) are also likely to have an anomalous extinction curve.

4. UV PHOTOLYSIS AND THE EVOLUTION OF GRAIN MANTLES

Greenberg and co-workers (1989) suggested, and tested in laboratory simulations, a possible scenario for mantle evolution. Sticking of atoms and molecules inside dense clouds proceeds simultaneously with surface reactions and processing by UV photolysis. As a result of this photoprocessing, chemically-active radicals are produced and frozen in the mantles. If the radical concentration is large enough, a triggering event, such as collisions between grains or interactions of grains with cosmic-ray particles, can ignite an explosive release of stored chemical energy capable of evaporating the volatile part of the mantle. Before the explosion, the irradiation can alter the chemical composition of the grain mantles.

Since the interstellar radiation field is mostly excluded from the inner regions of dense clouds, alternate sources such as the emission mechanism considered here may provide the needed UV photons.

The photolysis rate of the X molecular species is given by $\varphi = \sigma_X \xi N_{tot}$ where σ_X is the molecular absorption cross-section; ξ is the photolysis efficiency, and N_{tot} is the ambient photon flux.

For the sake of simplicity we assume that the mantles are composed of pure H_2O ice: σ_x is then 2 10^{-17} cm^2, $\xi \simeq 0.005$ and φ, at the cloud center, ranges from 5.8 10^{-16} to 1 10^{-15} s^{-1} for Case I and Case II, respectively. Assuming 5 10^7 years as the lifetime of dense clouds, we find that practically all mantle molecules can be photolysed during the course of the cloud lifetime by cosmic-ray-induced photons. In this situation the 1% concentration of radicals, a necessary condition for the mantle explosion, obtained in less than 10^6 years, is fulfilled many times during the cloud's lifetime.

Even allowing for uncertainties in our knowledge of the interstellar low-energy cosmic-ray spectrum and of the properties of interstellar dust in dense clouds, our results strengthen the previous conclusions that the cosmic-ray-induced photons can affect the chemistry of dense interstellar clouds.

REFERENCES

Draine, B. T.: 1985, private communication.

Gredel, R., Lepp, S., Dalgarno, A. (1987) 'The C/CO ratio in dense interstellar clouds', Ap.J.Letters, 323, L137.

Greenberg, J. M. (1989) Proc. CI SIF Int. School of Physics "Enrico Fermi", Varenna.

Lepp, S., Dalgarno, A., Sternberg, A. (1987), 'The abundance of H_3^+ in dense interstellar clouds, Ap.J., 321, 383.

Mathis, J.S., Perinotto, M., Patriarchi, P., Schiffer, F.M. (1981) The UV properties of dust in the Orion nebula', Ap.J., 249, 109.

Prasad, S.S., Tarafdar, S.P. (1983), 'UV radiation fields inside dense clouds', Ap.J., 267, 603.

Snow, T.P., York, D.G. (1975) 'Far UV extinction in σ Scorpii', Astrophys.Space Sci, 34, 19.

Sternberg, A., Dalgarno, A., Lepp, S. (1987) 'Cosmic-ray-induced photodestruction of interstellar molecules in dense clouds', Ap. J., 320, 676.

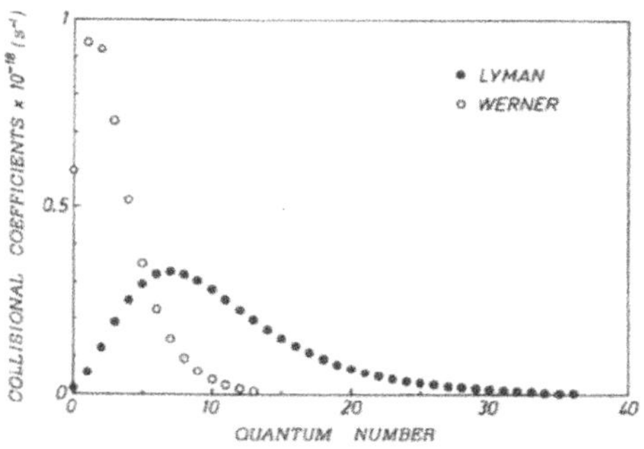

Figure 1. Excitation coefficients for vibrational levels of the states $B^{\prime}\sum_{u}^{+}$ and $C^{\prime}\prod u$ due to the excitation both by protons and electrons

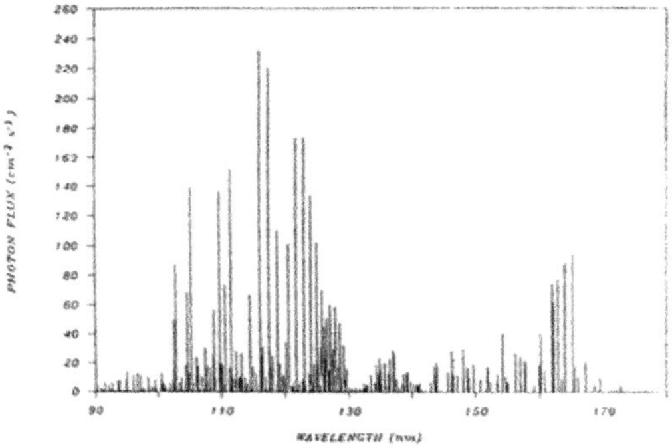

Figure 2. Cosmic-ray-induced emission spectrum in Lyman and Werner bands of molecular hydrogen (Case II)

ION-MOLECULE REACTIONS AT DENSE INTERSTELLAR CLOUD TEMPERATURES: THE CRESUS APPARATUS

J.-B. MARQUETTE
Laboratoire d'Aérothermique du CNRS
4ter Route des Gardes
92195 Meudon Cedex
France

ABSTRACT. The newly-developed CRESUS (Cinétique de Réactions en Ecoulement Supersonique Uniforme avec Sélection) apparatus allows mass-selected ion-molecule reaction rate coefficients to be measured at translational temperatures down to 20 K. The technique is briefly described and the first results are summarised.

Dalgarno [1] has recalled a few years ago the need of an extensive database containing reliable values of parameters caracterising molecular processes for achieving quantitative theories of interstellar chemistry. Among these parameters, particularly important are the rate coefficients of ion-molecule reactions. Such processes are indeed the majority of those involved in numerical models of dense interstellar clouds (see for example the recent work of Viala [2]). Thus, it is worth paying attention to the question of the influence of temperature on their efficiency.

During the last decade, a very large amount of data has been obtained down to 80 K [3] by different groups using the so-called selected ion flow drift tube (SIFT) apparatus which was developed by Smith and Adams [4] at the University of Birmingham. More recently (in the early 1980s), the CRESU ('Cinétique de Réactions en Ecoulement Supersonique Uniforme') apparatus was designed in Meudon for attaining lower temperatures (down to 8 K). This technique is based on the use of a uniform supersonic flow, generated by the expansion of a buffer gas through a suitably contoured converging-diverging nozzle, as a reactor [5]. Some results relevant to astrochemistry were then obtained, for example those concerning the validity of capture theories at very low temperatures [6]. However, the flexibility of the CRESU apparatus was not as great as that achieved in the SIFT technique because primary ions were produced by ionization of the buffer gas by a 20 keV electron beam and, if necessary, by addition of a convenient ion parent gas. To remove the limitations inherent in this ionization method, it became necessary to add an ion injector to the CRESU experiment for yielding mass-selected primary ions as in the SIFT apparatus. This new configuration has been called CRESUS, standing for 'CRESU avec Sélection'.

The ions are produced by electron impact in a low pressure source. After mass selection through a quadrupole filter they are injected in the nozzle nearly radially. Since diffusion is negligible in uniform supersonic flows, they are drifted toward the flow axis by a tranverse electric field whose intensity is set, according to the ion mobility, by adjusting the potentials applied to different parts of the nozzle wall [7]. The reaction zone, downstream the nozzle exit, is field free, so that rate coefficients are measured in the usual way, i.e. by monitoring the primary ion density decrease versus the neutral reactant flow rate or the distance from

J. M. Greenberg and V. Pirronello (eds.), Chemistry in Space, 433–434.

the nozzle exit [6].

The first data have been obtained at 20 and 70 K in helium buffer gas. The results concerning $He^+ + N_2$, $N^+ + H_2$, and $O_2^+ + CH_4$ are in very good agreement with previous CRESU measurements [6]. The reactions of Ar^+ with H_2, N_2, O_2, and CO exhibit specific behavior with temperature which are discussed in detail elsewhere [8]. Surprisingly, the rate coefficient of $N_2^+ + H_2$ is temperature dependent below room temperature. It is well represented by $k = 3.9 \times 10^{-10} T^{-0.29}$ cm^3 s^{-1}. On the other hand, third-body association reactions involving C^+ and CH_3^+ as primary ions have been examined. When H_2 is the neutral the rate coefficient at 20 K is $< 10^{-28}$ and 3.4×10^{-27} cm^6 s^{-1}, respectively. These values are significantly smaller than those deduced from the theoretical power law $k \propto T^{-n} (n > 0)$ currently used for both ternary and radiative association processes.

The perspectives for the near future include various aspects. Studies of processes relevant to astrochemistry, such as isotope fractionation or association reactions, are planned. It should also be desirable to develop other types of ion sources for gaining in flexibility, for example for examining the H_3^+ chemistry at very low temperatures.

ACKNOWLEDGMENTS. This work is has been supported in part by the CNRS through the GDR "Physico-Chimie des Molécules Interstellaires" and the GRECO "Dynamique des Réactions Moléculaires".

REFERENCES

[1] Dalgarno, A. (1985) 'Molecular Astrophysics', in G.H.F. Diercksen, W.F. Huebner et P.W. Langhoff (eds.) Molecular Astrophysics, State of the Art and Future Directions, D. Reidel Publishing Company, Dordrecht, pp. 3-22.

[2] Viala, Y. (1986) 'Chemical equilibrium from diffuse to dense interstellar clouds. I. Galactic molecular clouds', Astron. Astrophys. Suppl. Ser. 64, 391-437.

[3] Anicich, V.G. and Huntress, W.T. (1986) 'A survey of bimolecular ion-molecule reactions for use in modeling the chemistry of planetary atmospheres, cometary comae, and interstellar clouds', Astrophys. J. Suppl. 62, 553-672.

[4] Smith, D. and Adams, N.G. (1987) 'The selected ion flow tube (SIFT): studies of ion-neutral reactions', Adv. Atom. Mol. Phys. 24, 1-49.

[5] Dupeyrat, G., Marquette, J.B., and Rowe, B.R. (1985) 'Design and testing of axisymmetric nozzles for ion-molecule reaction studies between 20 K and 160 K', Phys. Fluids 28, 1273-1279.

[6] Rowe, B.R. and Marquette, J.B. (1987) 'CRESU studies of ion-molecule reactions', Int. J. Mass Spectrom. Ion Proc. 80, 239-254.

[7] Rowe, B.R., Marquette, J.B., and Rebrion, C. (1989) 'Mass-selected ion-molecule reactions at very low temperatures: the CRESUS apparatus', J. Chem. Soc. Faraday Trans. 2, in press.

[8] Rebrion, C., Rowe, B.R., and Marquette, J.B. (1989) 'Reactions of Ar^+ with H_2, N_2, O_2, and CO at 20, 30, and 70 K', J. Chem. Phys., to be published.

INTERSTELLAR H_2S: PROBE OF GRAIN SURFACE CHEMISTRY

Y. C. MINH and W. M. IRVINE
Five College Radio Astronomy Observatory
Department of Physics and Astronomy
University of Massachusetts, Amherst
MA 01003, U.S.A.

ABSTRACT. Interstellar hydrogen sulfide has been observed at fractional abundances ~ 10^{-9} relative to molecular hydrogen towards cold dark clouds (such as L134N and TMC-1: kinetic temperature T_{kin} ~ 10 K) and quiescent warm clouds (such as the Orion extended ridge: T_{kin} ~ 30-50 K), while it is enhanced by a factor of 1000 in the Orion hot core (T_{kin} ~ 200 K) and the plateau (T_{kin} ~ 150 K). We conclude that H_2S may be evaporating from the grain mantles in the hot core and the small, dense condensations in the plateau as a result of heating from IRc 2 or shocks. Even in the cold, dark clouds, grain surface reactions for the formation of H_2S may be responsible for the gas phase H_2S abundances.

Interstellar hydrogen sulfide has been suggested as a tracer of high-temperature chemistry and/or grain surface chemistry because crucial gas-phase reactions leading to the formation of H_2S are thought to be appreciably endothermic (Watson and Walmsley 1985; Duley, Millar, and Williams 1980). H_2S was first observed in the ISM via its 1_{10}-1_{01} transition by Thaddeus *et al.* (1972). This transition has recently been observed towards the cold, dark clouds L134N and TMC-1 (Minh, Irvine, and Ziurys 1989) and the Orion Molecular Cloud 1 (Minh *et al.* 1989).

Recent ion-molecule chemical models employ the radiative association reaction between SH^+ and H_2 as a principal formation route for the ion H_3S^+ in dense clouds, which is the assumed precursor to H_2S. Predicted abundances of H_2S in the ion-molecule chemical models are about ~ 10^{-10}, which is somewhat below the observed abundances in L134N and the Orion extended ridge or 3N position. The observed values are included in Table 1.

Standard ion-molecule chemistry clearly does not explain the observations in the Orion hot core and plateau, where the fractional abundances $f(H_2S) \geq 10^{-7}$ (see Table 1). Under high temperature conditions, the endothermicities or the activation energies of relevant reactions can be overcome easily. Although the high temperature chemistry models predict a large abundance of sulfur-containing

J. M. Greenberg and V. Pirronello (eds.), Chemistry in Space, 435–436.

TABLE 1. Abundances of Hydrogen Sulfide[a]

Source	RA(1950)	Dec(1950)	$N(H_2S)$	$N(H_2S)/N(H_2)$[b]
L134N	$15^h51^m30^s.0$	$-2°43'31''$	8(12) cm^{-2}	1(-9)
L134N(SO)	15 51 26.0	-2 43 31	3(13)	4(-9)
TMC-1	4 38 38.6	25 35 45	≤5(12)	≤5(-10)
TMC-1(NH_3)	4 38 22.0	25 41 45	7(12)	7(-10)
Orion(KL)	5 32 47.0	-5 24 23		
Extended ridge[c]			3(14)	~ 1(-9)
Hot Core[c]			4(17)	4(-7)
Plateau[c]			1-3(17)	~ 2(-6)
Orion 3N	5 32 51.0	-5 21 23	2(14)	2(-9)

[a] See details in Minh, Irvine, and Ziurys (1989) and Minh *et al.* (1989).; [b] Fractional abundance relative to hydrogen molecule.; [c] From gaussian fits for the KL spectra of H_2S and $H_2^{34}S$.

molecules after a passage of shocks (*e.g.* Hartquist, Oppenheimer, and Dalgarno 1980), H_2S formation in the shocked region may not be efficient (Neufield and Dalgarno 1989). Moreover, SH^+, the precursor of H_2S, has not been detected in shocked regions (Millar and Hobbs 1988).

Duley, Millar, and Williams (1980) have invoked grain surface reactions to synthesize H_2S. If sulfur has been depleted onto grain surfaces, ions like C^+ or H^+ are likely to find SH^- or S^- sites onto grain surfaces, which may lead to the formation of H_2CS or H_2S, respectively. Although this model contains many uncertainties and produces too much H_2S, it may be possible that in dense clouds, where sulfur is highly depleted, a variety of sulfur-containing molecules may be formed easily through reactions on the grains.

We conclude that although ion-molecule reactions cannot be ruled out entirely as a source of H_2S in the cold dark clouds and quiescent warm clouds, grain surface reactions may be a more promising pathway to explain the observed abundances. The similar fractional abundances of H_2S in both cold and warm clouds may imply that desorption processes in cold dark clouds may be as efficient as those in quiescent warm clouds. The large H_2S abundances in the Orion hot core and the plateau may result from the evaporation of grain mantles by UV radiation or shocks.

References

Duley,W.W.,Millar,T.J.,Williams,D.A. 1980, *M.N.R.A.S.*, 192, 945.
Hartquist,T.W.,Oppenheimer,M.,Dargarno,A. 1980, *Ap.J.*, 236, 182.
Millar,T.J.,Hobbs,L.M. 1988, *M.N.R.A.S.*, 231. 953.
Minh,Y.M.,Irvine,W.M.,Ziurys,L.M. 1989, *Ap.J.(Lett.)*, in press.
Minh,Y.M.,Irvine,W.M.,McGonagle,D.,Ziurys,L.M. 1989, *Ap.J.*, submitted.
Neufield,D.A.,Dalgarno,A. 1989, preprint.
Thaddeus,P.,Kutner,M.L.,Penzias,A.A.,Wilson,R.W.,Jefferts,K.B. 1972, *Ap.J.(Lett.)*, 176, L73.
Watson,W.D.,Walmsley,C.M. 1982, in *Regions of Recent Star Formation*, eds. R.S.Rogers and P.E.Dewdney (Dordrecht:Reidel), p. 357.

RESULTS OF A GALACTIC SURVEY FOR THE RING MOLECULE CYCLOPROPENYLIDENE (C_3H_2)

S. C. MADDEN
Five College Radio Astronomy Observatory (FCRAO)
University of Massachusetts,
Amherst, Massachusetts 01003

ABSTRACT. A survey for the organic ring molecule, cyclopropenylidene (C_3H_2) in 100 Galactic sources indicates that the molecule is widespread with abundances relative to H_2 as large as 10^{-8} in dark cloud environments. Gas phase chemical models are consistent with observations.

1. INTRODUCTION

Searches for cyclic species in space have been carried out for about 15 years with little success. The carbon ring molecule, SiCC, was detected only in the carbon star IRC+10216 (Thaddeus, Cummins and Linke 1984). Increasingly larger interstellar chain molecules were found, but even simple 3-carbon rings remained elusive until C_3H_2 (cyclopropenylidene) was finally detected in space (Matthews and Irvine 1985) and in the laboratory (Thaddeus, Vrtilek and Gottlieb 1985). More recently, the related cyclic molecule C_3H was detected in the laboratory and in space and found to be less abundant than and less widely distributed than C_3H_2 (Yamamoto *et al.* 1987). The results of our survey of C_3H_2 demonstrate that it is very widespread in a variety of environments throughout the Galaxy : giant molecular clouds, cold dark clouds, the diffuse spiral arm clouds in the direction of distant continuum sources, and the envelope of a carbon star (Madden *et al.* 1989). Extragalactic observations have revealed C_3H_2 in the nucleus of the radio galaxy Centaurus A (Seaquist and Bell 1986).

We are currently carrying out radiative transfer calculations for the sources we have surveyed in order to determine accurate abundances. Initial C_3H_2 calculations for dark clouds (Avery and Green 1989) suggest that a simple uniform density model may be inappropriate. We are refining these calculations to incorporate a dense core surrounded by a less dense halo and are applying this model to dark cloud sources.

2. CHEMISTRY

The data we have obtained thus far indicate that C_3H_2 is one of the more abundant organic constituents of dark cloud sources. In the dark cloud TMC-1, the fractional abundance of C_3H_2 relative to molecular hydrogen is estimated to be 1 to 3 x 10^{-8} (Madden *et al.* 1989), at least comparable to that of formaldehyde and HCN. For L134N the abundance is somewhat less, consistent with the lower abundance of species like the cyanopolyynes and C_3O relative to TMC-1. The abundance of C_3H_2 has been shown to correlate with the

J. M. Greenberg and V. Pirronello (eds.), Chemistry in Space, 437–438.

abundance of HC_3N and HC_5N in dark clouds suggesting a common chemical precursor to both the cyanopolyynes and C_3H_2 (Brouillet, Baudry and Daigne, 1988; Cox, Walmsley and Güsten 1988).

On the basis of measured laboratory rates for the production of $C_3H_3^+$ via three-body association, Herbst, Adams and Smith (1984) and Adams and Smith (1987) estimate that the radiative association reaction

$$C_3H^+ + H_2 \rightarrow C_3H_3^+ + h\nu \qquad (1)$$

is relatively efficient. The precursor C_3H can be produced from acetylene (C_2H_2) in a fast reaction with C^+. Dissociative recombination then yields

$$C_3H_3^+ + e \rightarrow C_3H_2 + H \qquad (2a)$$

$$\rightarrow C_3H + H_2. \qquad (2b)$$

If the branching ratio favors (2a) over (2b), it might explain the unexpectedly low abundances observed for C_3H, in both the linear and cyclic species (Thaddeus *et al.* 1985; Yamomoto *et al.* 1987) relative to theoretical predictions (*e.g.* Millar and Freeman 1984). The abundances of cyclic and linear C_3H in TMC-1 have been observed to be comparable and they are at least an order of magnitude lower than that of C_3H_2 and much less widespread in the Galaxy.

Steady state gas phase calculations have difficulty predicting sufficient abundances of C_3H_2 (Dalgarno, this volume). In a recent expansion of their pseudo-time-dependent gas-phase model, Herbst and Leung (1989) have included complex hydrocarbons in their reaction network. Their model calculations for dark clouds are successful in reproducing the necessary large abundances of C_3H_2 at the time of peak production, when the clouds are young - well before steady state has been reached.

Another model by Lepp and Dalgarno (1988) simulating steady state conditions introduces large molecules such as PAHs into the chemistry and enhances previous steady state calculations by 2 orders of magnitudes to agree with observed C_3H_2 abundances.

REFERENCES

Adams, N. G., and Smith, D. 1987, *Ap. J. (Letters)*, **317**, L25.

Avery, L. W., and Green, S. 1989, *Ap. J.*, **337**, 306.

Brouillet, N., Baudry, A., Daigne, C. 1988, *Astr. Ap.*, **199**, 312.

Cox, P., Walmsley, M., and Güsten, R. 1988, *Astr. Ap.*, **209**, 382.

Herbst, E., Adams, N.G., and Smith, D. 1984, *Ap. J.*, **285**, 618.

Herbst, E. and Leung, C. M. 1989, *Ap. J. Suppl.*, **69**, 271.

Lepp, S. and Dalgarno, A. 1988, *Ap. J.*, **324**, 553.

Madden, S. C., Irvine, W. M., Matthews, H. E., Friberg, P., and Swade, D., 1989, *A. J.*, **97**, 1403.

Matthews, H., and Irvine, W.M. 1985, *Ap. J. (Letters)*, **298**, L61.

Millar, T. J., and Freeman, A. 1984, *M.N.R.A.S.*, **207**, 405.

Seaquist, E. R., and Bell, M. B. 1986, *Ap. J. (Letters)*, **303**, L67.

Thaddeus, P., Cummins, S. E., and Linke, R. A. 1984, *Ap. J. (Letters)*, **283**, L45.

Thaddeus, P., Vrtilek, J. M., and Gottlieb, C. A. 1985, *Ap. J. (Letters)*, **299**, L63.

Yamamoto, S., Saito, S., Ohishi, M., Suzuki, H., Ishikawa, S., Kaifu, N., and Murakami, A. 1987, *Ap. J. (Letters)*, **322**, L58.

POSSIBLE ASSIGNMENT OF THE 11.3μm UIR FEATURE TO EMISSION FROM SOOT-LIKE MICROPARTICLES WITH INTERNAL HYDROGENS

S.P.Balm and H.W.Kroto

School of Chemistry and Molecular Sciences,
University of Sussex, Brighton, BN1 9QJ, UK.

The carriers of the UIR bands seen in many galactic and extragalactic objects have previously been assigned to emission from polycyclic aromatic hydrocarbon (PAH) molecules (Léger and Puget 1984) or hydrogenated amorphous carbon (HAC) particles (Duley and Williams 1983). However, the match between laboratory spectra of PAH's and HAC particles with the interstellar features in the structurally sensitive 11-15μm region remains unsatisfactory. In particular it has been difficult to reproduce the prominent 11.3μm feature found at the same position in nearly all sources. This consistency is most remarkable considering that a PAH-mixture is usually invoked. Between 11 and 15μm the out-of-plane C-H bending mode is found whose position is dependent on the number of adjacent hydrogen atoms on the aromatic periphery : for isolated hydrogens this feature occurs at 11.0-11.6μm while paired hydrogens are found in the 11.6-12.5μm and triplet hydrogens between 12.4 and 13.3μm (Clar *et al.* 1981).

Kekulene ($C_{48}H_{24}$) is a novel cyclic PAH molecule (Diederich and Staab 1978) possessing an unusual type of isolated hydrogen functionality - the internal hydrogen. It has an annulus made of twelve fused aromatic rings with six internal isolated C-H bonds pointing into the centre of the structure. On the outside it has alternating isolated and paired external hydrogens (Fig. 1, (a)). Comparison of its solid state (KBr disc) IR spectrum (Diederich 1988) in the 11.0-12.6μm region with the spectrum of the compact PAH ovalene ($C_{32}H_{14}$) which has two external isolated hydrogens in similar environments to the external isolated hydrogens of kekulene (Fig. 1, (b)) leads to the conclusion (Fig. 2, (a) and (c)) that external isolated (on the ring periphery) and internal isolated (in the centre) hydrogens absorb at different frequencies since kekulene has two main bands in the isolated region (11.0-11.6μm) at 11.25 and 11.42μm with the latter band close in position to the external isolated band of ovalene at 11.45μm. Thus the 11.25μm band would appear to be due to the internal isolated hydrogens. The band at 11.17μm is most probably a component of the main band at 11.25μm with the splitting possibly due to solid state interactions.

Comparison of the 11.25μm kekulene band with the latest observations of the 11.3μm feature from the Orion Bar (Roche *et al.* 1989) and the well known sources HD 44179 and IRAS 2182+5050 (Witteborn *et al.* 1989) in emission give a close match (Fig. 2, (b) and (c)).

J. M. Greenberg and V. Pirronello (eds.), Chemistry in Space, 439–441.

In an attempt to understand how carbon condenses in the cool circumstellar outflows of objects such as IRC+10216 a simulation experiment was undertaken in which graphite was laser vapourised and clusters formed and analysed in a molecular beam apparatus (Kroto *et al.* 1985). This resulted in the production of long carbon chains and higher mass clusters including a remarkably stable and inert C_{60} cluster ('Buckminsterfullerene') which is believed to be a graphitic network which has eliminated reactive edges by adopting a truncated icosahedral closed cage structure. This closed shell nature is only possible due the presence and regular disposition of pentagons in the hexagonal network. However, if during nucleation this criterion is not met then closure will not occur; instead the reactive edges will overlap and continue to grow to form quasi-icosahedral carbon particles with spiral structures (Kroto and McKay 1988).

In the cool carbon-rich envelopes around mass-losing red-giant stars like IRC+10216 these icospiral microparticles and closed shell 'fullerenes' are expected to form. However mass-loss generally occurs in a hydrogen-rich environment so that as microparticles are expelled they are likely to have some sections of their reactive edges annealed with hydrogen. In such an environment there is another way of producing the required curvature in which the pentagon carbon atoms are replaced by internal hydrogen 'holes'. These are expected to form during nucleation via hydrogenated edges in the growing spiral and are likely to be numerous in circumstellar carbon particles.

Once hydrogenated spirals and fullerenes are expelled they will be subsequently excited during the planetary nebula phase. On absorption of a UV photon the intermediate sized microparticles (25-100 atoms) for which internal energy conversion is fast should re-emit in the mid-IR at the fundamental group vibrational frequencies. Species of this type will have many internal hydrogen sites which will lead to a prominent 11.3μm feature. The peripheral groups on the outer edges of the spirals will probably contribute to the observed asymmetry of the 11.3μm band and will also produce the emission plateau features (Cohen *et al.* 1985). Closed shell species like C_{60} may contribute to the skeletal features at 6.2 and 7.7μm. The largest spiral particles should reach thermal equilibrium and re-radiate in the far-IR. Disruption of these larger species during star formation may be a source of carriers in HII regions and reflection nebulae.

It is important to note that the internal hydrogen functionalities will be similar in all particles so that a single, sharp feature is predicted and since these particles are likely to form whenever carbon condenses in circumstellar environments they are able to explain both the widespread observation of the UIR features and in particular the consistency of the 11.3μm feature.

References

Clar, E., Robertson, J.M., Schlogl, R. and Schmidt, W. (1981), *J. Am. Chem. Soc.*, **103**, 1320.

Cohen, M., Tielens, A.G.G.M. and Allamandola, L.J. (1985), *Ap. J.*, **299**, L93.

Diederich, F. (1988), *private communication.*

Diederich, F. and Staab, H.A. (1978), *Angew. Chem. Int. Ed. Engl.*, **17**, 372.

Duley, W.W and Williams, D.A. (1983), *Mon. Not. R. Astron. Soc*, **205**, 67P.

Kroto, H.W., Heath, J.R., O'Brien, S.C, Curl, R.F. and Smalley, R.E. (1985), *Nature (Lond.)*, **318**, 162.

Kroto, H.W. and M^c^Kay (1988), *Nature (Lond.)*, **331**, 328.

Léger, A. and Puget, J.L. (1984), *Astron. Astrophys.*, **137**, L5.

Roche, P.F., Aitken, D.K. and Smith, C.H. (1989), *Mon. Not. R. Astron. Soc.*, **236**, 485.

Witteborn, F.C., Sandford, S.A., Bregman, J.D., Allamandola, L.J., Cohen, M., Wooden, D.H. and Graps A.L. (1989), *Ap. J.*, **341**, 270.

SPECTROSCOPIC STUDIES OF POLYCYCLIC AROMATIC HYDROCARBONS (PAH) AND PAH MIXTURES IN THE INFRARED AND VISIBLE

P.EHRENFREUND, L.d'HENDECOURT
Groupe de Physique des Solides, University Paris VII, Tour 23,
2, Place Jussieu 75251 Paris Cedex 05 France

ABSTRACT. This short contribution presents some of our recent laboratory results. We have obtained the infrared absorption spectrum of isolated coronene ($C_{24}H_{12}$) in a neon matrix at 4 K. We have also performed spectroscopic studies in the visible of a natural mixture of PAHs which give us implications for the diffuse interstellar bands (DIBs).

1. MATERIALS, METHODS AND RESULTS

The experiments were performed with matrix isolation spectroscopy which allows us to study truly isolated molecules. We can simulate conditions which might dominate the environment of molecules in the interstellar medium like vacuum, low temperature and gas phase.

1.1. Infrared spectrum of isolated coronene

The infrared spectrum of isolated coronene in a neon matrix has been recently obtained in our laboratory. By comparing this spectrum with the infrared spectrum of solid coronene (microcristals of coronene dispersed in a CsI pellet), we measure (cf Fig.1) the same wavelength of the main coronene bands (1-6) within less than 1 %. Some additional features (S1, S2) appear in the spectrum, which are due to condensed CO_2 and H_2O. The oscillator strengths of the various bands, normalized to the 3.3 μm band, remain the same, with the exception of the 11.9 μm band which is reduced by a factor 2.5. The 11.9 μm band corresponds to the out of plane C-H bending mode. In the solid state, one might expect steric interactions due to neighbour atoms.

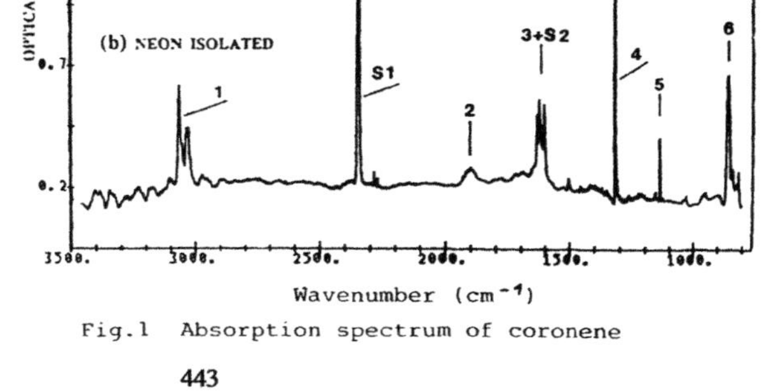

Fig.1 Absorption spectrum of coronene

J. M. Greenberg and V. Pirronello (eds.), Chemistry in Space, 443–444.

Since PAH molecules are supposed to be ionized in the astronomical objects where they are observed, another important goal is to obtain the infrared spectrum of the coronene ion.

1.2. Visible spectrum of coal pitch

It has been suggested that polycyclic aromatic hydrocarbon molecules or their ions could be attractive candidates for the carriers of the diffuse interstellar bands observed in the visible, due to their high abundance, their extreme stability against photodissociation, and their strong expected transitions in the visible. We present herewith spectroscopic studies in the visible range 3800-4300 Å and 4300-5400 Å of coal pitch, as a laboratory analogue mixture representative of an interstellar PAH mixture. The results in Fig.2 shows that the coal pitch mixture, even in the neutral state, gives few strong transitions in the visible, although at wavelengths shorter than the observed DIBs. We expect a shift of transitions to longer wavelengths by using coal pitch mixtures with increased size of molecules and conclude from our experiment that if PAHs are the carriers of the DIBs, they can be in neutral state. For a mixture of 300 molecules only a few lines appear in our spectra, which gives support to the idea, that only a few molecules out of many different species give strong signatures in the visible. In interstellar space these might be molecules, which provide the best conditions to survive in their specific environment, and are therefore also more abundant than other species. In the diffuse interstellar medium these PAH molecules might be of larger size and more compact. Therefore matrix isolation spectroscopy in the laboratory, on different coal pitch samples, may provide an interesting insight into the nature of the possible carriers of the diffuse interstellar bands. Laboratory studies shall be complemented by astronomical observations.

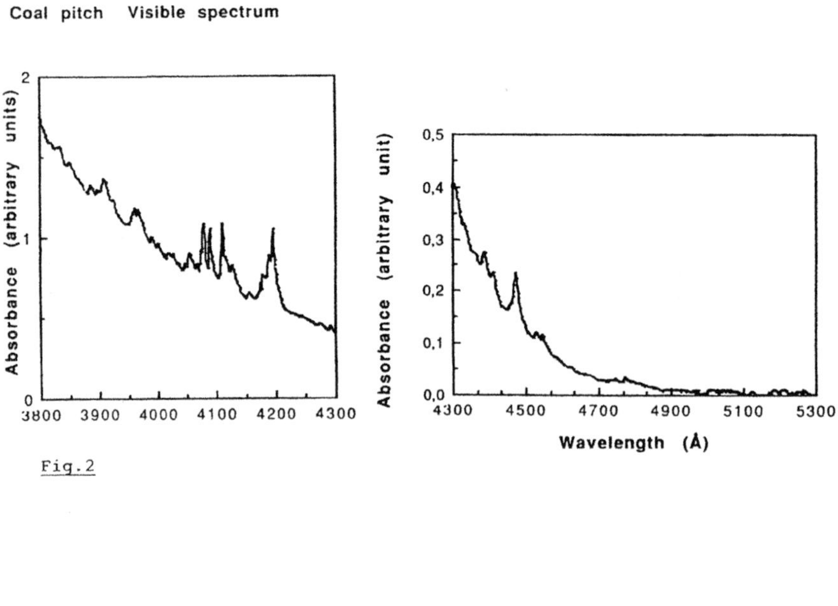

Fig.2

VERY SMALL PARTICLES IN THE INTERSTELLAR MEDIUM

F.-X. DÉSERT
Laboratory Astrophysics, Huygens Laboratorium
N. Borhweg 2, Postbus 9504
2300 RA Leiden
The Netherlands

ABSTRACT. We briefly review the evidence and the implications of the existence of very small grains and polycyclic aromatic hydrocarbon (PAH) molecules in the interstellar medium (ISM).

1. Existence of Very Small Solid Particles

A three-population model of IS dust is advocated (Désert *et al.* 1989), where alongside the traditional big grains (BGs: radius $\simeq 0.1\,\mu$m) made of silicate cores coated with hydrocarbons, there also are carbon-dominated very small grains (VSGs: $\simeq 7\,$nm) and polycyclic aromatic hydrocarbon (PAH) molecule ($\simeq 1\,$nm). Because a grain from these two latter species has a small heat capacity, its temperature fluctuates under the individual photon absorptions (see *e.g.* Désert *et al.* 1986). This non-equilibrium infrared (IR) emission has now been widely observed in The Galaxy and external galaxies (Sellgren 1981, 1984, Walterbos and Schwering 1987). PAHs and VSGs contribute for about 40 percent of the total power absorbed by dust in the solar neighborhood IS radiation field. Identified by a very characteristic set of near IR emission features (from 3.3 to $11.3\,\mu$m), the PAH molecules have been more studied (Léger and Puget 1984, Allamandola *et al.* 1985, Puget *et al.* 1985, Léger *et al.* 1989a) than the 3-dimensional VSGs for which no precise identification has yet been made (these grains could produce the bump at 220 nm in the IS extinction curve) (see the review by Puget and Léger 1989).

2. Abundance and Origin of Very Small Grains and PAHs

The model of IS dust has been built to fit both the average observed extinction curve and the IR emission spectrum of the local HI medium, the so-called "cirrus" clouds. It implies a major revision of early IS dust models (Mathis *et al.* 1977, Draine and Anderson 1985). Various observations, in particular some IR color variations, can be accounted by the model by simply varying the intensity and spectrum of the radiation field, and keeping the abundance of grains constant. However, there is some evidence that in certain galactic regions, one has to include some variation of the abundance of PAHs and VSGs, relative to BGs. PAHs are probably destroyed near strong UV sources (Ryter *et al.* 1987, Boulanger *et al.* 1988) and both components are also much depleted deep inside molecular clouds (Boulanger *et al.* 1989), probably condensed on BGs (Greenberg 1982). Both components are very abundant in some external layers of molecular clouds, usually the ones facing distant UV sources. It has already been suggested that VSGs could be the hydrocarbon mantle debris (De Groot *et al.* 1988) released from BGs when BGs go outside molecular clouds. PAHs could also have a similar origin: the smallest sizes of VSGs could be annealed by photon absorption into the very stable honeycomb structure of PAHs.

J. M. Greenberg and V. Pirronello (eds.), Chemistry in Space, 445–446.

3. Role of VSGs and PAHs in the Interstellar Chemistry

VSGs and especially PAHs are at the difficult frontier between molecules (line emitting and absorbing) and solids (having a continuum emissivity. Hence, the study of the chemistry of these large molecules or small particles is still in its infancy. However outlines of this new chemistry have already been made (Omont 1986). Photothermodissociation of PAHs have been studied in order to deduce the minimum size a PAH molecule must have to survive in a given IS radiation field (Léger *et al.* 1989b). PAHs and VSGs are likely to be the main source of heating of the interstellar medium, through the photoelectric effect (d'Hendecourt and Léger 1987). These large molecules also modify the gas-phase chemistry (Dalgarno 1989). Some important issues on this topic are obviously going to be addressed with new laboratory data and improved astrophysical observations (*e.g.* from IR satellites like *COBE* and *ISO*).

This work was done while the author held an ESA fellowship.

References

Allamandola, L. J., Tielens, A. G. G. M., and Barker, J. R. 1985, *Astrophys. J. (Letters)*, **290**, L25.
Boulanger, F., *et al.* 1988, *Astrophys. J.*, **332**, 328.
Boulanger, F., *et al.* , 1989, preprint.
Dalgarno, A, 1989, these proceedings and references therein.
De Groot, M. S., *et al.* 1988, in *Dust in the Universe: Manchester Symp.*, ed. Bailey, M. E., and Williams, D. A., Cambridge Press, p. 265.
Désert, F. X., Boulanger, F., and Shore, S. N. 1986, *Astron. Astrophys.*, **160**, 295.
Désert, F. X., Boulanger, F., and Puget, J. L. 1989, *Astron. Astrophys.*, to be submitted.
d'Hendecourt, L. B., and Léger 1987, *Astron. Astrophys.*, **180**, L9.
Draine, B. T., and Anderson, N. 1985, *Astrophys. J.*, **292**, 494.
Greenberg, J. M. 1982, in *Submillimetre Wave Astronomy*, ed. Beckman, J. E., and Phillips, J. P., Cambridge University Press.
Léger, A., and Puget, J. L. 1984, *Astron. Astrophys.*, **137**, L5.
Léger, A., d'Hendecourt, L. and Défourneau, D. 1989a, *Astron. Astrophys.*, **216**, 148 (LHD).
Léger, A., Désert, F. X., d'Hendecourt, L., and Boissel, P. 1989b, *Astron. Astrophys.*, **213**, 351.
Mathis, J. S., Rumpl, W., and Nordsieck, K. H. 1977, *Astrophys. J.*, **217**, 425.
Omont, A. 1986, *Astron. Astrophys.*, **164**, 159.
Puget, J. L., Léger, A., and Boulanger, F. 1985, *Astron. Astrophys.*, **142**, L19.
Puget, J. L., and Léger, A. 1989, *Ann. Rev. Astron. Astrophys.*, in press.
Ryter, C., Puget, J. L., and Pérault, M. 1987, *Astron. Astrophys.*,**186**, 312.
Sellgren, K. 1981, *Astrophys. J.*, **245**, 138.
Sellgren, K. 1984, *Astrophys. J.*, **277**, 623.
Walterbos, R. A. M., and Schwering, P. B. W. 1987, *Astron. Astrophys.*, **180**, 27.

THE EFFECT OF THE CHOICE OF THE GRAIN MODEL ON THE H_2 PRODUCTION BY COSMIC RAYS IN DENSE CLOUDS

D. Averna
Istituto di Astronomia
Universita' di Catania
Viale A. Doria, 6
Catania, ITALY

V. Pirronello
Dipartimento di Fisica
Universita' della Calabria
87036 Roges di Rende
ITALY

Since fifty years, the presence of dust in interstellar space was established and a lot of models of the structure of grains have been proposed, changing composition and size distribution, in order to describe correctly extinction and polarization curves.

In 1977, Mathis, Rumpl and Nordsiek suggested that the grains in diffuse clouds could be made of graphite or silicates and follow a size distribution going as $a^{-3.5}$.

Another model of grain, that is very interesting for its evolutionary meaning, is the Greenberg one (Hong and Greenberg, 1980). In his view a core-mantle grain, during its life in a dense cloud, can migrate into regions of high UV radiation level that induces photolytic reactions on the original mantle, that is then progressively transformed into a non volatile form.

When such grains enter or reenter a molecular cloud they can accrete a dirty ice layer.

We use these two models to compare how such a choice can affect the process of formation of molecular hydrogen induced by low energy (E<1 GeV) cosmic rays interacting with interstellar dust grains, a process recently proposed by us (Pirronello and Averna, 1988).

A computer code was built to simulate by means of the Monte Carlo method the bombardment of grains by 40.000 H^+ and He^+ ions inside a dense cloud to compute the rate of H_2 formation as a sputtering process, induced by the energy loss of cosmic rays into the grains.

The CR energy spectrum of Morfill, Volk and Lee (1976), distorted by their loss of energy in ionizing the hydrogen of the cloud before impinging on grains, has been used.

The simulation has been performed for grains placed at different depths in a Boland and de Jong (1984) cloud with central depth of $A_V = 8$ mag.

J. M. Greenberg and V. Pirronello (eds.), Chemistry in Space, 447–449.

For both grain models we considered a core made of olivine-like silicates ($MgFeSiO_4$) and the icy mantle consisting of a mixture of frozen water, ammonia, methane and carbon monoxide in the relative abundance 5:3:2:1.

Treating with Greenberg grains we represent the composition of the organic refractory layer as a polymer with the ratio 1:2 between "metals" (C+N+O) and hydrogen. In this case, as a size distribution function we use the following one proposed by Greenberg (1979).

$$n(a)=n_0 \exp [-5((a_m-a_c)/a_i)^3] \qquad (1)$$

where

n_0= normalization factor

a_c= core radius

a_m= thickness of the inner mantle

a_i= the maximum mantle dimension consistent with cosmic abundance (0.21 μm).

The radius of the core and the thickness of the mantle we use in both models are reported in table 1.

TABLE 1. Radius of the core and thickness of the mantles in MRN and Greenberg models

MODEL	CORE [μm]	INNER MANTLE [μm]	OUTER MANTLE [μm]
MRN	.02-.2		.02-.2
GREENBERG	.05	.06-.24	.01-.04

The experimental data for our Monte Carlo simulation are taken from Brown et al. (1982) for the production yield of H_2 by H_2O ice together with a suitable scaling of these data for CH_4 and NH_3, as in Pirronello and Averna (1988), while molecular hydrogen production from the organic refractory intermediate mantle of the Greenberg model have been calculated by us from measurements of Lanzerotti et al. (1985).

Here we should like to present only our first preliminary result relative to the central core of the cloud; it shows that the rate obtained using the model of Greenberg is roughly a thousand times larger than the one obtained using the MRN model. This result seems to us quite acceptable, considering that the inner mantle of Greenberg model can contribute considerably to the H_2 production.

REFERENCES

Boland, W., de Jong, T.: 1984, Astron. Astrophys. 134, 87-98

Brown, W.L., Augustyniak, W.M., Simmons, E., Marcantonio, K.J., Lanzerotti, L.J., Johnson, R.E., Boring, J.W., Reimann, C.T., Foti, G., Pirronello, V.: 1982, Nucl. Instr. and Methods 198, 1-8

Greenberg, J.M.: 1979, "Grain Mantle Photolysis: a Connection between the Grain Size Distribution Function and the Abundance of Complex Interstellar Molecules", in "Stars and Star Systems", B.E. Westerlund ed., p. 173-193, Reidel Publ. Co., Dordrecht, Holland

Hong, S.S., Greenberg, J.M.: 1980, Astron. Astrophys. 88, 194-202

Lanzerotti, L.J., Brown, W.L., Johnson, R.E.: 1985, "Laboratory Studies of Ion Irradiation of Water, Sulfur Dioxide and Methane Ices", in "Ices in the Solar System", J. Klinger, D. Benest, A. Dollfus and R. Smoluchowski eds., p. 316-325, Reidel Publ. Co., Dordrecht, Holland

Mathis, J.S., Rumpl, W., Nordsiek, K.H.: 1977, Astrophys. J. 217, 425-433

Morfill, G.E., Volk, K.J., Lee, M.A.: 1976, J.G.R. 181, 5841

Pirronello, V., Averna, D.: 1988, Astron. Astrophys. 196, 201-206

CHEMICAL ENRICHMENT OF THE HALO BY DUST EXPULSION

A.FERRARA
Dip. di Astronomia
Universita' di Firenze
Largo Fermi, 5
50125 Firenze ITALY

ABSTRACT. The dynamical behaviour of dust grains subjected to light and matter distribution of spiral galaxies is analyzed in a hot galactic corona, including gas drag and sputtering. Grains can travel large radial and vertical distances from their original position removing material from the inner parts of the disk and sending it to the outskirts of the disk. Some heavy elements can be injected in the gas phase during the evolution in the galactic halos.

1. INTRODUCTION

The presence of dust grains in the galactic halo has been recognized through the infrared observations made by the IRAS satellite. This dust is coupled to the diffuse and molecular clouds that reside in the halo environment distributed with a typical scale height of the order of 120 pc.

The possibility that some interstellar grains may be expelled from the disks of spiral galaxies by the radiation pressure of the stellar field has been studied in the recent years. In particular the problem must be considered in the two considerably different environments of the disk and the halo.

Franco (1989) suggests that small dusty clouds can be expelled from the disk by radiation pressure if the cloud has an efficient dust-gas coupling. This occurrence depends upon the cloud column density, on cloud-cloud velocity dispersions and on the vertical distribution of stars in the disk. The minimum gas column density for dust coupling, without considering the magnetic field is 10^{19} cm^{-2}. In this conditions the cloud is accelerated in the upward direction, and, given the drift of the dust grains inside the cloud, some naked grains may detach from the cloud and escape from the disk through low density "tunnels".

As soon as the grains have reached the basis of the corona ($z \sim 100$ pc), their behaviour is determined by the overall mass and luminosity distribution of the galaxy. The dynamical evolution of naked grains in the halo environment has been studied by Barsella et al. (1989b). The main difference with the expulsion inside the disk is the highly modified interaction between dust and gas. The gas drag in the disk is due to localized dense structures while in the halo the sputtering driven by the hot, but sustantially less dense, coronal gas becomes prominent.

The radiation pressure induced dust expulsion can be regarded as a common process among spiral galaxies as demonstrated by Ferrara et al. (1989) who calculated a static solution of the problem. The results show that the galaxies within the sample expell dust in a very similar way in what concerns the qualitative behaviour of the process regarding grain sizes and materials.

J. M. Greenberg and V. Pirronello (eds.), Chemistry in Space, 451–453.

One of the consequences of the removal of solid material from the disk and of its subsequent injection in the galactic halo is the chemical enrichment of the latter with heavy elements. As a matter of fact the very efficient sputtering process in the coronal environment may remove heavy elements from the grain surface, transferring them to the gaseous phase.

2. THE MODEL FOR THE EVOLUTION OF DUST GRAINS IN THE HALO

Following the work of Barsella et al. (1989b) we study the motions of grains starting from the edge of the galactic disk, where the radiation field is anisotropic. The main sources of the force fields acting on the grains, if we do not consider the magnetic field whose structure in this environment is too poorly known, are three: the radiation and gravitational fields and the gas drag. A spiral galaxy is modeled as a superposition of two luminous components (bulge and disk) and three massive components (bulge, disk and halo). The optical grain properties have been taken from the work of Draine and Lee (1984), while the mass and luminous distributions are simulated as in Barsella (1989a). The gas drag and the dynamical sputtering are treated following the work of Draine and Salpeter (1979), without considering the possible effects of grain evaporation due to hard radiation field expected in the halo.

The coronal gas has been modeled in the following way: we suppose, as a first approch, an uniform distribution of the gas in the whole halo with thermodynamical properties derived from observations (e.g. Pettini et al.(1982)). The typical values are $10^{-2} - 10^{-3}$ cm^3 for the density and a pressure such that $3 \leq \log(p/k) \leq 4$, where k is the Boltzmann constant.

We studied in more detail the case of the galaxy NGC 3198 because several authors agree about the fact that it is a normal typical spiral galaxy and that it can be taken in some way like a model galaxy. For this galaxy and for other two (NGC 300 and NGC 224) we used the parameters given by Kent (1987), while for NGC 598 we referred to Athanassoula et al. (1987).

The initial position of the grains has been chosen at an height of 100 pc over the mid-galactic plane and at various galactocentric distances from 0.1 to 2.0 Kpc with zero velocity relative to the gas.

3. RESULTS

Calculations have been made for astronomical silicate grains with an initial radius of 100 nm, the typical size of an interstellar grain. There is an appreciable similar behaviour among three galaxies, (namely NGC 3198, NGC 300 and NGC 598) while NGC 224 does not expell grains. The explanation for NGC 224 peculiarity resides in the higher mass content and lower luminosity of that galaxy.

We can summarize the main results in the following points:

a) The grains can reach considerably heights over the galactic plane($z = 0.3 - 0.5\ R_h$, where R_h is the halo radius) depending on the density value in a time scale of 2×10^8 yrs

b) The grains spread also over large galactocentric radii and attain a radial position of the order of the outer arm zone. This effect is greater for grains that start from a larger galactocentric radius. In this way some material of the inner part of the galaxy, where the abundances of the condensed material are greater, may be transported in the outern regions of the galaxy.

c) The grains suffer from a substantial sputtering in the case of higher gas density. As it is shown in Fig. 1 for NGC 3198, the grain radius is reduced of several tenths while it is travelling in the hot medium with typical velocities of 100 km s^{-1}, in a time scale of about 5×10^8 yrs.

Some uncertainty in the sputtering yield allows us to increase the efficiency of the latter within the uncertainties. This approach tries to represent the effect of the high energy part of the radiation spectrum incident on the particle. In this case the calculations show that the grain is completely destroyed in a time of the order of 4×10^8 yrs.

The only consequence we point out in this paper is the possibility that at least some of the heavy chemical species observed in the halo have been carried inside the dust grains and subsequently released in the gas phase. Furthermore a cycling of these elements can take place between the inner and outern parts of the disk affecting the galactic evolution and abundances. Finally the expulsion process may give a reasonable explanation of the metallicity gradients dependence from the galactic z-height discussed by Savage (1986).

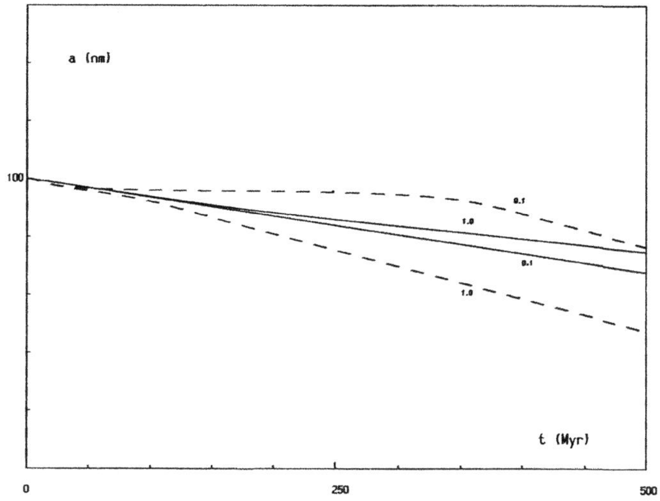

Figure 1 Grain radius reduction by sputtering for two different sets of thermodynamical properties of the halo gas: $n = 10^{-2}$, $\log(p/k) = 4$ (solid curve) and $n = 10^{-2}$, $\log(p/k) = 3$ (dashed); the starting radius is shown for each curve in Kpc units.

REFERENCES

Athanassoula, E., Bosma, A. and Papaioannu, S. 1987, *Astron. Astrophys.*,**179**, 23

Barsella, B., Ferrini, F., Greenberg, J.M., and Aiello, S. 1989a, *Astron. Astrophys.*, **209**, 349

Barsella, B., Franco, J., Ferrini, F., and Ferrara, A. 1989b, in preparation

Draine, B.T., and Lee, H.M. 1984, *Ap. J.*, **285**, 89

Draine, B.T., and Salpeter, E.E. 1979, *Ap. J.*. **231**, 77

Ferrara, A., Ferrini, F., Barsella, B., and Aiello, S. 1989, submitted to *Astron. Astrophys.*

Franco, J. 1989, preprint

Kent, S.M. 1987, *Astron. J.*,**93**, 816

Pettini, M., West, K. 1982, *Ap. J.*, **260**, 561

Savage, B., 1986, in "New Insights in Astrophysics", Proc. Joint NASA/ESA/SERC ESA-SP **263**

ASTROMACROMOLECULES: FORMATION OF VERY LARGE MOLECULES IN INTERSTELLAR SPACE

C.X. MENDOZA-GOMEZ*, J.M. GREENBERG*, G.B. EIJKEL** and J.J. BOON**
*Laboratory Astrophysics, University of Leiden, Postbus 9504
2300 RA Leiden, The Netherlands
**FOM-Institute for Atomic and Molecular Physics, Amsterdam
The Netherlands

The study of the physical and chemical processes taking place in the interstellar medium is made possible by the creation of laboratory conditions which simulate those relevant in space - low temperature, ultra-high vacuum and the radiation field. For details of the experimental methods see Hagen *et al.*(1979). The irradiation of the icy mantles in the molecular cloud medium leads to the formation of refractory organic material on interstellar dust. The more soluble part of the laboratory photoproduced residues has been analysed by GCMS and HPLC (see e.g. Agarwal *et al.*(1986)). Recently we have performed analyses of the non-soluble part of the organic residue by pyrolysis chemical ionization mass spectrometry (PY-CIMS). For a description of this method see Pouwels *et al.*(1989) and Westmore and Alauddin (1986). Two examples of the PY-CIMS spectra of one of the samples starting with labelled and unlabelled mixtures are shown in figures 1 and 2. It should be noted that the labelled sample (fig. 2) was a residue left over from a residue analysed by GCMS - therefore there was very little material. The peaks are quite reproducible, and comparing the labelled and unlabelled samples, the shifts obviously indicate that the peaks we obtain are real and do not come from contamination; e.g., 69 shifts to 74, 83 to 89, and so on, depending on the number of carbons and nitrogens we have. All the samples consist of compounds of very high molecular weight - many (especially those with CH_4 and C_2H_2) with more than 500 mass peaks well above the background noise level (see figs. 1, 2 and 3). The arrow in the three figures indicates the *maximum* intensity of the base peak in the blank, i.e. *all* the peaks above the arrow are almost certainly real, and probably some of the peaks below it are real as well. The peak at 391 is always high in the 'blank' spectrum and is therefore spurious. A significant number of main peaks are separated by mass number 14 and, following a main peak, decrease in intensity with increasing fragment weight. This clearly indicates hydrocarbon chains (CH_2 groups). With such a complex mixture -and without other studies- it is difficult to say precisely what the non-soluble part of the organic residue samples consists of. Nevertheless, we can say that the pyrolysate is a very complex mixture of unsaturated hydrocarbon chains, phenols, and aromatic compounds generated from a radical induced macromolecule.

J. M. Greenberg and V. Pirronello (eds.), Chemistry in Space, 455–457.

In order to characterize the composition of the laboratory photoproduced organic residues, they are also being analyzed by other methods. Residues examined for us by the group of Dr. Colin Pillinger using stepwise heating and oxidation, for isotopic analysis, indicated an evaporation sequence similar to that of meteoritic organics, i.e., the presence of aromatic kerogens (see Franchi *et al.*(1989)). This is yet to be confirmed, although the preliminary mass spectra obtained by Dr. Krueger, similar to the one performed for Comet Halley (see Kissel and Krueger (1987)) also gave such indications. This work is to be pursued further.

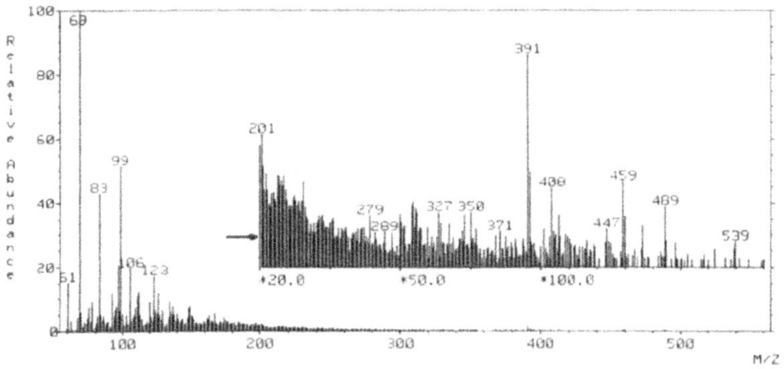

Figure 1. DCI-MS spectra of irradiated H_2O: CO: NH_3 = 5:5:1
The maximum level of the blank (marked with an arrow) is scaled to the amplification factor of 20.

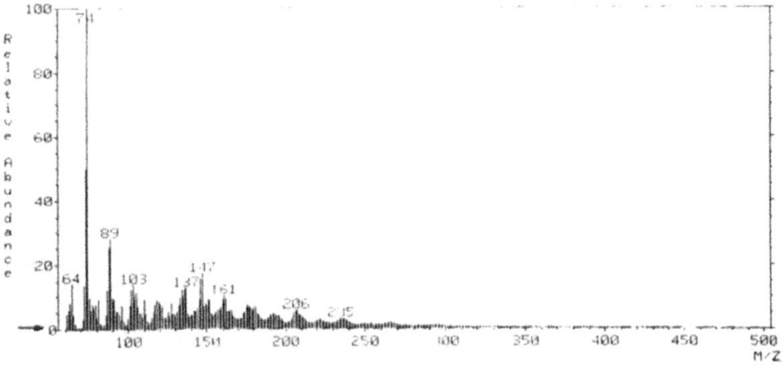

Figure 2. DCI-MS spectra of irradiated $H_2^{18}O$: ^{13}CO: $^{15}NH_3$ = 5:5:1

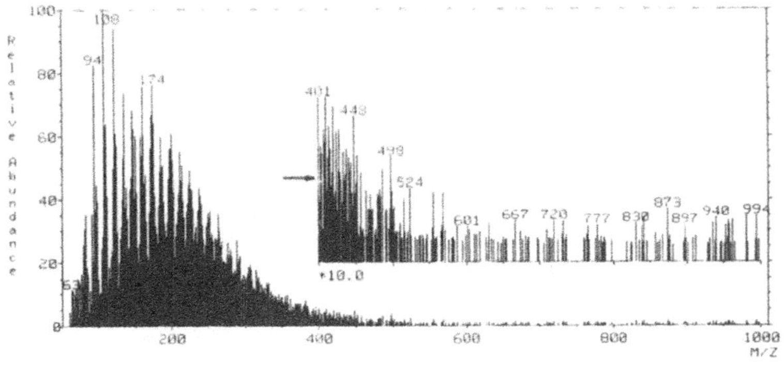

Figure 3. DCI-MS spectra of irradiated H_2O: CO: NH_3: C_2H_2 = 2:1:0.1:1

Aknowledgements

This work is supported by NASA-Grant nr. NGR 33-018-148, and one of us (C.X. Mendoza-Gómez) aknowledges a grant from DGAPA, University of Mexico (UNAM). The authors would like to thank Dr.C. Pillinger, Dr. I. Franchi and Dr. F. Krueger for their preliminary work on the organic residue samples.

References

Agarwal, V.K., Schutte, W., Greenberg, J.M., Ferris, J.P., Briggs, R., Connor, S., van de Bult, C.P.E.M. and Baas, F. (1985) in Origins of Life 16, D. Reidel Publ. Co., 21-40.

Franchi, I.A., Alexander, C.M.O., Pillinger, C., Mendoza-Gómez, C.X. and Greenberg, J.M. (1989) in Physics and Mechanics of Cometary Materials (ESA SP-302), 89-92.

Hagen, W., Allamandola, L.J. and Greenberg, J.M. (1979) in Astrophys. Sp. Sc. 65, 215-240.

Kissel, J., and Krueger, F.R. (1987) in Nature, 326, 755.

Pouwels, A.D., Eijkel, G.B., Arisz, P.W. and Boon, J.J. (1989) in J. Anal. Appl. Pyrol. 15, 71-84.

Westmore, J. and Alauddin, M. (1986) 'Ammonia Chemical Ionization', in Mass Spectrom. Rev., 5, 381.

SMALL-SIZE DUST PARTICLES NEAR HALLEY'S COMET

M.N.FOMENKOVA
Space Research Institute, USSR Academy of Sciences
117810, Profsojusnaya st., 84/32
Moscow
USSR

ABSTRACT. Dust-impact PUMA mass-analysers aboard the spacecrafts VEGA-1 and VEGA-2 allow to conduct the first direct measurements of mass spectra of comet Halley's dust envelope particles with masses higher than 10^{-17} g [1,2]. The analysis of spectra measured by the PUMA instruments showed that unindentified peaks in this spectra could be associated with very small particles (mass 10^{-17} - 10^{-20} g).

1. Introduction

The PUMA instruments are the time-of-flight mass-analysers. A collision of a dust particle with the target at the velocity of 80 km/s results in ion formation. Mass of the ions is determined from their time of flight via the optical system of the instrument. Ions are recorded by secondary electron multiplier so sensitive that even single ion could be detected. For the detailed description of the PUMA instruments see [3].

Generally the spectra of dust particles measured by the PUMA instruments point to the presence of two components: the light one (most probably, organic, consisted of non-volatile compounds) and heavy (silicates with other components added, e.g. FeS) [1,2]. Along with the peaks of major, most abundant elements, spectra of PUMA-1 show many unindentified peaks of smaller amplitude.

Quasi-random peaks are also seen in those spectra where peaks of major elements are absent. These quasi-noise events have been observed at large distances to the comet nucleus since the moment the instrument was switched on in the flyby session, but while both instruments have been switched on in "day before flyby" session quasi-noise pulses were not detected.

2. Experimental data

Changes of quasi-noise event rate along the flyby trajectory and dependance on the registration energy windows indicate that the observed peaks are caused by external factors associated with the comet, not by

J. M. Greenberg and V. Pirronello (eds.), Chemistry in Space, 459–460.

some instrumentation effects.
Number of peaks in the data frame (67 microseconds) grows with approach to the nucleus from several peaks per frame in the beginning to more than one hundred on the distance less than 10^5 km to the nucleus, where saturation on the number of peaks take place due to limit of instrument digitalisation system resolution.
Mean charge of peaks also increases with approach to the nucleus. On distances more than $3\ 10^5$ km before and more than 10^5 km after the nucleus amplitude of peaks is practically the same in both energy windows and corresponds to 1-2 elementary charges. In the inner region of trajectory amplitude value increases until the multiplier's saturation begins.

3. Discussion

Observed quasi-noise events can't be explained by influence of energetic suprathermal ions or vacuum ultraviolet photons. Events corresponding to small charges can in principle be explained by collision of neutral atoms with target but the appearance of peaks with large charges (up to tens) can be explained only by registration of very small dust particles of size about tens of angstroms. The measured pulses count rate results in estimation of tiny dust mass flux as 1-10% of total comet mass loss via dust particles.
Timing analysis of tiny dust particles data did not reveal the presence of correlated ion patterns. Out of our results it may be supposed that very small dust particles consist of one or two components (graphite or metal oxides) which does not contradict the indirect data on interstellar dust composition.

4. Conclusion

Thus the analysis performed of quasi-noise events shows that the most probable explanation of pulses observed is the combine influence of neutral particles, which may dominate in outer regions of flyby trajectory, and one of very small dust particles which prevail in inner regions.

References

1. Kissel, J., Sagdeev, R.Z., Bertaux, J.L., et al. (1986) 'Composition of comet Halley dust particles from Vega observations', Nature 321, 280-282.
2. Sagdeev, R.Z., Evlanov, E.N., Zubkov, B.V., et al. (1987) 'Analysis of comet Halley dust particles based on data obtained by PUMA instrument in zero mode', Kosmicheskie Issledovanija 25, 849-855.
3. Balebanov, V.M., Vorontzova, E.N., Scuridin, G.A. (eds.) (1984) Venus-Halley project, Lui-Jean, Paris.

THE POSSIBILITY OF FORMATION OF LAYER SILICATES IN HALLEY'S COMET NUCLEUS

A. D. GRECHINSKIY
Space Research Institute
117810, Profsoyuznaya 84/32
Moscow, USSR

Information on the elementary chemical composition of Halley's Comet nucleus dust component was obtained as a result of mass-spectrometric experiments performed during VEGA mission with the help of PUMA instruments. The study of 511 mass-spectra of dust particles, analyzed by the PUMA-2, shows that about 10% of particles contain Mg, Fe, Si and H in proportions allowing an assumption that layer silicates exist in the mineral component of comet dust. Rietmeijer et al. (1989) suggested that hydrotation reactions of silicates can take place directly in comet's nucleus - in the micron films of liquid water, that can exist inside a porous nucleus within temperatures from 200 K and higher. Let us consider several simple estimations, proving this assumption.

We shall assume that the cometary nucleus consists of porous ice covered with thin (several tens of cm) dark crust formed of the caked dust in which all volatiles were sublimated due to heating of the Solar radiation. Since the period of Halley's Comet nucleus rotation is $t_{rot} \simeq 53\ h \sim 4 \cdot 10^5$ s, that is, as will be shown later, much less than characteristic time of our problem, we can consider the nucleus to be uniformly heated by Solar radiation. The temperature on the mineral crust surface T_m, in this case, practically coincides with equilibrium black-body temperature and it is $\simeq 400$ K at the distance of 1 AU. On ice surface (under the mineral layer) the temperature T_{ic} reaches 255 K (Shulman (1987)).

The temperature distribution inside the nucleus is defined from the equation of thermoconductivity. The nucleus surface moves with a velocity V, depending on T_m and T_{ic} because of the permanent ice sublimation. We shall consider this velocity as constant, i.e., we shall make all estimations assuming that the comet is all the time at the distance 1 AU from the Sun. This is quite acceptable, for, as will be shown below, the characteristic time of our problem is close to the time during which Comet Halley was situated at the distance not father than 1 AU from the Sun, where hydrotation reactions take place very effectively.

Since the characteristic scale of our problem is several meters, we can do no consideration of the comet's nucleus curvature and concern ourselves with one-dimensional heat equation In the coordinate system, moving with the evaporation front, we have

J. M. Greenberg and V. Pirronello (eds.), Chemistry in Space, 461–462.

$$\frac{\partial T}{\partial t} = V \frac{\partial T}{\partial \xi} + a \frac{\partial^2 T}{\partial \xi^2}, \qquad (1)$$

where $\xi = x - Vt$, a is the thermal conductivity. As will be shown below, the characteristic evaporation time $t_{ev} = L/V$ is much longer than characteristic heat conduction time $t_{cond} = L^2/a$, therefore (1) can be solved in stationary approximation. Solution of (1), satisfying boundary conditions $T(\xi = 0) = T_{ic}$, $T(\xi \to \infty) = 0$, is

$$T = T_{ic} \exp(-V\xi/a). \qquad (2)$$

Thickness of the layer in which reactions of hydrotation of silicates are possible in the micron layers of liquid water (i.e., where $T > T_c \simeq 200$ K), is

$$L = a \ln(T_{ic}/T_c) / V. \qquad (3)$$

Thus, the characteristic evaporation time is

$$t_{ev} = L/V = a \ln(T_{ic}/T_c) / V^2. \qquad (4)$$

Substituting the values of a from Shulman (1987), where all the possible heat conduction mechanisms are taken into account and $v = Zm/\rho$, where gas production $Z = 8.8 \cdot 10^6\ s^{-1}\ cm^{-2}$ (calculated according to Huebner et al. (1986)), into (3) and (4), we obtain the following values:

$$L = 2m ; \quad t_{ev} = 2 \cdot 10^7\ s. \qquad (5)$$

Extrapolating the rate of hydrotation of silicates, adduced in Wood et al. (1983), to the low temperature region

$$k = 16 \exp(-6.85 - 2900/T)\ g/cm^2\ s, \qquad (6)$$

we obtain that the total mass of the formed layer silicates is $m = kt_{ev} \simeq 4\ g/cm^2$, or about 8% of the total mass in the column with the lenght L and the cross-section 1 cm^2, that is close to the experimental data.

REFERENCES

Huebner, W.F. et al. (1986) 'Dust-gas interaction deduced from Halley multicolour camera observations', in B. Battrick, E.J. Rolfe and R. Reinhard (eds.), 20th ESLAB Symposium on the Exploration of Halley's Comet, Heidelberg, Germany, vol. 2, pp. 363 -364.

Rietmeijer, J.M. et al. (1989) 'Layer silicate chemistry in P/Halley from PUMA-2 data', in Lunar and Planetary Science Conference XX, Houston, Texas, pp. 904 - 905.

Shulman, L.M. (1987) Comet's nuclei, Nauka, Moscow.

Wood, B.J. and Wather, J.V. (1983) 'Rates of hydrothermal reactions', Science, 222, 413 - 415.

INDEX

INDEX OF CHEMICAL SPECIES

GPSR Compliance

The European Union's (EU) General Product Safety Regulation (GPSR) is a set of rules that requires consumer products to be safe and our obligations to ensure this.

If you have any concerns about our products, you can contact us on ProductSafety@springernature.com

In case Publisher is established outside the EU, the EU authorized representative is:

Springer Nature Customer Service Center GmbH
Europaplatz 3
69115 Heidelberg, Germany

Batch number: 10755284

Printed by Printforce, the Netherlands